"Climate change is a reality, causing tangible changes to environments and landscapes where humans—and insects—live their lives. Insects represent key elements in biotas, in terms of biomass, diversity and ecological function. This book brings together these two important topics (climate change and insects), and examines the consequences from diverse perspectives. Chapters include views ranging from paleontology to physiology, and from genome and transcriptome to landscapes and distributions. This book should provide an excellent basis for many fascinating discussions, such as in a graduate student seminar, and should be an excellent means by which any researcher new to the area can get a panoramic perspective of the phenomenon."

A. Townsend Peterson
University Distinguished Professor
University of Kansas, USA

"Now more than ever we need to understand how climate change affects biological systems as it threatens the future not only of our species but also of our planet. This book brings together an internationally excellent team of researchers who synthesize the latest findings about the effect of climate change on the most diverse and ecologically important group of organisms—the insects. This volume ranges from fossil records of insect response to climate change to how this cascaded through the web of life. It promises to be widely read and a useful and updated piece of work for years to come."

Inara R. Leal
Universidade Federal de Pernambuco, Brazil

"Insects are essential to terrestrial ecosystem functioning but they now face the novel and rapid threat of human-induced climate change. How are they faring? Are some doing better than others? Why are some winners yet others are losers? What is it about their genetics, physiology, ecology and evolution that makes some survivors and others not? These critical questions are addressed in this fascinating and timely book. It explores what we are learning and how we should secure a future for them and us, bearing in mind that our mutual fate is entwined."

Michael Samways
Stellenbosch University, South Africa

"*Effects of Climate Change on Insects: Physiological, Evolutionary and Ecological Responses* brings together an authoritative collection of studies from a broad range of disciplines. This is a landmark book that delivers in depth evidence of how insects will respond to climate change and the important limitations in our knowledge. Insects are the most dominant organisms on the planet with critical roles in ecological function and food webs, hence the prediction of future declines in abundance offers a frightening prospect. This book will have a major impact on scientific thinking and will be a valuable resource to ecologists, entomologists and other researchers, and hopefully will raise public awareness to the potential cascading effects of climate change on our ecosystems."

Susan G.W. Laurance
James Cook University, Australia

Effects of Climate Change on Insects

Effects of Climate Change on Insects

Physiological, Evolutionary, and Ecological Responses

Edited by

DANIEL GONZÁLEZ-TOKMAN AND WESLEY DÁTTILO

Instituto de Ecología, A. C. Xalapa, México

OXFORD
UNIVERSITY PRESS

OXFORD
UNIVERSITY PRESS

Great Clarendon Street, Oxford, OX2 6DP,
United Kingdom

Oxford University Press is a department of the University of Oxford.
It furthers the University's objective of excellence in research, scholarship,
and education by publishing worldwide. Oxford is a registered trade mark of
Oxford University Press in the UK and in certain other countries

Published in the United States of America by Oxford University Press
198 Madison Avenue, New York, NY 10016, United States of America

British Library Cataloguing in Publication Data
Data available

Library of Congress Control Number: 2023946522

ISBN 9780192864161

DOI: 10.1093/oso/9780192864161.001.0001

Printed and bound by
CPI Group (UK) Ltd, Croydon, CR0 4YY

Cover image: Álvaro Hernández–Rivera

Links to third party websites are provided by Oxford in good faith and
for information only. Oxford disclaims any responsibility for the materials
contained in any third party website referenced in this work.

Contents

Preface

Our planet is experiencing unprecedented climate changes, largely caused by human activities. Insects, the most diverse and numerous organisms, are suffering the consequences of these changes. Insects play critical roles in ecosystems in nutrient cycling, pollination, seed dispersal, control of populations of other organisms, besides being a major food source for other taxa and vectors of many animal and human diseases. This makes insect responses to climate change a matter of global concern. *Effects of Climate Change on Insects: Physiological, Evolutionary and Ecological Responses* provides a thoughtful and stimulating contribution to debate about how and why climate change affects insects. In general, we bring together fifty-four leading experts from twelve countries in the fields of insect genomics, paleontology, physiology, ecology, evolution and conservation to provide a comprehensive overview of the impact of current and past climate changes on insects.

This volume comprises seventeen chapters starting by an overview where we highlight the causes and consequences of anthropogenic climate change and emphasize the need for urgent action and research to mitigate its effects. In Chapter 2, Ellen Currano explores the fossil record to reveal how insects have responded to climate changes in the past, providing valuable insights into their future responses. In Chapter 3, Zach Fuller and Maren Wellenreuther examine the role of omic sciences in advancing our understanding of insect responses to climate change. Chapter 4, by Daniel González-Tokman and Sebastián Villada-Bedoya, focuses on the physiological mechanisms of heat tolerance in insects, while Patrick Rohner discusses the genetic and plastic responses of insects to climate change in Chapter 5. In Chapter 6, Gang Ma and colleagues explore the effects of climate change on insect phenology. In Chapter 7, Bruno A. Buzatto, Daniel P. Silva and Paulo Enrique C. Peixoto consider the impacts of climate change on sexual selection. Chapter 8, by Rosa Ana Sánchez-Guillén and colleagues, examines the role of interspecific hybridization in insect adaptation to climate change. Carol Boggs discusses changes in insect population dynamics due to climate change in Chapter 9, while Kleber Del-Claro and colleagues explore the evidence of the effects of climate change on insect diversity in Chapter 10. In Chapter 11, Lucie Aulus-Giacosa and colleagues examine the effects of climate change on insect distributions and invasions. In Chapter 12, Genoveva Rodríguez-Castañeda and Anouschka R. Hof discuss insect communities' adaptation to climate change using species' trajectories along elevation gradients in tropical and temperate ecosystems. Chapter 13, by Laura A. Burkle and Shalene Jha, focuses on the impacts of climate change on insect pollinators and their ecological function. In Chapter 14, Berenice González-Rete and colleagues examine the implications of climate change for insect vectors of human pathogens. Pedro Luna and Wesley Dáttilo consider the cascading effects of climate change on insect biotic interactions in Chapter 15, while Guim Ursul and colleagues examine the role of refugia from climate change in shaping insect diversity and conservation in Chapter 16. Finally, in Chapter 17, together with Ornela De Gasperin, we conclude the book by recognizing main perspectives to improve our understanding of insect responses to climate change.

We hope that this book will serve as a valuable resource for students, researchers, policy-makers, conservationists and anyone interested in understanding the complex and multifaceted effects of climate change on insects. Overall, this landmark book offers an advance in the current literature by bringing together cutting-edge knowledge from diverse fields, which should stimulate new avenues of research and contribute to addressing the global challenge of climate change.

Daniel González-Tokman and Wesley Dáttilo
Editors

Acknowledgments

This book would have not been possible without the enthusiasm and valuable contribution of a large number of expert reviewers from different regions of the world who provided insightful comments and suggestions which helped improving book quality. We also appreciate the constructive discussions with Fernanda Baena, Alex Córdoba and Fabricio Villalobos which enriched the final content of the book. We wish to acknowledge Ricardo Madrigal for his technical and logistic support. We thank the chapter authors, who donated their time and their knowledge to produce this book.

Wesley Dáttilo (WD) would like to express his deepest gratitude to Daniel González-Tokman for the invitation to share the editorial work and the great teamwork and opportunity for fun projects. WD was supported by the Consejo Nacional de Ciencia y Tecnología (CONACyT, Mexico) in 2022 under Grant No. FOP16-2021-01-319227 while working on this book.

Wesley Dáttilo dedicates this book to his son Cauê Dáttilo whose smile has relaxed him greatly during hard and pandemic times and for brightening up his life in so many ways.

Abbreviations

20E	ecdysone
AKH	adipokinetic hormone
AO	Arctic Oscillation
ARTs	Alternative reproductive tactics
a.s.l.	above sea level
ATAC-seq	Assay for Transposase-Accessible Chromatin using sequencing
AUC	the area under the receiver operating characteristic curve
BA	biogenic amines
BAM	Biotic, Abiotic, Movement
bp	base pair
BP	before present
BS-seq	bisulfite sequencing
CAT	catalases
CC	climate change
CCVI	Climate Change Vulnerability Index
ChIP-seq	chromatin immunoprecipitation sequencing
CPI	community precipitation index
CS	Community Science
CTI	community temperature index
CTmax	critical thermal maximum
CTs	critical thermal limits
dnmt	DNA methyltransferases
DT	damage type
EECO	Early Eocene Climatic Optimum
EI	ecoclimatic index
EIP	extrinsic incubation period
ENMs	ecological niche models
ENSO	El Niño-Southern Oscillation
FFG	functional feeding group
FISH	fluorescence *in situ* hybridization
FOXO	forkhead
Gb	gigabyte
GBIF	Global Biodiversity Information Facility
GBS	genotyping-by-sequencing
GDP	gross domestic product
GEA	genome–environment association
GLMs	general linear models
GR	glutathione reductase
GSH	glutathione
GST	glutathione-S-transferase
GWAS	genome-wide association studies

Hi-C	high-throughput chromosome conformation capture
HSP	heat shock protein
IIS	insulin/insulin-like growth factor signaling
IPBES	Intergovernmental Science-Policy Platform on Biodiversity and Ecosystem Services
IPCC	Intergovernmental Panel on Climate Change
JH	juvenile hormone
ka	thousand years ago
kb	kilobyte
kyr	thousand years
LDH	lactate dehydrogenase
LGM	Last Glacial Maximum
LTER	Long-Term Ecological Research
Ma	million years
MAP	mean annual precipitation
MAT	mean annual temperature
MaxEnt	Maximum Entropy Modeling
miRNAs	microRNAs
MIROC	Model for Interdisciplinary Research on Climate
MMH	match-mismatch hypothesis
m.a.s.l.	meters above sea level
NABA	North American Butterfly Association
NCBI	National Center for Biotechnology Information
NGS	Next Generation Sequencing
NOS	nitric oxide synthase
NPP	net primary productivity
OCLTT	oxygen and capacity limited thermal tolerance
ONT	Oxford Nanopore Technology
PBDMs	physiological based demography models
PbTf	transferrin
PETM	Paleocene–Eocene Thermal Maximum
PgC	petagrams of carbon
PO	phenoloxidase
RAD	Restriction-site Associated DNA
RNAseq	RNA sequencing
ROS	reactive oxygen species
SDMs	species distribution models
SMRT	Single Molecule Real Time
SNP	single nucleotide polymorphism
SOD	superoxide dismutase
TPCs	thermal performance curves
TSS	true skill statistics
UKBMS	UK Butterfly Monitoring Scheme
ULT	upper lethal temperature
UN	United Nations
VCPG	ventilatory central pattern generator
VOCs	volatile organic compounds
WGR	Whole-genome resequencing
ZMW	zero-mode waveguide

Contributors

Diego Anjos, Universidade Federal de Uberlândia, Instituto de Biologia, Brazil.

Luis Rodrigo Arce-Valdés, Red de Biología Evolutiva, Instituto de Ecología A.C., Xalapa, Veracruz, Mexico.

Lucie Aulus-Giacosa, Department of Ecology and Evolution, University of Lausanne, 1015 Lausanne, Switzerland.

Andrea Viviana Ballén-Guapacha, Red de Biología Evolutiva, Instituto de Ecología A. C., Xalapa, Veracruz, Mexico.

Olivia K. Bates, Department of Ecology and Evolution, University of Lausanne, 1015 Lausanne, Switzerland.

Cleo Bertelsmeier, Department of Ecology and Evolution, University of Lausanne, 1015 Lausanne, Switzerland.

Carol Boggs, School of Earth, Ocean & Environment and Department of Biological Sciences, University of South Carolina, Columbia, SC, USA; Rocky Mountain Biological Laboratory, Crested Butte, CO, USA.

Aymeric Bonnamour, Department of Ecology and Evolution, University of Lausanne, 1015 Lausanne, Switzerland.

Jelena Bujan, Department of Ecology and Evolution, University of Lausanne, 1015 Lausanne, Switzerland.

Laura A. Burkle, Montana State University, Department of Ecology, Bozeman, MT USA.

Bruno A. Buzatto, College of Science and Engineering, Flinders University, Sturt Road, Bedford Park SA 5042, Australia.

Margarita Cabrera-Bravo, Posgrado en Ciencias Biológicas, Universidad Nacional Autónoma de México, Ciudad de México, México; Departamento de Microbiología y Parasitología, Facultad de Medicina, Universidad Nacional Autónoma de México, Mexico.

Eduardo S. Calixto, University of Florida, Gainesville, USA.

Juan Pablo Cancela, Centre for Ecology, Evolution and Environmental Changes (CE3C), Lisbon, Portugal.

Elliot Centeno de Oliveira, Universidade de São Paulo, Faculdade de Filosofia Ciências e Letras de Ribeirão Preto, Pós-Graduação em Entomologia.

Alex Córdoba-Aguilar, Departamento de Ecología Evolutiva, Instituto de Ecología, Universidad Nacional Autónoma de México, México.

Vitor Miguel da Costa Silva, Universidade de São Paulo, Faculdade de Filosofia Ciências e Letras de Ribeirão Preto, Pós-Graduação em Entomologia.

Ellen D. Currano,   Departments of Botany and Geology & Geophysics, University of Wyoming, USA.

Wesley Dáttilo,   Red de Ecoetología, Instituto de Ecología AC, Xalapa, Veracruz, Mexico.

Ornela De Gasperin,   Red de Ecoetología, Instituto de Ecología AC, Xalapa, Veracruz, Mexico.

Kleber Del-Claro,   Universidade Federal de Uberlândia, Instituto de Biologia, Brazil.

Gyda Fenn-Moltu,   Department of Ecology and Evolution, University of Lausanne, 1015 Lausanne, Switzerland.

Any Laura Flores-Villegas,   Departamento de Microbiología y Parasitología, Facultad de Medicina, Universidad Nacional Autónoma de México, Mexico.

José Antonio de Fuentes-Vicente,   Universidad de Ciencias y Artes de Chiapas, Mexico.

Zach Fuller,   Department of Biological Sciences, Columbia University, USA.

Jérôme Gippet,   Department of Ecology and Evolution, University of Lausanne, 1015 Lausanne, Switzerland.

Berenice González-Rete,   Posgrado en Ciencias Biológicas, Universidad Nacional Autónoma de México, Ciudad de México, México; Departamento de Microbiología y Parasitología, Facultad de Medicina, Universidad Nacional Autónoma de México, Mexico.

Daniel González-Tokman,   Red de Ecoetología, Instituto de Ecología AC, Xalapa, Veracruz, Mexico.

Anouschka R. Hof,   Wildlife Ecology and Conservation Group, Wageningen University & Research Droevendaalsesteeg 3, 6708PB, Wageningen, The Netherlands.

Shalene Jha,   University of Texas at Austin, Department of Integrative Biology, Austin, TX USA.

Jesús Guillermo Jiménez-Cortés,   Departamento de Microbiología y Parasitología, Facultad de Medicina, Universidad Nacional Autónoma de México, Mexico.

Tristan Klaftenberger,   Department of Ecology and Evolution, University of Lausanne, 1015 Lausanne, Switzerland.

Cécile Le Lann,   UMR 6553 ECOBIO, Centre National de la Recherche Scientifique, Universite de Rennes I, Rennes, Cedex, France.

Pedro Luna,   Red de Ecoetología, Instituto de Ecología A.C., Xalapa, Veracruz, Mexico.

Chun-Sen Ma,   State Key Laboratory for Biology of Plant Diseases and Insect Pests, Institute of Plant Protection, Chinese Academy of Agricultural Sciences, Beijing, China.

Gang Ma,   State Key Laboratory for Biology of Plant Diseases and Insect Pests, Institute of Plant Protection, Chinese Academy of Agricultural Sciences, Beijing, China.

Helena Maura Torezan-Silingardi,   Universidade Federal de Uberlândia, Instituto de Biologia, Brazil.

Mario Mingarro,   Museo Nacional de Ciencias Naturales (MNCN-CSIC), Madrid, Spain.

Renan Fernandes Moura,   Professor (Assistant) at University of Cincinnati, Cincinnati, USA.

Jesús Ordaz-Morales,   Red de Biología Evolutiva, Instituto de Ecología A. C., Xalapa, Veracruz, Mexico.

Paulo Enrique C. Peixoto,   Departamento de Genética, Ecologia e Evolução, Universidade Federal de Minas Gerais, Avenida Antônio Carlos, Belo Horizonte 6627, Brazil.

Iasmim Pereira,   Universidade de São Paulo, Faculdade de Filosofia Ciências e Letras de Ribeirão Preto, Pós-Graduação em Entomologia.

Genoveva Rodríguez-Castañeda,   The Burke Museum, Seattle, Washington, USA.

Patrick T. Rohner,   Department of Ecology, Behavior, and Evolution, University of California San Diego, 9500 Gilman Drive, La Jolla, CA 92093, USA.

Helena Romo,   Universidad Autónoma de Madrid y Centro de Investigación en Biodiversidad y Cambio Global (CIBC-UAM), Madrid, Spain.

Paz María Salazar-Schettino,   Departamento de Microbiología y Parasitología, Facultad de Medicina, Universidad Nacional Autónoma de México, Mexico.

Rosa Ana Sánchez-Guillén,   Red de Biología Evolutiva, Instituto de Ecología A. C., Xalapa, Veracruz, Mexico.

Daniel P. Silva,   Departamento de Ciências Biológicas, Instituto Federal Goiano, Rodovia Geraldo Silva Nascimento, Km 2.5, Urutaí, GO 75790-000, Brazil.

Miguel Stand-Pérez,   Red de Biología Evolutiva, Instituto de Ecología A. C., Xalapa, Veracruz, Mexico.

Guim Ursul,   Museo Nacional de Ciencias Naturales (MNCN-CSIC), Madrid, Spain.

Joan van Baaren,   UMR 6553 ECOBIO, Centre National de la Recherche Scientifique, Universite de Rennes I, Rennes, Cedex, France.

Sebastián Villada-Bedoya,   Instituto de Ecología A.C. Xalapa, Mexico.

Maren Wellenreuther,   Plant and Food Research, Nelson, New Zealand; University of Auckland, Auckland, New Zealand.

Robert J. Wilson,   Museo Nacional de Ciencias Naturales (MNCN-CSIC), Madrid, Spain.

CHAPTER 1

Anthropogenic climate change

Causes, consequences and a call to action and research

Wesley Dáttilo and Daniel González-Tokman

1.1 Introduction

One of the great interests of scientists from different areas is to understand how and why the planet we live on has changed over time. Some of the notable changes taking place on Earth are those related to climate. During the 4.5-billion-year history of our planet, there have been several radical climate changes in which long periods of stable climate were followed by glaciations and warm periods that caused, for example, desertification of large continental areas (Dodson 2012). These changes have been related to natural causes, such as long periods of volcanic activity, weathering of rocks, changes in the Earth's rotation and variations in the incidence of solar radiation (Hulme et al. 1999; Lockwood et al. 2010). However, in recent decades different scientists and organizations around the world have drawn attention to the permanent change of climatic conditions in the atmosphere due to the emissions of greenhouse gases by human activity (i.e., CO_2—carbon dioxide, N_2O—nitrous oxide, CH_4—methane, CFCs—chlorofluorocarbons, HFCs—hydrochlorofluorocarbons, PFCs—perfluorocarbons, SF_6—sulfur hexafluoride) (Al-Ghussain 2019; Wuebbles and Jain 2001).

The greenhouse effect is a natural phenomenon that maintains life on the planet where gaseous substances present in the atmosphere absorb part of the infrared radiation emitted mainly by the Earth's surface, therefore preventing excessive heat loss to space (i.e., keeping our planet heated) (Venkataramanan et al. 2011). However, in the last century there has been a progressive increase in the concentration of greenhouse gases that are directly related to human activities such as agriculture, burning of fossil fuels, solid waste disposal and deforestation, which together are responsible for most greenhouse gas emissions (Anderson et al. 2016). The increase in greenhouse gases in the atmosphere has potentiated this natural phenomenon after the Industrial Revolution and in the second half of the twentieth century, making the world warmer through the phenomenon called climate change (nowadays also known as climate breakdown) (Mitchell 1989). During this

---

* All authors contributed equally.

Wesley Dáttilo and Daniel González-Tokman, *Anthropogenic climate change*. In: *Effects of Climate Change on Insects*. Edited by: Daniel González-Tokman and Wesley Dáttilo, Oxford University Press. © Oxford University Press (2024). DOI: 10.1093/oso/9780192864161.003.0001

period there was an expansion of industrial production, which generated a large increase in greenhouse gas emissions into the atmosphere, often related to the exponential growth in the number of people and their level of personal consumption (Dow and Downing 2016). For instance, it is important to highlight that about 1 percent of the world's richest billionaires emit more CO_2 than 90 percent of humanity combined. Although a few groups in society question whether human activity is a sufficient driving force to generate climate change at a global level, this issue is almost unanimous in the scientific literature (reviewed by Jankó et al. 2017). Since the late 1800s, Eunice Newton Foote and Svante Arrhenius have been showing us the relationship between CO_2 and the greenhouse effect and speculating on how this gas might be able to contribute to global warming (Uppenbrink 1996). In addition, about fifteen years ago, data produced by thousands of scientists around the world that were analyzed by the Intergovernmental Panel on Climate Change (IPCC) of the United Nations (UN) showed that 97 percent of global warming is derived from the greenhouse effect generated by human activities (Anderegg et al. 2010). Moreover, other biophysical aspects of anthropogenic causes of the impacts of regional and global climate (not only the biogeochemical) include the alteration of the surface energy balance due to extensive and intensive land use change (Duveiller et al. 2020; Pielke et al. 2002). Thus, there is no doubt that we are living in a period of unprecedented climate change that affects us all.

The rapid increase in the Earth's global temperature and changes in the climate affect the natural dynamics of the planet in different ways, such as the frequency and activity of cyclones, severity and duration of heat waves, extreme dry periods or droughts, severe storms, rising temperatures of ocean water, melting glaciers and an increased chance of coastal flooding, all of which have direct influence on ecosystems and human life (Al-Ghussain 2019; Moritz et al. 2002). In other words, climate change is all climate-related changes that are happening on the planet and include the warming of the Earth over time (Letcher 2021). A point that needs to be highlighted is that populous and economically strong countries (e.g., China and the United States of America) are responsible for most of the greenhouse gas emissions (about 40 percent combined), mainly due to their economic system (Oreggioni et al. 2021). However, developing countries have also increased considerably their emissions and land use changes and related surface energy balance alterations in recent decades (Ahmed and Shuai 2022; Winkler et al. 2021). Thus, climate change is a phenomenon that must be fought globally to reduce greenhouse gas emissions and the effects on climate change (a topic that will be addressed at the end of this chapter) as it will persist for millennia and will continue to cause long-term changes to the planet's climate.

1.2 Projections of future climate change

In the recent report published by the IPCC (2021), researchers showed that while the average global temperature has risen by approximately 5 °C in the last 10,000 years, this rate has only been 1.1 degrees since the second half of the nineteenth century. Furthermore, studies have shown that since the 1970s we have seen a large increase in global aridity and that dry areas have increased by about 27 percent in the 2000s alone (Dai 2011). In fact, the last few years have been the hottest since temperature records began in 1880 (NOAA 2020). In the best-case scenario, the average global temperature is expected to reach an average warming of 1.5 °C between 2030 and 2052 if it continues to increase at the current rate (IPCC 2021) and approximately 4 °C (or more) in the next century (Meehl et al. 2007). Although this increase may seem small at first glance, this heat stress is a consequence of air temperature increases, and according to Vicedo-Cabrera et al. (2021), 37 percent of all

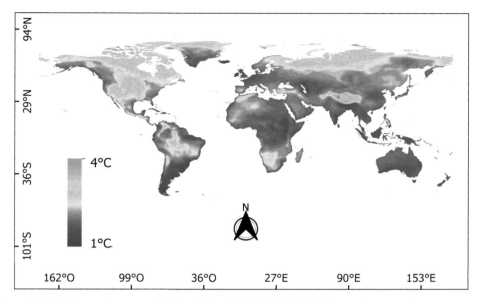

Figure 1.1 Difference between 2050 and historical (1961–1990) daily maximum surface temperatures according to the Japanese Model for Interdisciplinary Research on Climate (MIROC) based on the IPCC IV SRES Scenario A2 (Kriticos et al. 2012).

deaths of people due to heat between 1991 and 2018 in forty-three different countries can be attributed to anthropogenic climate change. Moreover, very small changes in global average temperature can have major impacts on the human environment, such as the spread of disease, loss of land due to rising sea levels and loss of agricultural productivity.

It is important to note that despite this alarming rise, there is considerable spatial heterogeneity of climate change impacts on biological and human systems (Figure 1.1). In other words, not all regions of the planet will be affected in the same way by climate change and some regions are more vulnerable than others (Li et al. 2018). The Arctic, Africa and Asian megadeltas, and small islands, mountains and coastal cities are regions that are likely to be especially affected by future climate change (IPCC 2007). Specifically for the Arctic, an alarming warming has been taking place, being two to three times above the annual global average (IPCC 2021) (e.g., that observed in the Thwaites Glacier region, also known as the "Doomsday Glacier"). Additionally, it is important to highlight that the importance of the ocean to global climate cannot be underestimated. This is because the oceans absorb around 23 percent of annual anthropogenic CO_2 emissions into the atmosphere and consequently act as a buffer against climate change (le Quéré et al. 2013; Sabine et al. 2004). However, the increase in the temperature of ocean waters has increased its acidification and decreased its ability to absorb and retain CO_2 (i.e., further increasing the concentration of CO_2 in the atmosphere) (le Quéré et al. 2013; Orr et al. 2005), but note that we still do not understand the ocean CO_2 sink via the biological pump very well and more research is required to develop a stronger empirical base (Crisp et al. 2022). All these factors affect marine species and ocean health, for example coral reefs, ocean current patterns and melting ice sheets (mainly in Greenland and Antarctica). Therefore, the impacts of climate change on terrestrial and marine ecosystems are innumerable, complex and interrelated.

When considering different scenarios for only terrestrial ecosystems' responses to climate change, most studies only consider variables related to temperature and precipitation in their projections. However, there are different variables that can mitigate or exacerbate the impacts of climate change on a particular region or population, which makes it difficult for us to understand the real situation. For example, social, economic and environmental factors, including government responsiveness, can affect climate change impact, adaptation and vulnerability (Al-Ghussain 2019; Fawzy et al. 2020). Maple Croft created the Climate Change Vulnerability Index (CCVI) in 2017 (https://www.maplecroft.com/risk-indices/climate-change-vulnerability-index/), which combines the risk of exposure of related extreme events (e.g., cyclones, landslides, flooding and sea-level rise, drought) with the level of current vulnerability to that exposure and the ability of a country to adjust and deal with climate change. According to the CCVI, Africa, Central America, South Asia, the Caribbean and Oceania are the most vulnerable regions to climate change, while India, South Sudan and Haiti are the most vulnerable countries. Thus, although some regions are more vulnerable than others, we need to think and act in a more integrative way and be aware of the causes and consequences of climate change on the planet to try to reduce the impact on the lives of all organisms.

1.3 Climate change research: A scientometric profile

Despite the problem of climate change at a global level, this topic has only recently occupied a central role in scientific literature, agendas of political institutions and public understanding (including newspapers and television). Using the keywords "Climate change" OR "Global Warming" in the Scopus database as a search source, we identified 52,916 papers that use these words in their title (i.e., studies directly related to the topic) between the years 1910 and 2021. The first article found is titled "Does the Indian Climate Change?" and was published in *Nature* in 1910 by William Lockyer. In this paper, the author drew attention to changes of great magnitude in the meteorological elements

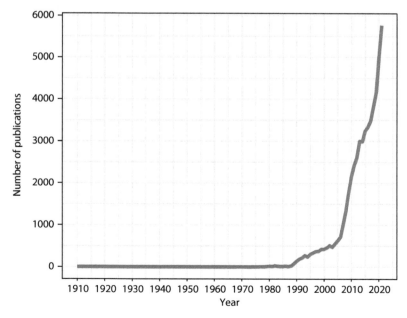

Figure 1.2 Number of articles published related to climate change and global warming between the years 1910 and 2021 available in the Scopus database.

of India. Despite the large number of articles published, there was a large increase from 2007 onwards (Figure 1.2). In fact, between 2007 and 2021, 87.08 percent (*n* = 46,081) of all articles found were published, and we identified that 160 different countries published literature on the topic, but the United States of America, United Kingdom and China published 51.63 percent (*n* = 27,324) (Figure 1.3). These articles fall within four main subject areas: Environmental Science; Earth, and Planetary Sciences; Agricultural and Biological Sciences; and Social Sciences, which together comprise 73.4 percent of all the published papers (Figure 1.4). We were also able to identify specialized journals on the topic, such as *Nature Climate Change, Global Change Biology, Climate Change, Global Environmental Change, Energy and Climate Change, Advances in Climate Change Research* and *Journal of Climate*, among others. According to Li et al. (2011), more than 2,000 different journals have been listed where papers on climate change have been published over time. These findings indicate that there is a large amount of technical knowledge available (mainly in recent decades) to assess the validity, quality and often the novelty of the published papers on this subject.

A quantitative analysis of research literature on climate change has also identified other important trends, biases and gaps. Although the first studies related to climate change were based on local observations looking for patterns at the global level, more recently the proportions of global, regional and local studies are very similar in the literature (Nabout et al. 2012). These scientometric studies have shown that about 60 percent of the studies carried out on climate change are sponsored by funding agencies (Venkatesan

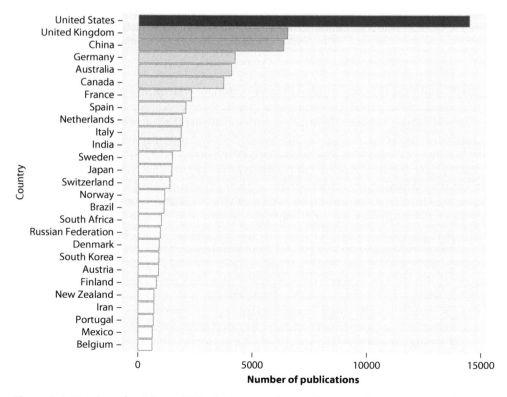

Figure 1.3 Number of articles published related to climate change and global warming by country between the years 1910 and 2021 available in the Scopus database. Note that only those countries with more than 500 publications are presented.

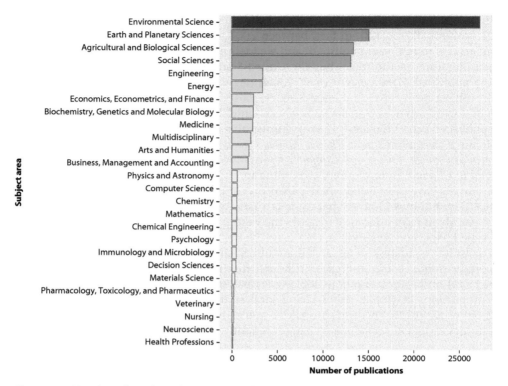

Figure 1.4 Number of articles published related to climate change and global warming by subject area between the years 1910 and 2021 available in the Scopus database.

et al. 2013) and that the largest subfield within climate change research includes continental biomass, climate modeling and impacts and effects of climate change (Haunschild et al. 2016). However, there is still little information available on and made by developing and underdeveloped countries, probably due to lack of funds, awareness agencies and a limited number of researchers and institutions (Venkatesan et al. 2013). Recently, Sharifi et al. (2021) used tools for science mapping and bibliometrics analysis on climate change impacts and identified some major thematic areas that have received little attention in the literature. Specifically, these authors showed that climate change impacts are also related to violent conflicts such as war and political tensions over climate policies (e.g., water resources and territories). This indicates that climate change is a complex problem with many facets and that it needs to be studied from different perspectives and using different approaches. Additionally, studies performed in culturally and politically diverse countries showed that the level of concern about climate change is positively related only to the level of knowledge about the causes of climate change (use of fossil fuels, deforestation, increasingly intensive agriculture, consumerism) but not to the physical characteristics of climate change (e.g., global temperature rise, glacier retreat, declines in Arctic sea ice and earlier spring flowering) (Shi et al. 2016). Thus, we need to improve how and what knowledge is disseminated in society since not all topics are related to public perceptions about climate change.

1.4 A call to action and research

Climate change threatens the future not only of our species but also of our planet. Thus, effective climate change mitigation will not be achieved if every person, institution, or country acts independently. In other words, we need collective action that encompasses different actors at different sectors of society to propose short-, medium- and long-term solutions. In this sense, after a series of international events on the subject and due to the great concern about the increase in the emission of greenhouse gases since the 1980s, in 1997 more than 180 countries joined in an international treaty called the Kyoto Protocol (Böhringer 2003). This treaty only came into force in 2005 and among the main goals, the protocol established a 5.2 percent reduction (in relation to 1990) in the emission of pollutants, mainly by industrialized countries.

Historically, the world focused its efforts on trying to limit CO_2 emissions to avoid the worst consequences of the climate crisis. However, the IPCC has recently diversified its efforts to push forward policies to mitigate (i.e., tackle the causes and minimize the possible impacts of) climate change. For example, increasing the use of renewable energy (i.e., improving energy efficiency), taxing the use of fossil fuels, CO_2 emission markets, carbon capture storage and utilization, conserving coastal wetlands because they protect us from increasing sea levels and store tonnes of carbon in their roots and soils, to achieve environmental justice for Indigenous peoples who manage almost 50 percent of the land on the planet, and to promote sustainability in mobility, industry, agriculture, fishing, livestock and more responsible consumption (i.e., reduce, reuse and recycle). Beyond mitigation, we need to reduce the negative consequences of climate change and seize the opportunities it can create. In this sense, we need, for example, to reforest and restore ecosystems, diversify crops to adapt better to more extreme climates, develop protocols for action in cases of climate emergencies and build safer and more sustainable buildings and infrastructure. However, funding these mitigation and adaptation activities remains an issue, particularly for the poorest individuals and countries. Advances have been proposed in recent years, since Global North pledged to give US$100 billion each year to poor nations for adaptation and mitigation, something they have so far failed to do. Thus, the sharing of economic resources and knowledge (i.e., synergy) is necessary to make the socio-economic system less carbon intensive and at the same time more resilient.

It is very important to highlight that most of the accumulated knowledge about the consequences of climate change is from an anthropomorphic perspective. However, we are just one of the species that inhabit this planet and we still have limited theoretical and empirical evidence on the effects of climate change on biodiversity. This limited knowledge is usually based on predictive models (but also field and laboratory experiments) that have alerted scientists and decision-makers to potential future risks and give us a scientific basis from which to reduce climate change impacts on biodiversity. Overall, climate change is expected to affect all levels of biodiversity, from genetic (e.g., mutation rates and natural selection) to biome (i.e., desertification) levels, in addition to reorganizing the way species interact, generating cascading effects. In addition, responses to climate change through phenotypic plasticity or evolutionary responses are expected, such as dispersing to areas with suitable habitat and altering gene flow, fitness and hybridization between species (Bellard et al. 2012; Parmesan 2006). Note that most of these studies involving climate change and biodiversity have focused on the consequences for vertebrates and plants, and only a few insect groups (e.g., butterflies, ants and beetles) have a relatively large amount of research in the literature (Halsch et al. 2021; Wilson and Maclean 2011). In fact, only approximately 3 percent of global climate change literature is based on invertebrates (Nabout et al. 2012). However, it is important to remember that insect biodiversity accounts for a large proportion of all known biodiversity on the planet with more than 1.5 million species described (Fox et al. 2019). Moreover, insects interact with many other species and play a major role in ecological and ecosystem services in both terrestrial and freshwater aquatic ecosystems, in addition to their influence on agriculture, human health and natural resources.

The new climate change report from the IPCC published in August 2021 shows that we are facing unprecedented changes in climate—some of which are irreversible. In this report, scientists have assessed that 6 percent of insect species will lose half of their climate-determined geographic range with global warming of 1.5 °C over the next few decades. In other words, at just 0.5 °C, there are a lot of potential effects of global warming and associated climate change on insects and their role in the functioning and maintenance of the Earth's ecosystems and biosphere. Despite the increasing knowledge on the effects of climate change on insects, it is not known in detail how and why some insect species are more vulnerable or resilient than others. Thus it is necessary to synthesize and interpret the available scientific information on climate change and insects to answer these and other questions and identify research directions to be pursued.

Key reflections

- We are living in a geologic phenomenon (i.e., Anthropocene) with a permanent change of climatic conditions in the atmosphere mainly due to human activity.
- The rapid increase in the Earth's global temperature and changes in the climate are affecting, for instance, the severity and duration of drought, frequency and activity of storms, sea levels, producing heat waves, melting glaciers and warming oceans, all of which have direct influence on ecosystems and human wellbeing.
- Despite the alarming scenario, some regions and people are more vulnerable to climate change than others.
- Climate change is expected to affect all species in the planet. However, the accumulated knowledge about the consequences of climate change is limited to the human perspective despite the effects of climate change on biodiversity are now evident.

Key further reading

- Chen, W. Y., Seiner, J., Suzuki, T., and Lackner, M. (2012). *Handbook of climate change mitigation*. New York: Springer US.
- Dunlap, R. E. and Brulle, R. J. (Eds.). (2015). *Climate change and society: Sociological perspectives*. Oxford University Press.
- Lovejoy, T. and Hannah, L. (2005). *Climate change and biodiversity*. Yale University Press, New Haven.
- Tziperman, E. (2022). *Global warming science: A quantitative introduction to climate change and its consequences*. Princeton University Press.

References

Ahmed, M., Shuai, C., Ahmed, M. 2023. Analysis of energy consumption and greenhouse gas emissions trend in China, India, the USA, and Russia. *International Journal of Environmental Science and Technology* 20: 2683–2698.

Al-Ghussain, L. 2019. Global warming: Review on driving forces and mitigation. *Environmental Progress & Sustainable Energy* 38 (1): 13–21.

Anderegg, W.R., Prall, J.W., Harold, J., and Schneider, S.H. 2010. Expert credibility in climate change. *Proceedings of the National Academy of Sciences* 107 (27): 12107–12109.

Anderson, T.R., Hawkins, E., and Jones, P.D. 2016. CO2, the greenhouse effect and global warming: from the pioneering work of Arrhenius and Callendar to today's Earth System Models. *Endeavour* 40 (3): 178–87.

Bellard, C., Bertelsmeier, C., Leadley, P., Thuiller, W., and Courchamp, F. 2012. Impacts of climate change on the future of biodiversity. *Ecology Letters* 15 (4): 365–77.

Böhringer, C. 2003. The Kyoto protocol: A review and perspectives. *Oxford Review of Economic Policy* 19 (3): 451–66.

Crisp, D., Dolman, H., Tanhua, T., McKinley, G.A., Hauck, J., Bastos, A., Sitch, S., et al. 2022. How well do we understand the land–ocean–atmosphere carbon cycle? *Reviews of Geophysics* 60 (2): e2021RG000736.

Dai, A. 2011. Drought under global warming: A review. *Wiley Interdisciplinary Reviews: Climate Change*, 2 (1): 45–65.

Dodson, J. 2012. "Climate change through time." In *Earth and Life, International Year of Planet Earth*, edited by E.F.J. de Mulder. Berlin: Springer Science+Business Media, 51–62.

Dow, K. and Downing, T.E. 2016. *The Atlas of Climate Change: Mapping the World's Greatest Challenge*. University of California Press, Berkeley: United States of America.

Duveiller, G., CapDuveiller, G., Caporaso, L., Abad-Viñas, R., Perugini, L., Grassi, G., Arneth, A., and Cescatti, A. 2020. Local biophysical effects of land use and land cover change: towards an assessment tool for policy makers. *Land Use Policy* 91: 104382.

Fawzy, S., Osman, A.I., Doran, J., and Rooney, D.W. 2020. Strategies for mitigation of climate change: A review. *Environmental Chemistry Letters* 18 (6): 2069–2094.

Fox, R., Harrower, C.A., Bell, J.R., Shortall, C.R., Middlebrook, I., and Wilson, R.J. 2019. Insect population trends and the IUCN Red List process. *Journal of Insect Conservation* 23 (2): 269–78.

Halsch, C.A., Shapiro, A.M., Fordyce, J.A., Nice, C.C., Thorne, J.H., Waetjen, D.P., and Forister, M.L. 2021. Insects and recent climate change. *Proceedings of the National Academy of Sciences* 118 (2): e2002543117.

Haunschild, R., Bornmann, L., and Marx, W. 2016. Climate change research in view of bibliometrics. *PloS One* 11 (7): e0160393.

Hulme, M., Barrow, E.M., Arnell, N.W., Harrison, P.A., Johns, T.C., and Downing, T.E. 1999. Relative impacts of human-induced climate change and natural climate variability. *Nature* 397 (6721): 688–91.

IPCC Intergovernmental Panel on Climate Change 2007. Climate Change 2007: Impacts, adaptation and vulnerability. Summary for policy makers. http://www.ipcc.cg/SPM13apr07.pdf

IPCC Intergovernmental Panel on Climate Change. 2021. Climate Change 2021: The Physical Science Basis. https://www.ipcc.ch/report/ar6/wg1/

Jankó, F., Papp Vancsó, J., and Móricz, N. 2017. Is climate change controversy good for science? IPCC and contrarian reports in the light of bibliometrics. *Scientometrics* 112 (3): 1745–1759.

Kriticos, D.J., Webber, B.L., Leriche, A., Ota, N., Macadam, I., Bathols, J., and Scott, J.K. 2012. CliMond: Global high-resolution historical and future scenario climate surfaces for bioclimatic modelling. *Methods in Ecology and Evolution* 3 (1): 53–64.

Le Quéré, C., Andres, R.J., Boden, T., Conway, T., Houghton, R.A., House, J.I., Marland, G., et al., 2013. The global carbon budget 1959–2011. *Earth System Science Data* 5 (1): 165–85.

Letcher, T.M. (ed). 2021. *Climate change: Observed impacts on planet Earth*. Amsterdam, Netherlands: Elsevier

Li, D., Wu, S., Liu, L., Zhang, Y., and Li, S. 2018. Vulnerability of the global terrestrial ecosystems to climate change. *Global Change Biology* 24 (9): 4095–4106.

Li, J., Wang, M.H., and Ho, Y.S. 2011. Trends in research on global climate change: A science citation index expanded-based analysis. *Global and Planetary Change* 77 (1–2): 13–20.

Lockwood, M. 2010. Solar change and climate: an update in the light of the current exceptional solar minimum. *Proceedings of the Royal Society A: Mathematical, Physical and Engineering Sciences* 466 (2114): 303–29.

Meehl, G.A., Stocker, T.F., Collins, W.D., Friedlingstein, P., Gaye, A.T., Gregory, J.M., Kitoh, A., et al. 2007. "Global climate projections. Climate change 2007: I physical science basis." In *Contribution of Working Group I to the Fourth Assessment Report of the Intergovernmental Panel on Climate Change*, edited by S. Solomon, D. Qin, M. Manning, Z. Chen, M. Marquis, K.B. Averty, M. Tignor, and H.L. Miller. Cambridge: Cambridge University Press, 747–845.

Mitchell, J.F. 1989. The "greenhouse" effect and climate change. *Reviews of Geophysics* 27 (1): 115–39.

Moritz, R. E., Bitz, C. M., and Steig, E. J. 2002. Dynamics of recent climate change in the Arctic. *Science* 297 (5586): 1497–1502.

Nabout, J.C., Carvalho, P., Prado, M.U., Borges, P.P., Machado, K.B., Haddad, K.B., Michelan, T.S., et al., 2012. Trends and biases in global climate change literature. *Natureza & Conservação* 10 (1): 45–51.

NOAA—National Centers for Environmental information 2020 Climate at a glance: Global Time Series. <http://www.ncdc.noaa.gov/cag/global/time-series/globe/land_ocean/ann/9/1880–2020>

Oreggioni, G.D., Ferraio, F.M., Crippa, M., Muntean, M., Schaaf, E., Guizzardi, D., Solazzo, E., et al. 2021. Climate change in a changing world: Socio-economic and technological transitions, regulatory frameworks and trends on global greenhouse gas emissions from EDGAR v. 5.0. *Global Environmental Change* 70: 102350.

Orr, J.C., Fabry, V.J., Aumont, O., Bopp, L., Doney, S.C., Feely, R.A., Gnanadesikan, A., et al. 2005. Anthropogenic ocean acidification over the twenty-first century and its impact on calcifying organisms. *Nature* 437 (7059): 681–86.

Parmesan, C. 2006. Ecological and evolutionary responses to recent climate change. *Annual Review of Ecology, Evolution and Systematics* 37: 637–69.

Pielke Sr, R.A., Marland, G., Betts, R.A., Chase, T.N., Eastman, J.L., Niles, J.O., Niyogi, D.D.S., et al. 2002. The influence of land-use change and landscape dynamics on the climate system: Relevance to climate-change policy beyond the radiative effect of greenhouse gases. *Philosophical Transactions of the Royal Society of London A: Mathematical, Physical and Engineering Sciences* 360 (1797): 1705–1719.

Sabine, C.L., Feely, R.A., Gruber, N., Key, R.M., Lee, K., Bullister, J.L., Wanninkhof, R., et al. 2004. The oceanic sink for anthropogenic CO2. *Science* 305 (5682): 367–71.

Sharifi, A., Simangan, D., and Kaneko, S. 2021. Three decades of research on climate change and peace: A bibliometrics analysis. *Sustainability Science* 16: 1079–1095.

Shi, J., Visschers, V.H., Siegrist, M., and Arvai, J. 2016. Knowledge as a driver of public perceptions about climate change reassessed. *Nature Climate Change* 6 (8): 759–62.

Uppenbrink, J. 1996. Arrhenius and global warming. *Science* 272 (5265): 1122–1122.

Venkataramanan, S. 2011. Causes and effects of global warming. *Indian Journal of Science and Technology* 4: 226–29.

Venkatesan, M., Gopalakrishnan, S., and Gnanasekaran, D. 2013. Growth of literature on climate change research: A scientometric study. *Journal of Advances in Library and Information Science* 2 (4): 236–42.

Vicedo-Cabrera, A.M., Scovronick, N., Sera, F., Royé, D., Schneider, R., Tobias, A., Astrom, C., et al. 2021. The burden of heat-related mortality attributable to recent human-induced climate change. *Nature Climate Change* 11 (6): 492–500.

Wilson, R.J. and Maclean, I.M. 2011. Recent evidence for the climate change threat to Lepidoptera and other insects. *Journal of Insect Conservation* 15: 259–68.

Winkler, K., Fuchs, R., Rounsevell, M., and Herold, M. 2021. Global land use changes are four times greater than previously estimated. *Nature Communications* 12 (1): 2501.

Wuebbles, D. J. and Jain, A. K. 2001. Concerns about climate change and the role of fossil fuel use. *Fuel Processing Technology* 71 (1–3): 99–119.

CHAPTER 2

Evidence from the fossil record on insect response to climate change

Ellen D. Currano

2.1 Introduction

Human civilization has set Earth up as a time machine that could take us back to a climate state not seen since the warmest sustained interval of the last 66 million years, the Early Eocene Climatic Optimum (EECO, 53–48 million years ago or Ma). By 2140, if we continue to burn fossil fuels at the current rate, we will reverse a 50-million-year cooling trend, and the climate across much of the globe will resemble the Eocene (Burke et al. 2018). Thus the fossil record has considerable potential to inform about how insects responded to global warming in the past and how they might respond in the future. Fossil studies are an integral component to understanding biotic response to climate change because they have different limitations than experiments or short-term ecological studies. Fossil assemblages record the net effects of climate change in natural ecosystems over thousands to millions of years, enabling research on long-term responses and feedback mechanisms. In this chapter, I review insect responses to climate change during the Cenozoic era, which includes intervals that were both warmer and cooler than today. I will first describe the insect fossil record and the paleoclimate record. Then, following the paleoclimatic framework of Zachos et al. (2001), I explore insect responses to long-term trends, cyclic variations and geologically abrupt aberrations in climate. Promising areas for new research are provided throughout the chapter.

2.2 The insect fossil record

2.2.1 *Insect body fossils*

Insect bodies perfectly preserved in amber, down to the tiniest hair, are among the most charismatic of all fossils, capturing public imagination as a fictional source of dinosaur DNA (Figure 2.1a). Unfortunately, amber deposits are rare, with few insect-producing

Ellen D. Currano, *Evidence from the fossil record on insect response to climate change*. In: *Effects of Climate Change on Insects.*
Edited by: Daniel González-Tokman and Wesley Dáttilo, Oxford University Press. © Oxford University Press (2024).
DOI: 10.1093/oso/9780192864161.003.0002

localities from the Cenozoic. Insects are more commonly preserved as compressions and impressions in lacustrine, pond, lagoonal and shallow marine settings (Figure 2.1b; Labandeira 1999). But even in these settings, it takes exceptional circumstances for insects to make it into the fossil record as they do not have mineralized hard parts. Insect bodies must be transported to an aquatic environment, break surface tension and sink to the bottom, get buried by sediments, and be collected and described by paleontologists (Karr and Clapham 2015; Smith 2012). How well different species are preserved is affected by body size, mass and shape as well as environmental variables like bathymetry, energy level and bottom-water oxygen content. Taxa that are of medium body size, heavy and lack large wings are most likely to have published descriptions, as exemplified by the diverse and abundant Coleoptera fossil record (Archibald and Makarkin 2006; Smith 2001). Even the very best fossil deposits, termed Konservat-Lagerstätten, do not capture

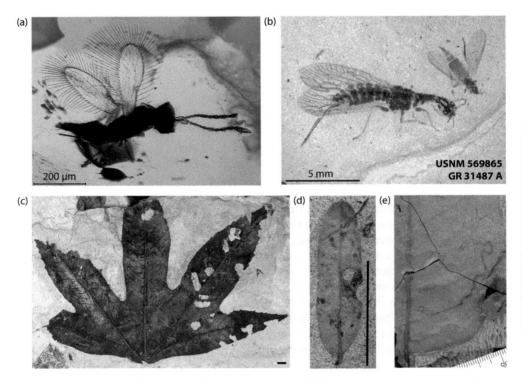

Figure 2.1 Example fossil insects and herbivory damage. a) *Macalpinia canadensis* preserved in Cretaceous amber. Photograph by George Poinar Jr. and John T. Huber, CC BY 3.0 <https://creativecommons.org/licenses/by/3.0>, via Wikimedia Commons. b) *Archiinocellia protomaculata* compression fossil from the Eocene Green River Formation. Photograph courtesy of Smithsonian National Museum of Natural History, CC0, via Wikimedia Commons. c) Eocene *Macginitiea* sp. leaf with hole and margin feeding damage (specimen UW-PB-232). d) PETM legume leaflet with hole feeding DT 2, margin feeding DT 12 and piercing and sucking DT 46 (specimen USNM 618006, reprinted from Currano et al. 2016). e) PETM leaf mine on Dicot sp. WW005 (specimen USNM 618007). Scale bars 200 um in a, 5 mm in b, 1 cm in c and d. Tick marks in mm in e.

the full diversity of living insect communities, and changes in fossil insect abundances probably have more to do with environmental changes (flash flooding, for instance) than with actual changes in insect population sizes (Schachat and Labandeira 2021). Therefore, most paleoentomological research has focused on taxonomic, morphologic, evolutionary and biogeographic analyses.

Considerable effort has been made to compile published taxon occurrences (Labandeira et al. 1993) and enter them into the Paleobiology Database (Clapham et al. 2016). These data are instrumental to our understanding of insect family diversification and extinction in the geologic past (e.g., Clapham et al. 2016; Condamine et al. 2016; Labandeira et al. 1993). Statistical analyses of these data demonstrate that the insect body fossil record is patchy across space and time, with occurrences heavily concentrated in ninety Konservat-Lagerstätten that range in age from the Carboniferous to the recent (Schachat and Labandeira 2021). The vast majority of these sites are at mid- to high-latitude in the Northern Hemisphere, and only Dominican and Tanzanian amber deposits preserve low-latitude Cenozoic insect faunas (Figure 2.2).

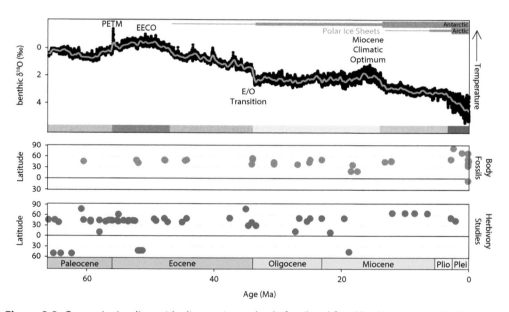

Figure 2.2 Cenozoic timeline with climate, insect body fossil and fossil herbivory records. Oxygen isotope values of benthic foraminifera are a proxy for global mean temperature, with lower values indicating warmer temperatures. Climate records, including $\delta^{18}O$ data, climate states and estimates of polar ice sheet extent, are from Westerhold et al. (2020). Polar ice sheet extent is indicated by gray bars, and climate states by colored bars, with Hothouse in red, Warmhouse in orange, Coolhouse in light and medium blue and Icehouse in dark blue. The insect body fossil record shown here includes Cenozoic Konservat-Lagerstätten and is modeled after Schachat and Labandeira (2021). Sites were selected using the Paleobiology Database and Smith (2012). The herbivory record consists of unbiased insect damage censuses published prior to 2021, and these data were previously compiled by Currano et al. (2021).

2.2.2 *Fossil herbivory records*

Leaf compression fossils preserve insect herbivore feeding traces, making the paleobotanical record an important source of information on insect response to ancient climate change. Today, plants make up ~80 percent of global biomass, and while leaves are estimated to compose just 7 percent of plant biomass, leaf biomass is still more than double total animal biomass (Bar-On et al. 2018). Further, trees shed and replace their leaves regularly, creating an abundance of potential fossils. And so unsurprisingly, fossil leaves have far greater spatial and temporal coverage than fossil insects and capture important time intervals and regions that lack insect body fossils (Figure 2.2). Each leaf fossil assemblage typically represents tens to thousands of years, and it is rare for leaves to be reworked into younger sediments (Wilf 2008). Leaf fossils are generally preserved in fluvial or lacustrine environments, constraining preservational biases and facilitating comparisons among fossil floras (Wing and DiMichele 1995). Last, the very same leaf fossils examined for herbivory can also be used to reconstruct climate, plant community composition and paleoecology.

Insect herbivore damage on leaves informs us about both the insects present in an ecosystem and provides direct evidence of fossil food webs. Herbivory damage can be easily distinguished from physical damage and post-mortem processes (e.g., detritivory or taphonomic alteration) as follows (Labandeira 1998; Labandeira et al. 2007). First, if leaf tissue is removed while the leaf is alive, cells along the damaged area become enlarged or multiply, creating a dark, thickened "reaction rim" that is readily observable in fossils. Second, insect herbivores frequently target particular plant tissues, organs, or species and make more consistent patterns and morphologies of damage than detritivores or physical processes. Last, insect-specific patterns of damage like cusp marks or pupal chambers can be observed. Insect damage is classified into morphologically distinct, diagnosable damage types (DTs) using *The guide to insect (and other) damage types on compressed fossil plant fossils* (Labandeira et al. 2007). DTs are grouped into the functional feeding groups (FFGs) hole feeding, margin feeding, skeletonization, surface feeding, mining, galling, piercing and sucking (Figure 2.1c–e). Although excluded from herbivory analyses, oviposition and fungal/pathogen DTs are also included in the guide.

DTs can be subdivided into generalized DTs, most often made by polyphagous insects, and specialized DTs, typically made by insects that eat only one or several closely related plant species. In some cases, morphologically distinct DTs, typically leaf mines and galls, can be attributed to an insect family or even genus, and these have proven important in dating insect clades and diversification events (e.g., Maccracken et al. 2021; Wilf et al. 2000). More often though, this is not possible as some DTs can be made by many unrelated insect species. Further complicating matters is the fact that some insects can make multiple DTs. However, important neontological work by Carvalho and colleagues (2014) showed a strong, positive correlation between the number of insect species captured feeding on a plant host and the number of DTs made by these insects. Thus, DT diversity, which can be examined at the level of individual plant host or bulk flora, is a proxy for insect herbivore diversity.

Herbivory is analyzed in the fossil record using unbiased insect damage censuses (Wilf and Labandeira 1999), in which every leaf from a fossil site that can be identified and is > 50 percent complete is scored for the presence/absence of every DT. If possible, multiple quarries should be excavated laterally along a fossil-bearing horizon to capture spatial heterogeneity in plant composition and herbivory (Burnham et al. 1992; Currano 2009), and a total of 1,000 leaves should be excavated. Herbivory damage is most often quantified using the metrics frequency (percentage of leaves with herbivory damage), diversity (number of DTs, standardized by sample size), composition (which DTs/FFGs occur at a site, often examined using ordinations), and intensity (percent of leaf area damaged) (Currano et al. 2021). Intensity data are extremely time-consuming to produce and, as of 2021, have only been reported for eleven Cenozoic fossil sites. Herbivory metrics can be examined on the bulk floras or on individual plant hosts, and values are often reported for total damage, specialized damage and mining and galling damage.

As of 2021, sixty-one fossil insect damage censuses have been conducted on Cenozoic angiosperm-dominated floras (Currano et al. 2021). The Paleocene and Eocene are the best-sampled time intervals, and two-thirds of all sites are located at mid-latitude in North America or Eurasia (Figure 2.2). The spatial and temporal distribution of fossil insect damage censuses does not reflect the availability of paleofloras that could be studied for herbivory. Rather, it is determined by the research interests and geographic locations of the small community of paleontologists who have undertaken these analyses. Paleobotanists are encouraged to add insect herbivory to the arsenal of paleoecological analyses conducted on fossil floras.

2.3 The Cenozoic paleoclimate record

The best record of global climate change during the Cenozoic comes from oxygen isotope analyses of bottom-dwelling, deep sea fossil foraminifera (Figure 2.2). The oxygen isotopic composition ($\delta^{18}O$) of their carbonate shells depends on the temperature in which they developed (higher $\delta^{18}O$ indicates cooler temperatures) and $\delta^{18}O$ of the seawater. Seawater $\delta^{18}O$ is influenced by polar ice volume; ^{16}O is preferentially incorporated into polar ice sheets, and so as ice volume increases, $\delta^{18}O$ also increases. Thus, during the Paleocene and early Eocene, prior to the formation of large continental ice sheets, the $\delta^{18}O$ record is indicative of changing ocean bottom-water temperatures, with the magnitude of changes corresponding to changes in globally averaged surface temperature. After large Antarctic ice sheets developed in the middle to late Eocene, the $\delta^{18}O$ record represents a mix of temperature and ice volume changes.

Since the first publication of a Cenozoic deep-sea foraminifer isotope record by Kennett and Shackleton (1975), paleoceanographers have produced and compiled additional records, increasing temporal resolution and documenting long- and short-term climate changes. Zachos et al. (2001) introduced the framework of trends, rhythms and aberrations in climate, a key advancement in our understanding of global climate dynamics. Climate trends are gradual, occurring over 10^5 to 10^7 years, and primarily driven by tectonics. The most recent compilation of deep-sea foraminiferal $\delta^{18}O$ has a temporal resolution of one sample every 2200 years for 0–34 Ma and one sample every 4400 years for 34–67 Ma (Westerhold et al. 2020). I will discuss this record in detail in section 2.4. Rhythms, or cyclic changes with durations of 10^4 to 10^6 years, are caused by regular changes in Earth's orbital parameters (Milankovitch cycles). Pleistocene and Holocene (2.588 Ma to the present) glacial-interglacial cycles were orbitally driven, and migrations of plants and insects have been well documented, particularly since the last glacial maximum (section 2.5). Climate aberrations are rapid (10^3 to 10^5 years), rare, and often accompanied by large perturbations to the global carbon cycle. The Paleocene–Eocene Thermal Maximum (PETM; 56 Ma) is the best-studied of these events and will be considered in section 2.6.

A variety of paleoclimate proxies have been developed to reconstruct terrestrial climate at the local to regional scale. In many cases, the very same fossil leaves analyzed in an insect damage census can be used to estimate past temperature and precipitation. Studies of modern forest ecosystems have demonstrated correlations between climate variables and morphological characters of woody dicot leaves (summarized in Peppe et al. 2018), which can then be used to extrapolate ancient climates from fossil leaf assemblages. For example, the percent of species with smooth vs. serrate margins correlates to mean annual temperature (MAT) in subtropical and temperate forests (Greenwood and Wing 1995), and average leaf size is related to mean annual precipitation (MAP; Jacobs 2002). Taxonomic composition can also be used to reconstruct climate for more recent floras (Neogene to recent), and a variety of methods have been proposed to use the climatic tolerances of nearest living relatives to estimate ancient climate parameters like MAT, MAP, growing season temperature and precipitation, growing season length, and mean cold month temperature and precipitation. Finally, geochemical proxies using pedogenic carbonate nodules, enamel, speleothems, leaf waxes and microbial biomarkers provide additional means to reconstruct past climate.

2.4 Trends: Cenozoic climate change

2.4.1 *Setting the scene: Climate and vegetation*

Building upon previous paleoclimate studies, Westerhold et al. (2020) divided the Cenozoic into major climate states (Figure 2.2). The first 10 million years of the Cenozoic (66–56 Ma) are characterized as a Warmhouse state, with temperatures > 5 °C warmer than today. Temperatures began to steadily rise around 60 Ma, and by the start of the Eocene (56 Ma), Earth shifted into a Hothouse state, with temperatures > 10 °C warmer than the present day and abrupt global warming events (aberrations) or hyperthermals, the largest of which is the PETM. The EECO marked the highest sustained temperatures of the Cenozoic. Temperatures gradually cooled through the remainder of the Eocene, first returning Earth to the Warmhouse from 47–34 Ma and then transitioning to the Coolhouse near the Eocene–Oligocene boundary (34 Ma). The large drop in $\delta^{18}O$ in the early Oligocene was likely due to a combination of cooling and major expansion of Antarctic ice sheets. Following the Miocene Climatic Optimum (17–14 Ma), $\delta^{18}O$ steadily decreased, representing a combination of cooling and ice sheet expansion. The current Icehouse state began around 3.3 Ma (middle Pliocene), with fully established ice sheets at both poles.

Closed-canopy, broad-leaved forests covered much of the globe during the warm, humid Paleocene and early Eocene, as evidenced by fossil plants and climatically informative geologic deposits like coals, evaporites and bauxites. Latitudinal vegetation bands corresponded to tropical everwet (~30°N to 30°S), subtropical summerwet (tropical everwet belt to ~45°N or ~60°S), warm temperate (extending above 60°N and covering coastal Antarctica) and warm/cool temperate forests (Willis and McElwain 2014). Cooling during the late Eocene and early Oligocene caused significant contraction of megathermal rainforests, particularly in the northern hemisphere (Morley 2007). By ~30 Ma, cool and cold temperate forests covered much of land above 45°N, southern South America and the ice-free parts of Antarctica (Willis and McElwain 2014). Continued cooling and aridification of continental interiors through the Oligocene and Miocene further restricted the distributions of moist forests, although the Miocene Climatic Optimum is likely to have allowed temporary expansion (Morley 2007). Grassland expansions during this time (reviewed in Strömberg 2011) further fragmented forests and created new habitats and food sources for insects.

2.4.2 *Insect diversity dynamics*

Molecular phylogenies, dated using both insect body fossils and diagnostic insect damage, are increasingly being used to understand links between insect diversity dynamics, vegetation and climate change. A review and synthesis of recent phylogenetic research suggests that Cenozoic climate trends affected plant distributions, abundances and diversities, which in turn affected speciation and extinction of associated herbivore groups (Nyman et al. 2012). The authors propose "diffuse cospeciation" and "resource abundance-dependent diversity patterns" as key processes. Under diffuse cospeciation, climate change causes fragmentation of plant geographic ranges, insect herbivores diversify in parallel with their plant hosts, and then, when geographic ranges fuse again, colonize the host's recently diverged relatives. Under resource abundance-dependent diversity patterns, insect speciation events are driven by increased abundance and distribution of the host plant, which is often driven by climate changes. Integration and expansion of insect fossil records, paleovegetation reconstructions and models and paleoclimate proxy data and models provide additional avenues for examining links between diversification events and climate change. Fossil studies also complement molecular phylogenetic work because they alone permit examination of extinction events, or lack thereof.

At the global scale, long-term temperature and vegetation trends may impact insect diversity, although caution must be taken due to uneven sampling in the fossil record. Analyses of fossil insect diversity are traditionally performed at the family level, rather than the genus- or species level, as sampling is more regular and less biased, and insect families are better established (Labandeira and Sepkoski 1993). Sampling-standardized analyses of insect family diversity indicate that richness peaked in the Early Cretaceous, ~125 Ma, declined through the Late Cretaceous, and recovered in the Early Paleogene (Clapham et al. 2016; Schachat et al. 2019). Modern family diversity is thought to be comparable to, or slightly lower than, the Early Cretaceous peak. Both insect diversity curves and phylogeny-based diversification analyses show relatively high origination rates and net diversification at the family-level during the globally warm Paleocene–Eocene and then declines near the Eocene–Oligocene boundary, as Earth shifts from Warmhouse to Coolhouse (Figure 2.3; Condamine et al. 2016; Schachat et al. 2019). No temporal trends in insect damage diversity or frequency have been observed globally across the entire Cenozoic (Figure 2.3; Currano et al. 2021), although it is important to note that the data are quite unevenly distributed in time and space. Additionally, lumping geographically distant sites may obscure regional diversification or extinction events.

2.4.3 *Herbivory records*

The fossil record clearly demonstrates that temperature changes affect herbivory in natural ecosystems over thousands to millions of years. Statistically significant relationships exist in the Cenozoic global herbivory database between reconstructed MAT and total, specialized and galling damage diversity (Figure 2.4; Currano et al. 2021), and correlations were also observed in an older dataset that included Paleozoic through Cenozoic sites (Pinheiro et al. 2016). The Cenozoic database encompasses 64 million years and sites span the globe, from Antarctica to the Arctic; it is remarkable that any relationship was

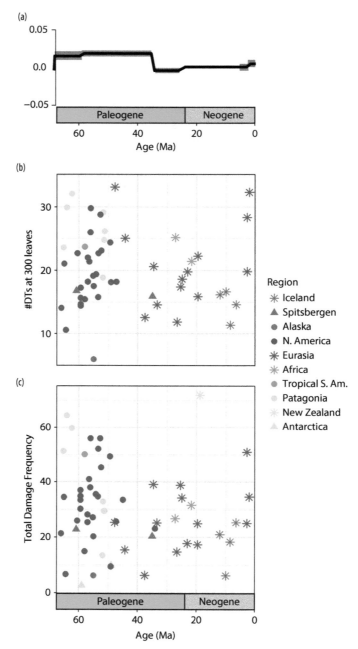

Figure 2.3 Cenozoic records of insect diversity. a) Net insect family diversification rate (origination minus extinction) calculated by Condamine et al. (2016) using insect body fossil data and a Bayesian model. b) Insect damage diversity at each Cenozoic, angiosperm-dominated site where an unbiased damage census was conducted. These data were previously compiled by Currano et al. (2021). Damage diversity is reported as the number of DTs observed on 300 fossil leaves, standardizing for unequal sample sizes. c) Total damage frequency, or the percentage of leaves in a damage census with any herbivory damage, at each damage census site. The color and shape of the points in B and C denotes geographic region, as shown in the figure legend. The Paleogene Period (66–23 Ma) consists of the Paleocene, Eocene and Oligocene Epochs, and the Neogene Period (23 Ma–11.7 ka) consists of the Miocene, Pliocene and Pleistocene Epochs.

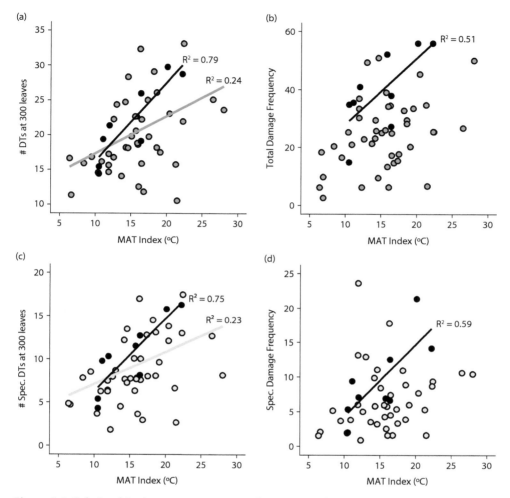

Figure 2.4 Relationships between reconstructed mean annual temperature (MAT index, °C) and herbivory frequency and diversity. Colored points are sites from the global insect damage census dataset of Currano et al. (2021). Black points denote sites from the Paleocene and Eocene of the Bighorn Basin, Wyoming, USA (Currano et al. 2010). Total (a) and specialized (c) damage diversity are reported as the number of DTs (or specialized DTs) observed on 300 fossil leaves, standardizing for unequal sample sizes. Total damage frequency (b) is the percent of leaves in a flora with any herbivory damage, and specialized damage frequency (d) is the percentage with specialized damage. Linear models with an R^2 value above 0.2 and a p-value less than 0.05 are shown for the global (colored lines) and Bighorn Basin (black lines) datasets.

observed. Much stronger correlations exist in regional mid- to high-latitude datasets. Nine sites in the Bighorn Basin, Wyoming, USA (paleolatitude ~50°N), that span 6 million years and include the PETM and the EECO, document strong, positive correlations between reconstructed MAT and damage diversity metrics (Figure 2.4; Currano et al. 2010). Weaker but still significant positive correlations were observed between reconstructed MAT and damage frequency (Figure 2.4; Currano et al. 2010). Similar results were obtained from a sequence of Neogene sites in Iceland, where long-term cooling caused a decline in damage diversity (Wappler and Grímsson 2016). Leaf area damaged data has only been collected at nine Eocene and two Paleocene sites, and these display a weak positive correlation with temperature ($R^2 = 0.36$, $p = 0.05$).

A variety of processes could drive increased herbivory at high temperature, many of which are discussed in greater detail in other chapters in this volume. Damage diversity is a proxy for insect herbivore diversity, and high damage diversity during intervals of global warmth could indicate either migration or speciation events. Given the Bighorn Basin's location in the mid-latitudes and the (albeit limited) documentation for plant taxa migrating northwards during the Eocene as temperatures warmed (Wing and Currano 2013), elevated damage diversity within the basin during warmer intervals has been attributed to northward migration of diverse, thermophilic insect herbivore species (Currano et al. 2010). Mid-latitude sites also dominate the global herbivory database; insect species could have migrated northwards, or higher in elevation, with warming and southwards, or down in elevation, with cooling. Insect speciation and extinction must also be considered as drivers of damage diversity trends. During the warm Early Paleogene, broad-leaved, closed canopy forests had their maximum geographic extent, and mountain ranges may have posed greater biogeographic barriers than they do in cooler climates (Janzen 1967). It is possible that global insect herbivore diversity, and thus damage diversity, is elevated during Warmhouse and Hothouse intervals due to a combination of the species-area effect and allopatric speciation. Insect herbivore extinctions, manifested in the fossil record as decreases in damage diversity, could have occurred as mesothermal forests contracted during cooling intervals.

Elevated damage frequency and intensity with warming is driven by direct effects of climate change on insect herbivores and by indirect effects via changes to plant nutritional content. Warming increases insect metabolic and growth rates, decreasing development time and larval mortality rates and permitting an increased number of generations per year, provided that the warming is not so extreme as to exceed a species' thermal tolerance (Bale et al. 2002). All these factors have the potential to increase population densities and thus herbivory frequency and intensity. However, plants grown experimentally at elevated CO_2 often have decreased C:N ratios (Robinson et al. 2012), which could negatively affect insect population densities but could also lead to compensatory feeding. The fossil record, though, demonstrates significant variations in plant species composition with climate change, and the Bighorn Basin paleofloras suggests that high temperature and CO_2 select for legumes and other plants that commonly have symbioses with N_2-fixing bacteria (Figure 2.5e). Today, legumes and other plants with N_2-fixing symbionts have higher leaf nitrogen contents than plants without N_2-fixing symbionts (Adams et al. 2016). Incorporation of high-nitrogen plant material into soils likely provides a source of biologically accessible nitrogen and supports larger and more diverse insect populations. Significant positive correlations are observed in the fossil record between prevalence of N_2-fixing plants and total and specialized damage diversity, total and specialized damage frequency and damage intensity (Figure 2.5; Currano et al. 2016; Currano and Jacobs 2021). Experimental and ecological studies are necessary to further disentangle the influences of the processes discussed in this paragraph, but the fossil record demonstrates that over long time intervals and in natural systems, the end result of warming and elevated CO_2 is increased herbivory.

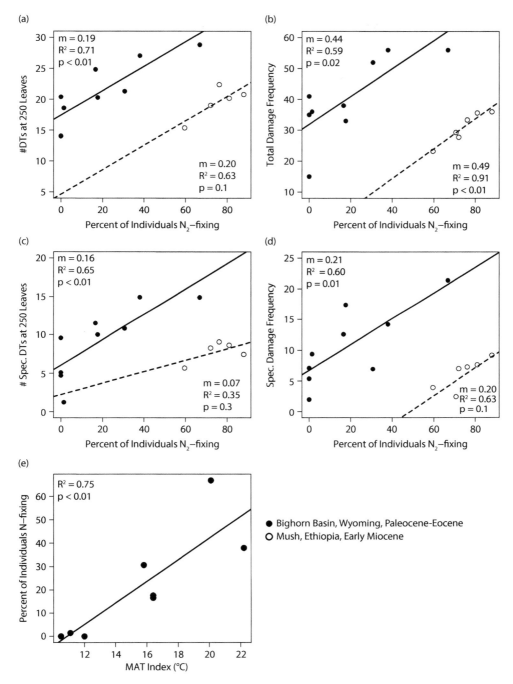

Figure 2.5 Herbivory on bulk floras from the Paleocene-Eocene Bighorn Basin, USA (filled symbols) and early Miocene Mush Valley, Ethiopia (open symbols) has significant, positive correlations with the prevalence of plants in the flora with nitrogen-fixing symbionts. Prevalence is given as the percent of leaves at a site that could be taxonomically placed within clades that today have symbioses with N_2-fixing bacteria (legumes and *Alnus*). Herbivory metrics include a) damage diversity (# DTs observed on 250 leaves; previously reported by Currano and Jacobs, 2021),

2.4.4 *Future research directions*

There are relatively few paleoecologists who study interactions among climate, plants and insects in deep time, and so there is tremendous potential for new agendas and new discoveries. Here are just three ideas. First, herbivory responses to water availability and drought should be examined in the fossil record, as these have been observed to accentuate modern biotic responses (Scherber et al. 2013). Application of new paleoclimate proxies as well as more qualitative comparisons of herbivory from wetter and drier locations would advance our understanding significantly. Second, evaluating links between climate, vegetation and herbivory in the fossil record may require data at broader spatial scales than that of plant macrofossil assemblages. According to both the global and Bighorn Basin herbivory datasets, plant diversity is not strongly affected by temperature, nor is herbivory affected by plant diversity (Currano et al. 2010, 2021). However, plant diversity was quantified in these datasets using only the leaves in the insect damage censuses, which may not be representative of regional plant diversity. Pollen samples capture a much greater source area than fossil leaf assemblages and therefore may constrain regional changes in diversity and vegetation type better. Last, only a few studies have considered herbivory over long timescales on individual plant clades (Leckey and Smith 2017). Work like this is essential to record colonization events and host-shifts (e.g., Winkler et al. 2009) and disentangle the effects of plant turnover from direct effects of climate change on insects.

2.5 Rhythms: Quaternary glacial-interglacial cycles

The past 2.6 million years (i.e., the Quaternary) are characterized by glacial-interglacial cycles driven by regular changes in Earth's orbital parameters (rhythms). Temperature shifts as large as 7 °C occurred over the course of decades and caused northern hemisphere ice sheets to expand, covering much of the mid-latitude continental landmasses, and retreat (Stauffer 1999). Often, these shifts in temperature and ice sheet extent were accompanied by large-scale changes in precipitation (e.g., Shuman and Serravezza 2017). Vegetation change in the Northern Hemisphere during the most recent deglaciation has been well-studied, and compilation of many hundreds of fossil pollen records allow scientists to map spatial distributions of individual plant species, plant associations and biomes at millennial-scale resolution for the past 21,000 years (e.g., Huntley and Birks, 1983; Overpeck et al. 1992; Webb 1993; Williams et al. 2004). This work has demonstrated large changes in biogeographic ranges, individualistic responses of plant taxa to climate change and the existence of "no-analog" communities, or combinations of plant taxa not observed today (Jackson and Overpeck 2000; Overpeck et al. 1992).

Figure 2.5 (*Continued*) b) damage frequency (the percent of leaves with any damage; previously reported by Currano and Jacobs 2021), c) specialized damage diversity (# specialized DTs observed on 250 leaves) and d) specialized damage frequency (the percentage of leaves with specialized damage). Regression lines were determined for each region, and the slope (m), R^2, and p value for each are given. e) A strong, positive correlation exists in the Bighorn Basin between reconstructed mean annual temperature (MAT index, °C) and prevalence of plants with nitrogen-fixing symbionts, reported as in a–d.

Quaternary insect fossils are relatively common in anoxic, water-logged deposits like peat bogs or lake-bottom sediments. Their original chitin exoskeletons are extremely well preserved, allowing for detailed comparison with living faunas. Perhaps surprisingly, given the sudden and severe climate and vegetation changes described above, the vast majority of Quaternary fossil specimens belong to extant species (Buckland and Coope 1991). Influential studies of beetle fossils from glacial and interglacial deposits demonstrate the capacity for species to shift their geographic ranges to track suitable climates (e.g., Coope 1994). Taxa from British glacial deposits occur today in northern Eurasia, northern North America, or high elevation regions of Asia; species from warmer, late glacial sites in Britain presently live in continental Europe, Turkey and Northern Africa (Coope 1994). Elevational and latitudinal range shifts associated with climate change have similarly been documented in New Zealand, using fossil beetle assemblages spanning the last 1.3 million years (Marra and Leschen, 2011). Coope (1994, 2004) hypothesized that the high frequency, intensity and rapidity of Quaternary climate changes, coupled with the high mobility of insect populations, allowed species to "maintain constancy for millions of generations" and suppressed origination and extinction. Populations would have regularly split up and rejoined as they migrated across complex terrains, resulting in well-mixed gene pools and stabilizing selection. Species that did go extinct had their mobility limited by either geography (e.g., occurrence on islands or equatorial mountains) or photoperiod requirements (Coope 1994).

As described above, the Quaternary insect fossil record supports the hypothesis that long-term climate trends are more likely to be "strong creative forces in macroevolution" than glacial-interglacial cycles (Jansson and Dynesius 2002). However, new research is needed to rigorously test this idea. Most Quaternary insect fossils are from mid- to high-northern latitudes (e.g., Abellán et al. 2011), and evolutionary and ecological patterns observed may differ at lower latitudes. DNA analyses document Quaternary insect speciation events in parts of Europe not affected by glaciation (e.g., Abellán et al. 2011), suggesting diversification events may also have occurred at lower latitudes. Comparisons of insect herbivore damage on *Quercus* section *Heterobalanus* leaves from 3 Ma and modern forests in southwestern China show a 50 percent increase in DTs observed in the modern and extinction of just one gall DT (Su et al. 2015). Future research should focus on fossil faunas and floras from subtropical and tropical latitudes, including the discovery of new localities. Another promising line of research would be to integrate Quaternary insect fossil records with the highly detailed paleobotanical record, as suggested by Jablonski (2008). Extensive pollen data are freely available in the Neotoma Paleoecology Database, and efforts are currently underway to add fossil insect data (https://www.neotomadb.org/). New research could address how ecological and evolutionary patterns differ among generalist herbivores, specialist herbivores, predators, scavengers and detritivores.

2.6 Aberrations: The PETM as an analog for modern-day climate change

The PETM (56 Ma) is widely regarded as the best geologic analog for modern, anthropogenic climate change. Over the course of less than 20,000 years, a large amount isotopically light carbon was released into the atmosphere and ocean, as evidenced by a global decrease in carbon isotope values in geologic records (reviewed in McInerney and

Wing 2011). The PETM occurs during an interval of sustained global warmth (Figure 2.2), and sedimentary sequences from shallow marine and continental settings document warming prior to the primary carbon release (Secord et al. 2012; Sluijs et al. 2007). The carbon source remains controversial, with possibilities including destabilization of methane hydrates on the ocean floor (Dickens et al. 1995), volcanic heating of organic-rich marine sediments in the North Atlantic (Svensen et al. 2004), melting of permafrost (DeConto et al. 2012), burning of peats and shallowly buried coals (Kurtz et al. 2003), or a combination of these sources (Sluijs et al. 2007). Carbon isotope values remained low for 70–100 thousand years (kyr) before gradually returning to background values, and the entire PETM is estimated to be ~170 kyr (Röhl et al. 2007).

PETM climate change is comparable in magnitude to that predicted for the twenty-first century if humans continue to burn fossil fuels at the current rate, but the rate of change was at least ten times slower than today. The rate of carbon emission during the PETM is difficult to constrain due to uncertainty in both the duration (near instantaneous to 20 kyr; Turner 2018 and references therein) and amount (4,500–10,000 petagrams of carbon (PgC), depending on the carbon source; Gutjahr et al. 2017 and references therein) of carbon release. Estimates range from < 1 PgC/yr to greater than the current anthropogenic rate of ~10 PgC/yr (Peters et al. 2012), with some consensus that PETM emission rates were about ten times slower than modern (Cui et al. 2011; Kirtland Turner 2018). Reconstructing atmospheric CO_2 levels before, during and after PETM interval is hindered by large uncertainties in paleo-CO_2 proxies, but modeling results propose that CO_2 doubled to quintupled during the PETM (Cui et al. 2011). Temperature changes have been successfully reconstructed using geochemical and paleobotanical proxies, and warming ranges from 4 °C to 8 °C depending on geographic location (McInerney and Wing 2011). If five degrees of warming occurred over 10 kyr, PETM warming was twenty to fifty times slower than projections for the next century (Wing and Currano 2013). Last, proxy data and models demonstrate that PETM warming intensified the hydrological cycle, with wet areas generally becoming wetter and dry areas becoming drier (Carmichael et al. 2017). Precipitation amount and seasonality, extreme storm events, evapotranspiration, and humidity were all affected.

The best terrestrial record of climate and ecosystem change during the PETM is from the Bighorn Basin. Here, MAT rose to ~20 °C during the PETM, an increase of ~5 °C (Currano et al. 2010; Wing et al. 2005; Secord et al. 2010), and summer temperatures may have been as high as 31–35 °C (Snell et al. 2013). Fossil soils and leaves document significant drying during the PETM, with MAP decreasing by as much as 40 percent (Kraus et al. 2013). Plant macrofossil assemblages indicate near-complete taxonomic turnover during the PETM, as plant species migrated in response to warming and drying (Wing and Currano 2013). Mesophytic species common in the late Paleocene, notably Cupressaceous conifers, were replaced by thermophilic and dry-tolerant species. Legumes (Fabaceae) were diverse and abundant during the PETM, and other important plant families include the Arecaceae, Lauraceae, Hernandiaceae and Salicaceae (Wing and Currano 2013). Floral composition following the PETM is very similar to the late Paleocene. The rarity of well-studied Paleocene–Eocene fossil plant sites makes it impossible to map biogeographic range changes, as was done for the Quaternary. However, Wing and Currano (2013) hypothesized that the Bighorn Basin Paleocene taxa migrated to higher latitudes and/or elevations during the PETM and returned when temperature declined and water availability increased. The PETM taxa are likely to have migrated northwards, into the Bighorn Basin, as temperatures warmed, retreated during the cooler part of the early Eocene, and

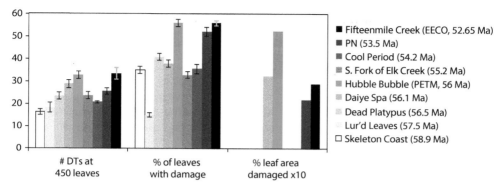

Figure 2.6 Damage diversity (number of DTs observed on 450 leaves), frequency (percentage of leaves with damage) and intensity (percentage leaf area damaged) during the Paleocene-Eocene Thermal Maximum (PETM, 56 Ma) compared to other times in the late Paleocene and early Eocene. All data are from the Bighorn Basin, USA, and were previously published by Currano et al. (2010, 2016). The Fifteenmile Creek site is within the Early Eocene Climatic Optimum (EECO, 53–48 Ma), the highest sustained temperatures of the Cenozoic. In order to show all three herbivory metrics on the same axis, percentage of leaf area damaged values were multiplied by 10.

then returned several million years after the PETM, when temperatures warmed to a sustained Cenozoic maximum 53–51 Ma (the EECO). In northwestern Wyoming at least, it appears that plant taxa were able to migrate, keeping pace with PETM climate change.

No insect body fossils are known from the PETM, but a series of nine insect damage censuses from before, during and after the PETM (59–52.65 Ma) reveals the response of Bighorn Basin insect herbivore communities (Currano et al. 2008, 2010, 2016). These are the same sites that demonstrated a strong positive correlation between MAT and damage diversity or frequency (see section 2.4; Currano et al. 2010). As might be expected given that result, frequency and diversity of total, specialized and mining damage on the PETM bulk flora is higher than at sites with cooler MAT and similar to the very warm EECO site (Figure 2.6; Currano et al. 2010). Percentage of leaf area damaged (damage intensity) on the PETM flora is much higher than at any other fossil site, and at least twice as high as at the EECO site (Currano et al. 2016). Leaf-chewing damage, particularly margin feeding, is responsible for the high PETM value. Patterns of damage frequency and diversity on individual species generally match those observed on bulk floras, with the highest mean values of individual species damage metrics in the PETM (Currano et al. 2010). PETM plant hosts were more extensively colonized and damaged, particularly by specialized herbivores (Currano et al. 2008). Floral turnover during the PETM largely precludes comparisons of damage on a single plant host through time, but *Populus cinnamomoides* occurs at both the PETM and EECO sites. Damage diversity, frequency and intensity are all at least 150 per cent higher during the PETM (Currano et al. 2016).

Floral diversity is highest in the PETM, but correlations between rarefied plant richness and damage metrics are weak and often non-significant (Currano et al. 2010). PETM plants do not appear to be more cheaply constructed than taxa from other sites, as leaf mass per area distributions show no significant differences (Currano et al. 2010). Rather, elevated herbivory during the PETM is attributed to a combination of increased temperature, CO_2 and nitrogen availability (Currano et al. 2008, 2010, 2016), as discussed in section 4.3. PETM warming likely permitted northward migration of speciose insect herbivore communities from lower latitudes (Currano et al. 2008, 2010), and high specialized damage diversity during the PETM suggests that insect species migrated in tandem with plant hosts. Warming may also have affected insect herbivores by increasing metabolic rates and decreasing generation times, which would manifest in the herbivory record as increased frequency and intensity. Although changes in food quality at high pCO_2 may mediate this increase, the Bighorn Basin PETM site is remarkable for its abundance of legumes (Figure 2.4). Incorporation of legume remains into PETM soils is likely to have provided a source of biologically accessible nitrogen, buffering ecosystem-wide decreases in plant nitrogen content at high CO_2 and supporting larger insect populations (Currano et al. 2016).

Discoveries of Bighorn Basin PETM plant species at late Paleocene or early Eocene sites outside the Bighorn Basin are desperately needed to improve our understanding of the processes driving increases in herbivory and how PETM results apply to the present day. How similar is the spectrum of DTs observed on a particular PETM host plant in the Bighorn Basin to that observed on the same host plant at pre- or post-PETM sites outside the Bighorn Basin? How does frequency and intensity on these host plants compare? High DT similarity indicates that insects migrated in synchrony with their hosts, whereas distinct DT compositions correspond to insect herbivore diversifications, extinctions, or host shifts. Changes in frequency and intensity inform about changes in insect physiology, insect abundances, or leaf chemistry at high temperature and CO_2.

2.7 Conclusions and future directions

The paleontological research summarized here shows that insects' primary response to climate trends, rhythms and aberrations is to migrate, tracking their optimal environment and food sources. The global record of Cenozoic herbivory damage, compiled by Currano et al. (2021), also shows a weak but significant correlation between herbivory and mean annual temperature, with stronger correlations observed at the regional level (e.g., Currano et al. 2010; Wappler and Grímsson 2016). Changing atmospheric CO_2 levels may also influence herbivory by affecting leaf C:N ratios, but selection for legumes and other plants with N_2-fixing symbionts at high CO_2 and temperature appears to mediate changes. There is a net increase in the frequency and intensity of herbivory as CO_2 levels and temperatures rise (Currano et al. 2010, 2016), suggesting that herbivory will increase in the future due to anthropogenic climate and atmospheric change. This prediction has since received support from models that used insect population growth and metabolic rates to predict increased crop losses to insect pests in a warming world (Deutsch et al. 2018).

Specific suggestions for future research on insect response to Cenozoic climate trends, rhythms and aberrations were provided at the end of this section. These fall within three main categories.

1. *Collection of new data from the tropics, southern hemisphere and polar regions.* As is true for most paleontological records, insect body fossil sites and fossil insect damage censuses are concentrated in the middle latitudes of the northern hemisphere. It is very likely that evolutionary and ecological responses were different in tropical and polar regions,

as these represent climate end-members. Additionally, no Cenozoic insect Konservat-Lagerstätten are known from Australia, nor have any insect damage censuses been conducted. How does Australia compare with larger, more connected continents?

2. *Better integration of paleobotanical and paleoentomological records.* Insect herbivores, particularly specialists, are dependent on their host plants. Thus, insect response to climate change must be examined in concert with paleobotanical research on taxonomic composition, diversity, abundance structure and canopy structure. Research on time periods when plant and insect taxonomy is well known will be especially productive.

3. *Analysis of herbivory on single plant hosts.* To date, most research has analyzed herbivory on bulk floras, as considerable floral turnover occurs within a region over the course of millions of years of climate change. Examination of plant lineages over large spatial and temporal scales will illuminate ecological and evolutionary dynamics.

The fossil record demonstrates the resilience of plants and insects to climate change, but never before has a species impacted the planet as much and as quickly as humans. Anthropogenic climate change is occurring at least ten times faster than even PETM climate change, and humans directly change biogeographic ranges of plants and insects by transporting organisms (knowingly and unknowingly) and by altering landscapes (e.g., deforestation, urbanization, agriculture). Perhaps the greatest lesson we can learn from the fossil record is this: we live in unprecedented times, and we must all work together to quickly reduce CO_2 emissions and protect natural spaces.

Acknowledgments

I am grateful to have had many productive and insightful conversations with mentors, colleagues and students about ancient climate change, plants and insect herbivores. I especially thank Lauren Azevedo-Schmidt, Bonnie Jacobs, Conrad Labandeira, Gussie Maccracken, Claudia Richbourg, Bryan Shuman, Anshuman Swain, Peter Wilf and Scott Wing. I appreciate the greater community studying fossil insect herbivory for their willingness to share data and work collaboratively. The University of Wyoming, where this chapter was written, is on the ancestral land of the Cheyenne, Arapaho, Očhéthi Šakówin, Crow and Shoshone; acknowledging ancestral land is an important first step in undoing generational trauma and historical (and present) levels of harm that academia has placed on Indigenous communities. Synthetic analyses presented here were supported by NSF EAR 145031.

Key reflections

- Insect response to long-term climate trends, cyclical variations due to changes in Earth's orbital parameters and geologically rapid climate events can be investigated using insect body fossils and herbivory damage on fossil leaves.
- Insects' primary response to ancient climate change has been to migrate, tracking their optimal environment and food sources.
- Herbivory damage on leaves increases with warming and elevated atmospheric CO_2 levels, mediated by nutrient content.

- Humans' impact on climate, plants and insects is likely unprecedented in Earth's history, emphasizing the need to act now to reduce CO_2 emissions and protect natural spaces.

Key further reading

- Labandeira, C.C., Wilf, P., Johnson, K.R., and Marsh, F. 2007. Guide to insect (and other) damage types on compressed plant fossils. Smithsonian Institution, National Museum of Natural History, Department of Paleobiology, Washington, DC. Photographs and descriptions of the first 150 DTs observed on fossil plant organs. A must have for anyone interested in conducting an insect damage census.
- Nyman, T., Linder, H.P., Peña, C., Malm, T. and Wahlberg, N. 2012. Climate-driven diversity dynamics in plants and plant-feeding insects. *Ecology Letters* 15: 889–98. Synthesis paper that examines evolutionary processes linking Cenozoic climate trends, vegetation, and insect lineages. Paleontologists are encouraged to read this paper and think about how their research can test hypotheses based on molecular phylogenies.
- Carvalho, M.R., Wilf, P., Barrios, H., Windsor, D.M., Currano, E.D., Labandeira, C.C., and Jaramillo, C.A. 2014. Insect leaf-chewing damage tracks herbivore richness in modern and ancient forests. *PloS One* 9: e94950. The only study to date that has tested the relationship between insect herbivore diversity and DT diversity. The authors also propose using insect damage censuses in modern forests to investigate herbivore and herbivory dynamics.
- Su, T., Adams, J., Wappler, T., Huang, Y., Jacques, F., Liu, Y., and Zhou, Z. 2015. Resilience of plant-insect interactions in an oak lineage through Quaternary climate change. *Paleobiology* 41: 174–86. One of the few studies to examine herbivory damage on a single plant host (*Quercus* section *Heterobalanus*) through time (~3 Ma to present-day) and across an interval of significant climate change (Pleistocene and Holocene glacial-interglacial cycles).
- Westerhold, T., Marwan, N., Drury, A.J., Liebrand, D., Agnini, C., Anagnostou, E., Barnet, J.S., Bohaty, S.M., De Vleeschouwer, D., Florindo, F., and Frederichs, T., 2020. An astronomically dated record of Earth's climate and its predictability over the last 66 million years. *Science* 369: 1383–1387. The most updated compilation of benthic foraminifer isotope data, which was used to define Cenozoic Hothouse, Warmhouse, Coolhouse, and Icehouse climate states. The authors use a variety of statistical analyses to examine forcing factors and predictability of climate dynamics.
- Currano, E.D., Azevedo-Schmidt, L.E., Maccracken, S.A., and Swain, A. 2021. Scars on fossil leaves: An exploration of ecological patterns in plant–insect herbivore associations during the Age of Angiosperms. *Palaeogeography, Palaeoclimatology, Palaeoecology* 582: 110,636. The most recent review paper on fossil insect herbivory, including a compilation of all unbiased insect damage censuses published before 2021 and a YouTube video entitled, "So you want to damage type a fossil flora."

References

Abellán, P., Benetti, C.J., Angus, R.B., and Ribera, I. 2011. A review of Quaternary range shifts in European aquatic Coleoptera. *Global Ecology and Biogeography* 20 (1): 87–100.

Adams, M.A., Turnbull, T.L., Sprent, J.I., and Buchmann, N. 2016. Legumes are different: Leaf nitrogen, photosynthesis, and water use efficiency. *Proceedings of the National Academy of Sciences* 113: 4098–4103.

Archibald, S.B., and Makarkin, V.N. 2006. Tertiary giant lacewings (Neuroptera: Polystoechotidae): Revision and description of new taxa from western North America and Denmark. *Journal of Systematic Palaeontology* 4 (2): 119–155.

Bale, J.S., Masters, G.J., Hodkinson, I.D., Awmack, C., Bezemer, T.M., Brown, V.K. Butterfield J. et al. 2002. Herbivory in global climate change research: Direct effects of rising temperature on insect herbivores. *Global Change Biology* 8 (1): 1–16.

Bar-On, Y.M., Phillips, R., and Milo, R. 2018. The biomass distribution on Earth. *Proceedings of the National Academy of Sciences* 115 (25): 6506–6511.

Buckland, P.C., and Coope, G.R. 1991. *A Bibliography and Literature Review of Quaternary Entomology.* JR Collis Publications, Department of Archaeology & Prehistory, University of Sheffield.

Burke, K.D., Williams, J.W., Chandler, M.A., Haywood, A.M., Lunt, D.J., and Otto-Bliesner, B.L. 2018. Pliocene and Eocene provide best analogs for near-future climates. *Proceedings of the National Academy of Sciences* 115 (52): 13288–13293.

Burnham, R.J., Wing, S.L., and Parker, G.G. 1992. The reflection of deciduous forest communities in leaf litter: implications for autochthonous litter assemblages from the fossil record. *Paleobiology* 18 (1): 30–49.

Morley, R.J. 2007. "Cretaceous and Tertiary climate change and the past distribution of megathermal rainforests." In Tropical Rainforest Responses To Climatic Change 2nd edn, edited by Bush, M.B., Flenley, J.R., and Gosling, W.D. Berlin, Germany: Springer Science & Business Media, 1–31. https://link.springer.com/chapter/10.1007/978-3-642-05383-2_1

Carmichael, M.J., Inglis, G.N., Badger, M.P., Naafs, B.D.A., Behrooz, L., Remmelzwaal, S., Monteiro, F.M., et al. 2017. Hydrological and associated biogeochemical consequences of rapid global warming during the Paleocene-Eocene thermal maximum. *Global and Planetary Change* 157: 114–38.

Carvalho, M.R., Wilf, P., Barrios, H., Windsor, D.M., Currano, E.D., Labandeira, C.C., and Jaramillo, C.A. 2014. Insect leaf-chewing damage tracks herbivore richness in modern and ancient forests. *PloS One* 9 (5): e94950.

Clapham, M.E., Karr, J.A., Nicholson, D.B., Ross, A.J., and Mayhew, P.J. 2016. Ancient origin of high taxonomic richness among insects. *Proceedings of the Royal Society B: Biological Sciences* 283: 20152476.

Condamine, F.L., Clapham, M.E., and Kergoat, G.J. 2016. Global patterns of insect diversification: Towards a reconciliation of fossil and molecular evidence? *Scientific Reports* 6 (1): 1–13.

Coope, G.R. 1994. The response of insect faunas to glacial-interglacial climatic fluctuations. *Philosophical Transactions of the Royal Society of London Series B: Biological Sciences* 344 (1307): 19–26.

Coope, G.R. 2004. Several million years of stability among insect species because of, or in spite of, Ice Age climatic instability? *Philosophical Transactions of the Royal Society of London. B: Biological Sciences* 359 (1442): 209–14.

Cui, Y., Kump, L.R., Ridgwell, A., Charles, A.J., Junium, C.K., Diefendorf, A.F., Freeman, K.H., Urban, N.M., and Harding, I.C. 2011. Slow release of fossil carbon during the Palaeocene–Eocene Thermal Maximum. *Nature Geosciences* 4: 481–85.

Currano, E.D. 2009. Patchiness and long-term change in early Eocene insect feeding damage. *Paleobiology* 35 (4): 484–98.

Currano, E.D., Azevedo-Schmidt, L.E., Maccracken, S.A., and Swain, A. 2021. Scars on fossil leaves: An exploration of ecological patterns in plant–insect herbivore associations during the Age of Angiosperms. *Palaeogeography, Palaeoclimatology, Palaeoecology* 582: 110636.

Currano, E.D., Labandeira, C.C., and Wilf, P. 2010. Fossil insect folivory tracks paleotemperature for six million years. *Ecological Monographs* 80 (4): 547–67.

Currano, E.D., Laker, R., Flynn, A.G., Fogt, K.K., Stradtman, H., and Wing, S.L. 2016. Consequences of elevated temperature and pCO 2 on insect folivory at the ecosystem level: perspectives from the fossil record. *Ecology and Evolution* 6 (13): 4318–4331.

Currano, E.D., Wilf, P., Wing, S.L., Labandeira, C.C., Lovelock, E.C., and Royer, D.L. 2008. Sharply increased insect herbivory during the Paleocene–Eocene Thermal Maximum. *Proceedings of the National Academy of Sciences* 105 (6): 1960–1964.

Currano, E.D. and Jacobs, B.F. 2021. Bug-bitten leaves from the early Miocene of Ethiopia elucidate the impacts of plant nutrient concentrations and climate on insect herbivore communities. *Global and Planetary Change* 207: 103655.

DeConto, R.M., Galeotti, S., Pagani, M., Tracy, D., Schaefer, K., Zhang, T., Pollard, D., and Beerling, D.J. 2012. Past extreme warming events linked to massive carbon release from thawing permafrost. *Nature* 484 (7392): 87–91.

Deutsch, C.A., Tewksbury, J.J., Tigchelaar, M., Battisti, D.S., Merrill, S.C., Huey, R.B. and Naylor, R.L. 2018. Increase in crop losses to insect pests in a warming climate. *Science* 361: 916–19.

Dickens, G.R., O'Neil, J.R., Rea, D.K., and Owen, R.M. 1995. Dissociation of oceanic methane hydrate as a cause of the carbon isotope excursion at the end of the Paleocene. *Paleoceanography* 10 (6): 965–71.

Greenwood, D.R., and Wing, S.L. 1995. Eocene continental climates and latitudinal temperature gradients. *Geology* 23 (11): 1044–1048.

Gutjahr, M., Ridgwell, A., Sexton, P.F., Anagnostou, E., Pearson, P.N., Pälike, H., Norris, R.D., et al. 2017. Very large release of mostly volcanic carbon during the Palaeocene–Eocene thermal maximum. *Nature* 548 (7669): 573–77.

Huntley, B., and Birks, H.J.B. 1983. *Atlas of past and present pollen maps for Europe, 0–13,000 years ago*. Cambridge: Cambridge University Press.

Jablonski, D. 2008. Biotic interactions and macroevolution: extensions and mismatches across scales and levels. *Evolution* 62 (4): 715–39.

Jackson, S.T., and Overpeck, J.T. 2000. Responses of plant populations and communities to environmental changes of the late Quaternary. *Paleobiology* 26 (S4): 194–220.

Jacobs, B.F. 2002. Estimation of low-latitude paleoclimates using fossil angiosperm leaves: examples from the Miocene Tugen Hills, Kenya. *Paleobiology* 28 (3): 399–421.

Jansson, R., and Dynesius, M. 2002. The fate of clades in a world of recurrent climatic change: Milankovitch oscillations and evolution. *Annual Review of Ecology and Systematics* 33 (1): 741–77.

Janzen, D.H. 1967. Why mountain passes are higher in the tropics. *The American Naturalist* 101 (919): 233–49.

Karr, J.A., and Clapham, M.E. 2015. Taphonomic biases in the insect fossil record: Shifts in articulation over geologic time. *Paleobiology* 41 (1): 16–32.

Kennett, J.P., and Shackleton, N.J. 1975. Laurentide ice sheet meltwater recorded in Gulf of Mexico deep-sea cores. *Science* 188 (4184): 147–50.

Kraus, M.J., McInerney, F.A., Wing, S.L., Secord, R., Baczynski, A.A., and Bloch, J.I. 2013. Paleohydrologic response to continental warming during the Paleocene–Eocene thermal maximum, Bighorn Basin, Wyoming. *Palaeogeography, Palaeoclimatology, Palaeoecology* 370: 196–208.

Kurtz, A.C., Kump, L.R., Arthur, M.A., Zachos, J.C., and Paytan, A. 2003. Early Cenozoic decoupling of the global carbon and sulfur cycles. *Paleoceanography* 18 (4). doi:10.1029/2003PA000908.

Labandeira, C. C. 1998. Early history of arthropod and vascular plant associations. *Annual Review of Earth and Planetary Sciences* 26: 329–77.

Labandeira, C.C. 1999. Insects and other hexapods. *Encyclopedia of Paleontology* 1: 603–24.

Labandeira, C.C. and Sepkoski Jr, J.J. 1993. Insect diversity in the fossil record. *Science* 261 (5119): 310–15.

Labandeira, C.C., Wilf, P., Johnson, K.R., and Marsh, F. 2007. Guide to insect (and other) damage types on compressed plant fossils. *Smithsonian Institution, National Museum of Natural History, Department of Paleobiology, Washington, DC*.

Leckey, E.H., and Smith, D.M. 2017. Individual host taxa may resist the climate-mediated trend in herbivory: Cenozoic herbivory patterns in western North American oaks. *Palaeogeography, Palaeoclimatology, Palaeoecology* 487: 15–24.

Maccracken, S.A., Sohn, J.C., Miller, I.M., and Labandeira, C.C. 2021. A new Late Cretaceous leaf mine *Leucopteropsa spiralae* gen. et sp. nov. (Lepidoptera: Lyonetiidae) represents the first confirmed fossil evidence of the Cemiostominae. *Journal of Systematic Palaeontology* 19 (2): 131–44.

Marra, M.J. and Leschen, R.A.B. 2011. Persistence of New Zealand Quaternary beetles. *New Zealand Journal of Geology and Geophysics* 54 (4): 403–13.

McInerney, F.A. and Wing, S.L. 2011. The Paleocene-Eocene Thermal Maximum: A perturbation of carbon cycle, climate, and biosphere with implications for the future. *Annual Review of Earth and Planetary Sciences* 39: 489–516.

Morley, R.J. 2007. "Cretaceous and Tertiary climate change and the past distribution of megathermal rainforests." In *Tropical Rainforest Responses to Climatic Change*, edited by M.B. Bush and J.R. Flenley. Berlin: Springer, 1–31.

Nyman, T., Linder, H.P., Peña, C., Malm, T., and Wahlberg, N. 2012. Climate-driven diversity dynamics in plants and plant-feeding insects. *Ecology Letters* 15 (8): 889–98.

Overpeck, J.T., Webb, R.S., and Webb III, T. 1992. Mapping eastern North American vegetation change of the past 18 ka: No-analogs and the future. *Geology* 20 (12): 1071–1074.

Peppe, D.J., Baumgartner, A., Flynn, A., and Blonder, B. 2018. "Reconstructing paleoclimate and paleoecology using fossil leaves." In *Methods in Paleoecology: Reconstructing Cenozoic Terrestrial Environments and Ecological Communities*, edited by Croft, D.A., Su, D.F., and Simpsion, S.W. Cham, Germany: Springer International, 289–317.

Peters, G.P., Marland, G., Le Quere, C., Boden, T., Canadell, J. G., and Raupach, M.R. 2012. Rapid growth in CO 2 emissions after the 2008–2009 global financial crisis. *Nature Climate Change* 2: 2–4

Pinheiro, E.R., Iannuzzi, R., and Duarte, L.D. 2016. Insect herbivory fluctuations through geological time. *Ecology* 97 (9): 2501–2510.

Robinson, E.A., Ryan, G.D., and Newman, J.A. 2012. A meta-analytical review of the effects of elevated CO2 on plant–arthropod interactions highlights the importance of interacting environmental and biological variables. *New Phytologist* 194 (2): 321–36.

Röhl, U., Westerhold, T., Bralower, T.J., and Zachos, J.C. 2007. On the duration of the Paleocene–Eocene thermal maximum (PETM). *Geochemistry, Geophysics, Geosystems* 8 (12).

Schachat, S.R., Labandeira, C.C., Clapham, M.E. and Payne, J.L., 2019. A Cretaceous peak in family-level insect diversity estimated with mark–recapture methodology. *Proceedings of the Royal Society B* 286: 20192054.

Schachat, S.R. and Labandeira, C.C. 2021. Are insects heading toward their first mass extinction? Distinguishing turnover from crises in their fossil record. *Annals of the Entomological Society of America* 114 (2): 99–118.

Scherber, C., Gladbach, D.J., Stevnbak, K., Karsten, R.J., Schmidt, I.K., Michelsen, A., Albert, K.R., et al. 2013. Multi-factor climate change effects on insect herbivore performance. *Ecology and Evolution* 3 (6): 1449–1460.

Secord, R., Gingerich, P.D., Lohmann, K.C., and MacLeod, K.G. 2010. Continental warming preceding the Palaeocene–Eocene thermal maximum. *Nature* 467 (7318): 955–8.

Secord, R., Bloch, J.I., Chester, S.G., Boyer, D.M., Wood, A.R., Wing, S.L., Kraus, M.J., et al. 2012. Evolution of the earliest horses driven by climate change in the Paleocene-Eocene Thermal Maximum. *Science* 335 (6071): 959–62.

Shuman, B.N. and Serravezza, M. 2017. Patterns of hydroclimatic change in the Rocky Mountains and surrounding regions since the last glacial maximum. *Quaternary Science Reviews* 173: 58–77.

Sluijs, A., Brinkhuis, H., Schouten, S., Bohaty, S.M., John, C.M., Zachos, J.C., Reichart, G.J., et al. 2007. Environmental precursors to rapid light carbon injection at the Palaeocene/Eocene boundary. *Nature* 450 (7173): 1218–1221.

Smith, D.M. 2001. Taphonomic bias and insect diversity: A lesson from the beetles and flies. *PaleoBios* 21 (2, Suppl.): 117.

Smith, D.M. 2012. Exceptional preservation of insects in lacustrine environments. *Palaios* 27 (5): 346–353.

Snell, K.E., Thrasher, B.L., Eiler, J.M., Koch, P.L., Sloan, L.C., and Tabor, N.J. 2013. Hot summers in the Bighorn Basin during the early Paleogene. *Geology* 41 (1): 55–8.

Stauffer, B. 1999. Cornucopia of ice core results. *Nature*, 399(6735): 412–13.

Strömberg, C.A. 2011. Evolution of grasses and grassland ecosystems. *Annual review of Earth and Planetary Sciences* 39: 517–44.

Su, T., Adams, J.M., Wappler, T., Huang, Y.J., Jacques, F.M., Liu, Y.S., and Zhou, Z.K. 2015. Resilience of plant–insect interactions in an oak lineage through Quaternary climate change. *Paleobiology* 41 (1): 174–86.

Svensen, H., Planke, S., Malthe-Sørenssen, A., Jamtveit, B., Myklebust, R., Rasmussen Eidem, T., and Rey, S.S. 2004. Release of methane from a volcanic basin as a mechanism for initial Eocene global warming. *Nature* 429 (6991): 542–45.

Turner, S.K. 2018. Constraints on the onset duration of the Paleocene–Eocene thermal maximum. *Philosophical Transactions of the Royal Society A: Mathematical, Physical and Engineering Sciences* 376 (2130): 20170082.

Wappler, T., and Grímsson, F. 2016. Before the "Big Chill": Patterns of plant-insect associations from the Neogene of Iceland. *Global and Planetary Change* 142: 73–86.

Webb, T. 1993. "Vegetation, lake levels, and climate in eastern North America for the past 18,000 years." In *Global Climates Since the Last Glacial Maximum*, edited by H.E. Wright Jr, J.E. Kutzbach,

T. Webb, III, W.F. Ruddiman, F.A. Street-Perrott, and P.J. Bartlein. Minneapolis: University of Minnesota Press, 415–67.

Westerhold, T., Marwan, N., Drury, A.J., Liebrand, D., Agnini, C., Anagnostou, E., Barnet, J.S., et al. 2020. An astronomically dated record of Earth's climate and its predictability over the last 66 million years. *Science* 369 (6509): 1383–1387.

Wilf, P. 2008. Insect-damaged fossil leaves record food web response to ancient climate change and extinction. *New Phytologist* 178 (3): 486–502.

Wilf, P. and Labandeira, C.C. 1999. Response of plant-insect associations to Paleocene-Eocene warming. *Science* 284 (5423): 2153–2156.

Wilf, P., Labandeira, C.C., Kress, W.J., Staines, C.L., Windsor, D.M., Allen, A.L., and Johnson, K.R. 2000. Timing the radiations of leaf beetles: Hispines on gingers from latest Cretaceous to recent. *Science* 289 (5477): 291–94.

Williams, J.W., Shuman, B.N., Webb III, T., Bartlein, P. J., and Leduc, P.L. 2004. Late-Quaternary vegetation dynamics in North America: Scaling from taxa to biomes. *Ecological Monographs* 74 (2): 309–34.

Willis, K., and McElwain, J. 2014. *The Evolution of Plants*. Oxford: Oxford University Press.

Wing, S.L., and Currano, E.D. 2013. Plant response to a global greenhouse event 56 million years ago. *American Journal of Botany* 100 (7): 1234–1254.

Wing, S.L. and DiMichele, W.A. 1995. Conflict between local and global changes in plant diversity through geological time. *Palaios* 10 (6): 551–64.

Wing, S.L., Harrington, G.J., Smith, F.A., Bloch, J.I., Boyer, D.M., and Freeman, K.H. 2005. Transient floral change and rapid global warming at the Paleocene-Eocene boundary. *Science* 310 (5750): 993–6.

Winkler, I.S., Mitter, C., and Scheffer, S.J. 2009. Repeated climate-linked host shifts have promoted diversification in a temperate clade of leaf-mining flies. *Proceedings of the National Academy of Sciences* 106 (43): 18103–18108.

Zachos, J., Pagani, M., Sloan, L., Thomas, E., and Billups, K. 2001. Trends, rhythms, and aberrations in global climate 65 Ma to present *Science* 292 (5517): 686–93.

CHAPTER 3

Studying climate change effects in the era of omic sciences

Zach Fuller and Maren Wellenreuther

3.1 Introduction

Climate change is imposing novel selection pressures on insect species by altering abiotic and biotic environmental conditions (Parmesan 2006). Climate parameters, such as increased temperatures, rising atmospheric CO_2 levels and changing precipitation patterns all have significant impacts on the distribution and abundance of insect species. Indeed, for many species, particularly those occupying habitats that are already close to their physiological maxima, increasing climate parameters pose a significant threat to their persistence (Scheffers et al. 2016). However, insect species do not respond uniformly to changes in climate parameters and recent reviews show that changes are often variable and ecologically complex across species (Lancaster et al. 2022; Lehmann et al. 2020; Wellenreuther et al. 2022). It is thus of utmost importance that we gain a better understanding of the species-specific responses that various insect species show to a changing climate. Filling these knowledge gaps is critical as some of the anticipated changes in insect biodiversity are likely to have a significant impact on ecosystem functioning (e.g., pollinators), insect–human disease transmission (e.g., insect disease vectors like *Anopheles* spp.) and conservation (e.g., invasive and rapidly expanding species). Consequently, predicting the response of insects to projected environmental change is a priority area for research, with large impacts on conservation and management (Bonebrake et al. 2018).

As with other challenges related to climate change, technological advances are likely to play a key role in improving our understanding of the mechanistic processes involved. In this context, the rapid development of affordable sequencing technologies, in combination with computational advancements, has begun to uncover some of the molecular patterns underlying the responses of insects to changes in climate parameters. Excitingly, these new developments also present the opportunity to track eco-evolutionary processes using a greatly expanded set of omic markers, allowing, for example, researchers to carry out high-powered tests for dispersal and demographic processes impacting connectivity (e.g., Balkenhol et al. 2017). Omic sequencing technologies refer to a field of study in life sciences that ends with "omics" (see Glossary) to describe projects that focus on large-scale data. Together with improvements in

Zach Fuller and Maren Wellenreuther, *Studying climate change effects in the era of omic sciences*. In: *Effects of Climate Change on Insects*. Edited by: Daniel González-Tokman and Wesley Dáttilo, Oxford University Press. © Oxford University Press (2024). DOI: 10.1093/oso/9780192864161.003.0003

tools to gather detailed environmental and phenotypic data, these omic technologies have accelerated our ability to detect drivers of variation and mechanistic processes across the genome-to-phenome spectrum in model and non-model insect species (Ellegren 2014).

To date, genomic technologies have been applied largely to investigate genome-wide responses of species to changing climate parameters, and to separate neutral from adaptive genomic variation (Aguirre-Liguori et al. 2021). Likewise, transcriptomic approaches, the study of genes expressed at a particular moment, have been used to unravel the relationships between environment, genotype and phenotype in natural populations of insects (Alvarez et al. 2015). Epigenomics, the study of the modification of DNA to control expression, is another rapidly growing but still young field where insights have been gained to understand how insect species can modify their DNA to respond to a changing climate (Anastasiadi et al. 2021; McGuigan et al. 2021). Importantly, epigenomic modification can happen rapidly, even within one generation, and as such, epigenomic change can take place uncoupled from the slower process of natural selection that is needed to change the frequency of genetic variants.

Either in separation or in combination with one another, these omic tools have all been applied to answer the central question in this research field: how can species cope with the accelerated rate of climatic change? In insects, the focus has been to apply these tools to gain a better understanding of climate change responses that facilitate their ability to exhibit plastic phenotypic changes, shifts in abundance and distribution, and ultimately, evolutionary adaptation. An increasing number of studies are revealing that climate change impacts are often species-specific, but also the geographic region itself and overall, that insect phenotypes, and the "omics" that determine them, are often far more complex in time and space than has been appreciated previously. This accumulating evidence suggests that a better understanding of variation at diverse levels will be crucial to decipher how species adapt to warming environments, to improve forecasting and managing distributions of pest and invasive species, and to preserve biodiversity around the globe.

In this chapter, we first review the sequencing tools and computational technologies that have facilitated a deepening of our understanding of the omic responses of insects in response to climate change. Second, we synthesize the recent literature and detail some of the omic evidence of how insects are responding to climate change via phenotypic plasticity, movement changes, and, finally, evolutionary adaptive shifts. Third, we end by describing some future topics related to insect conservation that deserve increased attention.

3.2 Tools and technologies in the omic era

3.2.1 *Omic sequencing*

Over the two decades since the sequence of the first insect genome *Drosophila melanogaster* was published (Adams et al. 2000), omic sequencing technologies have experienced rapid progress. The continuing development of massively parallel, high-throughput sequencing platforms, alongside the emergence of single-molecule sequencing technologies, has enabled researchers around the world to assemble high-quality genomes cheaply. In addition, these new developments have also enabled researchers to investigate genomic variation in large populations more easily, and profile complete patterns of gene expression and epigenetic modifications, among many other abilities. These developments have ushered an era of omic sciences to the study of climate change responses of insects.

3.2.2 Genomics

Genomic sequencing can be defined as the process of determining the complete (or nearly complete) DNA sequence of an organism (Ellegren 2014). Today, more than 600 insect species (representing over twenty orders) have their full genome sequence publicly available and a variety of consortia and initiatives have been established with the goal of substantially increasing this number (e.g., i5k, see i5K Consortium 2013). Having a "reference" genome sequence available for a species is critical to understanding its genomic architecture/organization, gene content, level of heterozygosity and a host of other fundamental biology aspects. Importantly, reference genomes also serve as a guide to compare the genomic sequences between multiple individuals and are often required for many other downstream omic analyses.

For instance, the genome sequence of *D. melanogaster* paved the way for an assembly strategy known as whole genome shotgun sequencing, the principles of which still form the basis of many contemporary approaches. The basic idea is to generate a large number of sequenced fragments from sheared genomic DNA and then to use overlapping information between these fragments to assemble them into larger *contigs*. Larger sequenced fragments that align to multiple contigs can then be used to determine their order and orientation to generate *scaffolds*. While early genome assemblies relied on expensive Sanger-sequenced fragments of genomic DNA segments (typically a few thousand base pairs (bps)) cloned into plasmid vectors, newer technologies have resulted in dramatically cheaper, albeit shorter, fragments of DNA that can be sequenced. The introduction of platforms such as 454 pyrosequencing, Ion Torrent semiconductor sequencing and Illumina sequencing-by-synthesis provided the ability to sequence massive numbers of short (of the order of a few bps or fewer) fragments of DNA called *reads* at a cost orders of magnitude cheaper than initial Sanger-based approaches. These technological developments are referred to as *Next Generation Sequencing* (NGS).

Because of the low-cost and high-throughput of NGS sequencing platforms, each nucleotide in the accessible genome could be sequenced multiple times. The number of times a bp is sequenced (on average) in the genome is known as *coverage*, with higher values typically needed for shorter reads to provide enough overlap information to assemble contigs successfully. For example, early Sanger-based paired-end assemblies used five to ten times mean coverage, while more contemporary high-quality Illumina short-read genome assemblies typically require upwards of forty times coverage. Today, by far the most popular platform for short-read sequencing is Illumina, with newer instruments such as the NextSeq series having the throughput to generate upwards of 300 gigabytes (Gb) worth of data in a single run. Illumina technology provides the ability to sequence both ends of a DNA fragment (known as "paired-end" reads), enabling improved accuracy for read alignment compared to "single-end" reads and allows for the detection of repetitive elements and structural rearrangements. A standard contemporary sequencing configuration generates paired-end reads, each 150 bp in length of a DNA fragment in the range of 200–800 bp (often referred to as the "insert size").

Alongside the development of short read sequencing over the last decade, several platforms have been brought to market with the goal of sequencing *long reads*. While short reads have the advantage of being highly accurate and cheap to generate, assembling an entire genome from fragments of only a couple hundred bps in length proves a significant challenge. Thus, long read sequencing platforms seek to fill this gap by generating reads of single DNA molecules thousands of bps or more in length. These long read sequencing platforms are sometimes referred to as "third-generation sequencing." The first major platform to be developed for long read sequencing was introduced by Pacific Biosciences (PacBio) which relies upon zero-mode waveguide (ZMW) technology to fluorescently detect individual nucleotides as they are incorporated with a polymerase. PacBio refers to its sequencing approach as Single Molecule Real Time (SMRT) technology, which uses a circular DNA template that can accommodate insert sizes ranging from 1 kilobyte (kb) to 100 kb or more. After the circular DNA libraries are constructed and bound with a polymerase, they are loaded on to SMRT cells, which contain millions of ZMWs for sequencing. While PacBio SMRT platforms could initially generate reads with an average length of > 10 kb, a major drawback was the significantly higher error rates (~15 percent) compared to Illumina short reads (< 0.1 percent). For this reason, genome assemblies using PacBio long reads typically required high coverage (> eighty times) and often relied on Illumina short reads for error correction and assembly polishing. However, recently PacBio unveiled their HiFi sequencing method, which boasts single read lengths averaging 10–25 kb and accuracies exceeding 99.5 percent, opening the door for more affordable, long-read only genome assemblies.

Despite the progress of PacBio SMRT technology towards longer and more accurate reads, cost still remains a prohibitive factor and samples typically need to be sent to core genomics facilities. Oxford Nanopore Technology (ONT) offers an alternative long read sequencing platform that can be run on local desktops and is small enough to fit in most research labs (or even field sites). While the platform remains a promising alternative, high error rates (~20 percent; Wang et al. 2020), inconsistent/lower throughput and the requirement of high-quality high-molecular weight DNA (fragment lengths > 50 kb) are current drawbacks. Other technologies seek to leverage the cheap cost of Illumina short reads while providing long range genomic linkage information, such as 10X Genomics Chromium linked reads. Here, molecular barcodes are attached to short reads that are derived from the same long DNA molecules, allowing for improved scaffolding and the possible resolution of both haplotypes in diploids. Hybrid assembly strategies incorporating both long reads and linked reads have been successful in constructing chromosome-scale reference genomes for several insect species, including the honeybee *Apis mellifera* (Wallberg et al. 2019), olive fruit fly *Bactrocera oleae* (Bayega et al. 2020), aphid *Myzus persicae* (Mathers et al. 2020) and mosquito *Culex tarsalis* (Main et al. 2021). Moreover, additional technological developments have recently been made that enable chromosome scale scaffolding of genomic sequencing data, such as high-throughput chromosome conformation capture (Hi-C) and optical mapping (e.g., BioNano). Both Hi-C and optical mapping approaches typically require high molecular weight DNA as input.

Once a reference genome is obtained, the nucleotide sequences of multiple individuals (both within and between populations) can be compared for downstream analyses. Several strategies currently exist to sequence and map multiple individuals for comparative genomic analyses. Whole-genome resequencing (WGR) approaches involve preparing and sequencing short-read libraries at moderate to high coverage and then aligning the reads for each individual sample sequenced back to the assembled reference genome. Because WGR preserves individual-level information and aims to sequence reads from the entire accessible genome, such approaches have been successful in insects for detecting candidate targets of selection (Fuller et al. 2015; Harpur et al. 2014; Soria-Carrasco et al. 2014; Wragg et al. 2016), identifying large-scale structural rearrangement polymorphism

(Fuller et al. 2017; Martin et al. 2013) and inferring demographic history (Martin et al. 2013). While WGR provides whole-genome information for individual samples and sequencing costs continue to decrease, it still remains prohibitively expensive in many applications. Thus, several alternative, cost-effective strategies have been developed to produce data for comparative and population genomic analyses. One such approach is Pool-Seq, where instead of sequencing or molecularly barcoding individual samples, reads are generated from a single library of DNA extracted from pools of individuals (Ferretti et al. 2013). While individual level information is lost, Pool-Seq is a cost-effective strategy to estimate allele frequencies in populations or groups of individuals, allowing for the estimation of population structure and genetic differentiation, for example. Alternatively, several approaches are based around the idea of using restriction enzymes to digest small subsets of the genome randomly that are then sequenced, thus dramatically reducing cost compared to sequencing the entire genome. Such strategies are known as genotyping-by-sequencing (GBS) or Restriction-site Associated DNA sequencing (RAD-seq) (e.g., Elshire et al. 2011), and while they can provide genomic data at a reduced price tag, they have the important limitation of being an incomplete (i.e., reduced) representation of the genome. Yet another set of strategies is centered on the notion of sequencing individual samples at low-fold coverage (as low as 0.1× per sample) and relying on genotype imputation algorithms to provide probabilities for common variants not sequenced (as applied here: Mérot et al. 2018). While similar approaches are extremely popular in human and agricultural genetics, the often rapid breakdown of linkage disequilibrium and high genetic diversity of many insect species complicates the application of standard imputation procedures (which depend on observed patterns of linkage disequilibrium).

Finally, there are several considerations and challenges unique to insects that any genomic project should be aware of before proceeding. First, while isogenic strains are ideal for genome assemblies to avoid issues with heterozygosity, inspect species have a diverse range of life histories, mating cycles, life cycles, etc., posing a challenge to maintaining inbred strains in the lab. Additionally, it is often difficult to avoid extracting DNA from gut bacteria, endosymbionts, or other contaminants alongside the species of interest. Bioinformatics tools exist to filter raw reads from contaminants by aligning with reference databases, clustering based on sequence composition, or grouping by sequence composition. Moreover, because many insect species are physically small, DNA extractions must be optimized to yield enough genomic material from a single individual for sequencing, or in some cases, highly related, inbred individuals can be pooled together (often requiring assemblers that are aware of heterozygous input). The genome sizes of insect species also span a range of several orders of magnitude (e.g., *Drosophila* spp. ~175 Mb to the short-horned grasshopper *Podisma pedestris* ~16 Gb), meaning that a successful assembly strategy for one species may not be successful in another.

3.2.3 *Transcriptomics*

While genomic sequencing aims to obtain the DNA sequence of the nuclear genome, *transcriptomic* sequencing aims to obtain the sequence of expressed RNA in the cells or tissue used in extraction. The most widely used approach for RNA sequencing (RNAseq) is based on the same Illumina short read platform described in the previous section. Here, the isolated and purified RNA is converted into cDNA by reverse transcription and constructed into a sequencing library for short read sequencing. If a reference genome is available, the reads can then be mapped and aligned, or alternatively, the reads can be assembled *de novo* to produce a transcriptome assembly. More recently, PacBio long read sequencing technology has also been adapted to RNAseq (termed IsoSeq), with the goal of capturing full length transcripts (including transcription start sites) with single reads.

RNAseq enables a number of downstream applications. Because a reference genome is not necessarily required and expressed transcripts typically represent a small fraction of the genome, *de novo* assembled transcriptomes provide a cost-effective option for comparative genomic analyses. When a reference assembly is available, RNAseq data can be used to annotate the genome and characterize the genes expressed in an organism. Because the number of reads sequenced of a given transcript relates to its expression level, perhaps most notably RNAseq is used to test for differential gene expression in samples/replicates across experimental conditions, developmental stages, populations, environments, etc. For instance, RNAseq has been applied to understand insect responses to climate change by testing for differential expression across diel thermal changes (Chou et al. 2020), extreme temperatures (Vatanparast et al. 2021), diapause and quiescenece (Poelchau et al. 2013) and humidity conditions (Ma et al. 2019).

Like genomic sequencing, there are also a number of considerations to be aware of when performing RNAseq in insects. First and foremost, the most important step is isolating high-quality RNA of sufficient quantity. Factors that may create a challenge for this step include a wide range of ribosomal RNA content among insect species (Kukurba and Montgomery 2015) and non-uniform degradation rates across tissues and transcripts. Additionally, it is important to sequence multiple biological and technical replicates for any differential expression analysis.

3.2.4 *Epigenomics*

Epigenetics refers to a broad category of heritable phenotypic changes that do not directly involve changes to the genomic DNA sequence. These changes can occur through DNA methylation, histone protein modification, or non-coding RNA activity which can all modify gene regulation, and can be transmitted through both mitotic and meiotic cell divisions (Anastasiadi et al. 2021). Epigenetic modifications can respond to both intracellular and extracellular environmental signals. In insects, such epigenetic mechanisms can allow for plastic phenotypic responses to changing environments, such as increased temperature (Main et al. 2021) and desiccation (Wang et al. 2021).

As in genomics and transcriptomics, the development of massively parallel, high-throughput sequencing has enabled a number of platforms to assay epigenetic modifications (Glastad et al. 2019). A popular method to analyse protein associations and interactions with DNA is chromatin immunoprecipitation sequencing (ChIP-seq). Here, antibodies specific to target proteins are used to immunoprecipitate the bound DNA, which can then be constructed into a library and sequenced on, for example, an Illumina platform. ChIP-seq approaches have been used to uncover genes involved in phenotypic plasticity of wing size in response to environmental changes in *Sogatella furcifera* (Gao et al. 2019), plastic reproductive capacity in *Apis mellifera* (Duncan et al. 2020) and dispersal polyphenism in aphids (Richard et al. 2021). Another method to analyze DNA-associated proteins is *Assay for Transposase-Accessible Chromatin using sequencing* (ATAC-seq), which is used to identify regions of open chromatin. In this technique, a modified Tn5 transposase is used to integrate sequencing adapters into open chromatin regions. These tagged DNA sequences, representing regions of accessible chromatin, can then be constructed into libraries and sequenced on a high-throughput platform.

Whole-genome bisulfite sequencing (BS-seq) is the most popular approach to investigate patterns of DNA methylation. In this technique, DNA is treated with bisulfite which converts unmethylated cytosine nucleotides into uracil while leaving 5-methylcytosine untouched. Thus, only methylated cytosines remain. The resulting DNA is then PCR amplified (which converts uracil to thymine) and sequenced on a high-throughput platform. Then, the reads can be aligned to a reference genome and used to determine the proportion of methylated cytosines by calling Single Nucleotide Polymorphisms (SNPs) generated from the bisulfite treatment. More recently, long read sequencing platforms have been adapted to study DNA methylation and approaches developed that do not require bisulfite treatment—a harsh treatment that destroys nearly 90 percent of genomic DNA input. For example, nanopore-based methods offer the capability for direct sequencing of 5-methylacytosine from single molecules (Jain et al. 2015, Li and Tollefsbol 2021).

A number of methods exist to distinguish non-coding RNA molecules from mRNAs, adapted from standard RNAseq approaches. For instance, long non-coding RNA (lncRNA) molecules often contain poly-A tails and 5′-caps like mRNAs and are thus usually sequenced in RNAseq experiments. Bioinformatic and computational tools exist to distinguish these lncRNAs based on sequence analysis (Housman and Ulitsky 2016). Moreover, size selection techniques can be used to enrich extracted total RNA for small regulatory noncoding RNA molecules, such as microRNAs (miRNAs) (Guo et al. 2017). Together, these epigenetic techniques have revealed important insights into the adaptive response to climate change and novel environments in a number of insect species, including the tiger mosquito *Aedes albopictus* (Oppold et al. 2015), the whitefly *Bemisia tabaci* (Dai et al. 2018) and bumblebees *Bombus* spp. (Maebe et al. 2021).

3.3 Omic insights into insect climate change response

3.3.1 *Plastic phenotypic responses*

The extent of phenotypic plasticity, the capacity of a single genotype to exhibit diverse phenotypes in response to the environment, can have a significant impact on how insects respond to climate change (Pigliucci 2003; Pigliucci et al. 2006; see Figure 3.1 for a diagram that shows the interaction of different processes). Importantly, phenotypic plasticity allows populations to respond to changing environmental conditions within very short timescales (i.e., within one generation) and because of that, phenotypic plasticity in important life history traits can play an important role in the survival of species (Gibert et al. 2016). For example, plastic responses that modulate the initiation and duration of diapause have been demonstrated to facilitate population persistence in some insect species, for example in *Drosophila* spp. (Sgrò et al. 2016). Likewise, ontogenetic plasticity in traits such as hardening as well as temperature acclimation can be critical for the persistence of insects, as these physiological responses are critical for survival when facing rapid changes in temperature (Sgrò et al. 2016).

Genomic approaches are starting to uncover the mechanistic genes involved in regulating the expression of important life history traits in insects in response to a warming climate (Lafuente and Beldade 2019). Study systems were this has been dissected in detail are still rare, with the best example coming from the exhaustive work on *D. melanogaster* and close relatives, where a powerful body of empirical data now allows assessment of the genomic basis of thermal plasticity and its repeatability (Kapun et al. 2020; Machado et al. 2016; Rolandi et al. 2018). Additionally, transcriptome studies in insects have documented thermal plasticity in gene expression levels, including how many and which

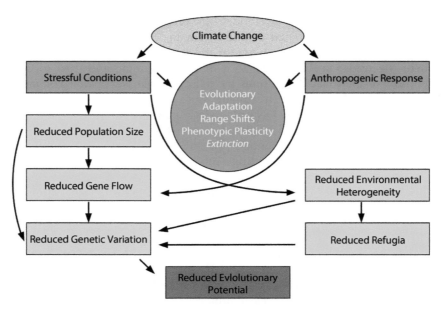

Figure 3.1 Schematic showing the effects and outcomes of climate change on different parameters of insect populations. Stressful conditions generated by climate change, as well as human responses to changing environments, press species to respond, either by evolutionary adaptation, range shifts, plastic phenotypic changes, or face extinction. These conditions can also reduce population sizes and environmental heterogeneity, ultimately leading to reduced evolutionary potential.

genes are differentially expressed between temperatures. Again, the effect of temperature on transcription has been particularly well studied in *D. melanogaster* (Chen et al. 2015), and generally studies document that this species harbours a diverse range of plastic responses to temperature. Indeed, the high plasticity potential in *D. melanogaster* and many other drosophilid species is what has probably facilitated the various climate-induced range shifts in this group (Lafuente and Beldade 2019; Sgrò et al. 2016). These studies indicate that the genetic basis of thermal plasticity is diverse and at times highly distinct with little overlap between the candidate loci underlying variation in plasticity (e.g., in body size: Lafuente et al. 2018).

Unlike the extensive standing genetic variation present in *Drosophila* spp., studies of the thermal transcriptional architecture in the African savannah butterfly *Bicyclus anynana* revealed pervasive gene expression differences only between seasonal phenotypes, and the almost complete absence of intra-population genetic variation for plasticity within season (Oostra et al. 2018). This indicates that a lack of genetic variation within populations may lead to catastrophic effects if there is a future mismatch between seasonal phenotypes and climate, something that can happen if the accuracy of seasonal cues deteriorates due to climate change, rendering dominant reaction norms maladaptive. Work on the range-expanding bluetailed damselfly *Ischnura elegans* showed that both the number of genes involved and levels of gene expression under heat stress have become attenuated during the expansion, consistent with theoretical work that predicts such a pattern when there is a release from selection (Lancaster et al. 2016; Figure 3.2). Moreover, the genes upregulated under cold stress differed between core and edge populations, which is consistent with a rapid plasticity response to cooler climates at the expansion front (Lancaster et al. 2015).

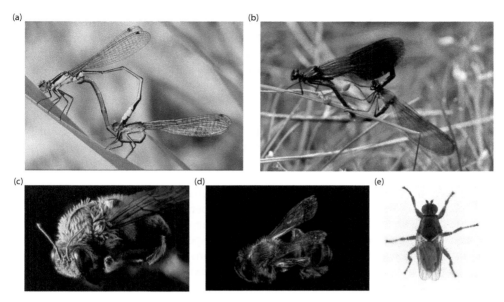

Figure 3.2 Photographs of some of the insect exemplar species reviewed in this chapter. a) bluetailed damselfly male and female in copula (*Ischnura elegans*); b) banded demoiselle male and female in copula (*Calopteryx splendens*); c) a carpenters bee (*Xylocopa* spp.); d) a honey bee (*Apis mellifera*); and e) the seaweed fly (*Coelopa frigida*). Photo credits for a, b and e: Maren Wellenreuther; and for c and d: Jeff Kerby.

Integrative genomic high-throughput approaches using individuals exposed to different thermal regimes have revealed that many phenotypically plastic responses are mediated by alterations in the epigenetic control of gene expression. Work on the freeze-tolerant goldenrod gall fly (*Eurosta solidaginis*) and the goldenrod gall moth (*Epiblema scudderiana*) suggests epigenetic regulation through the differential expression of DNA methyltransferases (dnmt), histone acetyltransferase and demethylation enzymes, which are all genes involved in freeze tolerance (Williamson 2017). Similarly, chromatin remodeling and differential expression of dnmt1 and dnmt3 was found to be essential for heat shock survival and chill coma recovery in the invasive white fly *Bemisia tabaci* (Ji et al. 2020). Future progress comes from the identification of genes that have both expression patterns as well as functions related to temperature, which are typically the best candidate genes that drive changes in thermally sensitive phenotypes (Rodrigues and Beldade 2020).

3.3.2 *Range shifts in response to a changing climate*

Climate change is leading to shifts in insects' geographical distributions around the globe. Indeed, in both the Northern and Southern hemispheres, meta-analyses have revealed extensive range shifts of different insect groups (Hickling et al. 2006a, 2006b; Parmesan and Yohe 2003; Parmesan et al. 1999), with widespread evidence of range expansions in warm-adapted species and range contractions in cold-adapted species. This expansion is leading to novel community compositions and species interactions; for example, an increased hybridization rate between the formerly allopatric species (Sánchez-Guillén et al. 2016). With the raising awareness that many insect species do show some degree of niche conservatism, such as demonstrated in the damselfly species *Calopteryx splendens* (Wellenreuther et al. 2012), it is important to understand the degree to which species can respond to a changing climate. As part of this effort, researchers have turned to landscape genomic analyses to improve the modeling of evolutionary processes that integrate measures of gene flow, population dispersal and genomic load, all of which are critical for predicting the fate of species across landscapes (Aguirre-Liguori et al. 2021). In addition, approaches such as genome–environment association (GEA) studies, allow for linking specific genomic variants to new environmental features and provide indirect evidence for adaptation to a new environment. For example, work on range-shifting damselflies through mapping of genomic variants to environmental conditions at range margins using GEA studies indicate that dispersal evolution is a strong component of the process (Dudaniec et al. 2018), and that some species may be pre-adapted to range shifts by exhibiting high dispersal propensities already at their range core (Swaegers et al. 2015b). Furthermore, genome-wide association studies (GWAS) have been applied to pinpoint the genomic regions responsible for climate related range shifts and a GWAS study on *D. melanogaster* populations uncovered twelve SNP changes associated with the critical thermal maximum (CTmax) (Rolandi et al. 2018). Interestingly, this study also revealed that in most of these SNPs, the minor allele increased the upper thermal limit, suggesting raw genetic variation for heat tolerance exists (Rolandi et al. 2018).

Studies on range shifts have also revealed that populations are likely to be adapting to environmental change at different rates across their range. Typically, species are expected to exhibit declines in genetic diversity but increases in genetic differentiation towards the edge of their range, where populations tend to be more isolated, recently colonized, and subject to stronger colonization–extirpation dynamics due to a mixture of drift and strong selection (Eckert et al. 2008; Hewitt 2001). Indeed, such a loss of genetic diversity and/or increase in genetic differentiation in populations near range margins is a common feature that has been found in multiple odonate species (Dudaniec et al. 2018; Iserbyt et al. 2010; Sánchez-Guillén and Ott 2018; Swaegers et al. 2014). However, the strength of the species-specific patterns are tempered by historic and ongoing range shift processes (Dudaniec et al. 2022).

3.3.3 *Adaptation to climatic changes*

Changes in environmental conditions are expressed as changes in the local selection regime acting on populations. Such altered selection regimes will eventually drive genomic changes and, ultimately, adaptation. Local adaptation occurs when organisms have higher average fitness in their local environment compared to individuals from elsewhere (Kawecki and Ebert 2004) and is seen as the most relevant scenario to ensure the long-term persistence of species under a warming climate (Merilä 2012). Caution is needed though when interpreting genomic population differentiation in key traits as being adaptive in cases where fitness experiments are lacking. Without clear evidence that genomic changes in key traits impact fitness, it is not possible to make that direct link, and future work will need to be conducted to quantify phenotypic fitness effects. Evidence for genetic shifts in important life history traits either over time, or across the species range, is manifold, including climate-driven genomic changes in phenology tolerances to high temperatures and increased metabolic costs of living in a warmer world (Ma et al. 2021; Macgregor et al. 2019). For example, a study on replicate populations of the range-expanding damselfly *Coenagrion scitulum* revealed that one SNP, associated with increased flight performance, to be under consistent selection at the expanding range edge (Swaegers et al. 2015a). Further, the authors also detected a genomic signature of adaptation to the newly encountered thermal regimes, both at a single SNP as well as in a set of covarying SNPs, using a combination of single-locus and a polygenic multilocus approach (Swaegers et al. 2015a). Whole-genome approaches are now increasingly applied to study insect responses to a changing climate, particularly in insect species that have small genomes. By studying thousands of *D. melanogaster* and exposing them to different thermal experiments followed by whole genome sequencing, hundreds of SNPs were identified that were associated with two divergent and thermally induced diapause phenotypes (Erickson et al. 2020). Together, these small-effect, clonally varying SNPs were able to explain latitudinal variation in diapause in North America consistent with local adaptation (Erickson et al. 2020).

These studies have also shown that in addition to SNPs, large inversion polymorphisms are responsible for part of the temperature responses and fluctuate predictably in frequency over seasonal time spans, and this has been a consistent finding across continents in *Drosophila* spp. (Kapun and Flatt 2019). Similar findings were also reported in a study on the seaweed fly *Coelopa frigida*, where a large inversion polymorphism was found to be linked to multiple environmental variables, including temperature (Mérot et al. 2018), and where different natural and sexual selection patterns on males and females lead to balancing selection and polymorphism maintenance (Mérot et al. 2020) across heterogeneous landscapes. Additionally, a comparative whole genome resequencing study on the spider mites *Tetranychus truncates* and *Tetranychus pueraricola* in China showed the faster spreading *Tetranychus truncates* exhibited a greater extent of local climatic adaptation with more genes (seventy-six versus seventeen) associated with precipitation than its sister species, including candidates involved in regulation of homeostasis of water and ions, signal transduction and motor skills (Chen et al. 2020). Therefore, comparative genomics is helping to identify changes to syntenic blocks and gene families to provide functional links between components of the genome and climate-related phenotypes (Moritz and Agudo 2013).

3.4 Conclusions and future research

The current revolution in several omics technologies is providing unprecedented insights into the eco-evolutionary processes of how insects are responding to climate change. It is now a well-accepted fact that we urgently need to improve our understanding of and more accurately predict how species respond to changing climatic conditions to inform management and implement conservation strategies. Most of the work to date has applied genomic tools to study insect responses to a changing climate, with a much lesser number using transcriptome approaches, and an even lesser number applying epigenomic technologies. Recent reviews have argued that genomic data can be used to quantify a species' evolutionary potential, for example, by measuring species genomic vulnerability, and then use this to improve prediction models to forecast how species will respond to changing climatic conditions across their distribution area (Layton and Bradbury 2022). What has become clear from the work to date is that we need to incorporate ecological as well as evolutionary genetic parameters in our models to account for a species' potential to compensate changing environmental conditions by either plasticity, range shifts, or adaptation. Of these, eco-evolutionary models appear to be promising for successfully integrating ecological and evolutionary genomic information, something that will only be achievable through the establishment of large consortia and engagement with local communities (Waldvogel et al. 2020).

Caution needs to be paid to the taxonomic basis of our knowledge. Genomic data are increasingly accumulating in public databases, such as National Center for Biotechnology Information (NCBI). However, there is a strong imbalance of sequencing model versus non-model species and data from laboratory versus natural populations, and this taxonomic imbalance may hamper general insights. In the future, efforts need also to focus on the sufficient resources required for covering different species, different geographical and temporal scales and different approaches. Such combined efforts will undoubtedly help shed much-needed light onto the omic responses of insects to climate change, to inform management and conservation.

Key reflections

- Climate change is already changing ecosystems worldwide, and a major challenge is to understand which species and populations have the genetic potential to adapt.
- Research on climate change adaptation needs to move beyond SNP approaches and include structural genomic variation as a potentially potent source of evolutionary adaptation.
- Epigenomic variation has a significant and mostly unexplored potential to increase the plastic responses of species to cope with a changing climate, and this needs to be considered more in future research.
- Multifaceted and integrated omics approaches are needed to investigate key species that may be impacted by a changing climate to inform mitigation measures in a timely manner.

Key further reading

- Aguirre-Liguori, J. A., et al. 2021. The evolutionary genomics of species' responses to climate change. *Nature Ecology & Evolution* 5:1–11.
- Lancaster, L. T., et al. 2022. Understanding climate change response in the age of genomics. *Journal of Animal Ecology* 91: 1056–1063.
- Wellenreuther, M., et al. 2022. The importance of eco-evolutionary dynamics for predicting and managing insect range shifts. *Current Opinion in Insect Science* 52: 100939.

References

Adams, M.D., Celniker, S.E., Holt, R.A., Evans, C.A., Gocayne, J.D., Amanatides, P.G., Scherer, S.E., et al. 2000. The genome sequence of Drosophila melanogaster. *Science* 287 (5461): 2185–2195.

Aguirre-Liguori, J.A., Ramírez-Barahona, S., and Gaut, B.S. 2021. The evolutionary genomics of species' responses to climate change. *Nature Ecology & Evolution* 5: 1–11.

Alvarez, M., Schrey, A.W., and Richards, C.L. 2015. Ten years of transcriptomics in wild populations: What have we learned about their ecology and evolution? *Molecular Ecology* 24 (4): 710–25.

Anastasiadi, D., Venney, C.J., Bernatchez, L., and Wellenreuther, M. 2021. Epigenetic inheritance and reproductive mode in plants and animals. *Trends in Ecology & Evolution* 36 (12): 1124–1140.

Balkenhol, N., Dudaniec, R.Y., Krutovsky, K.V., Johnson, J.S., Cairns, D.M., Segelbacher, G., Selkoe, K.A., et al. 2017. *Population Genomics: Concepts, Approaches and Applications*, 261–322.

Bayega, A., Djambazian, H., Tsoumani, K.T., Gregoriou, M.E., Sagri, E., Drosopoulou, E., Mavragani-Tsipidou, P., et al. 2020. De novo assembly of the olive fruit fly (Bactrocera oleae) genome with linked-reads and long-read technologies minimizes gaps and provides exceptional Y chromosome assembly. *BMC Genomics 21* (1): 1–21.

Bonebrake, T.C., Brown, C.J., Bell, J.D., Blanchard, J.L., Chauvenet, A., Champion, C., Chen, I.C., et al. 2018. Managing consequences of climate-driven species redistribution requires integration of ecology, conservation and social science. *Biological Reviews 93* (1): 284–305.

Chen, J., Nolte, V., and Schlötterer, C. 2015. Temperature-related reaction norms of gene expression: Regulatory architecture and functional implications. *Molecular Biology and Evolution 32* (9): 2393–2402.

Chen, L., Sun, J.T., Jin, P.Y., Hoffmann, A.A., Bing, X.L., Zhao, D.S., Xue, X.-F. 2020. Population genomic data in spider mites point to a role for local adaptation in shaping range shifts. *Evolutionary Applications 13* (10): 2821–2835.

Chou, H., Jima, D.D., Funk, D.H., Jackson, J.K., Sweeney, B.W., and Buchwalter, D.B. 2020. Transcriptomic and life history responses of the mayfly Neocloeon triangulifer to chronic diel thermal challenge. *Scientific Reports 10* (1): 1–11.

Dai, T.M., Lü, Z.C., Wang, Y.S., Liu, W.X., Hong, X.Y., and Wan, F.H. 2018. Molecular characterizations of DNA methyltransferase 3 and its roles in temperature tolerance in the whitefly, Bemisia tabaci Mediterranean. *Insect Molecular Biology 27* (1): 123–32.

Dudaniec, R.Y., Carey, A.R., Svensson, E.I., Hansson, B., Yong, C.J., and Lancaster, L.T. 2022. Latitudinal clines in sexual selection, sexual size dimorphism and sex-specific genetic dispersal during a poleward range expansion. *Journal of Animal Ecology 91* (6): 1104–1118.

Dudaniec, R.Y., Yong, C.J., Lancaster, L.T., Svensson, E.I., and Hansson, B. 2018. Signatures of local adaptation along environmental gradients in a range-expanding damselfly (Ischnura elegans). *Molecular Ecology 27* (11): 2576–2593.

Duncan, E.J., Leask, M.P., and Dearden, P.K. 2020. Genome architecture facilitates phenotypic plasticity in the honeybee (Apis mellifera). *Molecular Biology and Evolution, 37* (7): 1964–1978.

Eckert, C.G., Samis, K.E., and Lougheed, S.C. 2008. Genetic variation across species' geographical ranges: The central–marginal hypothesis and beyond. *Molecular Ecology 17* (5): 1170–1188.

Ellegren, H. 2014. Genome sequencing and population genomics in non-model organisms. *Trends in Ecology & Evolution 29* (1): 51–63.

Elshire, R.J., Glaubitz, J.C., Sun, Q., Poland, J.A., Kawamoto, K., Buckler, E.S., and Mitchell, S.E. 2011. A robust, simple genotyping-by-sequencing (GBS) approach for high diversity species. *PloS One 6* (5): e19379.

Erickson, P.A., Weller, C.A., Song, D.Y., Bangerter, A.S., Schmidt, P., and Bergland, A.O. 2020. Unique genetic signatures of local adaptation over space and time for diapause, an ecologically relevant complex trait, in Drosophila melanogaster. *PLoS Genetics 16* (11): e1009110.

Ferretti, L., Ramos-Onsins, S.E., and Pérez-Enciso, M. 2013. Population genomics from pool sequencing. *Molecular Ecology 22* (22): 5561–5576.

Fuller, Z.L., Haynes, G.D., Richards, S., and Schaeffer, S.W. 2017. Genomics of natural populations: Evolutionary forces that establish and maintain gene arrangements in Drosophila pseudoobscura. *Molecular Ecology 26* (23): 6539–6562.

Fuller, Z.L., Niño, E.L., Patch, H.M., Bedoya-Reina, O.C., Baumgarten, T., Muli, E., Mumoki, F., et al. 2015. Genome-wide analysis of signatures of selection in populations of African honey bees (Apis mellifera) using new web-based tools. *BMC Genomics 16* (1): 518.

Gao, X., Fu, Y., Ajayi, O.E., Guo, D., Zhang, L., and Wu, Q. 2019. Identification of genes underlying phenotypic plasticity of wing size via insulin signaling pathway by network-based analysis in Sogatella furcifera. *BMC Genomics 20* (1): 1–21.

Gibert, P., Hill, M., Pascual, M., Plantamp, C., Terblanche, J.S., Yassin, A., and Sgrò, C. M. 2016. Drosophila as models to understand the adaptive process during invasion. *Biological Invasions 18*: 1089–1103.

Glastad, K.M., Hunt, B.G., and Goodisman, M.A. 2019. Epigenetics in insects: Genome regulation and the generation of phenotypic diversity. *Annual Review of Entomology 64*: 185–203.

Guo, Y., Vickers, K., Xiong, Y., Zhao, S., Sheng, Q., Zhang, P., Zhou, W., et al. 2017. Comprehensive evaluation of extracellular small RNA isolation methods from serum in high throughput sequencing. *BMC Genomics* 18 (1): 1–9.

Harpur, B.A., Kent, C.F., Molodtsova, D., Lebon, J.M., Alqarni, A.S., Owayss, A.A., and Zayed, A. 2014. Population genomics of the honey bee reveals strong signatures of positive selection on worker traits. *Proceedings of the National Academy of Sciences* 111 (7): 2614–2619.

Hewitt, G.M. 2001. Speciation, hybrid zones and phylogeography—or seeing genes in space and time. *Molecular Ecology* 10 (3): 537–49.

Hickling, R., Roy, D.B., Hill J., and Thomas C. 2006b. A northward shift of range margins in British Odonata. *Global Change Biology* 11(3) 502–06.

Hickling, R., Roy, D.B., Hill, J.K., Fox, R., and Thomas, C.D. 2006. The distributions of a wide range of taxonomic groups are expanding polewards. *Global Change Biology* 12 (3): 450–55.

Housman, G. and Ulitsky, I. 2016. Methods for distinguishing between protein-coding and long noncoding RNAs and the elusive biological purpose of translation of long noncoding RNAs. *Biochimica et Biophysica Acta (BBA)—Gene Regulatory Mechanisms* 1859 (1): 31–40.

i5K Consortium. 2013. The i5K Initiative: Advancing arthropod genomics for knowledge, human health, agriculture, and the environment. *Journal of Heredity* 104 (5): 595–600.

Iserbyt, A., Bots, J., Van Gossum, H., and Jordaens, K. 2010. Did historical events shape current geographic variation in morph frequencies of a polymorphic damselfly? *Journal of Zoology* 282 (4): 256–65.

Jain, M., Fiddes, I.T., Miga, K.H., Olsen, H.E., Paten, B., and Akeson, M. 2015. Improved data analysis for the MinION nanopore sequencer. *Nature Methods* 12 (4): 351–356.

Ji, S.X., Wang, X.D., Shen, X.N., Liang, L., Liu, W.X., Wan, F.H., and Lü, Z.C. 2020. Using RNA interference to reveal the function of chromatin remodeling factor ISWI in temperature tolerance in Bemisia tabaci Middle East–Asia Minor 1 cryptic species. *Insects* 11 (2): 113.

Kapun, M., Barrón, M.G., Staubach, F., Obbard, D.J., Wiberg, R.A.W., Vieira, J., Goubert, C., et al. 2020. Genomic analysis of European Drosophila melanogaster populations reveals longitudinal structure, continent-wide selection, and previously unknown DNA viruses. *Molecular Biology and Evolution* 37(9): 2661–2678.

Kapun, M. and Flatt, T. 2019. The adaptive significance of chromosomal inversion polymorphisms in Drosophila melanogaster. *Molecular Ecology* 28 (6): 1263–1282.

Kawecki, T.J. and Ebert, D. 2004. Conceptual issues in local adaptation. *Ecology Letters* 7 (12): 1225–1241.

Kukurba, K.R. and Montgomery, S.B. 2015. RNA sequencing and analysis. *Cold Spring Harbor Protocols* 2015 (11): 951–69.

Lafuente, E. and Beldade, P. 2019. Genomics of developmental plasticity in animals. *Frontiers in Genetics* 10: 720.

Lafuente, E., Duneau, D., and Beldade, P. 2018. Genetic basis of thermal plasticity variation in Drosophila melanogaster body size. *PLoS Genetics* 14 (9): e1007686.

Lancaster, L.T., Dudaniec, R.Y., Chauhan, P., Wellenreuther, M., Svensson, E.I., and Hansson, B. 2016. Gene expression under thermal stress varies across a geographical range expansion front. *Molecular Ecology* 25 (5): 1141–1156.

Lancaster, L.T., Dudaniec, R.Y., Hansson, B., and Svensson, E.I. 2015. Latitudinal shift in thermal niche breadth results from thermal release during a climate-mediated range expansion. *Journal of Biogeography* 42 (10): 1953–1963.

Lancaster, L.T., Fuller, Z.L., Berger, D., Barbour, M.A., Jentoft, S., and Wellenreuther, M. 2022. Understanding climate change response in the age of genomics. *Journal of Animal Ecology* 91 (6): 1056–1063.

Layton, K.K. and Bradbury, I.R. 2022. Harnessing the power of multi-omics data for predicting climate change response. *Journal of Animal Ecology* 91 (6): 1064–1072.

Lehmann, P., Ammunét, T., Barton, M., Battisti, A., Eigenbrode, S.D., Jepsen, J.U., Kalinkat, G., et al. 2020. Complex responses of global insect pests to climate warming. *Frontiers in Ecology and the Environment* 18 (3): 141–50.

Li, S. and Tollefsbol, T.O. 2021. DNA methylation methods: Global DNA methylation and methylomic analyses. *Methods* 187: 28–43.

Ma, C.S., Ma, G., and Pincebourde, S. 2021. Survive a warming climate: Insect responses to extreme high temperatures. *Annual Review of Entomology* 66: 163–84.

Ma, W., Li, X., Shen, J., Du, Y., Xu, K., and Jiang, Y. 2019. Transcriptomic analysis reveals Apis mellifera adaptations to high temperature and high humidity. *Ecotoxicology and Environmental Safety 184*: 109599.

Macgregor, C.J., Thomas, C.D., Roy, D.B., Beaumont, M.A., Bell, J.R., Brereton, T., Bridle, J.R., et al. 2019. Climate-induced phenology shifts linked to range expansions in species with multiple reproductive cycles per year. *Nature Communications* 10 (1): 4455.

Machado, H.E., Bergland, A.O., O'Brien, K.R., Behrman, E.L., Schmidt, P.S., and Petrov, D.A. 2016. Comparative population genomics of latitudinal variation in Drosophila simulans and Drosophila melanogaster. *Molecular Ecology* 25 (3): 723–40.

Maebe, K., Hart, A.F., Marshall, L., Vandamme, P., Vereecken, N.J., Michez, D., and Smagghe, G. 2021. Bumblebee resilience to climate change, through plastic and adaptive responses. *Global Change Biology 27* (18): 4223–4237.

Main, B.J., Marcantonio, M., Johnston, J.S., Rasgon, J.L., Brown, C.T., and Barker, C.M. 2021. Whole-genome assembly of Culex tarsalis. *G3* 11 (2): jkaa063.

Martin, S.H., Dasmahapatra, K.K., Nadeau, N.J., Salazar, C., Walters, J.R., Simpson, F., Blaxter, M., et al. 2013. Genome-wide evidence for speciation with gene flow in Heliconius butterflies. *Genome Research* 23 (11): 1817–1828.

Mathers, T.C., Wouters, R.H., Mugford, S.T., Swarbreck, D., Van Oosterhout, C., and Hogenhout, S.A. 2020. Chromosome-scale genome assemblies of aphids reveal extensively rearranged autosomes and long-term conservation of the X chromosome. *Molecular Biology and Evolution 38* (3): 856–75.

McGuigan, K., Hoffmann, A.A., and Sgrò, C.M. 2021. How is epigenetics predicted to contribute to climate change adaptation? What evidence do we need?. *Philosophical Transactions of the Royal Society B: Biological Sciences* 376 (1826): 20200119.

Merilä, J. 2012. Evolution in response to climate change: In pursuit of the missing evidence. *BioEssays* 34 (9): 811–18.

Mérot, C., Berdan, E.L., Babin, C., Normandeau, E., Wellenreuther, M., and Bernatchez, L. 2018. Intercontinental karyotype–environment parallelism supports a role for a chromosomal inversion in local adaptation in a seaweed fly. *Proceedings of the Royal Society B: Biological Sciences* 285 (1881): 20180519.

Mérot, C., Llaurens, V., Normandeau, E., Bernatchez, L., and Wellenreuther, M. 2020. Balancing selection via life-history trade-offs maintains an inversion polymorphism in a seaweed fly. *Nature Communications* 11 (1): 670.

Moritz, C. and Agudo, R. 2013. The future of species under climate change: Resilience or decline? *Science* 341 (6145): 504–08.

Oostra, V., Saastamoinen, M., Zwaan, B.J., and Wheat, C.W. 2018. Strong phenotypic plasticity limits potential for evolutionary responses to climate change. *Nature Communications* 9 (1): 1005.

Oppold, A., Kreß, A., Bussche, J.V., Diogo, J.B., Kuch, U., Oehlmann, J., Vandegehuchte, M.B., et al. 2015. Epigenetic alterations and decreasing insecticide sensitivity of the Asian tiger mosquito Aedes albopictus. *Ecotoxicology and Environmental Safety* 122: 45–53.

Parmesan, C. 2006. Ecological and evolutionary responses to recent climate change. Annual Review of Ecology, Evolution, and Systematics 37: 637–69.

Parmesan, C., Ryrholm, N., Stefanescu, C., Hill, J.K., Thomas, C.D., Descimon, H., Huntley, B., et al. 1999. Poleward shifts in geographical ranges of butterfly species associated with regional warming. *Nature* 399 (6736): 579–83.

Parmesan, C. and Yohe, G. 2003. A globally coherent fingerprint of climate change impacts across natural systems. *Nature* 421 (6918): 37–42.

Pigliucci, M. 2003. Phenotypic integration: Studying the ecology and evolution of complex phenotypes. *Ecology Letters* 6 (3): 265–72.

Pigliucci, M., Murren, C.J., and Schlichting, C.D. 2006. Phenotypic plasticity and evolution by genetic assimilation. *Journal of Experimental Biology* 209 (12): 2362–2367.

Poelchau, M.F., Reynolds, J.A., Elsik, C.G., Denlinger, D.L., and Armbruster, P.A. 2013. RNA-Seq reveals early distinctions and late convergence of gene expression between diapause and quiescence in the Asian tiger mosquito, Aedes albopictus. *Journal of Experimental Biology* 216 (21): 4082–4090.

Richard, G., Jaquiéry, J., and Le Trionnaire, G. 2021. Contribution of epigenetic mechanisms in the regulation of environmentally-induced polyphenism in insects. *Insects* 12 (7): 649.

Rodrigues, Y.K. and Beldade, P. 2020. Thermal plasticity in insects' response to climate change and to multifactorial environments. *Frontiers in Ecology and Evolution* 8: 271.

Rolandi, C., Lighton, J.R., de la Vega, G.J., Schilman, P.E., and Mensch, J. 2018. Genetic variation for tolerance to high temperatures in a population of Drosophila melanogaster. *Ecology and Evolution* 8 (21): 10374–10383.

Sánchez-Guillén, R.A., Córdoba-Aguilar, A., Hansson, B., Ott, J., and Wellenreuther, M. 2016. Evolutionary consequences of climate-induced range shifts in insects. *Biological Reviews* 91 (4): 1050–1064.

Sánchez-Guillén, R.A. and Ott, J. 2018. Genetic consequences of range expansions along several fronts in Crocothemis erythraea. *International Journal of Odonatology* 21 (2): 81–91.

Scheffers, B.R., De Meester, L., Bridge, T.C., Hoffmann, A.A., Pandolfi, J.M., Corlett, R.T., Butchart, S.H.M., et al. 2016. The broad footprint of climate change from genes to biomes to people. *Science* 354 (6313): aaf7671.

Sgrò, C.M., Terblanche, J.S., and Hoffmann, A.A. 2016. What can plasticity contribute to insect responses to climate change? *Annual Review of Entomology* 61: 433–51.

Soria-Carrasco, V., Gompert, Z., Comeault, A.A., Farkas, T.E., Parchman, T.L., Johnston, J.S., Buerkle, C.A., et al. 2014. Stick insect genomes reveal natural selection's role in parallel speciation. *Science* 344 (6185): 738–42.

Swaegers, J., Mergeay, J., Van Geystelen, A., Therry, L., Larmuseau, M.H.D., and Stoks, R. 2015a. Neutral and adaptive genomic signatures of rapid poleward range expansion. *Molecular Ecology* 24 (24): 6163–6176.

Swaegers, J., Mergeay, J., St-martin, A.U.D.R.E.Y, De Knijf, G., Larmuseau, M.H., and Stoks, R. 2015b. Genetic signature of the colonisation dynamics along a coastal expansion front in the damselfly Coenagrion scitulum. *Ecological Entomology* 40 (4): 353–61.

Swaegers, J., Mergeay, J., Therry, L., Bonte, D., Larmuseau, M.H.D., and Stoks, R. 2014. Unravelling the effects of contemporary and historical range expansion on the distribution of genetic diversity in the damselfly Coenagrion scitulum. *Journal of Evolutionary Biology* 27 (4): 748–59.

Vatanparast, M., Puckett, R.T., Choi, D.S., and Park, Y. 2021. Comparison of gene expression in the red imported fire ant (Solenopsis invicta) under different temperature conditions. *Scientific Reports* 11 (1): 16476.

Waldvogel, A.M., Feldmeyer, B., Rolshausen, G., Exposito-Alonso, M., Rellstab, C., Kofler, R., Mock, T., et al. 2020. Evolutionary genomics can improve prediction of species' responses to climate change. *Evolution Letters* 4 (1): 4–18.

Wallberg, A., Bunikis, I., Pettersson, O.V., Mosbech, M.B., Childers, A.K., Evans, J.D., Mikheyev, A.S., et al. 2019. A hybrid de novo genome assembly of the honeybee, Apis mellifera, with chromosome-length scaffolds. *BMC Genomics* 20: 1–19.

Wang, L., Qu, L., Yang, L., Wang, Y., and Zhu, H. 2020. NanoReviser: An error-correction tool for nanopore sequencing based on a deep learning algorithm. *Frontiers in Genetics* 11: 900.

Wang, Y., Ferveur, J.F., and Moussian, B. 2021. Eco-genetics of desiccation resistance in Drosophila. *Biological Reviews* 96 (4): 1421–1440.

Wellenreuther, M., Dudaniec, R.Y., Neu, A., Lessard, J.-P., Bridle, J., Carbonell, J.A., Diamond, S.E., et al. 2022. The importance of eco-evolutionary dynamics for predicting and managing insect range shifts. *Current Opinion in Insect Science* 52: 100939.

Wellenreuther, M., Larson, K.W., and Svensson, E.I. 2012. Climatic niche divergence or conservatism? Environmental niches and range limits in ecologically similar damselflies. *Ecology* 93 (6): 1353–1366.

Williamson, S.M. 2017. *Epigenetic underpinnings of freeze tolerance in the goldenrod gall fly Eurosta solidaginis and the goldenrod gall moth Epiblema scudderiana* (Doctoral dissertation, Carleton University).

Wragg, D., Marti-Marimon, M., Basso, B., Bidanel, J.P., Labarthe, E., Bouchez, O., Le Conte, Y., et al. 2016. Whole-genome resequencing of honeybee drones to detect genomic selection in a population managed for royal jelly. *Scientific Reports* 6 (1): 1–13.

CHAPTER 4

Physiological mechanisms of heat tolerance in insects

Daniel González-Tokman and Sebastián Villada-Bedoya

4.1 Introduction

Global warming has caused insect extinctions in the past and is a major cause of extinction now (Chapter 2, this book; Cardoso et al. 2020). Survivors need to respond rapidly and the first step is a physiological machinery that has evolved over millions of years and allows insects tolerating heat. Such a machinery needs refined heat sensation and temperature interpretation by the nervous system, an effective but costly stress response, antioxidants, a well-orchestrated heat coma and a series of cellular, metabolic and thermoregulatory responses. These physiological mechanisms are responsible for insect heat tolerance (i.e., tolerance to high temperatures) but are also involved in heat acclimation (irreversible increased heat tolerance acquired during development of an individual) and hardening (reversible acquired tolerance) or adaptation (González-Tokman et al. 2020; Chapter 5, this book). In this chapter we describe such mechanisms in insects (Figure 4.1), that will allow individuals and species to persist in a warming world.

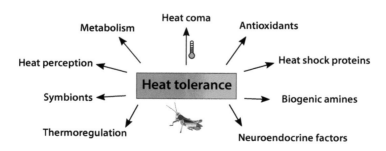

Figure 4.1 An appropriate response to heat involves several mechanisms that act coordinately and define heat tolerance in insects.

Daniel González-Tokman and Sebastián Villada-Bedoya, *Physiological mechanisms of heat tolerance in insects*. In: *Effects of Climate Change on Insects*. Edited by: Daniel González-Tokman and Wesley Dáttilo, Oxford University Press. © Oxford University Press (2024).
DOI: 10.1093/oso/9780192864161.003.0004

4.2 Heat tolerance and thermoregulation

Studies in insects have measured heat tolerance (i.e., tolerance to high temperatures) as either the maximum temperature that an insect is able to stand before dying (upper lethal temperature, or ULT) or before losing muscular control (critical thermal maximum, or CTmax) (Chown et al. 2015; Duffy et al. 2015). It can also be measured as the time an insect is able to stand upright at a given high temperature (heat knockdown time) (Bauerfeind et al. 2018). Either response results from a combination of physiological mechanisms that allows the insect to detect a stressfully high temperature and respond appropriately. Insects not only depend on internal protective physiological mechanisms to survive heat but also on evaporative cooling as well as external (cuticular) coloration which, combined with particular behavioral responses, allow them coping with high temperatures (Cooper et al. 1985; Rajpurohit et al. 2008). Heat tolerance is not constant across insect developmental stages, and also differs with age, sex, body size and condition and is sensitive to environmental conditions such as current and past environmental temperature, photoperiod, contaminants, symbionts, parasites and even the parental environment (González-Tokman et al. 2020).

Temperate regions of the world suffer the greatest effects of climate change on environmental temperature. However, insect diversity is more threatened in the tropics, first because more biodiversity exists at these latitudes, but mainly because of a biophysical principle of living at basal higher temperatures whereby even small increases in temperature need disproportionate increases in the metabolism (Dillon et al. 2010), potentially increasing energetic expenditure and oxidative stress, and thus even small increases can be fatal. However, tropical insect species remain poorly studied despite the potentially wide diversity of mechanisms involved in their heat responses. Insects are considered ectothermic and poikilothermic animals, meaning that they rely on external heat and that their body temperature is changeable. However, insects have the ability to thermoregulate with both behavioral and physiological mechanisms, thus insect body temperature rarely matches the ambient temperature (Heinrich 1993).

Behaviorally, insects exposed to heat thermoregulate by moving to cooler sites or changing perching orientation, such as *Enallagma doubledayi* damselflies or *Colias eurytheme* butterflies that consistently orient to minimize exposure to solar radiation and reduce energy loss at high temperatures (Mason 2017; Watt 1969). Insects can also thermoregulate by flapping appendages to circulate cool air when environmental temperature is high (Chapman 1998).

Physiologically, insects can thermoregulate by adjusting cuticular melanization during development, as dark phenotypes heat up and cool down faster than lighter phenotypes and are potentially more tolerant to desiccation given that melanin reduces cuticular permeability (Rajpurohit et al. 2008). As an adaptive response, insects at low altitudes or latitudes are generally lighter than conspecifics from cooler sites, in accordance with the thermal melanism hypothesis (Trullas et al. 2007; Zeuss et al. 2014). Evaporative cooling is another mechanism that is also common in insects and helps to decrease body temperature, as described at least in Diptera, Hemiptera, Orthoptera and Hymenoptera (González-Tokman et al. 2020). As an example, honeybees from the Sonoran desert regurgitate droplets of honey that refresh the body at extremely high environmental temperatures (Cooper et al. 1985). However, small insects are not expected to thermoregulate via evaporative cooling given the low amount of water they can store (Chapman 1998), even though dehydration and starvation apparently do not interfere with heat tolerance in *Drosophila melanogaster* (Overgaard et al. 2012). Finally, insects can regulate heat flow between different body sections from cooler to warmer regions via the tracheal and the circulatory systems (Heinrich 1973; Hillyer and Pass 2020).

4.3 Physiological mechanisms of heat tolerance

Heat tolerance needs a coordinated response of mechanisms first involving heat sensation, processing and interpretation of thermal information, and then responding with a variety of protective neural and physiological mechanisms, including heat shock proteins, biogenic amines and neuroendocrine factors, which together conform the stress response. Also, metabolic changes, antioxidants, thermoregulation and hormones are part of an integrated response to heat. We describe these physiological mechanisms involved in insect heat tolerance in the next section.

4.3.1 *Heat perception and coma*

The first step in the response to heat is an appropriate sensation of temperature. This occurs via neurons from the peripheral nervous system, which are connected with thermal projector neurons to the central nervous system that coordinates an appropriate response (Frank et al. 2015). Thermosensory neurons can be distributed across the whole insect surface, as in the kissing bug *Rhodnius prolixus*, or they can be restricted to some body parts, as in *Drosophila melanogaster*, which senses temperature via only six neurons in each antenna (Li and Gong 2017; Zermoglio et al. 2015). Some thermosensory neurons are sensitive to temperature changes of 0.5 °C, as in *D. melanogaster*, whereas others sense changes of 0.005 °C, as in the ant *Atta vollenweideri* (Gallio et al. 2011; Ruchty et al. 2010), indicating notably high heat sensitivity.

When facing extreme heat, the insect metabolism increases to a point that energetic expenditure becomes so high that death can occur, also because it becomes impossible to search for cooler habitats and regulate tracheal ventilatory movements (Robertson 2004). In response, insects enter heat coma, a reversible response involving neural and muscular shut down which prevents energetic depletion and cellular damage, increasing survival probabilities (Rodgers et al. 2010; Storey and Storey 2004). In insects adapted to extremely high temperatures, such as the desert locust *Locusta migratoria*, the central nervous system has a neuronal circuit, named the ventilatory central pattern generator (VCPG). This circuit, located in the metathoracic ganglion, delivers and removes O_2 and CO_2 via the tracheal system and promotes insect cooling via evaporation (Rodgers et al. 2010). Prior heat stress protects the VCPG by the action of the neurotransmitter serotonin, which increases in the heated insect and seems to have a fundamental role in thermoprotection during insect heat coma (Robertson 2004).

4.3.2 *Metabolic changes and the need for antioxidants*

Before entering heat coma, insect metabolism increases with increasing temperature (Gillooly et al. 2001; but see (Dinh et al. 2016). As resting metabolic rate (i.e., the rate at which organisms invest energy in basic biochemical and cellular processes) represents 50 percent of total energetic expenditure, increased metabolism becomes highly costly for insects exposed to heat because it demands increased ion pumping and protein turnover (Burton et al. 2011), leading to energetic depletion and death (Storey and Storey 2004). As temperature has a nonlinear effect on metabolic rate, insects from warm places, such as the tropics, are more threatened because baseline temperatures are higher and even small increases in temperature need an additional increase in metabolism (Dillon et al. 2010).

At high temperature, aerobic metabolism (respiration) increases but the capacity for circulation and ventilation may not be enough to supply the insect needs for oxygen, causing death (Pörtner 2002). The oxygen and capacity limited thermal tolerance (OCLTT) hypothesis states that animals die at high temperature because of insufficient oxygen supply, which is more limited in water where, in contrast to the air, oxygen decreases drastically with increasing temperature. Despite the fact that OCLTT does not seem to be the rule in insects, it is relevant underwater, where compensatory anaerobic metabolism could be used to extend lifespan by providing additional energy (Pörtner 2002; Verberk et al. 2016, 2021). In contrast, in air breathers, only severe hypoxia affects heat tolerance and compensatory anaerobic metabolism seems unlikely in the face of climate change (Verberk et al. 2016).

With increasing temperature, increased metabolism might result in higher production of free radicals such as peroxide anion (O_2^{2-}), hydroxyl radical (HO) and hydrogen peroxide (H_2O_2), which are highly reactive and may cause cell damage (Felton and Summers 1995). When the pro-oxidant effects cannot be counteracted by antioxidants, oxidative damage to biomolecules occurs, leading to cell death. Whereas some antioxidants are produced endogenously, others are obtained from the diet (Table 4.1; Felton and Summers 1995). In response to heat, insect antioxidant defenses include the enzymes superoxide dismutase (SOD), peroxidases, catalases (CAT) and glutathione-S-transferase (GST), but also include glutathione (GSH), glutathione reductase (GR), lactate dehydrogenase (LDH), transferrin (PbTf), ascorbic acid and proline. These antioxidants, as well as total antioxidant capacity, are upregulated in homogenates of heat-stressed Lepidoptera, Diptera,

Table 4.1 Different antioxidants and heat shock proteins have been described in insects in response to high temperatures. Most are upregulated but, in some cases, they are downregulated (down). Aldehyde dehydrogenase (ALDH), catalase (CAT), glutathione-S-transferase (GST), peroxidase (POD), superoxide dismutase (SOD), glutathione (GSH), glutathione reductase (GR), lactate dehydrogenase (LDH), transferrin (PbTf). na = no available information. References from González-Tokman et al. (2020).

Insect order	Antioxidants	Heat shock proteins
Coleoptera	ALDH, CAT, GST, POD, SOD, PbTf	HSP21, HSP23, HSP60, HSP70
Diptera	CAT, GST, POD, SOD	HSP20, HSP22, HSP23, HSP26, HSP27, HSP70, Hsc70, HSP83, HSP90
Hemiptera	CAT, GSH, GR, LDH, SOD	HSP40, HSP70, HSP90
Hymenoptera	CAT, GST, POD, SOD	HSP40, HSP70, hsc70-4h1, hsc70-4h2, HSP83
Lepidoptera	ALDH, ascorbic acid, CAT, GST, POD, SOD	HSP18.6, HSP19.2, HSP19.6, HSP19.7, HSP19.8, HSP19.9, HSP20, HSP20.1, HSP20.2, HSP20.4, HSP20.8, HSP21.3, HSP21.4, HSP21.5, HSP21.7, HSP22, HSP22.1, HSP23, HSP23.7, HSP23.9, HSP24.3, HSP25, HSP27, HSP28.4, HSP70 (down), HSP70, HSP71, HSP73, HSP74, HSP83, HSP84, HSP90
Odonata	na	HSP20, HSP70
Orthoptera	na	HSP90
Psocoptera	CAT, GST, POD, SOD	HSP23, HSP27
Thysanoptera	na	HSP60, HSP90

Coleoptera, Hemiptera and also in some particular organs such as testes of heat-exposed silkworms *Antheraea mylitta* (Cui et al. 2011; Jena et al. 2013; Jia et al. 2011; Zhang et al. 2015). Also, transcriptomic analysis of the whole body of heat-stressed Asian citrus psyllids (*Diaphorina citri*) reveals upregulation of oxidation resistance genes (Xiong et al. 2019).

Other antioxidants, including carotenoids, ommochromes (synthesized from the amino acid tryptophan) and melanins (synthesized from phenylalanine) (Dhinaut et al. 2017; Ushakova et al. 2019), are obtained from the diet but their role in heat resistance has not been studied. By affecting insect diets and therefore the availability of antioxidants for some insects, climate change may demand dietary adjustments in combination with effective endogenously produced antioxidants. Complementary mechanisms including uncoupling proteins, present in the mitochondrial membrane, might help preventing reactive oxygen species (ROS) formation and oxidative stress upon heat exposure, but evidence is still limited in insect heat responses (Slocinska et al. 2016).

Acclimation and adaptation to heat are also associated with antioxidant defenses in Lepidoptera, as heat-acclimated larvae induced more differentially expressed antioxidant genes in response to heat shock than the unacclimated larvae (Quan et al. 2020), but further experimental research is needed to find out the role of antioxidant responses in insect plastic and adaptive responses to heat (Figure 4.2).

4.3.3 *Heat shock proteins*

Exposure to heat may cause death due to protein denaturing. Heat shock proteins (HSP) are chaperone proteins varying in molecular weight that allow all living organisms, from bacteria to vertebrates, to cope with high environmental temperatures and other sources of stress (Feder and Hofmann 1999). The mode of action of HSP is by binding to other proteins to avoid denaturing at normally lethal temperature (King and MacRae 2015; Sørensen et al. 2003). In insects, HSP70 and HSP90 (with molecular weights of 70 and 90 kDa, respectively), have been the most studied in response to high temperatures, being upregulated in at least ten insect orders; other HSP (e.g.,

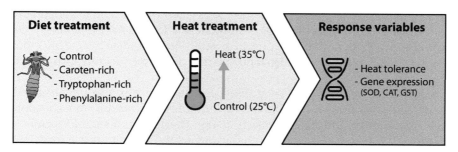

Figure 4.2 Elucidating the role of dietary and endogenously produced carotenoids on heat tolerance needs an experimental design manipulating diet and heat exposure. Heat tolerance (i.e., survival time at 35 °C) and differential gene expression could be evaluated as response variables. Ideally, full-sib or half-sib experiments could be used across the different treatments to estimate heritability in the response to combined stressors. Different "omic" approaches including genomic, transcriptomic, proteomic, etc., need to be decided carefully. Besides antioxidants, differential gene expression analysis (transcriptomic) could reveal other genes associated to heat tolerance, such as HSP. A tropical damselfly species is suggested as a study subject for being predator and aquatic, therefore being sensitive to food limitation and climate warming. Drawing by Yorleny Gil; SVB.

HSP19.8, HSP21) have also been described in response to heat (Table 4.1). Moreover, HSP are responsible for heat acclimation (Dahlhoff and Rank 2000) and adaptation (Sørensen 2010).

In heated insects, HSP protect cells from several tissues, including the nervous system (Armstrong et al. 2011; Gehring and Wehner 1995; Karunanithi et al. 1999), the gut (Benoit et al. 2011) and the fat body (Wang et al. 2014). In response to heat, most HSP are upregulated (González-Tokman et al. 2020), but in some particular cases they are downregulated, as in *Bombyx mori* (Lepidoptera) testis, ovary and fat body, where HSP70 expression is reduced (Li et al. 2012). As the heat shock response is energetically demanding, HSP upregulation is followed by a drop in its levels. In contrast, in sites with extremely high temperatures such as arid and semiarid regions of Africa, Asia and America, the desert locust *Locusta migratoria* and the fly *Drosophila arizonae* have high constitutive levels of HSP70, and do not respond to heat upregulating this HSP (Dehghani et al. 2011; Newman et al. 2005).

4.3.4 *Biogenic amines and neuroendocrine factors*

The neuroendocrine stress reaction includes the action of biogenic amines (BA) and hormones (Gruntenko and Rauschenbach 2008). Octopamine, dopamine, serotonin, epinephrine and norepinephrine, among other BA, are released from the insect fat body immediately after heat stress (Adamo 2012; Gruntenko et al. 2017; Hirashima et al. 1993). In nervous cells, BA are neurotransmitters and neuromodulators, but they can even be released into body fluids acting as neurohormones (Sinakevitch et al. 2018). Frequently, other sources of stress, such as cold, infection and starvation, also cause the release of BA (Lubawy et al. 2020). Biogenic amines regulate gene transcription and the Na+ K+ ATPase pump, which is fundamental in response to different environmental stimuli and the transmission of neural impulses (Alberts et al. 1997). Releasing BA needs several enzymes such as alkaline phosphatase or dopa-decarboxylase, among others (Sukhanova et al. 1996, 1997), which are sensitive to temperature, but the effects of climate warming on these enzymatic processes remain to be studied in insects (Table 4.1). After releasing BA, neuroendocrine factors (including peptides and proteins) such as adipokinetic hormone (AKH) mobilize energy reserves in insects, as glucocorticoids do in vertebrate stress responses (Adamo 2012). In honeybees, heat induces a drop in AKH receptors, reflecting a decrease in carbohydrate mobilization in stressed bees (Bordier et al. 2017), but further research is also needed to understand the role of neuroendocrine factors in insect responses to heat.

The endocrine stress reaction during insect heat tolerance is also understudied, but it is fundamental to survive heat. Involved hormones are insulin and gonadotropin hormones (Gruntenko and Rauschenbach 2018). The insulin/insulin-like growth factor signaling (IIS) pathway, involving the action of the transcription factors such as forkhead (FOXO, also present in vertebrates), regulates the synthesis of antioxidant enzymes such as SOD (see above) during insect diapause and provide tolerance to cold, oxidative stress and pathogenic infection (Sim and Denlinger 2013). In *Drosophila*, experimental evidence with mutants indicates that FOXO also regulates insulin signaling and energetic metabolism in response to heat stress (Eremina et al. 2021). Insulin signaling is also involved in the synthesis of the gonadotropin hormones juvenile hormone (JH) and ecdysone (20E), which complement the insect endocrine stress response against heat (Flatt et al. 2005; Gruntenko and Rauschenbach 2018). Non-model insects should be evaluated in the neuroendocrine response to heat, as the evidence of insulin signaling and gonadotropin hormones in insect heat tolerance is still incipient.

4.3.5 *The role of symbionts*

Several insect functions, including heat tolerance, are performed or facilitated by microbial endosymbionts, some of which are obligate and some facultative. However, climate change imposes a threat to these symbiosis, potentially reducing host performance at high temperature (Feldhaar 2011; Renoz et al. 2019). For example, gut symbioses collapse in the bug *Nezara viridula* exposed to simulated climate change, leading to severe fitness defects at temperatures 2.5 °C above the environment (Kikuchi et al. 2016). In aphids, probably the best studied system in this topic, the absence of a heat shock gene in the obligate endosymbiont bacteria *Buchnera* reduces heat tolerance in the host, indicating that shifts in bacterial heat sensitivity change heat tolerance in the insect host (Zhang et al. 2019). Also, in the pea aphid, facultative symbionts *Serratia symbiotica* protect the host and also *Buchnera* from heat stress (Burke et al. 2010). Facultative mutualisms also occur, as in *D. melanogaster* bearing certain strains of *Wolbachia*, which, instead of being pathogenic, confer increased heat tolerance by increasing dopamine metabolism (Gruntenko et al. 2017). As microbial communities within insect hosts are structured by both the evolutionary history and local environmental forces, microbes can be sex-, species- or population-specific (Parker et al. 2020). The role of the microbiome in heat tolerance remains an open question in most insects but it could be more important than previously envisioned.

4.4 Other mechanisms of heat tolerance

Transcriptomic analyses reveal the action of several physiological mechanisms that are involved in heat tolerance, but most of them have not been studied in detail. For example, transcriptomic analysis in *Apis mellifera*, revealed that exposure to 45 °C leads to the upregulation of 334 genes and the downregulation of 100 genes compared with a control temperature of 25 °C (Ma et al. 2019). In *Glyphodes pyloalis* (Lepidoptera: Pyralidae) exposure to 40 °C drives the upregulation of 1,275 genes and the downregulation of 1,222 genes compared with a control temperature of 25 °C (Liu et al. 2017). In the parasitoid wasp *Tamarixia radiata*, 476 genes are upregulated and 26 downregulated in response to 38 °C, also compared to 25 °C. These genes include HSP and antioxidants but also vitamin digestion and detoxification pathways, repressors of carbohydrate and lipid metabolism and even immune response genes, such as the Toll-like receptor signaling pathway. The differences in gene regulation among studied insects indicates particularities in the physiological response, but transcriptomic analyses of the heat shock response are still lacking for most insect orders. These mechanisms deserve attention in further studies. Appropriate experimental designs and gene co-expression network analyses are needed to identify modules of gene expression involved in heat tolerance (Torson et al. 2020). Other techniques such as fluorescence microscopy have revealed changes in the actin cytoskeleton of heat acclimated crickets *Gryllus pennsylvanicus* (Des Marteaux et al. 2018), but the role of the cytoskeleton in insect heat responses remains unexplored.

4.5 Limitations to heat tolerance: Life history trade-offs and synergistic effects

Survival and fecundity are highly compromised in insects exposed to high temperatures (Krebs and Loeschcke 1994). As mechanisms of heat tolerance demand energy and nutrients, there are risks of sacrificing fundamental body functions while tolerating heat.

This is evident in insects that lose body lipid reserves after heat exposure (Dinh et al. 2016). In our current period of global climate change, high temperatures inevitably interact with other environmental stressors, such as contaminants, food limitation and novel parasites and pathogens. The interactions between these stressors when combined can be additive (which means the sum of the effect of individual stressors) or synergistic, such that the effects are larger than the addition (Kaunisto et al. 2016). In general, it is expected that stressor effects will become more dramatic with climate warming (Noyes and Lema 2015). For example, in damselflies *Coenagrion puella*, exposure to a heat wave is only sublethal, causing immunosuppression (a reduction in the activity of the enzyme phenoloxidase) and metabolic depression (a reduction in cellular electron transport system) (Dinh et al. 2016). However, the effects of heat become lethal in larvae that are also exposed to the agricultural pesticide chlorpyrifos but only in combination with food limitation, revealing an important role of nutrition in the response to heat. The synergism between combined stressors is apparently regulated by the cellular electron transport system, while HSP70 or the enzyme acetylcholinesterase (fundamental in synapsis) do not seem to be involved in the response to the synergism (Dinh et al. 2016). In contrast, exposure to heat may have apparent beneficial effects when coping with other stressors, as in dung beetles *Euoniticellus intermedius*, that are more heat tolerant when artificially parasitized (González-Tokman 2021), or in *Culex pipiens* mosquitoes, that become less sensitive to chlorpyrifos when previously exposed to heat (Meng et al. 2020). If this finding results from cross-tolerance to both stressors or from selection of survivors to the first challenge (i.e., heat) remains to be studied, such as the physiological mechanisms involved (Figure 4.1).

Synergistic effects of heat with other stressors besides life-history trade-offs with heat tolerance will largely define species persistence in warm conditions. The underlying physiological mechanisms deserve special attention in survivors from natural conditions, where populations evolve facing multiple stressors and where the mechanisms involved in response to heat only may be quite different from the mechanisms involved in response to combined stressors (Ma et al. 2019).

4.6 Conclusions

Climate change imposes elevated temperatures that need to be counteracted by defensive physiological mechanisms. In insects, the most studied of these mechanisms are heat shock proteins, but antioxidants, biogenic amines and thermoregulation also play a fundamental role in heat tolerance. Less studied have been the roles of these mechanisms in heat acclimation and adaptation, opening an important avenue of research. Although evidence in non-model species has increased, our current knowledge of insect physiological responses to heat comes from a few orders. Omic analyses are still scarce in the search for the underlying physiological mechanisms of heat tolerance but are promising for studying non-model species. The role of symbionts in heat tolerance has also emerged as a relevant mechanism but evidence is still scarce in most insects. Right now, experiments in natural gradients and the laboratory, evaluating the physiological response to heat in combination with other stressors, are highly needed.

Key reflections

- Insects possess a battery of physiological mechanisms that allow them surviving and adapting to high temperatures

- Mechanisms of heat perception, processing and response are key for insect immediate survival but also for acclimation and adaptation
- The neuroendocrine stress reaction, antioxidant defenses and the microbiomes need to be more deeply explored in insect heat tolerance
- In times of global change, the insect physiological machinery needs to respond to multiple stressors simultaneously, but the response might be limited by life-history trade-offs

Key further reading

Heat perception

- Gallio, M., Ofstad, T.A., Macpherson, L.J., Wang, J.W., and Zuker, C. S. 2011. The coding of temperature in the *Drosophila* brain. *Cell* 144: 614–24.

Heat coma

- Robertson, R.M. 2004. Thermal stress and neural function: Adaptive mechanisms in insect model systems. *Journal of Thermal Biology* 29: 351–58.
- Rodgers, C.I., Armstrong, G.A., and Robertson, R.M. 2010. Coma in response to environmental stress in the locust: a model for cortical spreading depression. *Journal of Insect Physiology* 56: 980–90.

Metabolism

- Dillon, M.E., Wang, G., and Huey, R.B. 2010. Global metabolic impacts of recent climate warming. *Nature* 467: 704–06.

Antioxidants

- Zhang, S., Fu, W., Li, N., Zhang, F., and Liu, T. X. 2015. Antioxidant responses of *Propylaea japonica* (Coleoptera: Coccinellidae) exposed to high temperature stress. *Journal of Insect Physiology* 73: 47–52.

Heat shock proteins

- Feder, M., and Hofmann, G. 1999. Heat-shock proteins, molecular chaperones, and the stress response: evolutionary and ecological physiology. *Annual Review of Physiology* 61: 243–82.

Neuroendocrine response

- Gruntenko, N.E., and Rauschenbach, I.Y. 2018. The role of insulin signalling in the endocrine stress response in Drosophila melanogaster: A mini-review. *General and Comparative Endocrinology* 258: 134–39.

Symbionts

- Zhang, B., Leonard, S.P., Li, Y., and Moran, N.A. 2019. Obligate bacterial endosymbionts limit thermal tolerance of insect host species. *Proceedings of the National Academy of Sciences* 116: 24712–24718.

Other mechanisms

- Liu, Y., Su, H., Li, R., Li, X., Xu, Y., Dai, X., Zhou, Y. et al. 2017. Comparative transcriptome analysis of *Glyphodes pyloalis* Walker (Lepidoptera: Pyralidae) reveals novel insights into heat stress tolerance in insects. *BMC Genomics* 18: 1–13.

- Ma, W., Li, X., Shen, J., Du, Y., Xu, K., and Jiang, Y. 2019. Transcriptomic analysis reveals *Apis mellifera* adaptations to high temperature and high humidity. *Ecotoxicology and Environmental Safety* 184: 109599.

Combined stressors

- Dinh, K. V., Janssens, L., and Stoks, R. 2016. Exposure to a heat wave under food limitation makes an agricultural insecticide lethal: A mechanistic laboratory experiment. *Global Change Biology* 22: 3361–3372.
- Kaunisto, S., Ferguson, L.V., and Sinclair, B.J. 2016. Can we predict the effects of multiple stressors on insects in a changing climate? *Current Opinion in Insect Science* 17: 55–61.

References

Adamo, S.A. 2012. The effects of the stress response on immune function in invertebrates: An evolutionary perspective on an ancient connection. *Hormones and Behavior* 62 (3): 324–30.

Alberts, B., Johnson, A., Lewis, J., Raff, M., Roberts, K., and Walter, P. 2002. *Molecular Biology of the Cell*. New York: Garland Science.

Armstrong, G.A., Xiao, C., Krill, J.L., Seroude, L., Dawson-Scully, K., and Robertson, R.M. 2011. Glial Hsp70 protects K+ homeostasis in the Drosophila brain during repetitive anoxic depolarization. *PloS One* 6 (12): e28994.

Bauerfeind, S.S., Sørensen, J.G., Loeschcke, V., Berger, D., Broder, E.D., Geiger, M., Ferrari, M. et al. 2018. Geographic variation in responses of European yellow dung flies to thermal stress. *Journal of Thermal Biology* 73: 41–49.

Benoit, J.B., Lopez-Martinez, G., Patrick, K.R., Phillips, Z.P., Krause, T.B., and Denlinger, D.L. 2011. Drinking a hot blood meal elicits a protective heat shock response in mosquitoes. *Proceedings of the National Academy of Sciences* 108 (19): 8026–8029.

Bordier, C., Suchail, S., Pioz, M., Devaud, J.M., Collet, C., Charreton, M., LeConte, Y. et al. 2017. Stress response in honeybees is associated with changes in task-related physiology and energetic metabolism. *Journal of Insect Physiology* 98: 47–54.

Burke, G., Fiehn, O., and Moran, N. 2010. Effects of facultative symbionts and heat stress on the metabolome of pea aphids. *The ISME Journal* 4 (2): 242–52.

Burton, T., Killen, S.S., Armstrong, J.D., and Metcalfe, N.B. 2011. What causes intraspecific variation in resting metabolic rate and what are its ecological consequences?. *Proceedings of the Royal Society B: Biological Sciences* 278 (1724): 3465–3473.

Cardoso, P., Barton, P.S., Birkhofer, K., Chichorro, F., Deacon, C., Fartmann, T., Fukushima, C. et al. 2020. Scientists' warning to humanity on insect extinctions. *Biological Conservation* 242: 108426.

Chapman, R.F. 1998. *The insects: Structure and function*. Cambridge: Cambridge University Press.

Chown, S. L., Duffy, G. A., and Sørensen, J. G. 2015. Upper thermal tolerance in aquatic insects. *Current Opinion in Insect Science* 11: 78–83.

Cooper, P.D., Schaffer, W.M., and Buchmann, S.L. 1985. Temperature regulation of honey bees (Apis mellifera) foraging in the Sonoran desert. *Journal of Experimental Biology* 114 (1): 1–15.

Cui, Y., Du, Y., Lu, M., and Qiang, C. 2011. Antioxidant responses of Chilo suppressalis (Lepidoptera: Pyralidae) larvae exposed to thermal stress. *Journal of Thermal Biology* 36 (5): 292–97.

Dahlhoff, E.P. and Rank, N.E. 2000. Functional and physiological consequences of genetic variation at phosphoglucose isomerase: Heat shock protein expression is related to enzyme genotype in a montane beetle. *Proceedings of the National Academy of Sciences* 97 (18): 10056–10061.

Dehghani, M., Xiao, C., Money, T.G., Shoemaker, K.L., and Robertson, R.M. 2011. Protein expression following heat shock in the nervous system of Locusta migratoria. *Journal of Insect Physiology* 57 (11): 1480–1488.

Des Marteaux, L.E., Stinziano, J.R., and Sinclair, B.J. 2018. Effects of cold acclimation on rectal macromorphology, ultrastructure, and cytoskeletal stability in Gryllus pennsylvanicus crickets. *Journal of Insect Physiology* 104: 15–24.

Dhinaut, J., Balourdet, A., Teixeira, M., Chogne, M., and Moret, Y. 2017. A dietary carotenoid reduces immunopathology and enhances longevity through an immune depressive effect in an insect model. *Scientific Reports* 7 (1): 12429.

Dillon, M.E., Wang, G., and Huey, R.B. 2010. Global metabolic impacts of recent climate warming. *Nature* 467 (7316): 704–06.

Dinh, K.V., Janssens, L., and Stoks, R. 2016. Exposure to a heat wave under food limitation makes an agricultural insecticide lethal: A mechanistic laboratory experiment. *Global Change Biology* 22 (10): 3361–3372.

Duffy, G.A., Coetzee, B.W., Janion-Scheepers, C., and Chown, S.L. 2015. Microclimate-based macrophysiology: Implications for insects in a warming world. *Current Opinion in Insect Science* 11: 84–89.

Eremina, M.A., Menshanov, P.N., Shishkina, O.D., and Gruntenko, N.E. 2021. The transcription factor dFOXO controls the expression of insulin pathway genes and lipids content under heat stress in Drosophila melanogaster. *Vavilov Journal of Genetics and Breeding* 25 (5): 465.

Feder, M.E. and Hofmann, G.E. 1999. Heat-shock proteins, molecular chaperones, and the stress response: Evolutionary and ecological physiology. *Annual Review of Physiology* 61 (1): 243–82.

Feldhaar, H. 2011. Bacterial symbionts as mediators of ecologically important traits of insect hosts. *Ecological Entomology* 36 (5): 533–43.

Felton, G.W. and Summers, C.B. 1995. Antioxidant systems in insects. *Archives of Insect Biochemistry and Physiology* 29 (2): 187–97.

Flatt, T., Tu, M.P., and Tatar, M. 2005. Hormonal pleiotropy and the juvenile hormone regulation of Drosophila development and life history. *Bioessays* 27 (10): 999–1010.

Frank, D.D., Jouandet, G.C., Kearney, P.J., Macpherson, L.J., and Gallio, M. 2015. Temperature representation in the Drosophila brain. *Nature* 519 (7543): 358–61.

Gallio, M., Ofstad, T.A., Macpherson, L.J., Wang, J.W., and Zuker, C.S. 2011. The coding of temperature in the Drosophila brain. *Cell* 144 (4): 614–24.

Gehring, W.J. and Wehner, R. 1995. Heat shock protein synthesis and thermotolerance in Cataglyphis, an ant from the Sahara desert. *Proceedings of the National Academy of Sciences* 92 (7): 2994–2998.

Gillooly, J.F., Brown, J.H., West, G.B., Savage, V.M., and Charnov, E.L. 2001. Effects of size and temperature on metabolic rate. *Science* 293 (5538): 2248–2251.

González-Tokman, D., Córdoba-Aguilar, A., Dáttilo, W., Lira-Noriega, A., Sánchez-Guillén, R.A., and Villalobos, F. 2020. Insect responses to heat: Physiological mechanisms, evolution and ecological implications in a warming world. *Biological Reviews* 95 (3): 802–21.

González-Tokman, D., Gil-Pérez, Y., Servín-Pastor, M., Alvarado, F., Escobar, F., Baena-Díaz, F., García-Robledo, C. et al. 2021. Effect of chemical pollution and parasitism on heat tolerance in dung beetles (Coleoptera: Scarabaeinae). *Journal of Economic Entomology* 114 (1): 462–67.

Gruntenko, N.E. and Rauschenbach, I.Y. 2008. Interplay of JH, 20E and biogenic amines under normal and stress conditions and its effect on reproduction. *Journal of Insect Physiology* 54 (6): 902–08.

Gruntenko, N.E. and Rauschenbach, I.Y. 2018. The role of insulin signalling in the endocrine stress response in Drosophila melanogaster: A mini-review. *General and Comparative Endocrinology* 258: 134–39.

Gruntenko, N.E., Ilinsky, Y.Y., Adonyeva, N.V., Burdina, E.V., Bykov, R.A., Menshanov, P.N., and Rauschenbach, I.Y. 2017. Various Wolbachia genotypes differently influence host Drosophila dopamine metabolism and survival under heat stress conditions. *BMC Evolutionary Biology* 17 (2): 15–22.

Heinrich, B. 1973. "Mechanisms of insect thermoregulation". In *Effects of Temperature on Ectothermic Organisms: Ecological Implications and Mechanisms of Compensation*, Edited by Wieser, W. Effects of temperature on ectothermic organisms. New York: Springer-Verlag, 139–50.

Heinrich, B. 1993. "Social thermoregulation." In *The Hot-Blooded Insects: Strategies and Mechanisms of Thermoregulation*, Edited by Heinrich, B. Berlin: Springer Berlin Heidelberg, 447–509.

Hillyer, J.F. and Pass, G. 2020. The insect circulatory system: Structure, function, and evolution. *Annual Review of Entomology* 65: 121–43.

Hirashima, A., Nagano, T., and Eto, M. 1993. Stress-induced changes in the biogenic amine levels and larval growth of Tribolium castaneum Herbst. *Bioscience, Biotechnology, and Biochemistry* 57 (12): 2085–2089.

Jena, K., Kar, P.K., Kausar, Z., and Babu, C.S. 2013. Effects of temperature on modulation of oxidative stress and antioxidant defenses in testes of tropical tasar silkworm Antheraea mylitta. *Journal of Thermal Biology* 38 (4): 199–204.

Jia, F.X., Dou, W., Hu, F., and Wang, J.J. 2011. Effects of thermal stress on lipid peroxidation and antioxidant enzyme activities of oriental fruit fly, Bactrocera dorsalis (Diptera: Tephritidae). *Florida Entomologist* 94 (4): 956–63.

Karunanithi, S., Barclay, J.W., Robertson, R.M., Brown, I.R., and Atwood, H.L. 1999. Neuroprotection at Drosophila synapses conferred by prior heat shock. *Journal of Neuroscience* 19 (11): 4360–4369.

Kaunisto, S., Ferguson, L.V., and Sinclair, B.J. 2016. Can we predict the effects of multiple stressors on insects in a changing climate? *Current Opinion in Insect Science* 17: 55–61.

Kikuchi, Y., Tada, A., Musolin, D.L., Hari, N., Hosokawa, T., Fujisaki, K., and Fukatsu, T. 2016. Collapse of insect gut symbiosis under simulated climate change. *MBio* 7 (5): e01578–16.

King, A.M. and MacRae, T.H. 2015. Insect heat shock proteins during stress and diapause. *Annual Review of Entomology* 60: 59–75.

Krebs, R.A. and Loeschcke, V. 1994. Costs and benefits of activation of the heat-shock response in Drosophila melanogaster. *Functional Ecology* 730–37.

Li, J., Moghaddam, S.H.H., Du, X., Zhong, B.X., and Chen, Y.Y. 2012. Comparative analysis on the expression of inducible HSPs in the silkworm, Bombyx mori. *Molecular Biology Reports* 39: 3915–3923.

Li, K. and Gong, Z. 2017. Feeling hot and cold: Thermal sensation in Drosophila. *Neuroscience Bulletin* 33: 317–22.

Liu, Y., Su, H., Li, R., Li, X., Xu, Y., Dai, X., Zhou, Y. et al. 2017. Comparative transcriptome analysis of Glyphodes pyloalis Walker (Lepidoptera: Pyralidae) reveals novel insights into heat stress tolerance in insects. *BMC Genomics* 18 (1): 1–13.

Lubawy, J., Urbański, A., Colinet, H., Pflüger, H.J., and Marciniakm P. 2020. Role of the insect neuroendocrine system in the response to cold stress. *Frontiers in Physiology* 11: 1–11.

Ma, W., Li, X., Shen, J., Du, Y., Xu, K., and Jiang, Y. 2019. Transcriptomic analysis reveals Apis mellifera adaptations to high temperature and high humidity. *Ecotoxicology and Environmental Safety* 184: 109599.

Mason, N.A. 2017. Effects of wind, ambient temperature and sun position on damselfly flight activity and perch orientation. *Animal Behaviour* 124: 175–81.

Meng, S., Delnat, V., and Stoks, R. 2020. Mosquito larvae that survive a heat spike are less sensitive to subsequent exposure to the pesticide chlorpyrifos. *Environmental Pollution* 265: 114824.

Newman, A.E., Xiao, C., and Robertson, R.M. 2005. Synaptic thermoprotection in a desert-dwelling Drosophila species. *Journal of Neurobiology* 64 (2): 170–80.

Noyes, P.D. and Lema, S.C. 2015. Forecasting the impacts of chemical pollution and climate change interactions on the health of wildlife. *Current Zoology* 61 (4): 669–89.

Overgaard, J., Kristensen, T.N., and Sørensen, J.G. 2012. Validity of thermal ramping assays used to assess thermal tolerance in arthropods. *PLoS One* 7 (3): e32758.

Parker, E.S., Newton, I.L., and Moczek, A.P. 2020. (My microbiome) would walk 10,000 miles: Maintenance and turnover of microbial communities in introduced dung beetles. *Microbial Ecology* 80 (2): 435–46.

Pörtner, H.O. 2002. Climate variations and the physiological basis of temperature dependent biogeography: Systemic to molecular hierarchy of thermal tolerance in animals. *Comparative Biochemistry and Physiology Part A: Molecular & Integrative Physiology* 132 (4): 739–61.

Quan, P.Q., Li, M.Z., Wang, G.R., Gu, L.L., and Liu, X.D. 2020. Comparative transcriptome analysis of the rice leaf folder (Cnaphalocrocis medinalis) to heat acclimation. *BMC Genomics* 21 (1): 1–12.

Rajpurohit, S., Parkash, R., and Ramniwas, S. 2008. Body melanization and its adaptive role in thermoregulation and tolerance against desiccating conditions in drosophilids. *Entomological Research* 38(1): 49–60.

Renoz, F., Pons, I., and Hance, T. 2019. Evolutionary responses of mutualistic insect–bacterial symbioses in a world of fluctuating temperatures. *Current Opinion in Insect Science* 35: 20–26.

Robertson, R.M. 2004. Thermal stress and neural function: Adaptive mechanisms in insect model systems. *Journal of Thermal Biology* 29 (7-8): 351–58.

Rodgers, C.I., Armstrong, G.A., and Robertson, R.M. 2010. Coma in response to environmental stress in the locust: A model for cortical spreading depression. *Journal of Insect Physiology* 56 (8): 980–90.

Ruchty, M., Roces, F., and Kleineidam, C.J. 2010. Detection of minute temperature transients by thermosensitive neurons in ants. *Journal of Neurophysiology* 104 (3): 1249–1256.

Sim, C. and Denlinger, D.L. 2013. Insulin signaling and the regulation of insect diapause. *Frontiers in Physiology* 4: 189.

Sinakevitch, I.T., Wolff, G.H., Pflüger, H.J., and Smith, B.H. 2018. Biogenic amines and neuromodulation of animal behavior. *Frontiers in Systems Neuroscience* 31: 1–3.

Slocinska, M., Barylski, J., and Jarmuszkiewicz, W. 2016. Uncoupling proteins of invertebrates: A review. *IUBMB life* 68 (9): 691–99.

Sørensen, J.G. 2010. Application of heat shock protein expression for detecting natural adaptation and exposure to stress in natural populations. *Current Zoology* 56 (6): 703–13.

Sørensen, J.G., Kristensen, T.N., and Loeschcke, V. 2003. The evolutionary and ecological role of heat shock proteins. *Ecology Letters* 6 (11): 1025–1037.

Storey, K.B. and Storey, J.M. 2004. Metabolic rate depression in animals: Transcriptional and translational controls. *Biological Reviews* 79 (1): 207–33.

Sukhanova, M., Grenback, L., Gruntenko, N., Khlebodarova, T., and Rauschenbach, I. 1996. Alkaline phosphatase under heat stress in Drosophila. *Journal of Insect Physiology* 42: 161–65.

Sukhanova, M.J., Shumnaya, L.V., Grenback, L.G., Gruntenko, N.E., Khlebodarova, T.M., and Rauschenbach, I.Y. 1997. Tyrosine decarboxylase and dopa decarboxylase in Drosophila virilis under normal conditions and heat stress: Genetic and physiological aspects. *Biochemical Genetics* 35: 91–103.

Torson, A.S., Dong, Y.W., and Sinclair, B.J. 2020. Help, there are "omics" in my comparative physiology! *Journal of Experimental Biology* 223 (24): jeb191262.

Trullas, S.C., van Wyk, J.H., and Spotila, J.R. 2007. Thermal melanism in ectotherms. *Journal of Thermal Biology* 32 (5): 235–45.

Ushakova, N., Dontsov, A., Sakina, N., Bastrakov, A., and Ostrovsky, M. 2019. Antioxidative properties of melanins and ommochromes from black soldier fly Hermetia illucens. *Biomolecules* 9 (9): 408.

Verberk, W.C., Atkinson, D., Hoefnagel, K.N., Hirst, A.G., Horne, C.R., and Siepel, H. 2021. Shrinking body sizes in response to warming: Explanations for the temperature–size rule with special emphasis on the role of oxygen. *Biological Reviews* 96 (1): 247–68.

Verberk, W.C., Overgaard, J., Ern, R., Bayley, M., Wang, T., Boardman, L., and Terblanche, J.S. 2016. Does oxygen limit thermal tolerance in arthropods? A critical review of current evidence. *Comparative Biochemistry and Physiology Part A: Molecular & Integrative Physiology* 192: 64–78.

Wang, H., Fang, Y., Wang, L., Zhu, W., Ji, H., Wang, H., Xu, S. et al. 2014. Transcriptome analysis of the Bombyx mori fat body after constant high temperature treatment shows differences between the sexes. *Molecular Biology Reports* 41: 6039–6049.

Watt, W.B. (1969). Adaptive significance of pigment polymorphisms in Colias butterflies, II. Thermoregulation and photoperiodically controlled melanin variation in Colias eurytheme. *Proceedings of the National Academy of Sciences* 63 (3): 767–74.

Xiong, Y., Liu, X.Q., Xiao, P.A., Tang, G.H., Liu, S.H., Lou, B.H., Wang, J.J. et al. 2019. Comparative transcriptome analysis reveals differentially expressed genes in the Asian citrus psyllid

(Diaphorina citri) upon heat shock. *Comparative Biochemistry and Physiology Part D: Genomics and Proteomics* 30: 256–61.

Zermoglio, P.F., Latorre-Estivalis, J.M., Crespo, J.E., Lorenzo, M.G., and Lazzari, C.R. 2015. Thermosensation and the TRPV channel in Rhodnius prolixus. *Journal of Insect Physiology*, 81: 145–56.

Zeuss, D., Brandl, R., Brändle, M., Rahbek, C., and Brunzel, S. 2014. Global warming favours light-coloured insects in Europe. *Nature Communications* 5 (1): 3874.

Zhang, B., Leonard, S.P., Li, Y., and Moran, N.A. 2019. Obligate bacterial endosymbionts limit thermal tolerance of insect host species. *Proceedings of the National Academy of Sciences* 116 (49): 24712–24718.

Zhang, S., Fu, W., Li, N., Zhang, F., and Liu, T.X. 2015. Antioxidant responses of Propylaea japonica (Coleoptera: Coccinellidae) exposed to high temperature stress. *Journal of Insect Physiology* 73: 47–52.

CHAPTER 5

Genetic and plastic responses of insects to climate change

Patrick T. Rohner

5.1 Introduction

Ever since insects emerged around 480 million years ago, they have been a key component of terrestrial ecosystems and have become one of the most speciose groups of animals (Misof et al., 2014). This evolutionary success story speaks to their resilience and ability to evolve in response to environmental change. However, the speed at which environments have been changing in the last 150 years or so has imposed major challenges on insect populations. Indeed, there is mounting evidence that insect populations have been plummeting in the last few decades and that even once common species now face the threat of extinction (Hallmann et al. 2017; Wagner et al. 2021; Warren et al. 2021). Because changes in insect abundances have immediate ecosystem-level consequences, including ecosystem services (e.g., Losey and Vaughan 2006), these observations are reason for concern. Consequently, the mechanisms that allow insects to cope with changing environments are of great interest to entomologists, evolutionary biologists, and ecologists alike.

In general, insects have three options to cope with climatic changes. The first option, if possible, is to track changes in environmental conditions through *shifts in geographic distribution ranges*. These shifts are not due to active migration but primarily a consequence of some populations experiencing increasingly unsuited conditions facing extinction, while others disperse and establish in areas where environmental conditions were previously unsuited. Although climate change-mediated distribution shifts have long been recognized and are common (Parmesan and Yohe 2003), relocation is unlikely to be an option for most insects due to habitat fragmentation, as well as natural and anthropogenic barriers to dispersal. In cases where relocation is not possible, insects may rely on *phenotypic plasticity*. Insects have evolved mechanisms to deal with environmental variation within their lifetime and can adjust various aspects of their morphology, physiology, or behavior to deal with climatic challenges. If adaptive, such within-generation responses may buffer against rapid environmental change and allow populations to persist. Lastly, insects may also withstand rapid environmental changes through *adaptation*. That is, insects can evolve in response to novel conditions through genetic (i.e., heritable) changes. Although adaptation is often a very slow process, there is ample evidence that populations can

Patrick T. Rohner, *Genetic and plastic responses of insects to climate change*. In: *Effects of Climate Change on Insects.*
Edited by: Daniel González-Tokman and Wesley Dáttilo, Oxford University Press. © Oxford University Press (2024).
DOI: 10.1093/oso/9780192864161.003.0005

evolve on the scale of just a few decades (so-called 'ecological timescales'; Carroll et al. 2007; Hendry 2017).

While Chapters 6 and 11 of this book focus on changes in phenology and geographic distribution ranges, this chapter focuses on genetic (i.e., evolutionary) and plastic (i.e., environmental) responses of insects to climate change. Sections 5.2 and 5.3 provide a brief quantitative genetic overview on genetic variation, heritability and adaptation. In sections 5.4–5.7, we turn to plasticity and why plastic changes are so abundant. In the remainder of the chapter, we discuss how plasticity, once evolved, shapes further evolutionary change and how genetic accommodation may contribute to rapid evolutionary responses to climate change.

5.2 Adaptation: Genetic responses to selection

In its widest sense, adaptation refers to the process of organisms becoming better suited to their habitats through evolutionary processes (Dobzhansky 1968). This process is driven by the joint action of heredity (i.e., the passing on of characteristics from one generation to the next) and natural selection (i.e., the non-random survival and reproduction of individuals with regard to their phenotype). Together, these two processes allow organisms that are better suited to their environment not only to contribute more offspring to the next generation but to also pass on the genes (and possibly other factors) to their offspring that made them better at thriving in their habitat. Over time, this leads to populations becoming better able to perform in their environment. Adaptation is often regarded as a slow process but it can be very fast and is thus an important mechanism that can allow insects to adapt to novel environments on short timescales (Hendry 2017). This includes rapid evolution of resistance to insecticides (e.g., in mosquitos; Weetman et al. 2015), adaptation to urban habitats (reviewed in Diamond et al. 2022) and rapid adaptation in recently introduced invasive species (e.g., Gibert et al. 2016; Olazcuaga et al. 2022). Below, we start by providing a brief quantitative genetic perspective on adaptation, discuss how evolutionary potential can be quantified and then provide several examples of how insects may adapt to changing climates.

5.2.1 *Identifying and predicting adaptation: An evolutionary quantitative genetic perspective*

Evolutionary quantitative genetics provides a powerful framework to predict evolutionary change in most continuous polygenic traits (Walsh and Lynch 2018). Although quantitative genetics largely ignores the molecular developmental basis of trait variation, it is a very successful discipline that continues to lay the foundation of modern evolutionary and ecological genetics (Hill and Kirkpatrick 2010). One of the cornerstones of the field is the so-called *breeder's equation* and its derivatives. In essence, it predicts that the response to selection (i.e., change in mean trait value, $\Delta\mu$) is not only dependent on the magnitude and direction of selection (S) but also on the degree to which traits are inherited from parents to their offspring (h^2) (Falconer 1960).

$$\Delta\mu = h^2 S$$

Heritability (and in particular, as we will see below, genetic variation) is thus a major determinant of responses to selection and is likely to play a key role in shaping the magnitude and speed of evolutionary responses to climate change. Below, we provide a brief primer on why genetic variation is an important estimate of evolvability and how the amount of heritable variation can be estimated.

5.2.2 *Quantifying heritable variation*

Heritability can be assessed by quantifying the extent to which phenotypic variation is explained by genetic relatedness. Shared variation among relatives can be estimated, for instance, by rearing groups of individuals that vary in their degree of relatedness (e.g., clones, inbred lines, full-sib or half-sib families) under standardized conditions in a laboratory or greenhouse. These approaches are often referred to as 'common garden experiments'. Using statistical approaches (e.g., standard analysis of variance (ANOVA) or mixed model procedures (Walsh and Lynch 2018)), the total phenotypic variance (V_P) observed among animals reared in common garden conditions can be decomposed into a genetic (i.e., heritable) component (V_G) and a residual (or error) term (V_ε) that subsumes non-genetic sources of variation as:

$$V_P = V_G + V_\varepsilon.$$

The genetic variance component V_G captures the amount of variation explained by mean differences in trait expression between genotypes (or other genetic groupings, depending on the experimental design). In this simplistic example, the non-genetic variance component V_ε consists of measurement error and environmental variation (see section 5.4) and various other factors. Importantly, the term '*genetic* variation' is used to indicate the amount of *heritable* variation. This entity does *not* indicate whether the development of a trait *per se* is related to genes or gene products in any way. For example: like many scarabs, *Onthophagus* dung beetles develop shovel-like foretibiae with four tooth-like structures that are used to dig through soil. When beetles are collected in the field, the number, shape and size of these tibial teeth varies greatly (Figure 5.1A,B, and C), indicating large amounts of phenotypic variation. However, when rearing these beetles in the laboratory, each individual develops exactly four pointed tibial teeth (Figure 5.1D and E). It turns out that the variation observed in the field is caused by wear of tibial teeth when digging through compact soil. Most of the variation among individuals we observe in nature is thus driven by wear and tear and has no heritable basis. Consequently, the genetic variation is nil (or at least very small). It is important to note that, at the same time, the number of tibial teeth is under tight genetic control. For instance, knocking down the embryonic patterning gene *mex3* leads to the disappearance of two out of the four tibial teeth (Linz et al. 2019). This suggests that while genes and gene products play a role in the development of tibial teeth, there is just no heritable (i.e., genetic) variation in these developmental processes that selection could act on.

The heritability that features prominently in the breeder's equation can be calculated by dividing the variation attributable to genetic factors by the total phenotypic variation. Heritability comes in two different flavors. *Broad-sense* heritability (H^2) is the ratio between total genetic variation and total phenotypic variation calculated as:

$$H^2 = V_G/V_P$$

This proportion indicates the degree to which phenotypic variation among individuals is due to shared heritable factors. However, the genetic component V_G is composed of several different types of genetic effects and it is useful to differentiate between them. The

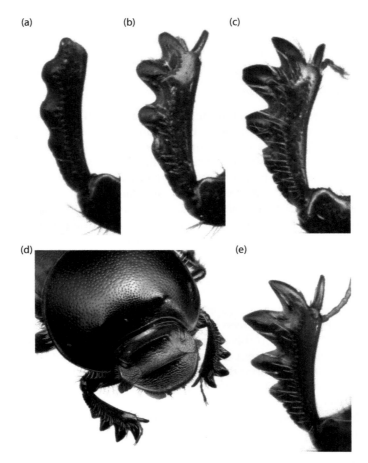

Figure 5.1 The forelegs (specifically the tibiae) of the bull-headed dung beetle *Onthophagus taurus* are equipped with four exaggerated tibial teeth that are used to construct underground breeding tunnels. In natural populations, the morphology of the forelegs differs strongly among individuals (a–c), indicating large amount of phenotypic variation. However, when females (d) from various families and populations are reared under laboratory conditions, they all develop a very similar morphology (e). This is because the variation in the field is caused by the wear of the tibia when digging through hard soil. The large phenotypic variation observed in the field thus does not necessitate to be heritable.

most important one is the variance attributable to additive genetic variance V_A; that is, the variance due to the additive effects of segregating alleles. In contrast to non-additive sources of variance (such as dominance and epistasis), additive effects take center stage in evolutionary quantitative genetics because they are primarily responsible for the resemblance among relatives and respond to selection the fastest (Hill et al. 2008; Walsh and Lynch 2018). Most of the quantitative literature thus focuses on the *narrow-sense* heritability (h^2) that is calculated by dividing the additive genetic variance by the total phenotypic variance.

$$h^2 = V_A/V_P$$

Calculating the additive genetic variance component requires more complex breeding designs, such as full-sib/half-sib designs or so-called animal models (Wilson et al. 2010) which require pedigree information. Because V_G is only one component of V_G, h^2 is always smaller than H^2. Note that most quantitative genetic studies in insects are performed in heavily controlled and often artificial (laboratory) environments. As genetic (and phenotypic) variance components are contingent on the environment they are measured in, it is often unclear how findings made in the laboratory translate into the wild. This continues to be a major caveat, especially because laboratory conditions rarely resemble the complex and fluctuating conditions in nature (Rodrigues and Beldade 2020).

5.2.3 Genetic variation, heritability, and evolvability

As evident from the breeder's equation, the amount of additive genetic variation relative to the total phenotypic variance (i.e., h^2) is a key determinant of the magnitude of the response to selection. However, while heritability estimates measure how much of the *variance* in a trait is heritable, they do not tell us how V_A compares to the *mean* phenotypic value. It is thus useful to also consider genetic variance relative to the mean trait values. A useful measure in this regard is *evolvability (I_A)* (Hansen and Pélabon 2021; Houle 1992). I_A is calculated by dividing the additive genetic variance by the square of the mean trait value (m):

$$I_A = V_A/m^2$$

The value of I_A is the expected percentage change in a trait under a unit strength of selection. I_A is thus a more intuitive measure of evolutionary potential than heritability because it focuses on changes in mean trait values rather than changes in variances. Even though evolvability and heritability are just different ways of standardizing genetic variances, they may lead to very different conclusions. For instance, while morphological traits tend to have higher heritability compared to many life-history traits, their evolvability tends to be smaller. These inconsistencies lead to the lack of a correlation between h^2 and I_A (Hansen and Pelabon 2021). Consequently, comparing V_A across species, environments and traits is not necessarily straightforward and the method of standardization (mean- or variance) should be chosen with care.

5.2.4 Testing for adaptation

Per definition, adaptation refers to the process of organisms becoming better able to perform in their habitats through evolutionary processes (Dobzhansky 1968). In order to infer adaptation, investigators at minimum need to demonstrate two things: First, it needs to be shown that populations or species differ due to evolutionary change (as opposed to, for example, environmental effects). This can, for instance, be achieved by estimating heritable differences between populations using common garden experiments described above. Secondly, it must be demonstrated that the observed heritable differences increase fitness in the respective environment. This point is often challenging but remains crucial to understand the adaptive value of heritable trait differences (Gould and Lewontin 1979). It is thus preferable to directly estimate fitness in different environmental conditions. One way of doing so is to subject populations to reciprocal transplant experiments (Johnson et al. 2022). If populations perform better in their own environments relative to all other populations, this indicates strong support for local adaptation. For instance, using reciprocal transplants, Via (1991) showed that pea aphid strains collected either on alfalfa or red clover performed best when reared on the host plant on which they were collected. Similar findings are common across study systems, indicating that populations are often adapted to their local environment (Hereford 2009; Olazcuaga et al. 2022). In the next section (5.3), we briefly review some examples of how insects adapt to their environment (also see Chapter 12).

5.3 Examples of adaptation

5.3.1 *Physiological adaptations*

A key factor with regard to climate change is physiological resistance to temperature or desiccation stress. Populations and species often differ in their thermal limits (Blanckenhorn et al. 2021; Garbuz et al. 2003; Sgrò et al. 2010), suggesting that these traits can evolve. Indeed, thermal tolerance thresholds often have additive genetic variation and thus, in principle, have the capacity to respond to selection (Diamond 2017). For instance, *Drosophila* species occurring in cooler habitats evolved a lower critical thermal minimum indicating that physiological resistance to cool environments play a major role in local adaptation (Figure 5.2; MacLean et al. 2019). Similar responses are also found within species. For example, Hangartner and Hoffmann (2016) show that the fruit fly *Drosophila melanogaster* can evolve higher heat resistance in the laboratory (+0.5 °C) (also see e.g., Mesas et al. 2021), suggesting that populations—in principle—can evolve in response to heat stress. However, Hangartner and Hoffmann (2016) also report that the response to selection reached a plateau after eight to ten generations, probably due to the erosion of V_A. This indicates that while initial responses to selection may be rapid, it remains unclear whether responses will be sustained in the long term. In addition, heritability estimates often vary among species. Kellermann et al. (2009) showed that, in contrast to cosmopolitan generalists, tropical *Drosophila* species have very little genetic variation for

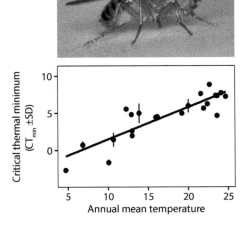

Figure 5.2 MacLean et al. (2019) demonstrate that *Drosophila* species occurring in different thermal habitats differ in their critical thermal minima indicating that physiological resistance to temperature plays a major role in local adaptation (picture of female *Drosophila melanogaster* by Hanna Davis CC-BY-SA-4.0).

desiccation and cold tolerance. Species adapted to narrow thermal niches may thus be severely constrained by low heritabilities which are expected to constrain the speed at which populations can evolve in response to environmental change.

5.3.2 *Morphological adaptations*

In addition to physiology, morphology also plays a major role in adaptation to climate change. This is especially true for traits that facilitate physiological or behavioral thermoregulation. For instance, desert ants of the genus *Cataglyphis* and *Ocymyrmex* evolved disproportionately long legs compared to other ant species. The increase in leg length not only reduces heat exposure by increasing the distance between the body and the hot ground but also by increasing convective cooling and foraging time (Sommer and Wehner 2012).

Another example of how morphology can facilitate thermoregulation is the evolution of relative wing size. Flight is a main contributor to thermoregulation in winged insects because it facilitates microhabitat choice. However, flight is also heavily limited at cool temperatures due to energetic constraints. One way to circumvent these biophysical constraints it to increase relative wing size, which allows insects to take off at cooler temperatures (Dudley 2002; Neve and Hall 2016). Such patterns have evolved convergently in two species of sepsid flies *Sepsis punctum* and *Sepsis fulgens* (Rohner et al. 2019), European populations of *Drosophila subobscura* (Gilchrist and Huey 2004) and the dung beetle *Onthophagus taurus* (Rohner and Moczek 2020). Similar patterns are also found across species of drosophilids (Rohner et al. 2018a). Evolutionary changes in relative wing size may thus contribute to adaptation to changing climates. Other aspects of wing morphology, such as coloration, have also been shown to relate to local adaptation to climatic differences (e.g., Ellers and Boggs 2004).

These examples highlight how populations and species can adapt to changing environments (also see Chapter 12 in this book). However, in addition to genetic responses, insects also have the capacity to adjust their phenotype using plastic within-generation mechanisms. In section 5.4, we discuss how plasticity arises and how it can be estimated.

5.4 Phenotypic plasticity

Phenotypic plasticity refers to the capacity of a single genotype to produce different phenotypes depending on the environment (West-Eberhard 2003). Plasticity is exceptionally common in insects (and organisms in general; see de Jong and van der Have 2009; Pfennig 2021; Sultan 2015). Examples include developmental adjustments to cuticular pigmentation to match substrate coloration in caterpillars (Figure 5.3; Noor et al. 2008), size-dependent development of many male secondary sexual structures (Rohner and Blanckenhorn 2018; Figure 5.3), or immune responses to the presence of pathogens (Rolff et al. 2009).

Plastic responses can be complex and non-linear. For instance, locusts that develop in a crowded environment develop into gregarious (i.e., swarming) adults while individuals experiencing low population densities develop into a solitary morph (Verlinden et al. 2009). In this case, plasticity to crowding results in a polyphenism—the presence of multiple discrete alternative morphs that are induced by environmental conditions. Similar polyphenisms are common in species with alternative reproductive tactics where large 'fighter' morphs develop elaborate secondary sexual traits while small 'sneaker'

Figure 5.3 Phenotypic plasticity can have major effects on phenotypic variation. This includes a) plastic adjustments of cuticular coloration to substrate coloration as in *Biston betularia*; b) seasonal polyphenism in *Papilio xuthus* (spring form on top, summer form on the bottom); and c) differences in ornament morphology between small (d) and large (e) sepsid flies (Rohner 2022; Rohner and Blanckenhorn 2018) (image credit: a) *Biston betularia* by M. A. F. Noor, R. S. Parnell, and B.S. Grant 2008, CC-BY-2.5; b) *Papilio xuthus* by S. Komata and T. Sota 2017, CC-BY-4.0; c) *Sepsis pyrrhosoma* male by K. Schulz 2015, CC-BY-2.0; d) and e) by P. Rohner).

morphs invest in clandestine mating behaviors (e.g., in horned and horn-less dung beetles (Moczek and Emlen 2000)). Similarly, there are many instances of seasonal morphs in butterfly wing coloration patterns (Brakefield and Reitsma 1991; Komata and Sota 2017; Figure 5.3). The induction of hibernation behavior can also be regarded as a polyphenism if individuals switch between a diapause and a direct development path (Kivelä et al. 2017).

Whether plasticity takes a continuous or discrete form, it is often a major contributor to phenotypic variation and can confound genetic differences among populations or species. That is, if environmental conditions are not accounted for, differences in population or species means across geographic regions (e.g., warmer versus cooler climates) or temporal samples (e.g., historical museum collections versus contemporary populations) are unsuited to infer *evolutionary* changes and adaptation. Consequently, this has major impacts on how we assess insects' responses to ongoing climate change in the field in that relatively few studies are able to disentangle plastic from genetic responses in nature (e.g., Blanckenhorn 2015; Bradshaw and Holzapfel 2001). Plasticity has thus been traditionally considered a confounding factor when studying genetics and evolution. However, the relative contribution of plasticity to phenotypic variation can be assessed in the laboratory. Below, we briefly outline a classic quantitative genetic approach.

5.4.1 *Quantifying phenotypic plasticity*

The contribution of phenotypic plasticity to the total phenotypic variation can be assessed by rearing closely related individuals (e.g., clones, inbred lines, or full-sib/half-sib families) in different environment. This allows to decompose the total phenotypic variation V_P into variance components due to genetic factors, the environment, and genotype-by-environment interactions (see Figure 5.4).

$$V_P = V_G + V_E + V_{G \times E} + V_\varepsilon$$

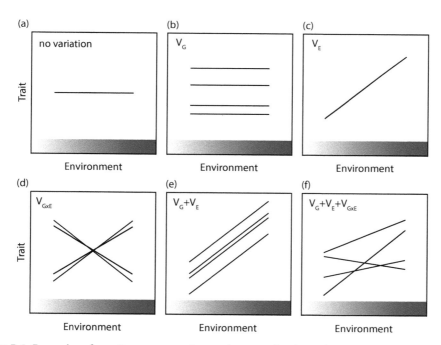

Figure 5.4 Examples of reaction norms under varying contributions of genetic differentiation, plasticity and G × E interactions. Each line represents a single genotype's reaction norm. a) No variation; b) only genetic differences among genotypes; c) plasticity in the absence of genetic variation; d) only genotype-by-environment interactions; e) genetic differences and a common plastic response; f) genetic differences, plasticity and G × E interactions.

V_E captures variation explained by the shared plastic response while $V_{G\times E}$ captures variation explained by heritable differences in plastic responses among related individuals. The latter is generally regarded as a component of plasticity and can account for a significant amount of the phenotypic variance. For instance, measuring temperature-specific thorax length in 196 *Drosophila melanogaster* lines, Lafuente et al. (2018) computed that the variance component associated to the genotype-by-temperature interaction component ($V_{G\times E} = 6.0 \times 10^{-4}$) was three times larger than V_G (2.0×10^{-4}, Figure 5.5). For abdomen length, which the authors measured as well, $V_{G\times E}$ (6.2×10^{-3}) was even six times larger than V_G (1.0×10^{-3}). This suggests substantial heritable variation for thermal plasticity in body size, a phenomenon found in many species (Rodrigues and Beldade 2020; DeWitt and Scheiner 2004).

Although often documented, the contribution of phenotypic plasticity (V_E and $V_{G\times E}$) to phenotypic variation is difficult to predict and depends strongly on the type of environmental variable and the range at which it is investigated. However, some environmental variables most likely to be affected by climate change, such as temperature, humidity, and food availability, often have strong and predictable effects on many life-history and morphological traits. Below we briefly highlight common forms of plasticity as a response to temperature.

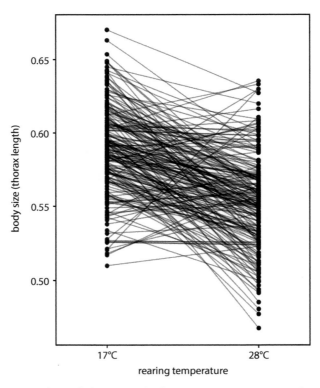

Figure 5.5 Differences in thermal plasticity in body size among 196 isogenic *Drosophila melanogaster* strains. Each line represents the thermal response of a genetic strain. The finding that lines differ in their response to rearing temperature indicates significant levels of genetic variation for plasticity (Data from Lafuente et al. 2018).

5.4.2 *Examples of common plastic responses: Thermal reaction norms*

The development and physiology of insects is strongly dependent on ambient temperature. This is especially true for egg-to-adult development time and growth rate (de Jong and van der Have 2009; Hochachka and Somero 2014). Plastic responses of these traits to temperature are so widespread that they can be expected *a priori*. As an example, Buckley et al. (2017) found that contemporary climate changes cause phenological advancements and an increase in the number of generations per year almost universally across the globe. Such plastic increases in developmental rates thus have major effects on entire ecosystems.

In addition to growing slower, insects (and other ectotherms) also tend to grow to larger sizes at cold temperatures. This response is so strong that the phenomenon has been dubbed the 'temperature size rule' (Atkinson 1994). The physiological mechanisms underpinning the temperature size rule diverge across insects. In the tobacco hornworm *Manduca sexta*, the temperature size rule is caused by temperature-dependent growth occurring late in development (Davidowitz and Nijhout 2004). In contrast, in *Drosophila*, it is caused by temperature-dependent plasticity in the size at which metamorphosis is initiated (Ghosh et al. 2013). The observation that different insects utilize different mechanisms to achieve the temperature size rule suggests that the temperature size rule may be adaptive. However, the precise mechanisms and their adaptive value remain disputed (Angilletta and Dunham 2003; Horne et al., 2015; Verberk et al. 2021).

Although most insects follow the temperature size rule, there are many exceptions (e.g., Walters and Hassall 2006). In addition, thermal plasticity in growth-related traits is usually nonlinear. Specifically, thermal performance curves take a typical asymmetric bell shape (Angilletta 2006; David and Clavel 1967). Estimating thermal plasticity using only two or just a few temperatures can be misleading and is insufficient to locate a species' thermal niche. Estimating full thermal performance curves is difficult and often challenging but provides important insights into the effect of environmental variation on development, morphology and fitness.

5.4.3 *Is plasticity adaptive?*

From an evolutionary genetic perspective, plasticity is expected to evolve if organisms occupy variable but predictable habitats, if selection favors alternative phenotypes in different environments, and if no phenotype is best suited across all levels of the environmental variable (DeWitt and Scheiner 2004; Ghalambor et al. 2007). Such adaptive plasticity enables an insect to take advantage of cues in its current environment to make predictions about future conditions, enabling adaptive adjustments to developmental trajectories. For instance, the flesh fly *Sarcophaga bullata* takes advantage of daylength as a cue to initiate diapause (i.e., hibernation) whenever there are less than around thirteen hours of light per day (Denlinger 1972). This is most likely to be an adaptive response that is found in many insects (Tauber et al. 1986). Similarly, the mayfly *Drunella coloradensis* takes advantage of chemical cues indicating the presence of fish predators to develop longer caudal filaments—the development of which reduces predation risk (Dahl and Peckarsky 2002). Any plastic response that relates to environment-specific survival (e.g., predator avoidance, immune responses, etc.), are most likely adaptive responses to selection. However, our increasing understanding of development and physiology demonstrates that developmental systems are generally sensitive to environmental conditions (Bateson and Gluckman 2011; Nijhout et al. 2021). Plasticity to variables that affect developmental processes directly, such as temperature, can thus be regarded as the default, not the exception. For instance, in a classic experiment, Waddington (1953) exposed *Drosophila* pupae to heat shock and found that some adults developed wings with unusual morphology (i.e., missing cross-veins). This is certainly a form of plasticity, but probably of little adaptive value. Hence, while many plastic responses to climatic variables are probably under selection and therefore have the potential to be adaptive, it needs to be demonstrated that the phenotypic changes have fitness consequences. In section 5.8, we will consider how plasticity, once evolved, impacts future evolutionary change.

5.5 Evolution of (and through) plasticity

Plasticity decouples an organisms' phenotype from its genotype. Consequently, plasticity impacts evolutionary changes in a variety of ways (Crispo 2008; Ghalambor et al. 2007; West-Eberhard 2003). For instance, adaptive plastic responses to climate change are thought to facilitate subsequent adaptation by maintaining a population's fitness until novel, beneficial mutations emerge (e.g., Corl et al. 2018). Plasticity can thus 'buy time'—which is of the essence because adaptation is often slow. However, plasticity may also hamper adaptation, for instance if plastic responses buffer the phenotypic effects of deleterious mutations. By decoupling phenotypes visible to selection from an organism's genotype, plasticity can prevent selection from removing maladaptive alleles (e.g., Huey et al. 2003). Plasticity thus has complex effects on evolutionary trajectories. In addition, plasticity itself can evolve, potentially changing the phenotypic variation visible to selection and influencing direction and magnitude of adaptive responses. Below, we first outline how the evolution of plasticity can contribute to local adaptation. Next, we discuss how robustness shapes evolutionary capacitance, and lastly, we touch on plasticity's ability to precede and 'lead' future evolution.

5.5.1 *Evolution of plasticity*

Plastic responses often differ between species, ecotypes, sexes and traits, demonstrating that plasticity has large potential to evolve (e.g., Foquet et al. 2021; Rohner et al. 2018b). While evolution of plasticity is often documented both across and within species, *how* plasticity evolves is still poorly understood, especially under natural conditions and on 'ecological timescales' (Fox et al. 2019). This is because to study its evolution, plasticity must first be quantified in the ancestral population and then be contrasted to patterns of plasticity after evolution has taken place. This can be done in the laboratory using artificial selection or laboratory evolution experiments (Mallard et al. 2020; Suzuki and Nijhout 2006; Waddington 1952), but it has been exceedingly difficult to investigate how plasticity evolves in nature.

The evolution of plasticity can be studied in the field using longitudinal approaches. For instance, Bradshaw and Holzapfel (2001) sampled pitcher plant mosquitos repeatedly over a time span of around thirty years. Rearing wild-caught populations under various controlled laboratory conditions, the authors were able to show that populations evolved a modified photoperiod threshold for diapause induction to match more southern day lengths, a finding consistent with the hypothesis that plasticity evolved rapidly and adaptively in response to climate change. Such longitudinal studies are very insightful and allow us to track adaptation to climate change in real time. However, these approaches are logistically and experimentally challenging. An alternative is to study species that success-fully and rapidly invaded new habitats that are climatically different from the ancestral range. Studying invasions offers an exceptional opportunity to study how organisms cope with and adapt to rapidly changing environments (Gilchrist et al. 2001; Kingsolver and Buckley 2017), and, provided that the ancestral source population (or a proxy thereof) is known and still accessible, even allows to investigate the role of ancestral plasticity therein (Moczek 2007).

One example is the invasion of the dung beetle *Onthophagus taurus* in the eastern United States. This species, which is native to the Mediterranean region, was acciden-tally introduced in Florida in the early 1970s. Upon its introduction, it rapidly expanded its range towards the North, and within forty years—which corresponds to about eighty to one hundred generations—reached the Canadian border (Figure 5.6; Rohner and Moczek 2020). A common garden experiment revealed that this rapid invasion coincided with a reduction in development time in the North when beetles were reared at 19 °C (Figure 5.6). This is likely to be adaptive because northern climates are too short for typ-ical southern populations to complete their reproductive cycle. However, no differences were found when rearing populations at 27 °C which represents the average temperature during the breeding season at the southern range edge (Figure 5.6). This indicates that *O. taurus* adapted to short seasons in the North by an evolutionary reduction of thermal plas-ticity in development time (Rohner and Moczek 2020). Such countergradient variation (or genetic compensation; Grether 2005) allows northern populations to complete their reproductive cycle despite shorter seasons. Similar responses are expected to be common in general and contribute to climate change (Kelly 2019).

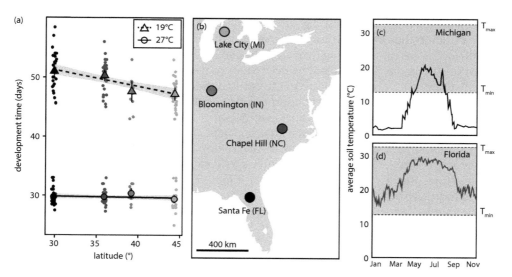

Figure 5.6 The invasion and rapid range expansion of *O. taurus* in the eastern United States involved an evolutionary reduction of thermal plasticity in development time, allowing northern populations to complete their reproductive cycle despite shorter seasons (Rohner and Moczek, 2020). Panels on the right show average soil temperature at a depth of 9 cm throughout the year 2020 in Michigan (c) and Florida (d). The shaded area indicates the temperature range between the minimal and maximal thermal limits of *O. taurus* (soil temperature data from the National Ecological Observatory Network).

5.5.2 *Robustness, cryptic genetic variation, and evolutionary potential*

Before we continue discussing how plasticity shapes evolution, it may be useful to discuss the complex relationship between plasticity and robustness. Robustness (i.e., the apparent insensitivity of phenotype expression to environmental variation) is often interpreted as the absence of plasticity. However, robustness and plasticity are in a very complex relationship because the mechanisms that make insects robust are often themselves plastic and the mechanisms that make a developmental system plastic are often surprisingly robust. Robustness and plasticity are thus not just flip sides of the same phenomenon but are in a reciprocal relationship (see: Bateson and Gluckman 2011; Schwab et al. 2019).

Robustness is important for evolution because it enables organisms to withstand environmental as well as genetic (e.g., mutational) perturbations. For instance, individuals within a population may carry various non-synonymous mutations but can still be indistinguishable on the phenotypic level. Such robustness is rooted in the way genes affect phenotypes through development. Genes don't affect phenotypes directly but act through developmental genetic networks that are riddled with redundant interactions and feedback loops such that one component can compensate for another (Bateson and Gluckman 2011; Nijhout 2002; Gursky et al. 2012). This allows developmental systems to buffer against deleterious variation and prevents mutational perturbations from affecting phenotypes. Because some of these mutations are not visible to selection, they can accumulate and form so-called *cryptic genetic variation*. This variation is referred to as *cryptic* because it usually has no effects on phenotypes and thus remains invisible.

Cryptic genetic variation can have major evolutionary implications. That is because the capacity to which developmental systems can buffer against perturbations is not limitless. If systems are disturbed too much (e.g., through the exposure to a novel environment), the cryptic variation that was previously buffered against can be released and hit phenotypes with full force. This can lead to an increase in heritable variation in the new environment. Most of this heritable variation will be deleterious or neutral but some decrypted effects are likely to bring an organism closer to the optimal trait value in the new environment. The release of previously cryptic variation can thus fuel adaptation to novel habitats or changing environments (Paaby and Rockman 2014).

One example where cryptic genetic variation shapes phenotypic variation in a novel climatic environment was documented in the yellow dung fly *Scathophaga stercoraria*. Females of this species typically develop three sperm storage organs (i.e., spermathecae), but sometimes they develop a fourth one. In nature, the frequency of females developing a fourth spermatheca (the so-called 4S phenotype) is nearly zero (Berger et al. 2011). However, when offspring of wild-caught 3S females are reared in the laboratory, the frequency of the 4S phenotype increases strongly. This increase is especially pronounced at warm temperatures (Figure 5.7) outside the species' preferred temperature range. Crucially, high temperatures not only increase the total phenotypic variance in 4S phenotype expression but also the relative amount of genetic variance. This leads to an increase of heritability with rearing temperature, suggesting that genetic variation that remains cryptic at low temperatures close to the species' optimum is released when encountering thermal stress. In the case of the yellow dung fly, this increase in heritability could fuel adaptation if favored by selection. Indeed, some studies suggest that females with four spermathecae have a fitness advantage via genetic benefits through female choice despite a fecundity cost (Ward 2007; Ward et al. 2008; but see Walters et al.

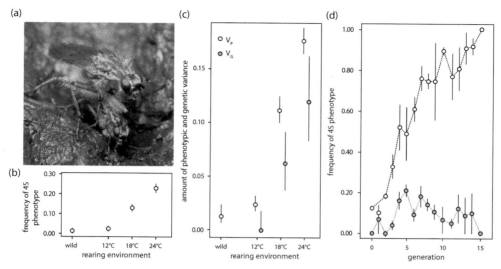

Figure 5.7 Female yellow dung flies (a) develop either three (3S) or four (4S) spermathecae. b) The frequency of the 4S phenotype is almost nil in nature but increases when females are reared in the laboratory, especially at warm temperatures. c) The increase in 4S phenotype expression with temperature not only increases the total phenotypic variance but also the relative amount of the genetic variance. This leads to an increase of heritability with rearing temperature, suggesting that genetic variation that remains cryptic at optimal low temperatures is released when developing flies encounter thermal stress. Interestingly, this released genetic variation can respond rapidly to artificial selection (d). (data from Berger et al. 2011 and War, 2000; picture of *Scathophaga stercoraria* by David Evans CC-BY-2.0).

2022). The context-dependent expression of spermatheca number in the yellow dung fly exemplifies how plasticity and robustness can shape genetic variation and adaptive potential. Next, we explore how plasticity may generally facilitate evolution through genetic accommodation.

5.5.3 *Plasticity-led evolution: Genetic accommodation and assimilation*

A large body of literature documents how plasticity emerges as a product of evolution. However, once evolved, plasticity may also precede and 'lead' subsequent evolutionary change. These ideas are relatively old (e.g., Baldwin 1896; Morgan 1896; Waddington 1942) but received reinvigorated interest and scrutiny since the early 2000s (Pfennig et al. 2010; West-Eberhard 2003). Most commonly, plasticity-led evolution is discussed in the context of genetic accommodation and genetic assimilation—two similar but distinct mechanisms. Let's first consider a hypothetical example of how genetic accommodation is thought to 'lead' evolution.

We begin with a hypothetical insect population that is adapted to a particular environment (Figure 5.8a). Because selection has removed unfit genotypes over time, most genotypes produce a phenotype close to the fitness optimum (in Figure 5.8a). Exposure to a novel environment—for example, a much hotter environment or unprecedented desiccation stress—will induce plastic responses. However, not all genotypes will respond equally to the environmental change. Some may be totally unaffected by environmen-

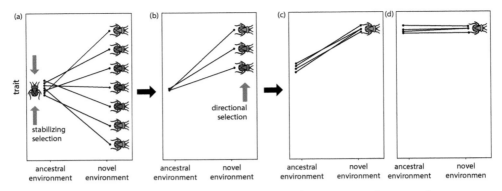

Figure 5.8 a) A hypothetical population adapted to its habitat is exposed to a novel environment. Exposure to this new environment induces plastic responses and releases cryptic genetic variation that is now visible to selection. b) Directional selection in the new environment favors an increase in the trait value and favors genotypes with an increased plastic response. c) The result is a population that adapted to a novel environment through genetic modification of an ancestral plastic response (i.e., genetic accommodation). d) If genetic accommodation leads to environment-insensitive trait expression, this process is referred to as genetic assimilation.

tal variation, while others may vary in the strength and direction of the change. The differences in the reaction norms among genotypes are due to previously cryptic genetic variation. The release of this cryptic variation manifests in increased phenotypic variation that is now visible to selection. In the example shown in Figure 5.8, selection will favor an increase in the trait value and favors genotypes with an increased plastic response. The result is a population that adapted to a novel environment through genetic modification of an ancestral plastic response. In other words, the trait has undergone *genetic accommodation*.

The process of genetic accommodation contrasts with classic neo-Darwinian frameworks in that new selectable variants initially arise through plasticity in a novel environment without the need for novel mutations (Moczek 2007). In this sense, plasticity precedes and has the potential to bias future genetic change (note, however, that plasticity-first evolution can be approached using standard quantitative genetics; e.g., Lande 2009). Genetic accommodation is also thought to be much faster than 'typical' neo-Darwinian modes of evolution which depend on new genetic variants entering the population one at a time through mutation and gene flow. That is because selection in a new environment can act on potentially large amounts of 'decrypted' variation that have accumulated over many generations in an entire population (as opposed to novel alleles entering the population at low frequency). To provide an empirical example of how fast responses to selection can be when fueled by cryptic genetic variation, we shall revisit variation in spermatheca number in the yellow dung fly. Ward (2000) selected on the 4S phenotype (which is almost completely cryptic in the field) and was able to fix the 4S phenotype within just fifteen generations (see Figure 5.8d). This highlights cryptic genetic variation's potential as an evolutionary capacitor.

Crucially, selection in the new environment can also change the phenotype produced when genotypes are again exposed to the ancestral environment. Because selection only 'sees' phenotypes in the new environment, mutations that increase trait values will be favored. Some of these will do so constitutively and consequently produce genotypes that in turn produce an overall higher level of trait expression irrespective of the environmental conditions. These responses can be so strong that traits become completely environment insensitive (Figure 5.8d). In that case, this process is referred to as *genetic assimilation*. Genetic assimilation is simply a form of accommodation where previously plastic traits evolve to become expressed constitutively across environments. In that sense, the concept of genetic accommodation can be seen as a generalization of genetic assimilation (Braendle and Flatt 2006).

Genetic accommodation has been demonstrated under laboratory conditions, but there are very few cases where it has been demonstrated in the field. This may suggest that genetic accommodation is not important. However, plasticity-led evolution via genetic accommodation is expected to be rapid and the products of plasticity-first and neo-Darwinian evolution cannot be distinguished once evolution has taken place. Based on these challenges, Levis and Pfennig (2016) provide four key criteria that need to be demonstrated in order to demonstrate plasticity-first evolution in nature. First, plasticity must have been present in the ancestral population (in the case of thermal plasticity this is very likely). Exposing an ancestral population to the new environment should thus induce the novel phenotype, although not necessarily to the same degree. Secondly, release of cryptic genetic variation must have occurred in the derived environment. Thirdly, it needs to be demonstrated that plasticity evolved, and, lastly, the evolved form of plasticity should increase fitness in the derived environment relative to ancestral plasticity. Based on these promising criteria, future work in natural populations will hopefully reveal how much plasticity-first evolution contributes to adaptation to climate change.

5.6 Conclusions

Although climate change is only one of many factors contributing to the rapid decline of insect populations across the globe, understanding how insects respond to changes in climate is crucial. Laboratory studies document that many key physiological, morphological or life-history traits harbor some degree of genetic variation. This indicates that insect populations have the potential to respond adaptively to rapid environmental changes. Many fitness-related traits also show plastic responses to climatic changes. Some (but certainly not all) of these plastic responses are adaptive and will be able to buffer against the potentially negative effects of climate change. Furthermore, many studies demonstrate genetic variation for plasticity, indicating that plasticity itself may be able to evolve in response to selection. Some studies even indicate that this is possible on ecological timescales. In addition, plasticity may facilitate adaptive evolution through genetic accommodation, a process potentially faster than classical neo-Darwinian evolution. Taken together, there are good reasons to believe that many insects will be able to adapt to novel climatic conditions—either via plasticity, genetic changes, or adaptation through the evolution of plasticity. However, whether these responses are going to be sufficiently strong and rapid enough remains unclear.

Key reflections

- There is accumulating evidence that insect populations can adapt to rapidly changing environments. However, our understanding of the evolutionary potential insect populations (e.g., the amount of genetic variation) is mostly limited to a few heavily studied groups of insects (e.g., drosophilids). Predicting adaptive responses more broadly will require more research in diverse insect groups.
- Environmental responses to temperature, humidity, and nutrient availability are so common that plastic responses to climate change can be expected *a priori*. This implies that changes in trait means observed in nature over time or space are unsuited to infer *evolutionary* diversification.
- Laboratory studies often estimate plastic and genetic responses under constant conditions and one variable at a time. However, natural environments are complex and often fluctuating. Plastic and evolutionary responses to climatic variability remain poorly understood, especially when several variables change simultaneously. The degree to which laboratory studies reflect the natural situation thus often, unfortunately, remains unclear.
- Plasticity can both hamper and facilitate adaptation to climate change. In addition, plasticity itself can evolve, potentially changing the phenotypic variation visible to selection and influencing direction and magnitude of adaptive responses. Finally, plasticity can precede and 'lead' future genetic changes. Although there are many examples that are consistent with a plasticity-first scenario, there is little unambiguous evidence for genetic accommodation in the field. Future research in diverse species will be necessary to reveal how often plasticity contributes to, hampers, or leads evolution.

Key further reading

- Angiletta, M.J. 2009. *Thermal adaptation: A theoretical and empirical synthesis*. Oxford: Oxford University Press.
- Kellermann, V., and van Heerwaarden, B. 2019. Terrestrial insects and climate change: adaptive responses in key traits. *Physiological Entomology* 44: 99–115.
- Nijhout, H.F., Kudla, A.M., and Hazelwood, C.C. 2021. "Genetic assimilation and accommodation: Models and mechanisms." In *Current topics in developmental biology*, edited by S.F. Gilbert. San Diego: Academic Press.
- Pfennig, D.W. (Ed) 2021. *Phenotypic plasticity and evolution: Causes, consequences, controversies*. Boca Raton, FL: CRC Press.
- Sultan, S.E. 2015. *Organism and environment: Ecological development, niche construction, and adaption*. Oxford: Oxford University Press.
- Diamond, S.E., Prileson, E., and Martinv, R.A. 2022. Adaptation to urban environments. *Current Opinion in Insect Science* 51: 100893.

References

Angilletta, M.J. 2006. Estimating and comparing thermal performance curves. *Journal of Thermal Biology* 31: 541–45.

Angilletta, M.J. and Dunham, A.E. 2003. The temperature-size rule in ectotherms: Simple evolutionary explanations may not be general. *The American Naturalist* 162(3): 332–42.

Atkinson, D. 1994. Temperature and organism size-a biological law for ectotherms? *Advances in Ecological Research* 25: 1–58.

Bateson, P. and Gluckman, P. 2011. *Plasticity, Robustness, Development and Evolution*. New York: Cambridge University Press.

Baldwin, J.M. 1896. A new factor in evolution. *The American Naturalist* 30: 441–51.

Berger, D., Bauerfeind, S.S., Blanckenhorn, W.U., and Schäfer, M.A. 2011. High temperatures reveal cryptic genetic variation in a polymorphic female sperm storage organ. *Evolution* 65 (10): 2830–2842.

Blanckenhorn, W.U. 2015. Investigating yellow dung fly body size evolution in the field: Response to climate change?. *Evolution* 69 (8): 2227–2234.

Blanckenhorn, W.U., Berger, D., Rohner, P.T., Schäfer, M.A., Akashi, H., and Walters, R.J. 2021. Comprehensive thermal performance curves for yellow dung fly life history traits and the temperature-size-rule. *Journal of Thermal Biology* 100: 103069.

Bradshaw, W.E. and Holzapfel, C.M. 2001. Genetic shift in photoperiodic response correlated with global warming. *Proceedings of the National Academy of Sciences* 98 (25): 14509–14511.

Braendle, C. and Flatt, T. 2006. A role for genetic accommodation in evolution? *Bioessays* 28 (9): 868–73.

Brakefield, P.M. and Reitsma, N. 1991. Phenotypic plasticity, seasonal climate and the population biology of *Bicyclus* butterflies (Satyridae) in Malawi. *Ecological Entomology* 16: 291–303.

Buckley, L.B., Arakaki, A.J., Cannistra, A.F., Kharouba, H.M., and Kingsolver, J.G. 2017. Insect development, thermal plasticity and fitness implications in changing, seasonal environments. *Integrative and Comparative Biology* 57 (5): 988–98.

Carroll, S.P., Hendry, A.P., Reznick, D.N., and Fox, C.W. 2007. Evolution on ecological time-scales. *Functional Ecology* 21 (3): 387–93.

Corl, A., Bi, K., Luke, C., Challa, A. S., Stern, A. J., Sinervo, B., and Nielsen, R. 2018. The genetic basis of adaptation following plastic changes in coloration in a novel environment. *Current Biology* 28 (18): 2970–2977.

Crispo, E. 2008. Modifying effects of phenotypic plasticity on interactions among natural selection, adaptation and gene flow. *Journal of Evolutionary Biology* 21 (6): 1460–1469.

Dahl, J. and Peckarsky, B.L. 2002. Induced morphological defenses in the wild: Predator effects on a mayfly, *Drunella coloradensis*. *Ecology* 83 (6): 1620–1634.

David, J. and Clavel, M.F. 1967. Influence de la température subie au cours du développement sur divers caractères biométriques des adultes de *Drosophila melanogaster* Meigen. *Journal of Insect Physiology* 13 (5): 717–29.

Davidowitz, G. and Nijhout, H.F. 2004. The physiological basis of reaction norms: The interaction among growth rate, the duration of growth and body size. *Integrative and Comparative Biology* 44 (6): 443–49.

de Jong, G. and van der Have, T.M. 2009. "Temperature dependence of development rate, growth rate and size: From biophysics to adaptation." In *Phenotypic Plasticity of Insects: Mechanisms and Consequences*, edited by D.W. Whitman and T.N. Ananthakrishnan. Enfield: Science Publishers, Inc, 461–526.

Denlinger, D.L. 1972. Induction and termination of pupal diapause in Sarcophaga (Diptera: Sarcophagidae). *The Biological Bulletin* 142 (1): 11–24.

DeWitt, T.J. and Scheiner, S.M. (eds). 2004. *Phenotypic Plasticity: Functional and Conceptual Approaches*. Oxford: Oxford University Press.

Diamond, S.E. 2017. Evolutionary potential of upper thermal tolerance: Biogeographic patterns and expectations under climate change. *Annals of the New York Academy of Sciences* 1389 (1): 5–19.

Diamond, S.E., Prileson, E., and Martinv, R.A. 2022. Adaptation to urban environments. *Current Opinion in Insect Science* 51: 100893.

Dobzhansky, T. 1968. "On some fundamental concepts of Darwinian biology." In *Evolutionary biology*, edited by T. Dobzhansky, M.K. Hecht, and W.C. Steere. Boston: Springer US, 1–34.

Dudley, R. 2002. *The Biomechanics of Insect Flight: Form, Function, Evolution*. Princeton: Princeton University Press.

Ellers, J. and Boggs, C.L. 2004. Functional ecological implications of intraspecific differences in wing melanization in *Colias* butterflies. *Biological Journal of the Linnean Society* 82 (1): 79–87.

Falconer, D.S. 1960. *Introduction to quantitative genetics*. London: Oliver and Boyd.

Foquet, B., Castellanos, A.A., and Song, H. 2021. Comparative analysis of phenotypic plasticity sheds light on the evolution and molecular underpinnings of locust phase polyphenism. *Scientific Reports* 11 (1): 11925.

Fox, R.J., Donelson, J.M., Schunter, C., Ravasi, T., and Gaitán-Espitia, J.D. 2019. Beyond buying time: The role of plasticity in phenotypic adaptation to rapid environmental change. *Philosophical Transactions of the Royal Society B: Biological Sciences* 374 (1768): 20180174.

Garbuz, D., Evgenev, M.B., Feder, M.E., and Zatsepina, O.G. 2002. Evolution of thermotolerance and the heat-shock response: evidence from inter/intraspecific comparison and interspecific hybridization in the *virilis* species group of *Drosophila*. I. Thermal phenotype. *The Journal of Experimental Biology* 206: 2399–2408.

Ghalambor, C.K., McKay, J.K., Carroll, S.P., and Reznick, D.N. 2007. Adaptive versus non-adaptive phenotypic plasticity and the potential for contemporary adaptation in new environments. *Functional Ecology* 21 (3): 394–407.

Ghosh, S.M., Testa, N.D., and Shingleton, A.W. 2013. Temperature-size rule is mediated by thermal plasticity of critical size in *Drosophila melanogaster*. *Proceedings of the Royal Society B: Biological Sciences* 280 (1760): 20130174.

Gibert, P., Hill, M., Pascual, M., Plantamp, C., Terblanche, J.S., Yassin, A., and Sgrò, C.M. 2016. *Drosophila* as models to understand the adaptive process during invasion. *Biological Invasions* 18: 1089–1103.

Gilchrist, G.W. and Huey, R.B. 2004. Plastic and genetic variation in wing loading as a function of temperature within and among parallel clines in *Drosophila subobscura*. *Integrative and Comparative Biology* 44 (6): 461–70.

Gilchrist, G.W., Huey, R.B., and Serra, L. 2001. Rapid evolution of wing size clines in *Drosophila subobscura*. *Genetica* 112–113:273–86.

Gould, S.J. and Lewontin, R.C. 1979. The spandrels of San Marco and the Panglossian paradigm: A critique of the adaptationist programme. *Proceedings of the Royal Society of London B: Biological Sciences* 205: 581–98.

Grether, G.F. 2005. Environmental change, phenotypic plasticity, and genetic compensation. *The American Naturalist 166* (4): E115–E123.

Gursky, V.V., Surkova, S.Y., and Samsonova, M.G. 2012. Mechanisms of developmental robustness. *Biosystems 109* (3): 329–35.

Hallmann, C.A., Sorg, M., Jongejans, E., Siepel, H., Hofland, N., Schwan, H., Stenmans, W., et al. 2017. More than 75 percent decline over 27 years in total flying insect biomass in protected areas. *PLoS One* 12 (10): e0185809.

Hangartner, S. and Hoffmann, A. A. 2016. Evolutionary potential of multiple measures of upper thermal tolerance in *Drosophila melanogaster*. *Functional Ecology* 30 (3): 442–52.

Hansen, T.F. and Pélabon, C. 2021. Evolvability: A quantitative-genetics perspective. *Annual Review of Ecology, Evolution, and Systematics* 52: 153–75.

Hendry, A.P. 2017. *Eco-evolutionary dynamics*. Princeton: Princeton University Press.

Hereford, J. 2009. A quantitative survey of local adaptation and fitness trade-offs. *The American Naturalist* 173 (5): 579–88.

Hill, W.G., Goddard, M.E., and Visscher, P.M. 2008. Data and theory point to mainly additive genetic variance for complex traits. *PLoS Genetics* 4 (2): e1000008.

Hill, W.G. and Kirkpatrick, M. 2010. What animal breeding has taught us about evolution. *Annual Review of Ecology, Evolution, and Systematics* 41: 1–19.

Hochachka, P.W. and Somero, G.N., 2014. *Biochemical Adaptation*. Princeton: Princeton University Press.

Horne, C.R., Hirst, A.G., and Atkinson, D. 2015. Temperature-size responses match latitudinal-size clines in arthropods, revealing critical differences between aquatic and terrestrial species. *Ecology Letters* 18 (4): 327–35.

Houle, D. 1992. Comparing evolvability and variability of quantitative traits. *Genetics* 130 (1): 195–204.

Huey, R.B., Hertz, P.E., and Sinervo, B. 2003. Behavioral drive versus behavioral inertia in evolution: A null model approach. *The American Naturalist 161* (3): 357–66.

Johnson, L.C., Galliart, M.B., Alsdurf, J.D., Maricle, B.R., Baer, S.G., Bello, N.M., Gibson, D.J. et al. 2022. Reciprocal transplant gardens as gold standard to detect local adaptation in grassland species: New opportunities moving into the 21st century. *Journal of Ecology* 110 (5): 1054–1071.

Kellermann, V., Van Heerwaarden, B., Sgrò, C.M., and Hoffmann, A.A. 2009. Fundamental evolutionary limits in ecological traits drive *Drosophila* species distributions. *Science* 325 (5945): 1244–1246.

Kelly, M. 2019. Adaptation to climate change through genetic accommodation and assimilation of plastic phenotypes. *Philosophical Transactions of the Royal Society B: Biological Sciences* 374 (1768): 20180176.

Kingsolver, J.G. and Buckley, L.B. 2017. Evolution of plasticity and adaptive responses to climate change along climate gradients. *Proceedings of the Royal Society B: Biological Sciences* 284 (1860): 20170386.

Kivelä, S.M., Friberg, M., Wiklund, C., and Gotthard, K. 2017. Adaptive developmental plasticity in a butterfly: Mechanisms for size and time at pupation differ between diapause and direct development. *Biological Journal of the Linnean Society* 122 (1): 46–57.

Lafuente, E., Duneau, D., and Beldade, P. 2018. Genetic basis of thermal plasticity variation in *Drosophila melanogaster* body size. *PLoS Genetics* 14 (9): e1007686.

Lande, R. 2009. Adaptation to an extraordinary environment by evolution of phenotypic plasticity and genetic assimilation. *Journal of Evolutionary Biology* 22 (7): 1435–1446.

Levis, N.A. and Pfennig, D. W. 2016. Evaluating "plasticity-first" evolution in nature: Key criteria and empirical approaches. *Trends in Ecology & Evolution* 31 (7): 563–74.

Linz, D.M., Hu, Y., and Moczek, A.P. 2019. The origins of novelty from within the confines of homology: The developmental evolution of the digging tibia of dung beetles. *Proceedings of the Royal Society B: Biological Sciences* 286 (1896): 20182427.

Losey, J.E. and Vaughan, M. 2006. The economic value of ecological services provided by insects. *Bioscience* 56 (4): 311–23.

MacLean, H.J., Sørensen, J.G., Kristensen, T.N., Loeschcke, V., Beedholm, K., Kellermann, V., and Overgaard, J. 2019. Evolution and plasticity of thermal performance: An analysis of variation in thermal tolerance and fitness in 22 *Drosophila* species. *Philosophical Transactions of the Royal Society B: Biological Sciences* 374 (1778): 20180548.

Mallard, F., Nolte, V. and Schlotterer, C. 2020. The evolution of phenotypic plasticity in response to temperature stress. *Genome Biology and Evolution* 12: 2429–2440.

Mesas, A., Jaramillo, A., and Castañeda, L.E. 2021. Experimental evolution on heat tolerance and thermal performance curves under contrasting thermal selection in *Drosophila subobscura*. *Journal of Evolutionary Biology* 34 (5): 767–78.

Misof, B., Liu, S., Meusemann, K., Peters, R.S., Donath, A., Mayer, C., Frandsen, P.B., et al. 2014. Phylogenomics resolves the timing and pattern of insect evolution. *Science* 346 (6210): 763–67.

Moczek, A.P. 2007. Developmental capacitance, genetic accommodation, and adaptive evolution. *Evolution & Development* 9 (3): 299–305.

Moczek, A.P. and Emlen, D.J. 2000. Male horn dimorphism in the scarab beetle, *Onthophagus taurus*: Do alternative reproductive tactics favour alternative phenotypes? *Animal Behaviour* 59 (2): 459–66.

Morgan, C.L. 1896. On modification and variation. *Science* 4: 733–40.

Neve, G. and Hall, C.R. 2016. Variation of thorax flight temperature among twenty Australian butterflies (Lepidoptera: Papilionidae, Nymphalidae, Pieridae, Hesperiidae, Lycaenidae). *European Journal of Entomology* 113: 571–78.

Nijhout, H.F. 2002. The nature of robustness in development. *Bioessays* 24 (6): 553–63.

Nijhout, H.F., Kudla, A.M., and Hazelwood, C.C. 2021. Genetic assimilation and accommodation: Models and mechanisms. *Current Topics in Developmental Biology* 141: 337–69.

Noor, M.A., Parnell, R.S., and Grant, B.S. 2008. A reversible color polyphenism in American peppered moth (*Biston betularia cognataria*) caterpillars. *PloS One* 3 (9): e3142.

Olazcuaga, L., Foucaud, J., Deschamps, C., Loiseau, A., Claret, J.L., Vedovato, R., Guilhot, R. et al. 2022. Rapid and transient evolution of local adaptation to seasonal host fruits in an invasive pest fly. *Evolution Letters* 6 (6): 490–505.

Paaby, A.B. and Rockman, M.V. 2014. Cryptic genetic variation: Evolution's hidden substrate. *Nature Reviews Genetics* 15 (4): 247–58.

Parmesan, C. and Yohe, G. 2003. A globally coherent fingerprint of climate change impacts across natural systems. *Nature* 421 (6918): 37–42.

Pfennig, D.W. (Ed) 2021. *Phenotypic Plasticity & Evolution: Causes, Consequences, Controversies*. Boca Raton, FL: CRC Press.

Pfennig, D.W., Wund, M.A., Snell-Rood, E.C., Cruickshank, T., Schlichting, C.D., and Moczek, A.P. 2010. Phenotypic plasticity's impacts on diversification and speciation. *Trends in Ecology & Evolution* 25 (8): 459–67.

Rodrigues, Y.K. and Beldade, P. 2020. Thermal plasticity in insects' response to climate change and to multifactorial environments. *Frontiers in Ecology and Evolution* 8: 271.

Rohner, P.T. 2022. Secondary sexual trait melanization in "black" scavenger flies: Nutritional plasticity and its evolution. *The American Naturalist*, 199 (1): 168–77.

Rohner, P.T. and Blanckenhorn, W.U. 2018. A comparative study of the role of sex-specific condition dependence in the evolution of sexually dimorphic traits. *The American Naturalist* 192 (6): E202–E215.

Rohner, P.T. and Moczek, A.P. 2020. Rapid differentiation of plasticity in life history and morphology during invasive range expansion and concurrent local adaptation in the horned beetle *Onthophagus taurus*. *Evolution* 74 (9): 2059–2072.

Rohner, P.T., Pitnick, S., Blanckenhorn, W.U., Snook, R.R., Bächli, G., and Lüpold, S. 2018. Inter-relations of global macroecological patterns in wing and thorax size, sexual size dimorphism, and range size of the Drosophilidae. *Ecography* 41 (10) 1707–1717.

Rohner, P.T., Teder, T., Esperk, T., Lüpold, S., and Blanckenhorn W.U. 2018. The evolution of male-biased sexual size dimorphism is associated with increased body size plasticity in males. *Functional Ecology* 32: 581–91.

Rohner, P.T., Roy, J., Schäfer, M.A., Blanckenhorn, W.U., and Berger, D. 2019. Does thermal plasticity align with local adaptation? An interspecific comparison of wing morphology in sepsid flies. *Journal of Evolutionary Biology* 32 (5): 463–75.

Rolff, J. and Reynolds, S. (eds). 2009. *Insect Infection and Immunity: Evolution, Ecology, and Mechanisms*. Oxford: Oxford University Press.

Schwab, D.B., Casasa, S., and Moczek, A.P. 2019. On the reciprocally causal and constructive nature of developmental plasticity and robustness. *Frontiers in Genetics* 9: 735.

Sgrò, C.M., Overgaard, J., Kristensen, T.N., Mitchell, K.A., Cockerell, F.E., and Hoffmann, A.A. 2010. A comprehensive assessment of geographic variation in heat tolerance and hardening capacity in populations of *Drosophila melanogaster* from eastern Australia. *Journal of Evolutionary Biology* 23 (11): 2484–2493.

Sommer, S. and Wehner, R. 2012. Leg allometry in ants: Extreme long-leggedness in thermophilic species. *Arthropod Structure & Development* 41 (1): 71–77.

Sultan, S.E. 2015. *Organism and Environment: Ecological Development, Niche Construction, and Adaptation*. New York: Oxford University Press.

Suzuki, Y. and Nijhout, H.F. 2006. Evolution of a polyphenism by genetic accommodation. *Science* 311 (5761): 650–52.

Tauber, M.J., Tauber, C.A., and Masaki, S. 1986. *Seasonal Adaptations of Insects*. Oxford: Oxford University Press.

Verberk, W.C., Atkinson, D., Hoefnagel, K.N., Hirst, A.G., Horne, C.R., and Siepel, H. 2021. Shrinking body sizes in response to warming: Explanations for the temperature–size rule with special emphasis on the role of oxygen. *Biological Reviews* 96 (1): 247–68.

Verlinden, H., Badisco, L., Marchal, E., Van Wielendaele, P., and Broeck, J.V. 2009. Endocrinology of reproduction and phase transition in locusts. *General and Comparative Endocrinology* 162 (1): 79–92.

Via, S. 1991. The genetic structure of host plant adaptation in a spatial patchwork: Demographic variability among reciprocally transplanted pea aphid clones. *Evolution* 45 (4): 827–52.

Waddington, C.H. 1942. Canalization of development and the inheritance of acquired characters. *Nature* 150: 563–65.

Waddington, C.H. 1952. Selection of the genetic basis for an acquired character. *Nature* 169 (4302): 625–26.

Waddington, C.H. 1953. Genetic assimilation of an acquired character. *Evolution* 7: 118–26.

Wagner, D.L., Grames, E.M., Forister, M.L., Berenbaum, M.R., and Stopak, D. 2021. Insect decline in the Anthropocene: Death by a thousand cuts. *Proceedings of the National Academy of Sciences* 118 (2): e2023989118.

Walsh, B. and Lynch, M. 2018. *Evolution and Selection of Quantitative Traits.* Oxford: Oxford University Press.

Walters, R.J., Berger, D., Blanckenhorn, W.U., Bussière, L.F., Rohner, P.T., Jochmann, R., Thüler, K., et al. 2022. Growth rate mediates hidden developmental plasticity of female yellow dung fly reproductive morphology in response to environmental stressors. *Evolution & Development* 24 (1–2): 3–15.

Walters, R.J. and Hassall, M. 2006. The temperature-size rule in ectotherms: May a general explanation exist after all? *The American Naturalist* 167: 510–23.

Ward, P.I. 2000. Cryptic female choice in the yellow dung fly *Scathophaga stercoraria* (L.). *Evolution* 54 (5): 1680–1686.

Ward, P.I. 2007. Postcopulatory selection in the yellow dung fly *Scathophaga stercoraria* (L.) and the mate-now-choose-later mechanism of cryptic female choice. *Advances in the Study of Behavior 37*: 343–69.

Ward, P.I., Wilson, A.J., and Reim, C. 2008. A cost of cryptic female choice in the yellow dung fly. *Genetica 134*: 63–67.

Warren, M.S., Maes, D., van Swaay, C.A., Goffart, P., Van Dyck, H., Bourn, N.A., Wynhoff, I., et al. 2021. The decline of butterflies in Europe: Problems, significance, and possible solutions. *Proceedings of the National Academy of Sciences 118* (2): e2002551117.

Weetman, D., Mitchell, S.N., Wilding, C.S., Birks, D.P., Yawson, A.E., Essandoh, J., Mawejje, H.D., et al. 2015. Contemporary evolution of resistance at the major insecticide target site gene Ace-1 by mutation and copy number variation in the malaria mosquito *Anopheles gambiae. Molecular Ecology* 24 (11): 2656–2672.

West-Eberhard, M.J. 2003. *Developmental Plasticity and Evolution.* Oxford: Oxford University Press.

Wilson, A.J., Reale, D., Clements, M.N., Morrissey, M.M., Postma, E., Walling, C.A., et al., 2010. An ecologist's guide to the animal model. *Journal of Animal Ecology* 79: 13–26.

CHAPTER 6

Effects of climate change on insect phenology

Gang Ma, Chun-Sen Ma, Cécile Le Lann and Joan van Baaren

6.1 Introduction

Phenology refers to the timing of cyclical life history events that occur at certain times of the year or under specific environmental conditions (Chmura et al. 2019). After long-term evolutionary adaption, species synchronize their life-history events with local environmental (including biotic and abiotic) conditions, which is of great significance to ensure individual survival and population persistence (Abarca and Spahn 2021; Marshall et al. 2020). As ectotherms, the life processes of insects are determined by ambient temperature (Bale and Hayward 2010; Buckley 2022). As such, the changes in temperature-dependent biological rates such as metabolic and development rates can lead to substantial shifts in insect phenology (Chmura et al. 2019). However, insects are highly variable in their life-history traits (e.g., multivoltine vs univoltine; diapause vs non-diapause), which may result in different phenological responses to temperature change (Teder 2020). In this sense, the changes in insect phenology are a combination of temperature-dependent and life-history trait-associated responses to climate change, leading to considerable variation in the direction and magnitude of phenological shifts within and across taxa (Chmura et al. 2019; Marshall et al. 2020).

Insect phenological events, such as spring emergence, swarming, migration and overwintering diapause, are changing in response to climate warming (Altermatt 2010; Buckley 2022; Vitasse et al. 2021). For example, warmer temperatures can advance spring first flight in butterflies (Vitasse et al. 2021) and spring swarming in bark beetles (Jakoby et al. 2019), postpone aphid autumn migration (Bell et al. 2019) and lead to delayed diapause induction in some UK butterfly species (Marshall et al. 2020). Temporal shifts in these events with climate change have been widely documented at both regional (Altermatt 2010; Duchenne et al. 2020, Teder 2020; Vitasse et al. 2021) and global (Cohen et al. 2018) scales. Phenological shifts can lead to changes in voltinism (generations per year) and abundance at the population and species levels (Altermatt 2010; Kerr et al. 2020; Mitton and Ferrenberg 2012). Importantly, these changes can scale up to alter species interactions at the community and ecosystem levels, not only through direct phenological mismatch between interacting species (Abarca and Spahn 2021) but also indirectly by modifying species abundance (Kerr et al. 2020; van Dyck et al. 2015) and distribution

Gang Ma et al., *Effects of Climate Change on Insect Phenology*. In: *Effects of Climate Change on Insects*. Edited by: Daniel González-Tokman and Wesley Dáttilo, Oxford University Press. © Oxford University Press (2024). DOI: 10.1093/oso/9780192864161.003.0006

(Jakoby et al. 2019; Macgregor et al. 2019). As a consequence, climate change-associated insect phenological shifts can influence biodiversity conservation and pest management, thereby impacting ecosystem functioning and stability.

6.2 Phenological shifts: Drivers and responses

6.2.1 *Environmental and organismal drivers*

Many abiotic and biotic factors can determine or affect insect phenological shifts under climate change and these factors can be divided into two types, environmental drivers and organismal drivers (Chmura et al. 2019). On the one hand, phenological shifts can be driven by the changes in environmental cues such as temperature, precipitation, photoperiod, etc., that can influence insect phenology. If the same species experience different climates at various latitudes and altitudes (correlated with environmental cues), they may differ in phenological shifts due to the different warming rates across latitudinal and elevational gradients, even if they have similar responses to temperature change. For example, the advancement for spring swarming of the European spruce bark beetle *Ips typographus* is expected to be greater at higher than lower elevations (Jakoby et al. 2019). The predictive result can be explained by the greater temperature increase at higher than lower elevations. On the other hand, phenological mismatches or shifts can also be driven by differences in sensitivity to environmental change among species or populations, that is, organismal drivers. The species or populations that are more sensitive to temperature changes are expected to show greater phenological shifts compared to those that are less sensitive to temperature changes, even if they experience the same warming rate. Differences in temperature sensitivities between sympatric species with different trophic levels can lead to increased or decreased phenological synchronization between herbivorous insects and their host plants and/or between pest insects and their natural enemies (Jakoby et al. 2019).

6.2.2 *Plastic and evolutionary responses*

As with many other ectothermic species, insects cannot regulate their internal body temperatures. Thus, the development and consequent seasonal timing (phenology) of insects show high plasticity in response to external temperature change (Chapter 5, this book). On the one hand, climate warming may determine the direction of phenological shifts. According to the thermal performance curves (TPCs), both the metabolic and development rates are temperature dependent. As such, these rates may show either positive or negative responses to temperature increase, depending on if and to what extent the amplitude of warming exceeds the optimal performance temperature (Figure 6.1a). Warm-adapted and cold-adapted populations/species are likely to express advanced and delayed phenologies, respectively, in response to similar climate warming (Figure 6.1b). Generally, warmer temperatures may accelerate insect development and lead to advanced spring phenology. By contrast, however, some insect species require prolonged exposure to chilling temperatures to terminate diapause, suggesting that warming is also likely to prolong the diapause duration, thereby resulting in delayed spring phenology (Abarca and Spahn 2021; Davies 2019).

On the other hand, warming can lead to differences in the magnitude of phenological shifts due to the high variations in phenological flexibility among taxa (Belitz et al. 2021) and species (Colom et al. 2022; Hassall et al. 2017). Overall, two existing patterns are found to contribute to these differences (Figure 6.1c and d). First, the plasticity in response to temperature change (i.e., thermal sensitivity) may determine the extent to which climate warming affect phenological shifts, with more sensitive populations/species exhibiting stronger phenological shifts (Figure 6.1c). For example, the temperature sensitivity pattern shows that organisms more sensitive to changes in temperatures have greater or faster phenological shifts. Cohen and colleagues synthesized 127 published studies globally concerning the animal phenologies across time series. They find that (i) ectotherms have stronger phenological shifts than endotherms because they are more sensitive to temperature change due to their limited ability to regulate body temperatures; (ii) invertebrates with smaller body size are found to advance their phenology faster than larger vertebrates because thermal inertia is positively associated with body size (Cohen et al.

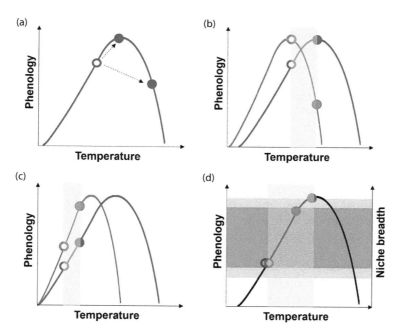

Figure 6.1 Existing patterns of the (a–b) direction and (c–d) magnitude of phenological shifts under climate change. Open and solid circles represent the insect (relative) phenology under current and future climate scenarios, respectively. The gray square within each subpanel indicates an equivalent increase of temperature experienced by different species/populations (red vs blue). (a) The direction of phenological shift can be determined by the extent to which warmer temperatures may alter insect phenology. (b) Similar temperature increase may result in different directions of phenological shifts between species/populations, depending on if warming exceeds their optimal temperatures. (c) The magnitude of phenological shift can be determined by internal physiology of insects, e.g., temperature sensitivity, with the species/populations more sensitive to temperature change showing greater phenological shifts. (d) Even assuming similar sensitivity to temperature change, the magnitude of phenological shift can also depend on some external factors such as the niche breadth, with the species/populations having wider niche breadth may exhibit greater phenological advancement.

2018). Generally, the species with more generations per year have shorter life cycles and thus are expected to be more sensitive to warming than that with fewer yearly generations (Ma et al. 2021). For instance, by analyzing multiple long-term monitoring datasets, the hoverfly species with greater voltinism flexibility are found to have a general trend of stronger phenological shifts (Hassall et al. 2017). Second, the niche breadth pattern indicates that generalist species are more likely to have stronger phenological shifts than specialist species (Figure 6.1d). For example, butterflies with a greater number of host plants are found to exhibit greater phenological advancement than species with fewer host plants (Brooks et al. 2017). Importantly, the flexibility of phenology is likely to determine the demographic responses and population abundances to climate change. For example, insect species/populations with greater phenological flexibility in response to different interannual climates may enable to maintain synchrony with their resources, which may allow them to thrive or be less impacted in abundance under recent climate change (Colom et al. 2022; Moraiti et al. 2014). Phenological flexibility, such as flexibility in voltinism and active period, may lead to positive demographic responses of multivoltine Lepidoptera species to climate change (Macgregor et al. 2019).

Many insect species have the capacity to evolutionarily adjust their phenological responses to rapid climate change. For example, the mosquito *Wyeomyia smithii* shows delayed diapause induction, which is genetically controlled by photoperiodic response, to adapt to the extended growing season caused by climate warming (Bradshaw and Holzapfel 2001). The UK butterflies with less dispersal abilities advanced their spring phenologies (emergence dates) to catch up with the temperature changes over their geographic ranges, suggesting extensive local adaptations (Roy et al. 2015). In addition, there are frequent examples that populations within species differ in their phenology such as the timing for diapause induction (Haugen and Gotthard 2015; Pruisscher et al. 2018) and annual number of generations (Braune et al. 2008; Lindestad et al. 2019) along climatic gradients, indicating the capacity to evolve to various climate conditions and thus widespread local adaptation. However, local adaptation is usually expected to be associated with limited gene flow and genetic divergence between populations, implying considerable differences of their genomic offset to future environments, which is in turn likely to cause local maladaptation under the strong directional selection of rapid climate warming.

Some insect species have evolved bet-hedging strategies to adapt to the variations in interannual climates or in lengths of growing/dormant season. These strategies can enable a certain proportion of the populations to track synchrony with their feeding/mating resources in case of missing a prior phenological window. For example, the univoltine fruit fly *Rhagoletis cerasi* experiences an obligatory pupal diapause during autumn-winter seasons and requires sufficient chilling for diapause termination (Moraiti et al. 2014). When failing to meet the chilling requirements for their diapause termination (insufficient or extended chilling), the individuals may adopt two alternative ways to cope with the particular climatic conditions. Warmer winters may cause insufficient chilling and thereby leading to prolonged dormancy (delayed diapause termination). By contrast, colder winters may cause extended chilling, resulting in another (facultative) diapausing cycle enabling adults to emerge a year later (Moraiti et al. 2014). The spruce beetle *Dendroctonus rufipennis* has an adult obligate diapause. Cool temperatures can induce a facultative prepupal diapause, leading to a semivoltine cycle. However, warm temperatures may enable to complete a univoltine life cycle, avoiding to generate a facultative prepupal diapause (Schebeck et al. 2017). Semivoltine individuals of the pine processionary moths *Thaumetopoea pityocampa* emerging one year later than their own cohort are found to synchronize with univoltine individuals emerging the same year, rather than with their own cohort, to optimize mating probability, which implies an adaptive response to the new climatic conditions (Martin et al. 2022). However, the prolonged dormancy is often found to incur fitness costs such as mortality risk and reduced adult fertility (Moraiti et al. 2012), implying certain constraints in the evolvability of bet-hedging strategies.

6.3 Effects of phenological shifts on populations

6.3.1 *Diapause and emergence*

Photoperiod is usually the main cue for insect diapause while other cues such as temperature, precipitation/drought, host plant quality, can affect this process (Marshall et al. 2020; Takagi and Miyashita 2008). As such, by modifying the temperature and/or precipitation/drought, climate change may shift the onset and/or termination of insect diapause and thereby lead to advanced or delayed phenology (Abarca and Spahn 2021; Ma et al. 2017), although the photoperiod will remain stable at a given latitude. Thus, new environmental conditions will bring selective pressures and result in rapid plastic or evolutionary diapausing responses in insect phenology (Bale and Hayward 2010). Generally, warmer autumn temperatures may result in postponed autumn phenology such as diapause induction (Marshall et al. 2020) while warmer winter or spring temperatures can trigger diapause termination earlier (Stuhldreher et al. 2014) and/or accelerate postdiapause development (Davies 2019) and thus advance spring phenology (Abarca and Spahn 2021). However, winter warming in temperate areas can also lead to delayed spring emergence due to insufficient chilling (short duration or warmer temperatures) to terminate diapause (Abarca and Spahn 2021; Marshall et al. 2020; Moraiti et al. 2014).

On the one hand, climate change has led to advances in spring phenologies, including spring migration (Bell et al. 2019; Luquet et al. 2019) and first observed activity (adult emergence) (Duchenne et al. 2020; Vitasse et al. 2021), in many insect species. For example, the spring migration of aphids has advanced by nearly one month during 1978–2015 in France (Luquet et al. 2019). However, the magnitudes of phenological shifts are variable, depending not only on the environmental drivers such as different warming rates across spatial and temporal scales but also on the organismal drivers such as various taxon-, species- and population-specific temperature sensitivities. For example, it has been found that the mean flight date of European pollinators (more than 2,000 species of Hymenoptera, Diptera, Lepidoptera and Coleoptera) advanced at a rate of 1.0 days/decade during past 60 years (Duchenne et al. 2020). However, in the European Alps with much faster warming rate, the first spring activity of butterflies has advanced at an average rate of 6.0 days/decade (Vitasse et al. 2021). Species with early season activities usually have a greater rate of phenological advancement than the species with later activities (Bell et al. 2019; Brooks et al. 2017). This may be caused either by the higher warming rate during winter and spring than that in summer and autumn (Luquet et al. 2019) or because early-season species are sensitive to temperature increases compared to those occurring later (Bell et al. 2019; Brooks et al. 2017; Hassall et al. 2017). Nevertheless, the causal experimental studies would be more welcomed than existing correlations in further studies.

On the other hand, however, advanced spring phenology is not universal; many insect species also show delayed spring emergence due to insufficient winter chilling for diapause termination (Davies 2019; Duchenne et al. 2020; Hällfors et al. 2021; Ward et al. 2019;

Zhang et al. 2016). As a consequence, seasonal warming may result in different or even contrasting shifts in phenologies (Abarca and Spahn 2021; Marshall et al. 2020). In this case, the phenological outcomes of some species may depend not only on the intensity of temperature increase but also on the time of the year that is subject to temperature change (Davies 2019; Ward et al. 2019). For example, warming in winter may lead to delayed spring phenology in the univoltine butterfly *Anthocharis cardamines* because the pupae require accumulated winter chilling to break obligate pupal diapause and higher winter temperatures thus delay the onset of diapause termination (Davies 2019). On the contrary, higher spring temperatures after postdiapause can advance adult emergence due to the acceleration of postdiapause development (Davies 2019).

6.3.2 *Voltinism and abundance*

Climate warming can advance spring phenology while delaying autumn phenology, thereby prolonging growing seasons for insect activity and causing increases in yearly generations (Figure 6.2). A recent study involving 101 insect species shows that warm temperatures have led to extended adult appearance periods in both univoltine and multivoltine species (Belitz et al. 2021). The longer period of insect adult activity has the potential to cause additional generations per year in both multivoltine and functionally univoltine species (Altermatt 2010; Kerr et al. 2020; Ma et al. 2021; Mitton and Ferrenberg 2012; van Dyck et al. 2015). For example, the frequency of second and subsequent generations of multivoltine butterfly and moth species in Central Europe was found to have

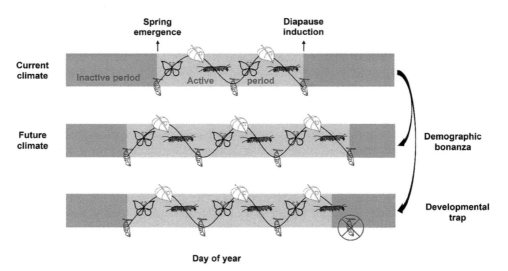

Figure 6.2 Phenological shifts and the demographic consequences of insects under climate change. In general, warming can lead to advanced spring emergence, postponed overwintering diapause induction and thereby prolonged active seasons, resulting in increased generations per year. However, the demographic consequences (demographic bonanza or developmental trap) of warming may depend on whether the additional generation can complete an extra lifecycle and overwinter successfully or not during the extended active period. Note that here the butterfly species is assumed to diapause as an obligate pupal-specific stage.

increased substantially since 1980, with 44 out of 263 species showing an increased number of generations per year (Altermatt 2010). The functionally univoltine mountain pine beetle *Dendroctonus ponderosae* does not diapause. Thus, warmer spring and summer temperatures can advance their flight season by more than one month earlier, and accelerate their temperature-dependent development and prolong their activity seasons, resulting in an unprecedented summer generation in the Colorado Front Range (Mitton and Ferrenberg 2012). However, many species still maintain obligately univoltine despite of warmer temperatures (Schebeck et al. 2017; Teder 2020). For example, a recent study shows that almost half of the 731 moth and butterfly species still have a single generation across their entire European ranges, implying that there are some constraints for enabling additional generations in obligately univoltine species under climate warming (Teder 2020).

Increases in voltinism associated to phenological shifts can lead to complex demographic consequences at the population level (Figure 6.2). Although the photoperiod will remain unchanged under climate change, warmer temperatures are able to modify the temperature-dependent development rates. This may reinforce or decouple the synchrony between diapause-sensitive development stages and stage-specific critical photoperiods for diapause induction (Bale and Hayward 2010). On the one hand, if multivoltine insects can produce yearly additional generations which may enter diapause successfully at a certain development stage and survive harsh winters, warmer temperatures could result in increased abundance (demographic bonanzas) (Altermatt 2010; Jakoby et al. 2019; Kerr et al. 2020; Mitton and Ferrenberg 2012). On the contrary, when the yearly additional generations cannot complete their development during the extended growing season, warming may cause ineffective reproduction and excessive energy consumption before entering diapause, leading to high mortality and thereby decreasing population abundance or even lead to extinction (developmental trap) (Kerr et al. 2020; van Dyck et al. 2015; Zhang et al. 2016). For example, warmer autumn temperatures can lead to a third generation in the wall brown butterfly *Lasiommata megera*. However, such autumn warming may not enable these individuals to develop into the third instar where they can enter the age-specific larval diapause, showing a high mortality (van Dyck et al. 2015). This species has been found to go extinct in areas where all the individuals were found to develop into the third generation without larval diapause. However, the population-level consequences will be variable in insects, depending on the biological (species-specific) responses (Marshall et al. 2020; Schebeck et al. 2017) and spatial (local climatic conditions) (Kerr et al. 2020; Moraiti et al. 2014) and temporal (interannual fluctuating weathers) (Martin et al. 2022; Moraiti et al. 2014) environmental cues. For example, under future climate warming, the butterfly *Pieris oleracea* which is originally from North America is predicted to experience a demographic bonanza in its southern ranges whereas it is expected to undergo a developmental trap in its northern ranges (Kerr et al. 2020).

6.4 Effects of phenological mismatches on interspecific interactions

6.4.1 *Host–plant/host and parasitoid/symbiont interactions*

In ecosystems, multitrophic interactions often result from a long co-evolutionary process specific to a particular environment and relatively stable climatic conditions (Visser and Both 2005; Visser and Gienapp 2019). Temperature changes may differentially affect the biology of each of the component species of a system as, for example, the herbivores, their natural enemies (parasitoids, predators and pathogens) and hyperparasitoids

(Abarca and Spahn 2021; van Baaren et al. 2010). Temperature increases can either accelerate or delay development: during the growing season, a mean increase of temperature generally accelerates ectotherm development while heat stress may decrease it, but not at the same pace for all species of a trophic web. During the unfavorable season (e.g., winter), the diapause of different species belonging to the same communities can be differentially affected, with some species still having a substantial proportion of diapausing individuals but some others becoming fully active (Abarca and Spahn 2021; Tougeron et al. 2020). Moreover, most of the insect species harbor obligatory and/or facultative bacterial symbionts whose functions can be altered or modified by temperature increase, with either deleterious consequences on insect survival or with an enhancement of their resistance to temperature stress by direct symbiont-mediated host protection (as observed in hot climatic areas) or indirect effects of temperature on the symbiont itself (Heyworth et al. 2020). These different effects could possibly result in phenological mismatch between at least two different species of the trophic web. The effect of climatic changes is likely to be even more important in higher trophic levels that depend on the capacity of the lower trophic levels to adapt to these changes (van Baaren et al. 2010).

There are two types of interspecific interactions that are differentially affected by phenological mismatches. In antagonistic trophic interactions, like plant–herbivore host–parasitoid or prey–predator interactions, mismatch will generally have positive fitness consequences for the plant or host/prey, which will escape at least partially to their consumer and therefore negative impacts for the herbivore/parasitoid/predator (Damien and Tougeron 2019). In mutualistic interactions, mismatch will have generally negative consequences for both species. In both cases, trophic mismatch is expected to last for short periods on an evolutionary point of view, because the species either adapt or disappear in a few generations (Renner and Zohner 2018). Several factors can explain the difficulty to detect phenological mismatch due to climate change, including their short duration (on an evolutionary point of view), and the necessity to have long-term ecological data (Renner and Zohner 2018). In particular, it is necessary to have a pre-climate change baseline, that is, an evaluation of the degree of synchrony that a given interaction had before recent climate change, to be able to detect the real phenological mismatches due to climate change (Abarca and Spahn 2021; Kharouba and Wolkovich 2020).

For antagonist relationships, the match-mismatch hypothesis (MMH) (Cushing 1990) provides a conceptual framework to explain the effects of phenological changes on population dynamics (Revilla et al. 2014). When the phenology of a consumer is close to that of its resource, the growth rate of the consumer is maximal and it decreases as the mismatch increases. This model also shows that consumer viability is constrained by the magnitude of the phenological mismatch with their resources. Nevertheless, a perfect synchronization between two species can induce an overexploitation of the resource, and by consequence, a long-term decrease of both species. In Revilla et al.'s (2014) model, which considers this possibility of the overexploitation of the resource, consumers display two abundance maxima, one large for slightly early recruiter consumers and one smaller for late recruiter consumers, who have less time to use resources to increase their progeny. These small mismatches allow the persistence of consumer–resource relationships. This highlights the role of feedbacks in the response of populations towards changes in phenology as the abundance of consumers is higher when present before or after their resources because this reduces the long-term effects of overexploitation that occurs under closer synchrony (Figure 6.3). Shorter phenologies promote oscillations and cycles (Revilla et al. 2014) and are less vulnerable to climate change. Some authors point out that until now,

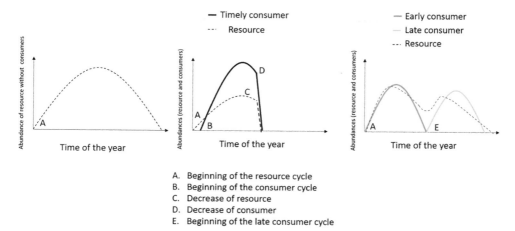

A. Beginning of the resource cycle
B. Beginning of the consumer cycle
C. Decrease of resource
D. Decrease of consumer
E. Beginning of the late consumer cycle

Figure 6.3 Effect of the phenology of a consumer on the abundances of its resource and feedback consequences on the abundance of the consumer. The first graph presents the abundance of a resource when the consumer is absent. The second presents the abundance of a resource in the presence of a timely consumer resulting in the overexploitation of the resource and the consumer's collapse. The third graph shows the persistence of consumer-resource relationships when early and late arriving consumers exploit a resource. The letters A, B, C, D and E indicate the different steps of the cycles of the resource–consumer relationship.

no study has collected appropriate data to test Cushing's MMH and most of the studies have fail to define a pre-climate change baseline (Kharouba and Wolkovich 2020). However, in some studies, this baseline has been calculated and asynchrony between insects and their host plants may be adaptive (Singer and Parmesan 2019).

The consequences of phenological mismatches differ between univoltine and multivoltine organisms. Voltinism is an important aspect of consumptive interspecific interactions and a modification of voltinism can affect the overall amount of phenological overlap (Revilla et al. 2014). Moreover, according to multiple studies, univoltine species may suffer from phenological mismatch and experience continual selection to adapt to changing seasonality (e.g., by adding a new generation), whereas at least the later generations in the season in multivoltine species may be less affected and experience reduced selection (Knell and Thackeray 2016).

Non-consumptive interspecific interactions in insects include mutualist interactions (pollination, seed dispersal, or ant–plant mutualisms) (Kiers et al. 2010) and interference or exploitative competition. For mutualist interactions, the likelihood of phenological mismatch following climate change depends mainly on the amount of interspecific temporal overlap, which is representative of the interaction strength (Rafferty et al. 2015; Revilla et al. 2014). Other factors of influence are the binding nature of the interaction and its specificity (Rafferty et al. 2015). Rafferty et al. (2015) showed that mismatch has a higher probability to occur when mutualism occurs during seasonal, facultative and generalized interactions and this situation is more probable at higher latitude and altitude characterized by a strong seasonality. By contrast, interacting species for which the relationship is reciprocally obligate and symmetrically specialized have been shaped by natural selection to respond in the same way to the same abiotic cues and should be more resistant to phenological mismatch. Generalist species are under weaker selection to respond to similar cues to their resources, but the mismatches could have less impact on their fitness if they can compensate for the loss of some interactions by other ones. Indeed, the degree of specialization is sometimes asymmetrical. For example, a specialized plant flowering earlier could suffer from a failure of pollination following a mismatch with its pollinator, but this mismatch could have no effect for the pollinator if it is able to compensate with other floral resources (Rafferty et al. 2015).

6.5 Plant–herbivore/pollinator/seed disperser interactions

In plant–herbivore interactions, climate change can either protect the plant from herbivores by inducing a phenological mismatch or to the contrary synchronize the plant to its herbivore, inducing herbivore outbreaks. Plants can escape herbivore damages due to phenological mismatch. For example, in the system between the European larch (*Larix decidua*) and the large bud moth (*Zeiraphera diniana*), a survey of more than 1,000 generations of the moth showed outbreaks every 9.3 years on average, during a period of 1,173 years. However, in the period between 1981 and 2018 no outbreak was observed, even though four outbreaks would have been expected. The exceptional warming in this area allows the plant to escape to the herbivory because the leaves develop earlier, inducing a phenological mismatch between the two species (Büntgen et al. 2020).

On the contrary, climate can also synchronize the herbivore with its plant, inducing herbivore outbreaks. In the system including the spruce budworm (*Choristoneura fumiferana*) and its host plant, the white spruce (*Picea glauca*), the timing of budburst is critical for young larvae because the suitability of foliage declines quickly after budburst. As a defensive strategy, the tree defoliates, which induces earlier budburst, that is, a mismatch with herbivores called defoliation-induced mismatch. Until now, the plant was protected by this phenological mismatch, but winter warming suppresses this mismatch and the synchrony increases the probability of outbreaks (Ren et al. 2020).

For plant–pollinator interactions, Rafferty (2017) reviewed that bees, butterflies and moths generally advanced their phenologies. However, while bees advance their phenology at similar rates as their flowers, this is not the case for butterflies. Until now, mismatches rarely lead to complete decoupling, probably because plant–pollinator interactions are constrained by natural selection to be synchronized (Gérard et al. 2020).

Phenological mismatches can also induce evolutionary changes in insects. In the winter moth *Operophtera brumata*, winter warming causes caterpillars to hatch before foliage becomes available, causing population decline. However, after twenty years, this selection pressure led to a decrease of the asynchrony by a delay in the hatching time of the caterpillar, showing that some species can adapt quickly to climate change (Van asch et al. 2013). Few studies have analyzed the effect of mismatch on both demographic rates and natural selection on fitness traits. This limits our ability to predict eco-evolutionary consequences of phenological mismatch reliably and hence the likelihood of successful adaptation to climate change (Visser and Gienapp 2019).

Few situations with phenological mismatch induce decoupling of mutualistic interactions. For example, according to Kudo and Ida (2013), ephemeral spring wildflowers from the northern forest ecosystem suffer frequently from phenological mismatch with their pollinators. Over fourteen years they studied the inter-annual variation in seed production of the spring ephemeral species *Corydalis ambigua* (Papaveraceae) in northern Japan. They showed that this plant has a short flowering period (ten days) and is pollinated by overwintering bumblebees at the end of the snowing period. The years with an early disappearance of snow correlate with a large decrease in seed production, as flowering tended to be ahead of first pollinator detection. In this system, the frequency of unusual events (warmer and dryer winters) will particularly affect this plant, but not the bumblebees which can use later floral resources. Similar mismatches apply to spring-fruiting plants and the ants that disperse their seeds, as ant optimal foraging activity occur within a limited range of temperatures allowing seed dispersal (Rafferty et al. 2015).

It is not only phenological mismatches that can alter mutualistic interactions but also mismatches of their life-history traits and behaviors that are often directly related to the fitness of individuals. Global warming may create a risk of a morphological mismatch. For example, the increase of temperature induces a reduced size for many insect species, whereas the quality of pollination is often linked to the length of the tongue of the pollinator (Gérard et al. 2020). The foraging behavior of the pollinator can be affected, as well as the attractiveness of the plant or the quality and quantity of the nectar reward. All these changes in life-history traits can induce a disruption of an interaction between a specific plant species and its pollinator (Gérard et al. 2020).

The fitness consequences of mismatches are commonly evaluated for plants by quantifying effects on seed production but remain poorly understood for pollinators (Rafferty 2017). However, phenological mismatches between plant and pollinators can have strong negative effects on pollinator fitness and populations. At the species level, males and females can be differentially affected. For example, in the solitary bees from the *Osmia* genus, males responded to warmer temperatures whereas females did not, inducing male emergences during the flowering period and the female emergences only after, potentially leading to a reduction in mating and fertilized eggs (Slominski and Burkle 2019). Moreover, specialist pollinators are generally more vulnerable than generalist ones (Abarca and Spahn 2021; Burkle et al. 2013; Gérard et al. 2020) as they rely on a single plant species or family to feed their progeny.

6.6 Community and ecosystem levels

Phenological mismatch between two species may have repercussions on multiple trophic networks in a given area as two species in strong interactions also interact with other species and these interactions can be reinforced or disappear (Renner and Zohner 2018; Tougeron et al. 2018). For instance, in a system including plants (two fern species), their herbivorous insect (caterpillar of the moth *Herpetogramma theseusalis* (Walker) (Lepidoptera: Crambidae)), its principal primary parasitoid wasp (*Alabagrus texanus* (Cresson) (Hymenoptera: Braconidae)) and hyperparasitoids (*Aprostocetus* sp.), studied since 2004, Morse (2021) showed that each species changed its phenology rapidly, and that the rhythm of change varies among species, inducing significant mismatches at all trophic levels. The higher trophic levels (parasitoids and hyperparasitoids) changed more and with a higher inter-annual variability than the ferns, being therefore more vulnerable. In another system, Meineke et al. (2014) showed that in the warmest habitats of a city, the scale insects *Parthenolecanium quercifex* (Hemiptera: Coccidae) were twelve times more abundant on oak trees than in the cooler habitats. In warmer areas, the scale insect oviposits earlier, whereas the main parasitoid species does not advance egg laying at the same rate, inducing a phenological mismatch. This does not modify the proportion of scale insect parasitized, but parasitized insects from warmer areas produced twice as many eggs as individuals from cooler areas, while the number of eggs produced by unparasitized individuals was not different between both habitats. This study provides evidence that a pest develops earlier due to urban warming but that phenology of its parasitoid community does not similarly advance. This mismatch is associated with greater egg production that likely leads to increased pest abundance. Rafferty et al. (2015) proposed that at the community level, season length is an important determinant of network resilience, with communities with shorter interaction seasons being potentially the most vulnerable.

Mismatches in the interactions between plants and phytophagous insects or pollinators can lead to the emergence of new biological networks and communities (Rafferty 2017; Tougeron et al. 2018). Though some communities will experience the loss of existing interactions, other communities could gain novel interactions (Gérard et al. 2020; Rafferty 2017). For example, in the cereal aphid-parasitoid community of Western France, Tougeron et al. (2018) showed that during the winter, part of the population of the aphid species *Metopolophium dirhodum* is now reproducing parthenogenetically in cereals and part of the parasitoids *Aphidius ervi* and *Aphidius avenae* are no longer diapausing, which completely modifies the winter network that is now richer and closer to what is observed in the spring (Jeavons et al. 2021; Tougeron et al. 2022).

Phenological mismatches between interacting species could also impact nutrient cycling (Beard et al. 2019). Beard et al. (2019) proposed one example with antagonistic interactions, between oak (*Quercus robur*), great tit (*Parus major*) and caterpillars. If the migration of birds does not match the peak of caterpillar biomass, the increase of caterpillar populations at the beginning of the season could result in increased oak herbivory resulting in a decrease of above ground leaf biomass, reducing the carbon (C) and nitrogen (N) sink strength of these trees. For mutualistic interactions, Beard et al. (2019) proposed that if apple trees are not pollinated, then the C a tree would dedicate to fruits may be shunted to growth and storage, making the plant a greater C sink.

At the evolutionary time scale (within a few decades), more and more studies highlighted the existence of mutualism breakdowns linked to climatic changes (Gerard et al. 2020; Kiers et al. 2010). Three possibilities were mentioned: shifts from mutualism to antagonism, switch to novel partners and mutualism abandonment. Examples including insects comprised plant–pollinators and seed dispersion interactions, which are disrupted by habitat loss and fragmentation or by phenological mismatches. Indeed, it was shown that mutualisms are dynamic at ecological and evolutionary time scales, as natural selection favors individuals that give up a mutualist interaction when it becomes costly. These switches have been observed in the past but seem to be more frequent following recent global changes (Kiers et al. 2010). As all mutualisms are embedded within larger ecological interactions, a loss or gain of one species, which can be induced by phenological mismatch, can drive an evolutionary mutualism breakdown. In the case of the pollination, the inclusion of nectar or pollen robbers in a trophic web may modify the cost-benefit balance of the interaction, although there is no concrete example yet (Kiers et al. 2010). An analysis of a fifty-year dataset for four *Prunus* species and their butterfly pollinator showed that the plant flowered earlier whereas the butterfly phenology was not modified, and this phenological mismatch can originate the mutualism breakdown except in areas where the butterfly shows rapid evolutionary responses (Doi et al. 2008).

6.7 Resilience of communities

In some situations, even in antagonistic interactions, phenological mismatches can still have no consequence at the community level due to several mechanisms. For example, Senior et al. (2020) studied a plant–aphid–parasitoids community. In this system comprising one tree (the sycamore *Acerpseudoplatanus*), two aphid species (*Drepanosiphum platanoidis* and *Periphylus testudinaceus*) and several braconid parasitoid species, phenological mismatches were demonstrated between the tree and the aphids with earlier budburst of the tree and delayed emergence of the aphids probably due to chill requirement to end the diapause, and between the aphids and their parasitoids (with earlier arrival of the parasitoids probably due to shorter diapause). However, the analysis of twenty years of data showed that these mismatches have no effect on population growth rates of both aphid species, contrary to what was expected. Indeed, in many tree–phytophagous interactions, there is a rapid seasonal increase in induced volatile chemicals that allow plants to defend themselves from insect herbivores. In this system, the lack of impact on aphid populations seems to be due to strong buffering effects of density dependence (Senior et al. 2020). Mechanisms such as density-dependent compensation can occur when asynchrony-triggered mortality is balanced by a subsequent increase in fecundity of the survivors due to reduced competition, resulting in a stable population size (Abarca and Spahn 2021). This highlights the strong resilience of aphid populations to both mismatches with their host plant and with their natural enemies (Senior et al. 2020).

Some taxa have characters or life histories that predispose them to resist phenological mismatch. For example, migratory insect species, such as a large number of hoverflies species, are highly mobile, have a high reproductive rate and a generalist alimentation. These characteristics allow them to exploit different food resources in case their usual resources are no more available following phenological mismatches. For these reasons, they decline at a lower rate than most of the other insect species (Doyle et al. 2020). These examples are at the species level. However, it can have an impact at the community level because the different species of a community have different reactions and that happens when resilience can be achieved by functional redundancy. Indeed, in community assemblages, functional complementarity and redundancy between species can be modified by phenological mismatches (Gérard et al. 2020). Duchenne et al. (2020) analyzed phenological shifts of over 2,000 species of European pollinators and showed a six days advance of the mean flight date and a shortening of the flight period of two days over the last sixty years, including a reduction of pollinator species overlap in European assemblages and an increase of extinction risks. However, a community or an ecosystem can be resilient to these phenological mismatches as predicted by the biodiversity insurance hypothesis (Bartomeus et al. 2013). For example, Bartomeus et al. (2013) studied a community of pollinators of apple trees using fifty-six years of data on the flowering phenology combined to historical data of bee pollinators. They showed that although the flowering time of apple trees and the phenology of most bee species is affected differently according to the species, pollination is still achieved, due to the functional redundancy between species that maintain synchrony at the community level. Biodiversity can then enhance and stabilize phenological synchrony at the community level.

6.8 Effects of phenological mismatches on ecosystem services

Very few studies analyzed precisely the impact of phenological mismatches on ecosystem services. In most studies, it is only mentioned that as the phenological mismatches are affecting the interactions between species, there should be detrimental consequences on ecosystem functioning and services, but the studies did not quantify this decrease. Moreover, the decrease in ecosystem service provision is generally linked to several causes and it is therefore difficult to assess how much a reduction in ecosystem provisioning is due to phenological mismatches. Some studies suggested solutions for managing the environment to limit the impact of phenological mismatches and maintain ecosystem services. Although insects are involved in several other ecosystem services such as decomposition and wildlife provisioning, we focused this paragraph on two major ecosystem services involving insects of agrosystems, that is, pollination and pest control for which a few publications are available.

6.9 Ecosystem services of pollination and pest control

Entomophilic pollination participates in the production of around 70 percent of crops across the world, and their productivity has an estimated annual economic value of US\$ 200–800 billion. Additionally, the value of insect pollination to native plant species is considered as essential but not measurable (Dalsgaard 2020; Doyle et al. 2020). Pollination also contributes to 9.5 percent of global food production (Gérard et al. 2020). The most important pollinators are wild and managed bees, but "non-bee insects" also play a large role. Hoverflies (Diptera: Syrphidae) are the second most important pollinators. Overall, 1,800–3,800 species of pollinators visit 72 percent of crops across the world (Doyle et al. 2020).

For pest control, it was evaluated that potential and actual losses due to arthropod pests for the major crops (i.e., wheat, rice, maize, potatoes, soybeans and cotton) reached 8–15 percent between 2001 and 2003 worldwide. These values could reach 9–37 percent in the absence of natural biological control or pesticides. Natural enemies can save up to US\$ 4.5 billion by their pest regulation service. The main providers are flies and parasitic wasps, which lay eggs on or inside the body of the pests, killing them and reducing their spread (Gutierrez-Arellano and Mulligan 2018). Mutualistic relationships can also contribute to pest control. For example, the mutualistic relationships between the Aztec ant and a coccid reduced the number of the coffee borer beetles, the main pest of coffee plants (Gutierrez-Arellano and Mulligan 2018).

Very few studies included both pollination and pest regulation services together. One of the most complete studies on this matter was conducted by Kőrösi et al. (2018). They analyzed the consequences of phenological mismatches on the arthropod community of apple trees in orchards in which they manipulated the degree of phenological mismatch by cold storing potted apple trees or keeping them in a greenhouse before transferring them into orchards. By doing so, they created two situations where plant phenology is either advanced or delayed relative to arthropod communities. They sampled the abundance and diversity of the arthropod community (including pollinators, generalist and specialist herbivores and predatory arthropods) throughout the season for the three types of trees (advanced, control and delayed phenology). The results showed that the communities of pollinators displayed changes in species composition between the three categories of trees, but that in general, the pollination service is maintained.

For herbivore abundances, one main result was a very high variability both between and within orchards, probably linked to their smaller dispersal capacity than bees. In general, the species diversity was lower in manipulated trees (with either advanced or delayed phenology). For wild bees and aphidophagous beetles, the reduced specific diversity is mostly due to the dominance of one or a few species, whereas for spiders and phytophagous bugs, only few species were found on manipulated trees, without increase of abundance of some species. Therefore, altered phenology is detrimental for most species except for a few that benefit from it. Specialist species were the most affected and the mutualistic interactions were more stable than antagonistic ones. Moreover, it was shown that apple crop yield and quality strongly depend on both pollination and pest control ecosystem services (Kőrösi et al. 2018). This manipulative experimental design is particularly interesting for future research aiming at evaluating the impact of phenological mismatches between organisms supporting main ecosystem services such as pest control and pollination, as it allows identifying the most affected species sustaining those services. The apple pollinator community accounts for 16 percent of the EU's total economic gains attributed to insect pollination, apple trees being the most important insect-pollinated crop plant (Kőrösi et al. 2018) and these trees are attacked by several species of phytophagous pests, resulting in crop loss. In the context of climate change, phenological mismatches between species may thus have serious consequences for human population and economy. However, so far, there is still no study that has measured the effects of phenological mismatches for both types of interspecific interactions on crop yields.

6.10 How to manage the environment to maintain ecosystem services in a changing world

Studying the impact of phenological mismatches on plant–insect and insect–insect interactions can allow to measure the decrease of ecosystem services (which is still rarely done) and to be able to propose practical applications to limit these losses (Dalsgaard 2020). A spatial approach is one method to evaluate the impact of climate changes and land-use on plant–pollinators or plant–pest–natural enemy interactions. By studying the linear distance to a patch of a given habitat type, it will be possible to define the minimum patch size to maintain ecosystem services, using spatial distances in accordance with the dispersal abilities of studied organisms (Dalsgaard 2020). The second solution is the buffer approach, which will allow the definition of the measures that should be taken at the landscape scale to optimize the pollination ecosystem services (Dalsgaard 2020; Tougeron et al. 2022). From a temporal point of view, it is possible to investigate the impact of historical changes in climate on continental gradients or landscape/field scale gradients, which focuses on recent variation in climate (Dalsgaard 2020). Combining both spatial and temporal approaches on multiple taxa allows us to understand the elements of the landscape which can maximize the potential of ecosystem services. For example, Tougeron et al. (2022) showed by studying eighty fields distributed in three regions between the western part of France to Belgium, in three different landscape types along a temporal gradient that (i) mild winter areas are favorable to parasitoid populations due to their activity during the winter inducing a precocious control of aphid populations; (ii) at the field scale, natural enemy communities were less diverse and had lower abundances in landscapes with high crop and wooded continuities, contrary to the studied pests (aphids and slugs); (iii) field boundaries with grass strips were more favorable to spiders and carabids than boundaries formed by hedges, while the opposite was found for crop pests, with the latter being less abundant towards the centre of the fields. These results reveal that it may be possible to both reduce pest pressure and promote natural enemies by accounting for taxa-specific antagonistic responses to multi-scale environmental characteristics. This study highlights the need for multitaxon approaches at the landscape scale to produce indicators of diversity and service at the scale of the territories.

Some studies evaluated the needs of pollinators or pest natural enemies, which can help to manage the habitat to aid species to resist climate change and its effects, including phenological mismatches (Dalsgaard 2020; Doyle et al. 2020; Jeavons et al. 2020). For example, Doyle et al. (2020) highlighted that hoverfly richness and abundance are favored in complex agricultural landscapes providing a temporarily stable supply of resources. Therefore, planting or maintaining semi-natural habitats such as hedgerows or grasslands, adapting agricultural practices, for instance by delaying mowing, sowing flower strips, especially those containing Umbelliferae, and using multicultural crops are concrete management practices to maintain pollination and biological control separately. Future studies should test which practices could favor both services simultaneously.

The phenological shifts may also offer promising opportunities for implementing engineering ecological methods for improving these ecosystem services. For example, in the western parts of France, winter covers, which are sown at the end of the summer and remain until maize is sown at the beginning of the spring, have been blooming all winter for the past decade. These flowers offer a new nectar source for the parasitoids of cereal aphids (Damien et al. 2017) which have recently given up their diapause strategy and remain active during the whole winter (Tougeron et al. 2018). Therefore, implanting designed flower strips during the whole year around crops is a developing strategy to sustain the pest regulation ecosystem service, reducing the impact of phenological mismatch.

6.11 Conclusion

Climate change has led to profound effects on insect phenology worldwide. Although there have been many research reports, some knowledge gaps still exist and limit our understanding of how to take appropriate strategies to cope with the ecological consequences. First, most of the existing studies focus mainly on the past phenological trends or correlation predictions of long-term field observations. Therefore, the causal mechanisms of climate change on insects should receive more attention in future research. This may help us to understand the physiological and ecological mechanisms of insects' responses to climate change as well as eco-evolutionary feedbacks more generally (Visser and Gienapp 2019). Second, most previous research on phenological mismatch caused by climate change fails to provide a baseline for comparison and quantitative assessment of the impact of climate change (Kharouba and Wolkovich 2020; Renner and Zohner 2018). Nevertheless, the assessment of the degree of phenological synchronization in a given species interaction before climate change can help to detect the real phenological mismatch caused by climate change. Long-term ecological data will offer a way to solve this problem (Renner and Zohner 2018). Finally, many factors can contribute to influencing insects and their involving communities and ecosystems (Chmura et al. 2019). It will be helpful to predict the changes of ecosystem functions by distinguishing the effects of climate factors from other threats that affect communities and ecosystems functioning, like invasive species or changes in land use and by further quantifying the ecological consequences caused by climate change compared to the ecological consequences caused by the other factors. This has important theoretical and practical significance for taking appropriate measures to improve ecosystem services and resilience in the context of ongoing climate change.

Key reflections

- Climate warming is modifying key phenological events such as spring emergence and overwintering diapause of insects, causing phenological shifts.
- At the population and species levels, phenological shifts can lead to changes in voltinism (generations per year) and abundance in insects.
- The direction and magnitude of phenological shifts can be taxon- and species-specific. Interacting species often differ in their phenological responses to climate warming, resulting in strengthened or mismatched phenological synchrony.
- At the community and ecosystem levels, phenological mismatches between interacting species can alter the structure and dynamics of communities, leading to community succession and thereby modifying ecosystem functioning.

Key further reading

- Ma, G., Le Lann, C., van Baaren, J., and Ma, C.S. 2020. "Night Warming Affecting Interspecific Interactions: Implications for Biological Control." In *Integrative Biological Control. Progress in Biological Control*, edited by Y. Gao, H. Hokkanen, and I. Menzler-Hokkanen. Cham: Springer, vol 20, 9–53. doi:10.1007/978-3-030-44,838-7_3
- Forrest, J.R. 2016. Complex responses of insect phenology to climate change. *Current Opinion in Insect Science 17*: 49–54. doi:10.1016/j.cois.2016.07.002

References

Aalberg Haugen, I.M., and Gotthard, K. 2015. Diapause induction and relaxed selection on alternative developmental pathways in a butterfly. *Journal of Animal Ecology* 84 (2): 464–72.

Abarca, M. and Spahn, R. 2021. Direct and indirect effects of altered temperature regimes and phenological mismatches on insect populations. *Current Opinion in Insect Science* 47: 67–74.

Altermatt, F. 2010. Climatic warming increases voltinism in European butterflies and moths. *Proceedings of the Royal Society B: Biological Sciences* 277 (1685): 1281–1287.

Bale, J.S. and Hayward, S.A.L. 2010. Insect overwintering in a changing climate. *Journal of Experimental Biolog* 213 (6): 980–94.

Bartomeus, I., Park, M.G., Gibbs, J., Danforth, B.N., Lakso, A.N., and Winfree, R., 2013. Biodiversity ensures plant–pollinator phenological synchrony against climate change. *Ecology Letters* 16: 1331–1338, doi: 10.1111/ele.12170.

Beard, K.H., Kelsey, K.C., Leffler, A.J., and Welker, J.M. 2019. The missing angle: Ecosystem consequences of phenological mismatch. *Trends in Ecology and Evolution* 34 (10): 885–88.

Belitz, M.W., Barve, V., Doby, J.R., Hantak, M.M., Larsen, E.A., Li, D., Oswald, J., et al. 2021. Climate drivers of adult insect activity are conditioned by life history traits. *Ecology Letters* 24 (12): 2687–2699.

Bell, J.R., Botham, M.S., Henrys, P.A., Leech, D.I., Pearce-Higgins, J.W., Shortall, C.R., Brereton, T.M., et al. 2019. Spatial and habitat variation in aphid, butterfly, moth and bird phenologies over the last half century. *Global Change Biology* 25: 1982–1994. doi:10.1111/gcb.14592

Bradshaw, W.E. and Holzapfel, C.M. 2001. Genetic shift in photoperiodic response correlated with global warming. *Proceedings of the National Academy of Sciences* 98 (25): 14509–14511.

Braune, E., Richter, O., Söndgerath, D., and Suhling, F. 2008. Voltinism flexibility of a riverine dragonfly along thermal gradients. *Global Change Biology* 14 (3): 470–82.

Brooks, S.J., Self, A., Powney, G.D., Pearse, W.D., Penn, M., and Paterson, G.L. 2017. The influence of life history traits on the phenological response of British butterflies to climate variability since the late-19th century. *Ecography* 40 (10): 1152–1165.

Buckley, L.B. 2022. Temperature-sensitive development shapes insect phenological responses to climate change. *Current Opinion in Insect Science* 52: 100897.

Büntgen, U., Liebhold, A., Nievergelt, D., Wermelinger, B., Roques, A., Reinig, F., Krusic, P.J., et al. 2020. Return of the moth: rethinking the effect of climate on insect outbreaks. *Oecologia* 192 (2): 543–52. doi:10.1007/s00442-019-04585-9.

Burkle, L.A., Marlin, J.C., and Knight, T.M. 2013. Plant-pollinator interactions over 120 years: Loss of species, co-occurrence, and function. *Science* 339: 1611–1616. doi:10.1126/science.1232728.

Chmura, H.E., Kharouba, H.M., Ashander, J., Ehlman, S.M., Rivest, E.B., and Yang, L.H. 2019. The mechanisms of phenology: The patterns and processes of phenological shifts. *Ecological Monographs* 89 (1): e01337.

Cohen, J.M., Lajeunesse, M.J., and Rohr, J.R. 2018. A global synthesis of animal phenological responses to climate change. *Nature Climate Change* 8: 224–28. doi:10.1038/s41558-018-0067-3.

Colom, P., Ninyerola, M., Pons, X., Traveset, A., and Stefanescu, C. 2022. Phenological sensitivity and seasonal variability explain climate-driven trends in Mediterranean butterflies. *Proceedings of the Royal Society B: Biological Science* 289 (1973): 20220251.

Cushing, D.H. 1990. Plankton production and year-class strength in fish populations: An update of the match/mismatch hypothesis. *Advances in Marine Biology* 26: 249–93. doi:10.1016/S0065-2881(08)60202–60203.

Dalsgaard, B. 2020. Land-use and climate impacts on plant–pollinator interactions and pollination services. *Diversity* 12 (5): 168.

Damien, M., Le Lann, C., Desneux, N., Alford, L., Al Hassan, D., Georges, R., and Van Baaren, J. 2017. Flowering cover crops in winter increase pest control but not trophic link diversity. *Agriculture, Ecosystems & Environment* 247: 418–25.

Damien, M., Le Lann, C., Desneux, N., Alford, L., Al-Hassan, D., Georges, R., and Van Baaren, J. 2017. Change in plant phenology during winter increases pest control but not trophic link diversity. *Agriculture Ecosystems and Environment* 247: 418–25. doi:10.1016/j.agee.2017.07.015.

Damien, M. and Tougeron, K. 2019. Prey–predator phenological mismatch under climate change. *Current Opinion in Insect Science* 35: 60–68.

Davies, W.J. 2019. Multiple temperature effects on phenology and body size in wild butterflies predict a complex response to climate change. *Ecology* 100 (4): e02612.

Doi, H., Gordo, O., and Katano, I. 2008. Heterogeneous intra-annual climatic changes drive different phenological responses at two trophic levels. *Climate Research* 36 (3): 181–90.

Doyle, T., Hawkes, W.L., Massy, R., Powney, G.D., Menz, M.H., and Wotton, K.R. 2020. Pollination by hoverflies in the Anthropocene. *Proceedings of the Royal Society B: Biological Sciences* 287 (1927): 20200508.

Duchenne, F., Thébault, E., Michez, D., Elias, M., Drake, M., Persson, M., Rousseau-Piot, J.S., et al. 2020. Phenological shifts alter the seasonal structure of pollinator assemblages in Europe. *Nature Ecology & Evolution* 4 (1): 115–21.

Gérard, M., Vanderplanck, M., Wood, T., and Michez, D. 2020. Global warming and plant–pollinator mismatches. *Emerging Topics in Life Sciences* 4 (1): 77–86.

Gutierrez-Arellano, C. and Mulligan, M. 2018. A review of regulation ecosystem services and dis-services from faunal populations and potential impacts of agriculturalisation on their provision, globally. *Nature Conservation* 30: 1–39. doi: 10.3897/natureconservation.30.26989.

Hällfors, M.H., Pöyry, J., Heliölä, J., Kohonen, I., Kuussaari, M., Leinonen, R., Schmucki, R. et al. 2021. Combining range and phenology shifts offers a winning strategy for boreal Lepidoptera. *Ecology Letters* 24 (8): 1619–1632.

Hassall, C., Owen, J., and Gilbert, F. 2017. Phenological shifts in hoverflies (Diptera: Syrphidae): Linking measurement and mechanism. *Ecography* 40: 853–63. doi:10.1111/ecog.02623

Heyworth, E.R., Smee, M.R., and Ferrari, J. 2020. Aphid facultative symbionts aid recovery of their obligate symbiont and their host after heat stress. *Frontiers in Ecology and Evolution* 8: 56.

Jakoby, O., Lischke, H., and Wermelinger, B. 2019. Climate change alters elevational phenology patterns of the European spruce bark beetle (Ips typographus). *Global Change Biology* 25: 4048–4063. doi:10.1111/gcb.14766

Jeavons, E., van Baaren, J., and Le Lann, C. 2020. Resource partitioning among a pollinator guild: A case study of monospecific flower crops under high honeybee pressure. *Acta Oecologica* 104: 103527.

Jeavons, E., van Baaren, J., Le Ralec, A., Buchard, C., Duval, F., Llopis, S., Postic, E., Le Lann, C. 2021. Third and fourth trophic level composition shift in an aphid-parasitoid-hyperparasitoid food web limits aphid control in an intercropping system. *Journal of Applied Ecology* 59 (1): 300–313. doi:10.1111/1365-2664.14055. Dryad, Dataset, doi:10.5061/dryad.d51c5b049. (hal-03464364)

Kerr, N.Z., Wepprich, T., Grevstad, F.S., Dopman, E.B., Chew, F.S., and Crone, E.E. 2020. Developmental trap or demographic bonanza? Opposing consequences of earlier phenology in a changing climate for a multivoltine butterfly. *Global Change Biology* 26 (4): 2014–2027.

Kharouba, H.M. and Wolkovich, E.M. 2020. Disconnects between ecological theory and data in phenological mismatch research. *Nature Climate Change* 10 (5): 406–15.

Kiers. E.T. Palmer, T.M., Ives, A.R., Bruno J.F., and Bronstein, J.L. 2010. Mutualisms in a changing world: An evolutionary perspective. *Ecology Letters* 13: 1459–1474. doi:10.1111/j.1461-0248.2010.01538.x.

Knell, R.J. and Thackeray, S.J. 2016. Voltinism and resilience to climate-induced phenological mismatch. *Climatic Change* 137: 525–539.

Kőrösi, Á., Markó, V., Kovács-Hostyánszki, A., Somay, L., Varga, Á., Elek, Z., Boreux, V., et al. 2018. Climate-induced phenological shift of apple trees has diverse effects on pollinators, herbivores and natural enemies. *PeerJ* 6: e5269.

Kudo, G. and Ida, T.Y. 2013. Early onset of spring increases the phenological mismatch between plants and pollinators. *Ecology* 94 (10): 2311–2320.

Lindestad, O., Wheat, C.W., Nylin, S., and Gotthard, K. 2019. Local adaptation of photoperiodic plasticity maintains life cycle variation within latitudes in a butterfly. *Ecology* 100 (1): e02550.

Luquet, M., Hullé, M., Simon, J.C., Parisey, N., Buchard, C., and Jaloux, B. 2019. Relative importance of long-term changes in climate and land-use on the phenology and abundance of legume crop specialist and generalist aphids. *Insect Science* 26 (5): 881–96.

Ma, C.S., Ma, G., and Pincebourde, S. 2021. Survive a warming climate: Insect responses to extreme high temperatures. *Annual Review of Entomology* 66: 163–84. doi:10.1146/annurev-ento-041520-074454.

Ma, G., Tian, B. L., Zhao, F., Wei, G.S., Hoffmann, A. A., and Ma, C.S. 2017. Soil moisture conditions determine phenology and success of larval escape in the peach fruit moth, Carposina sasakii (Lepidoptera, Carposinidae): Implications for predicting drought effects on a diapausing insect. *Applied Soil Ecology* 110: 65–72.

Macgregor, C.J., Thomas, C.D., Roy, D.B., Beaumont, M.A., Bell, J.R., Brereton, T., Bridle, J.R., et al. 2019. Climate-induced phenology shifts linked to range expansions in species with multiple reproductive cycles per year. *Nature Communications* 10: 4455. doi:10.1038/s41467-019-12479-w.

Marshall, K.E., Gotthard, K., and Williams, C.M. 2020. Evolutionary impacts of winter climate change on insects. *Current Opinion in Insect Science* 41: 54–62.

Martin, J.C., Mesmin, X., Buradino, M., Rossi, J.P., and Kerdelhué, C. 2022. Complex drivers of phenology in the pine processionary moth: Lessons from the past. *Agricultural and Forest Entomology* 24 (2): 247–59. doi:10.1111/afe.12488

Meineke, E.K., Dunn, R.R., and Frank, S.D. 2014. Early pest development and loss of biological control are associated with urban warming. *Biology Letters* 10 (11): 20140586.

Mitton, J.B. and Ferrenberg, S.M. 2012. Mountain pine beetle develops an unprecedented summer generation in response to climate warming. *The American Naturalist* 179 (5): E163–E171.

Moraiti, C.A., Nakas, C.T., and Papadopoulos, N.T. 2012. Prolonged pupal dormancy is associated with significant fitness cost for adults of Rhagoletis cerasi (Diptera: Tephritidae). *Journal of Insect Physiology* 58: 1128–1135. doi:10.1016/j.jinsphys.2012.05.012.

Moraiti, C.A., Nakas, C.T., and Papadopoulos, N.T. 2014. Diapause termination of Rhagoletis cerasi pupae is regulated by local adaptation and phenotypic plasticity: Escape in time through bet-hedging strategies. *Journal of Evolutionary Biology* 27 (1): 43–54.

Morse, D.H. 2021. Rapid phenological change differs across four trophic levels over 15 years. *Oecologia* 196 (2): 577–87.

Pruisscher, P., Nylin, S., Gotthard, K., and Wheat, C.W. 2018. Genetic variation underlying local adaptation of diapause induction along a cline in a butterfly. *Molecular Ecology* 27 (18): 3613–3626.

Rafferty, N.E. 2017. Effects of global change on insect pollinators: Multiple drivers lead to novel communities. *Current Opinion in Insect Science* 23: 22–27. doi:10.1016/j.cois.2017.06.009

Rafferty, N.E., CaraDonna, P.J., and Bronstein, J.L. 2015. Phenological shifts and the fate of mutualisms. *Oikos* 124 (1): 14–21.

Ren, P., Néron, V., Rossi, S., Liang, E., Bouchard, M., and Deslauriers, A. 2020. Warming counteracts defoliation-induced mismatch by increasing herbivore-plant phenological synchrony. *Global Change Biology* 26 (4): 2072–2080.

Renner, S.S. and Zohner, C.M. 2018. Climate change and phenological mismatch in trophic interactions among plants, insects, and vertebrates. *Annual Review of Ecology, Evolution, and Systematics* 49: 165–82.

Revilla, T.A., Encinas-Viso, F., and Loreau, M. 2014. (A bit) Earlier or later is always better: Phenological shifts in consumer–resource interactions. *Theoretical Ecology* 7 (2): 149–62.

Roy, D.B., Oliver, T.H., Botham, M.S., Beckmann, B., Brereton, T., Dennis, R.L., Harrower, C., et al. 2015. Similarities in butterfly emergence dates among populations suggest local adaptation to climate. *Global Change Biology* 21 (9): 3313–3322.

Schebeck, M., Hansen, E.M., Schopf, A., Ragland, G.J., Stauffer, C., and Bentz, B.J. 2017. Diapause and overwintering of two spruce bark beetle species. *Physiological Entomology* 42: 200–10. doi:10.1111/phen.12200.

Senior, V.L., Evans, L.C., Leather, S.R., Oliver, T.H., and Evans, K.L. 2020. Phenological responses in a sycamore-aphid-parasitoid system and consequences for aphid population dynamics: A 20 year case study. *Global Change Biology* 26 (5): 2814–2828. doi:10.1111/gcb.15015.

Singer, M.C. and Parmesan, C. 2019. Butterflies embrace maladaptation and raise fitness incolonizing novel host. *Evolutionary Applications* 12: 1417–1433. doi:10.1111/eva.12775

Slominski, A.H. and Burkle, L.A. 2019. Solitary bee life history traits and sex mediate responses to manipulated seasonal temperatures and season length. *Frontiers in Ecology and Evolution* 7: 314.

Stuhldreher, G., Hermann, G., and Fartmann, T. 2014. Cold-adapted species in a warming world—an explorative study on the impact of high winter temperatures on a continental butterfly. *Entomologia Experimentalis et Applicata* 151 (3): 270–79.

Takagi, S. and Miyashita, T. 2008. Host plant quality influences diapause induction of byasa alcinous (Lepidoptera: Papilionidae). *Annals of the Entomological Society of America* 101: 392–96.

Teder, T. 2020. Phenological responses to climate warming in temperate moths and butterflies: Species traits predict future changes in voltinism. *Oikos* 129 (7): 1051–1060.

Tougeron, K., Brodeur, J., Le Lann, C., and van Baaren, J. 2020. How climate change affects the seasonal ecology of insect parasitoids. *Ecological Entomology* 45 (2): 167–81.

Tougeron, K., Couthouis, E., Hecq, F., Barascou, L., Baudry, J., Boussard, H., Burel, F., et al. 2022. Where and when the biological control service can be maximized? *Science of the Total Environment* 822: 153569.

Tougeron, K., Damien, M., Le Lann, C., Brodeur, J., and van Baaren, J. 2018. Changes in host-parasitoid communities over the years in cereal crops of Western France: Does climate warming matters? *Frontiers in Ecology & Evolution* 6: 173. doi:10.3389/fevo.2018.00173

Van Asch, M., Salis, L., Holleman, L.J., Van Lith, B., and Visser, M.E. 2013. Evolutionary response of the egg hatching date of a herbivorous insect under climate change. *Nature Climate Change* 3 (3): 244–48.

Van Baaren, J., Le Lann, C., and van Alphen, J. 2010. "Consequences of climate change for aphid-based multi-trophic systems." In *Aphid biodiversity under environmental change*, pp. 55–68. In *Patterns and Processes* edited by P. Kindlmann, A.F.G. Dixon, and J.P. Michaud. The Netherlands: Springer Dordrecht. doi:10.1007/978-90-481-8601-3.

Van Dyck, H., Bonte, D., Puls, R., Gotthard, K., and Maes, D. 2015. The lost generation hypothesis: Could climate change drive ectotherms into a developmental trap? *Oikos* 124 (1): 54–61.

Visser, M.E. and Both, C. 2005. Shifts in phenology due to global climate change: the need for a yardstick. *Proceedings of The Royal Society B-Biological Sciences*, 272 (1581): 2561–2569. https://doi-org.inee.bib.cnrs.fr/10.1098/rspb.2005.3356.

Visser, M.E. and Gienapp, P. 2019. Evolutionary and demographic consequences of phenological mismatches. *Nature Ecology & Evolution* 3 (6): 879–85.

Vitasse, Y., Ursenbacher, S., Klein, G., Bohnenstengel, T., Chittaro, Y., Delestrade, A., Monnerat, C., et al. 2021. Phenological and elevational shifts of plants, animals and fungi under climate change in the E uropean Alps. *Biological Reviews* 96 (5): 1816–1835.

Ward, S.F., Moon, R.D., and Aukema, B.H. 2019. Implications of seasonal and annual heat accumulation for population dynamics of an invasive defoliator. *Oecologia* 190: 703–14.

Zhang, B., Zhao, F., Hoffmann, A., Ma, G., Ding, H.M., and Ma, C.S. 2016. Warming accelerates carbohydrate consumption in the diapausing overwintering peach fruit moth Carposina sasakii (Lepidoptera: Carposinidae. *Environmental Entomology* 45 (5): 1287–1293.

CHAPTER 7

Sexual selection in insects in times of climate change

Bruno A. Buzatto, Daniel P. Silva and Paulo Enrique C. Peixoto

7.1 Introduction: Climate change and the macroecology of sexual selection

This chapter explores how climate change (CC) can have extreme effects on the environmental factors that influence sexual selection on insects directly or indirectly, ultimately affecting secondary sexual traits, such as ornaments and weapons. Importantly, our goal was not to review all relevant studies on this topic systematically but rather to use case studies that illustrate the state of knowledge in the field. We also point to promising future perspectives in the macroecology of sexual selection in times of severe change.

7.1.1 *A brief historical context*

Sexual selection was first recognized as a separate mechanism from natural selection by Charles Darwin (1871) and was used as an explanation for sexual ornaments in males. Darwin's ideas on the topic generated controversy in the first decades that followed but the field of sexual selection eventually matured as a prolific area of evolutionary biology. The very definition of sexual selection has been updated and amended several times. Perhaps the most widely used definition is one first published in the 1990s, which states that sexual selection results from the differential reproduction of individuals of the same sex due to mating and fertilization success (Andersson 1994). As recently as in 2021, however, further debate has resulted in a refinement of this definition as "any selection that arises from fitness differences associated with nonrandom success in the competition for access to gametes for fertilization" (Shuker and Kvarnemo 2021). This latest definition also sparked debate, placing a stronger focus on competition for access to gametes more directly, to the exclusion of competition for reproductive resources. Despite our support for these recent developments, while endeavoring to cover the potential effects of CC on sexual selection, we adopt a more general definition of sexual selection. This was so that we could be more inclusive with the studies we covered and, at the same time, conveniently avoided the challenge of disentangling access to gametes from access to reproductive resources, which is a theoretical advance but still difficult in practice.

Bruno A. Buzatto, Daniel P. Silva and Paulo Enrique C. Peixoto, *Sexual selection in insects in times of climate change*. In: *Effects of Climate Change on Insects*. Edited by: Daniel González-Tokman and Wesley Dáttilo, Oxford University Press. © Oxford University Press (2024). DOI: 10.1093/oso/9780192864161.003.0007

Understanding sexual selection and mating system evolution is usually seen as a behavioral ecology problem dealt with on the scale of populations, in contrast to studies about CC that typically focus on how environmental traits may affect species richness, abundance and distribution. Nonetheless, mating system theory has long recognized the environment as having a significant role in sexual selection and mate choice (Emlen and Oring 1977). Environmental factors can mediate the interactions between mates and individuals of the same sex competing for access to mates. Accordingly, there is plenty of evidence that the intensity of sexual selection is influenced by environmental conditions, including the availability of resources and population density, which fluctuate with other environmental factors (García-Roa et al. 2020; Punzalan et al. 2008). A recent framework has been put forward for studying large-scale variation in reproductive traits, and the resulting field was named *"macroecology of sexual selection"* (Machado et al. 2016). This framework sheds light on how long-recognized geographical patterns of reproductive traits evolve. This is the context in which we attempt to understand the relationship between CC and sexual selection.

7.1.2 *The potential for environmental effects on sexual selection*

Rather than fully covering the effects of various environmental variables on sexual selection, we structured this chapter according to different components of sexual selection and discussed studies that exemplify how CC can affect ornaments, weapons, mating tactics and dimorphisms. The connection we make with CC is sometimes rather indirect. When studies that directly linked CC with a particular aspect of sexual selection were scarce or absent, we focused on the known effects of climate on sexually selected traits and inferred the potential for CC to disturb that connection or operate in any way that alters the reproductive biology and mating systems of insect species. This somewhat reflects the state of knowledge in this incipient field in insects—there is ample evidence suggesting that CC should affect sexual selection. Still, macroecological studies focusing on CC (Parmesan 2006) seldomly target sexual selection.

Climate offers an extensive set of environmental variables that can affect natural and sexual selection, consequently playing a role in different elements of mating systems. Temperature, for example, is an essential environmental dimension in the niche of a species. It can vary significantly in different timescales, from circadian changes that oscillate dramatically between day and night to seasonal changes that cause extreme differences between summer and winter, especially in high latitudes. García-Roa et al. (2020) recently proposed a conceptual framework to distinguish the direct and indirect effects of temperature on sexual selection. Direct effects occur when temperature variation causes immediate changes in sexually selected traits. In contrast, indirect effects occur when temperature variation affects phenotypes, demography, trade-offs and sex-specific reproductive costs and benefits (García-Roa et al. 2020), so sexual selection pressures are secondarily affected.

García-Roa et al. (2020) also emphasized the difference between effects within the range of temperatures that organisms are adapted to and effects outside that range, which might trigger stress responses and have maladaptive outcomes. Unfortunately, our understanding of how sexual selection is affected by environmental variables (including but not limited to temperature) seems very biased toward cases within the natural range of these variables to which animals are adapted. Meanwhile, CC has clearly been exposing species to environmental conditions outside these ranges, and the maladaptive outcomes and stress responses likely to result are still poorly understood.

7.2 Intersexual selection in times of climate change

Since the origin of sexual selection as a field of study, this evolutionary force has been split into two components. The first results from direct competition between members of one sex for reproductive access to receptive individuals of the other sex. Termed intrasexual selection (Andersson 1994), this process is more readily observed in the form of male–male competition. However, there are many well-known cases of female–female competition in nature (Rosvall 2011). The second component of sexual selection is derived from individuals of one sex choosing individuals of the other sex with whom to mate, termed intersexual selection (Andersson 1994). This process can also be called mate choice (Andersson and Simmons 2006) and it is more often observed as females choosing between different potential mates. However, cases of choosy males are also well known, especially in insects (Bonduriansky 2001).

Mate choice has historically been the most controversial component of sexual selection. This was probably because at the time, it was perceived that female animals would have needed an aesthetic taste to exert preference for different candidate males. This idea encountered resistance from historical sexism but it is now widely accepted and empirically corroborated that mate choice is a pervasive component of sexual selection and a powerful evolutionary mechanism (Andersson and Simmons 2006). Mate choice has been responsible for the evolution of diverse ornaments (Figure 7.1) and copulatory behaviors across animal taxa and different modalities of animal senses. In insects, mate choice has driven the evolution of spectacular visual ornaments (Sivinski and Wing 2008; Figure 7.1), chemically complex olfactory signals (Steiger and Stökl 2014) and elaborate acoustic and vibrational displays (Cocroft and Rodríguez 2005; Song et al. 2020).

7.2.1 *Mate choice, periods of activity and sexual conflict*

Insect reproduction is usually seasonal, and restricting mating to a particular time of the year is a universal solution to avoid extreme temperatures and drought periods that do not allow mating activity. In the smaller timescale of a single day, restricting mating activities to a specific part of the day also achieves a similar goal. In some insect orders, adults will benefit from the heat of the day, and sometimes even direct exposure to sunlight, to warm up their flying muscles and allow sexual behavior that demands highly energetic

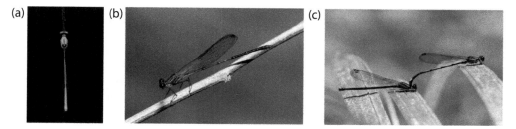

Figure 7.1 Sexual ornamentation in male damselflies from the genera (a) *Neoneura* and (b) *Hetaerina*. In (c), a couple of the genus *Argia*, where the male (on the right) has colorful sexual ornamentation that the female (left) lacks (Photos by João Gabriel Lacerda de Almeida).

courtship and territorial behaviors. In butterflies and the Odonata (dragonflies and damselflies, Figure 7.1), male courtship and copulations typically occur only during the day and are influenced by the access of adults to sunny spots (Corbet 1999; Vande Velde and Van Dyck 2013). This illustrates how environmental variables affecting the warming-up of flying muscles will influence how clustered in time mating activity will become.

The temporal and spatial clustering of females is a critical factor in modulating the intensity of sexual selection. It is reasonable to expect that when breeding seasons are short and mating activity clustered, individuals of the choosing sex should be able to assess a more extensive set of candidate mates. This creates a higher opportunity for mate choice to operate and has been linked to the incidence and geographic patterns of extra-pair paternity in birds, although not without some controversy (Spottiswoode and Møller 2004). Unlike birds, monogamy is rare in insects, where multiple mating by males and females is usually more common (Arnqvist and Nilsson 2000). There are plenty of examples of polygamous mating systems in insects, where males fight for territories and access to females (Thornhill and Alcock 1983). In such systems, the spatial and temporal clustering of female breeding activity also seems to affect the degree to which males monopolize access to females.

We are unaware of studies that directly connect the length of breeding seasons to the potential for mate choice in insects, but harvestmen (the arachnid order Opiliones) provide an interesting example. In this order, the length of breeding seasons is mainly determined by the number of warm months where species occur, and precipitation also plays a role. Interestingly, species with long breeding seasons are more likely to have mating systems based on resource defense by males. This type of mating system also increases the magnitude of sexual dimorphism in the group (Machado et al. 2016). It is quite likely that similar patterns exist in insects. If some males can monopolize access to breeding females, stopping other males from courting and mating with females in their territories, the opportunity for mate choice by females should be reduced. The potential connection with CC is that under warmer conditions, the length of insect breeding seasons is changing—the complexity of the different effects of CC in different parts of the globe means that breeding activity can become less synchronous in some places. Still, we expect more synchrony (shorter breeding seasons) in most cases. In shorter breeding seasons, polygamous mating systems should be unfavored, with a concomitant decrease in intrasexual (male–male competition) selection. The opposite is expected in areas where the breeding season is likely to be extended by CC.

Postcopulatory sexual selection in sperm competition (Parker 1970) and cryptic female choice (Eberhard 1996) may also be affected by changes in breeding seasons. However, the effects of CC on these postcopulatory processes in insects are still mostly unknown and are a fascinating avenue for future research. Another promising area for further studies is the relationship between climate and sexual conflict, which is the evolutionary arms race between males and females when the interests of the different sexes are not aligned. It is expected that environmental stresses on populations should reduce sexual conflict as they should increase the selective forces of natural selection, which affects the different sexes in a more similar way than sexual selection (Connallon 2015). As a result, the pressures and responses to selection should become more aligned between males and females, thus decreasing the conflict between the sexes. For example, in the fruit fly *Drosophila melanogaster* Meigen, 1830, an experimental evolution study manipulating the temperature and intensity of sexual selection demonstrated that sexual dimorphism decreases in populations of flies evolving under increasing temperatures, showing that environmental stress results in more sexually concordant selection (Gómez-Llano et al. 2021). Nonetheless, climate effects on sexual conflict are still very poorly studied in insects.

7.2.2 Temperature, resources and visual ornaments

Typical examples of insect mate choice are based on male ornaments, which are traits that seem to have evolved chiefly under the pressure of intersexual selection. Environmental conditions can affect the cost-benefit ratio of ornaments. Sexual signals and courtship usually significantly increase the conspicuousness of males (Figure 7.1). Therefore, predation risk should suppress the investment of males in secondary sexual traits. In fact, in the butterfly *Pieris rapae* (Linnaeus, 1758), more chromatic males are more attractive to females but also more conspicuous to avian predators (Morehouse and Rutowski 2010). If CC affects the abundance and activity of potential predators, this will indirectly affect sexual ornamentation via elevated predation risk in such a scenario, potentially to the point that increased predation could reduce and even eliminate sexual ornamentation via natural selection.

In contrast to the indirect effect of CC on sexual ornamentation mentioned above, a more direct link between these two processes can manifest in various ways. For example, in the African butterfly *Bicyclus anynana* (Butler, 1879), where there is mutual sexual ornamentation in the sexes, seasonal differences in the relative costs and benefits of mating for males and females drive a reversal of sexual roles and ornamentation. Wet season males are more ornamented and court females more often than dry season males, whereas dry season females are more ornamented and court males more often than wet season females. This seems linked to the fact that females who mate with dry season males enjoy an increase in longevity and fecundity as a result of receiving a more beneficial nuptial gift from these males (Prudic et al. 2011). In this system, seasonal differences in temperature (and presumably correlated resource availability) influence visual ornamentation, courtship behavior and mate choice between sexes. CC can impact this dynamic if temperature changes affect the duration of the seasons, especially if it leads to a shortening of the rainy period.

From a mechanistic perspective, CC can affect environmental conditions in a way that disrupts mate choice. This can occur at different stages of the mate choice process, starting from the production of traits evaluated by choosers, the transmission of information, the reception and processing of information and the resulting mate choice (Candolin 2019). The first type of disruption, on the production of an ornament, is better understood. It can be illustrated with the stalk-eyed fly *Teleopsis dalmanni* (Wiedemann 1830), where male eyestalks are extremely elongated and are under direct female choice (Wilkinson et al. 1998). By manipulating the amount of food available to developing larvae of this species, David et al. (1998) showed that eyestalks are disproportionately sensitive to condition. If CC affects environmental conditions, such that resource availability is decreased for larvae, the development of exaggerated male eyestalks would be directly affected in food-limited individuals, for example.

7.2.3 Other modalities of sexual signals affected by changing environments

The study of sexual selection and mate choice has historically been biased toward visual ornaments, which comes as no surprise due to the heavy reliance of humans on vision.

Most insects have reasonable visual acuity and the ability to detect an extensive range of the color spectrum, associated with their complex compound eyes (Chapman et al. 2013). However, some insect orders also have powerful chemical (Steiger and Stokl 2014) and vibrational senses (Cocroft and Rodríguez 2005), and courtship and mate choice operate via those senses in many groups.

In the red mason bee *Osmia bicornis* (Linnaeus, 1758), males use thorax vibrations to court females. Experimental temperature manipulation demonstrated that females preferentially mate with males that can maintain the frequency of their vibrations across various temperatures (Conrad et al. 2017). Females do not accept males who cannot consistently vibrate their thoraxes independently of environmental temperature, so this resilience to temperature effects on courtship is under sexual selection. This species occurs across continental Europe and Great Britain. CC is likely to expose males to temperatures exceeding their ability to sustain vibrational courtship in some parts of its distribution during the breeding season. Unfortunately, the effects of temperature on courtship and female choice in these bees have only been explored within a range of temperatures to which the species is already adapted. Responses beyond those temperatures are yet to be studied.

Chemical communication is vital for the mating systems of many insects, as females often exert mate choice based on male pheromones. Our understanding of how rapid environmental changes might affect sexual chemical communication in animals is still scarce. Still, a few key studies illustrate how CC might affect the synthesis of sexual pheromones and their detection by females (Groot and Zizzari 2019). In the beewolf *Philanthus triangulum* (Fabricius, 1775), the amount and composition of pheromones produced by adult males depend on the temperature experienced by them as larvae (Roeser-Mueller et al. 2010), showing that pheromone synthesis is temperature dependent. Meanwhile, the response specificity of male Oriental fruit moths *Grapholita molesta* (Busck, 1916) and pink bollworm moths *Pectinophora gossypiella* (Saunders, 1844) to pheromone levels are also affected by higher temperatures (Linn et al. 1988).

Acoustic communication is also widespread in insects (Chapman et al. 2013). Perhaps the most striking example of the role of acoustic signals in insect courtship comes from the Orthoptera (crickets, grasshoppers, katydids and relatives). In this order, sexual selection plays a central role in the coevolution of sound-producing and hearing structures (Song et al. 2020). In a classic experiment with tree crickets, Walker (1957) showed that species-specific female preferences for the pulse rate of male mating calls shifted with temperature, tracking the same temperature-related changes in the male calls. This has been termed thermal coupling, a process that might be widespread in insects with acoustic sexual communication. Thermal coupling in male song (and female preference for it) has been demonstrated in the moth *Achroia grisella* (Fabricius, 1794; Greenfield and Medlock 2007) and the katydid *Neoconocephalus triops* (Linnaeus, 1758; Beckers and Schul 2008). However, no evidence of this phenomenon was found in the fruit fly *Drosophila montana* (Stone, Griffen and Patter 1942; Ritchie et al. 2001). Interestingly, thermal coupling was demonstrated for the vibrational mating signals of males of the treehopper *Enchenopa binotata* Say (Jocson et al. 2019), extending the application of thermal coupling beyond acoustic communication.

Even though thermal coupling is not ubiquitous in insects, it could provide mate choice with some resilience to CC in some groups. Here again, however, the role of thermal coupling is only well understood within a range of temperatures to which the organisms are adapted. We still need to extend that framework to include responses beyond those temperatures. Thermal coupling has limitations, and CC might push mating systems in insects to the point where thermal coupling breaks down and mate choice ceases to work. It would be fascinating for future studies to aim at these temperature extremes in systems with thermal coupling between male signals and female preferences.

7.3 Intrasexual selection in times of climate change

Intrasexual selection occurs when there is direct competition between individuals of one sex for preferential access to potential mating partners (Andersson 1994). As for sexual selection in general, intrasexual selection may occur both before (precopulatory intrasexual selection) and after (postcopulatory intrasexual selection) copulation. Here, we will focus on precopulatory intrasexual selection (from now on, intrasexual selection), a mechanism often associated with events in which there is some type of agonistic interaction between individuals of the same sex.

Two additional processes are often integrated into intrasexual selection studies: (i) factors that determine the cost-benefit relationship of fights for access to mating partners (Ord 2021; Weir et al. 2011), and (ii) individual traits that may be favored due to their effect in reducing fighting costs or increasing winning chances (Pinto et al. 2019; Vieira and Peixoto 2013). Therefore, in a more general sense, studies on intrasexual selection may focus on the ecological and social factors that affect the cost-benefit relationship of fights, in order to understand when fighting evolves (Emlen and Oring 1977; Weir et al. 2011); the contest dynamics, in order to understand which traits increase the winning chances and how rivals make decisions during the contests (Chapin et al. 2019); and the consequences of fights on the evolution of individual traits, such as male weapons (Rico-Guevara and Hurme 2019).

Similar to what is known for intersexual selection studies, virtually no insect study has directly evaluated the potential consequences of CC on intrasexual selection processes. To our knowledge, the few studies focused on such relationships are restricted to fishes and crustaceans (Gherardi et al. 2013; Matthews and Wong 2015). For this reason, most of the potential effects of CC on intrasexual selection processes in insects discussed here are based on inferences from studies that presented some information on how climate may affect competition between individuals of the same sex. Below, we provide some key examples and discuss the potential implications of CC on male–male competition (including factors that may favor the occurrence of fights) and male weapons, which represent extreme cases of trait evolution in response to intrasexual selective pressures.

7.3.1 *Male–male fights*

In many insect species, males fight for direct access to females or to territories located in mating sites that provide a high chance of encountering females (Thornhill and Alcock 1983). There is significant variation (both within and between species) in the frequency and intensity of male–male fights (e.g., Peixoto et al. 2014). Theory aimed at understanding such variation relies on the evolution of mating systems and the economics of territorial defense (Ord 2021).

Generally, agonistic interactions between males should be favored when sexually receptive females are scarce but their location is predictable in time and space (Emlen and Oring 1977). However, the number of males competing for access to females may also affect the occurrence of fights. If the number of competing males is very high, the costs of fighting may surpass the benefits of mating, and agonistic behaviors should be abandoned (Weir et al. 2011). Based on this rationale, one of the most critical connections between CC and male–male fights is through the effect of temperature on population dynamics (Kokko and Rankin 2006). For example, seasonal variation in temperature and humidity may determine when territorial mating systems are favored. This is expected because temperature and humidity determine the periods suitable for growth and reproduction, and therefore the synchrony in which males and females become sexually receptive (Macías-Ordóñez et al. 2013). If the period suitable for reproduction is short and unpredictable, the ratio of sexually receptive males to females should approach equality, while the adult population size will tend to increase due to synchronous emergence (Peixoto and Mendoza-Cuenca 2013). Such conditions often hinder the evolution of territorial mating systems and agonistic behavior in males (Emlen and Oring 1977; Weir et al. 2011). Since seasonal variation is expected to become more extreme and unpredictable with climate warming (AghaKouchak et al. 2020), agonistic behaviors may become rarer in species that do not migrate to higher latitudes.

Another critical point in male–male fights is the dynamics of cost accrual during fights. When fighting, males pay costs related to energy depletion and injury accumulation (Ord 2021). The rate of energy depletion, in particular, may be affected by the behaviors adopted during the fights and the environmental conditions in which the fight takes place (e.g., deCarvalho et al. 2004; Vande Velde and Van Dyck 2013). For ectothermic species, such as most insects, high temperatures often increase energy consumption (Huey and Kingsolver 2019). Therefore, the increase in mean temperature due to global warming may elevate the costs paid by males during agonistic interactions. However, it will be essential to understand whether such changes in the costs of fights really occur and the consequences of potential changes in cost accrual during fights to fight dynamics. Changes in fight dynamics, in turn, may have significant implications for the reproductive patterns in a population. For instance, fights between males may select individuals in better condition as they concentrate most matings (e.g., Tsubaki et al. 2010). However, if an increase in the energetic costs of fights affects the chances of winning, this may increase the opportunity for males to adopt alternative non-aggressive tactics to increase mating success and change the evolutionary route of a population (e.g., López-Sepulcre and Kokko 2005; see also section 7.4). Nevertheless, despite the many potential consequences of climate warming for agonistic interactions, initial studies on how temperature variation may affect the adoption of aggressive behaviors by males and the dynamics of contests are still extremely needed.

7.3.2 *Male weapons*

In many species, males have morphological structures collectively called weapons (Figure 7.2), which are primarily used in aggressive interactions to gain an advantage against other males in competition for female access (McCullough et al. 2016, Eberhard et al. 2018). In general, the size of animal weapons is hyper-allometric in relation to body size, representing a significant portion of individual total length in some cases (e.g., McCullough et al. 2015). This pattern indicates that weapon development and maintenance are very costly for males (Somjee et al. 2018). In fact, the expression of weapons strongly depends on male nutritional condition, which in insects is mainly affected by the amount of energy accumulated during development (Johns et al. 2014). Therefore, changes in the environmental conditions in which individuals develop may profoundly

(a)

(b)

(c)

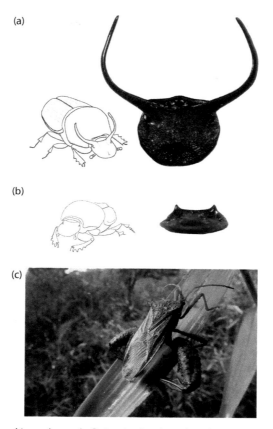

Figure 7.2 Weapons used in male–male fights. In the dung beetle *Onthophagus taurus* (Schreber, 1759), whereas fighter males (a) bear long horns used in fights, smaller males (b) are sneakers (an alternative reproductive tactic) and only develop vestigial horns (drawings by Rachel Werneck). In leaf-footed bugs (family Coreidae), the hind legs are often modified into strong and exaggerated weapons, which in (c) *Acanthocephala femorata* (Fabricius, 1775) males can represent 27% of the mass of this large species (photo by Ummat Somjee).

impact the expression of male weapons. For example, in the horned beetle *Allomyrina dichotoma* (Linnaeus, 1771), male larvae that had low food availability during development become adults with proportionally smaller horns than males raised in places with high food availability (Karino et al. 2004). Since male weapons are important determinants of the winning chances and mating success of males (Kelly 2008), changes in their expression due to environmental changes should also have significant consequences at the population level.

CC is predicted to have marked consequences in every habitat globally (Mantyka-Pringle et al. 2012). Such changes may affect the expression of animal weapons in two different ways. The first is through the direct effect of temperature on insect development. Due to increases in mean temperature, global warming may affect metabolic rates that cascade to individual growth (Huey and Kingsolver 2019). For example, a long-term study on the yellow dung fly *Scathophaga stercoraria* (Linnaeus, 1758) indicated that body size decreased with increasing temperature in response to global warming (Blanckenhorn 2015). Since weapon size is often hyper-allometrically related to body size, weapons in insects will likely become relatively smaller over the years if the increasing temperature trend is not halted. A strong reduction in weapon size relative to body size may also affect the dynamics of male–male interactions, increasing the chances that males with smaller weapons win contests (against rivals with greater weapons) for access to females (Palaoro and Peixoto 2022). This would disrupt a vital filter that allows strong males to concentrate most mating in the population. In fact, not only weapons but also other critical reproductive traits, such as male fertility and sperm competitiveness, may be compromised when temperatures increase (Sales et al. 2018).

The second effect of CC on animal weapons is indirect, through CC effects on habitat conditions. For example, CC may reduce local resource availability, impacting food intake (Huey and Kingsolver 2019). If juvenile forms of insects have reduced food consumption, the expression of weapons is compromised, leading to the predominant emergence of adults without weapons or with vestigial forms of such traits (Warren et al. 2013; Figure 7.2b). The mating dynamics in these populations may be changed entirely, since weaponless individuals tend to adopt alternative mating strategies (often determined by furtive tactics) and may be less preferred by females (Oliveira et al. 2008). Therefore, CC may have significant consequences for the expression of animal weapons, affecting the predominant mating system of species and impacting population dynamics. Nonetheless, since the effects of global warming on weapon expression may occur in many different ways, studies aimed at identifying such relationships are extremely necessary.

7.4 Alternative reproductive tactics

Insects are the most diverse group of multicellular organisms, and their evolutionary success and diversification are partially due to the remarkable phenotypic plasticity of their adaptive responses to environmental change (Simpson et al. 2011). Plasticity is also ubiquitous in insects' reproductive strategies and mating systems. This is especially evident in species where individuals may obtain fertilizations in two or more different ways, a phenomenon called alternative reproductive tactics (ARTs; Oliveira et al. 2008). ARTs are widespread across at least eleven insect orders (Buzatto et al. 2014). In many cases, the tactics are associated with the evolution of intrasexual polymorphisms, where there is more than one typical morphology (called "*morph*") within the same sex.

7.4.1 *Genetic polymorphisms versus phenotypic plasticity*

From an evolutionary perspective, the existence of two or more behaviorally and morphologically different morphs within the same population raises the question of how they can coexist without the morph that returns the highest fitness on average completely replacing the other one. In game theory (Maynard Smith 1982), three different models explain this conundrum: "*alternative strategies*", "*mixed strategy*", and "*conditional strategy*" (Gross 1996). An alternative strategy is characterized by the occurrence in a population of a stable mixture of different genetically based decision rules, which are completely insensitive to the environment. Meanwhile, a conditional strategy is a decision rule containing a conditional clause, so individuals employing such a strategy will express a different tactic (such as defending a territory or looking for unguarded females) depending on environmental or social conditions. Finally, the mixed strategy is a decision rule with a probabilistic basis, where an individual expresses each tactic with a fixed probability, regardless of any external influence. Despite the theoretical validity of mixed strategies, they do not seem common in nature, probably because any adjustment of the phenotypic expression to an individual's condition or external environment will have greater adaptive value than a probabilistic decision rule that is insensitive to the circumstances. Meanwhile, alternative (genetically fixed) strategies and conditional (plastic) strategies abound in nature, the latter being especially common in insects (Buzatto et al. 2014).

Alternative strategies containing two or more genetically fixed phenotypes have been recorded at least in the Diptera, Odonata and Orthoptera. The phenotypes that arise from these strategies are insensitive to the social and physical environments (Hazel et al. 2004), have a Mendelian pattern of inheritance and are maintained by frequency-dependent selection (Maynard Smith 1982; Gross 1996). In the damselfly *Mnais costalis* Selys, 1869, males mature as orange-winged territorial fighters or clear-winged non-territorial sneakers. Male morph expression is genetically controlled by a single-locus, two-allele autosomal gene (Tsubaki 2003). The thoracic temperature of territorial males in this species, which is vital for flying performance, varies with the insolation of their territories. In contrast, non-territorial males have less variable and higher thoracic temperatures. Accordingly, the proportion of orange-winged males seems higher in populations with more sunny spots (Samejima and Tsubaki 2010). It is reasonable to expect that if CC influences the patterns of insolation of populations of *M. costalis*, or significantly disturbs the air temperature where males occur, the proportions of territorial and nonterritorial males should respond to these changes. Given the genetic basis for this polymorphism, one of the morphs could go locally extinct, but the ecological consequences of losing one of the morphs are not well understood.

Meanwhile, more than 95% of male ARTs in insects are conditional strategies (Tomkins and Hazel 2007), where social or environmental cues will determine which reproductive tactic is expressed. The body size and condition of males, usually influenced by their diet, are the most common cues determining the expression of ARTs in the group, followed by population density and seasonality. However, these are only a few of the at least fourteen different types of cues that trigger the development of distinct morphs identified in insects with ARTs (Buzatto et al. 2014). In most insect orders, ARTs are dynamic and reversible, such as in the case of the butterfly *Lycaena hippothoe* (Linnaeus, 1761), where males can perch and defend a territory or conduct patrol flights looking for females, and the fluctuating weather conditions influence which tactic males adopt each time (Fischer and Fiedler 2001). These ARTs might respond quickly to extreme and unpredictable weather, so high reversibility would make such systems more resilient to CC. However, in the Coleoptera and Hymenoptera, conditional strategies are more commonly associated with irreversible adult phenotypes that are strongly dimorphic. Dung beetles provide many examples of striking male dimorphism (Figure 7.2a and b), usually expressed in the shape and size of thoracic and head horns (Emlen et al. 2007). Despite the irreversibility and morphological disparity between male phenotypes, the diet of dung beetle larvae is the main determinant of adult male morphology, and therefore mating tactic. Such systems with irreversible male dimorphism are probably more susceptible to the loss of one of the morphs under severe climatic change. In the case of dung beetles, the availability of dung from large mammals is potentially the main way in which CC might affect the persistence of male dimorphisms, given the vulnerability of mammals to habitat loss in future climatic scenarios.

7.4.2 *Alternative phenotypes and resilience to change*

The oceanic field cricket *Teleogryllus oceanicus* (Le Guillou, 1841) is native to Australia and has colonized the Hawaiian Islands, where the North American parasitoid fly *Ormia ochracea* (Bigot, 1889) also currently occurs. The typical call of male crickets attracts the parasitoid, which poses a deadly cost to male sexual signaling (Zuk et al. 1998). Interestingly, some populations in Hawaii contain a single locus mutation that affects male wings and makes calling impossible (Tinghitella et al. 2009; Zuk et al. 2006). Silent males use a satellite mating tactic that existed in the population before the male wing mutation appeared (Tinghitella et al. 2009) and can achieve matings while completely avoiding the parasitism risk from *O. ochracea*. This system illustrates how the existence of a plastic conditional tactic of mating without calling (satellite) facilitated the evolution of an alternative strategy associated with a genetically fixed male morphological dimorphism. Future CC can uniquely affect this system via the interaction with the parasitic fly. If future climate scenarios affect the abundance of that species positively or negatively, this will certainly have an influence on the proportions of silent and calling males of *T. oceanicus* in Hawaii. Here again, whether future changes will drive one of the morphs to extinction remains to be seen in this fascinating and dynamic system.

So far we have primarily focused on the risk of CC driving one of the male mating tactics extinct, which usually seems more likely in genetic alternative strategies than plastic conditional ones. However, it is also quite possible that alternative mating tactics in a system increase its resilience to CC in the first place. From a conservation perspective, the worst outcome of the impact of CC on sexual selection would be the risk of a population having its sexual reproduction completely obliterated due to extreme climatic conditions. We are unaware of any studies showing this but it seems to be a genuine concern. If mate choice is strong and females will not accept mating without the display of typical courtship by the male, and if new environmental conditions hinder the males' ability to perform such courtship, could sexual reproduction completely cease in a population? We believe that whenever (and if) this pessimistic scenario takes place, the existence of more than one way of achieving matings means that the complete cessation of sexual reproduction is less likely in the system. Even if alternative mating tactics are not yet present, the inability of males to perform the typical mating tactic in a new environmental condition would create the strongest selective pressure for novel alternative mating tactics to evolve. Is this the type of situation the oceanic field cricket faced in Hawaii—a novel environment where performing the typical courtship of calling was close to a death sentence? That pressure seems to be associated with more than one case of new mutations that affected male courtship and spread throughout the population (Tinghitella et al. 2021). Has this saved sexual reproduction and perhaps the persistence of *T. oceanicus* in Hawaii?

7.5 Climate change, sexual dimorphism and species recognition

As mentioned previously, temperature has a strong effect on the development of insects. Depending on the temperatures faced during their life cycles, insects may have varying developmental rates, affecting their adult sizes (Jaramillo et al. 2009). Such effects then carry over to the reproductive success of insects in their environments. The impact of CC on the physiological regulation of plasticity in the body sizes of males and females can ultimately affect the morphological differences between the sexes (sexual dimorphism; Stillwell et al. 2010). For example, Baroni and Masoero (2018) studied the alpine grasshopper *Stenobothrus ursulae* Nadig, 1986 (Coleoptera: Stenobothrini), using a species distribution model to predict the species range in high altitudes and sampling individuals in both high and low altitudes, to evaluate temperature effects on this species. They observed that the same environmental predictors describing the distribution of the species might also affect their morphology. Specifically, high altitudes (low temperatures), which represent better habitat suitability for the species, are also associated with declines in the sizes of individuals in general and an additional decrease in sexual dimorphism. With CC, temperatures are expected to increase dramatically in mountainous areas, and this relationship between habitat suitability and sexual dimorphism may also change significantly. The distribution and morphology of males and females in this species may change while adapting to a new climate. Still, this process may represent a severe concern for alpine species, such as *S. ursulae*, which generally have limited dispersal abilities.

Differences in the body sizes of males and females are perhaps the more straightforward form of sexual dimorphism. However, more complex dimorphisms are common in specific structures used differently between sexes. In ambrosia beetles, *Xylebrous afinis* Eichhoff, 1868 (Curculionidae: Scolytinae), where the sex ratio is heavily skewed toward females (Ospina-Garces et al. 2021), females are wood borers and cultivate ambrosial fungi, which is thought to have resulted in sexual dimorphism in mandibular features throughout the evolutionary history of the species. Females are also the only sex capable of flying and colonizing new areas, where they will mate with parthenogenetically produced males who do not participate in developing their galleries. Therefore, males are reduced in size, lack wings and have smaller mandibles. A recent study evaluated the effects of temperature rise on ontogeny and sexual dimorphism in this species. After establishing different experimental temperature treatments (17°C, 23°C, 26°C and 29°C), Ospina-Garcés et al. (2021) found that the sex differences in shape and size of mandibles were smallest at the highest temperature, with male and female mandibula becoming more similar at that temperature, and consequently decreasing overall sexual dimorphism. This illustrates another apparent effect of changing temperatures on sexual dimorphism, but the implications of such an effect are still poorly understood.

The effects of climate on sexual dimorphism can also have implications for species recognition through environmental impacts on the secondary sexual traits of one of the sexes. In the examples mentioned above, natural selection may be the driver of sexual dimorphism via ecological causes, primarily when the niche is partitioned between the sexes, and males and females specialize in different foraging methods or explore different resources (Shine 1989). However, sexual selection seems to be a more robust force in shaping sexual dimorphism, as males typically compete for mating opportunities (intrasexual competition). In contrast, females often select males in the best condition, or that can best advertise their quality as mates (intersexual selection). These processes make up sexual selection as we know it and are the primary drivers of morphological differences between males and females. Generally, only males bear the costly secondary sexual traits that evolve under sexual selection, including ornaments (Figure 7.1), weapons (Figure 7.2) or both. Given the previously documented and predicted effects of CC on such traits, both explored earlier in this chapter, it is unsurprising that sexual dimorphism and the processes of species recognition between the sexes will also subsequently be affected.

So far we have focused on sexual dimorphisms in the body sizes of males and females or the relative sizes of body parts across the sexes. If we move our focus to secondary sexual traits, another meaningful connection between CC and sexual dimorphism involves coloration via melanization in insects. Melanization processes may function quite differently between males and females. For instance, in odonates, sexual dimorphism frequently results from males showing more prominent and intense color patches than females (Figure 7.1). In ten different dragonflies with wing melanization that differs between the sexes, Moore et al. (2021) showed that more wing melanin means higher body temperatures. As a result, species consistently adapted to warm climates by evolving less male melanin ornamentation. Interestingly, female selection for more melanized males counters natural selection for less melanin due to rising temperatures in this system. More extreme temperatures may further decrease male melanization, causing males and females to be more morphologically similar and potentially affecting female–male recognition and mate preferences in future climate scenarios (Moore et al. 2021). According to the study's authors, wing melanization is expected to decline by 2070, and sexual selection may be severely affected by CC in odonates, with sexual dimorphism reducing in the future. This might affect mate choice and mating system dynamics in these species. Still, the degree to which it would affect species recognition is hard to predict, as it also depends on the future degree of sympatry of morphologically similar species.

There is a high risk that CC will significantly affect species recognition patterns in some cases. Temperature may have an entirely different effect on insect species with populations at lower and higher altitudes or species with broad pole-equator ranges (Larson et al. 2019). Gene flow between populations may be affected, resulting in local evolutionary processes, individual recognition difficulties and changes in species distribution boundaries. In such populations, conspecifics from different lineages may rapidly change in response to CC and exhibit different phenological features, even in just a few generations. Temperature changes may also affect species developmental rates throughout the species range, causing phenological mismatches between populations of a single species and increasing the possibility of hybridization between previously stable and isolated species (Larson et al. 2019).

CC may even result in new speciation events. For instance, Moore et al. (2019) observed that the wing coloration of males of the dragonfly *Pachydiplax longipennis* (Burmeister, 1839) was dramatically reduced in the hottest portions of the species' range. In colder areas, males with darker wing coloration have their flight ability improved, as they heat up faster than males with less pigmented wings. Meanwhile, males bearing light-colored wings have greater reproductive output than their dark-wing counterparts in hotter portions of the species' range. Since there is a biogeographic pattern related to wing coloration, where pigmented species tend to occur more frequently in temperate regions of the northern hemisphere (Svensson and Waller 2013), species from these regions may be more affected by CC.

It is important to note that many insect species rely on sounds for male–female recognition. Temperature changes may affect how these sounds are emitted and received by males and females, consequently affecting sexual selection (Larson et al. 2019). For instance, sound frequency and pulse/interpulse lengths are temperature dependent. In warmer environments, these features may differ from colder scenarios. Although male signaling may change due to temperature effects, female choosiness is usually temperature independent. Under potential future CC scenarios, mismatches between male signaling and female response may increase, resulting in poorer reproductive outputs (Larson et al. 2019). Greenfield and Medlock (2007) evaluated the effects of temperature on the emission of sounds by males and how females responded to such changes. They observed that while male song rates increased with rising temperatures, the female acceptance threshold of males also changed (thermal coupling, as discussed in section 7.2.3) but was much less variable, which could influence species recognition between the sexes.

Finally, we highlight that the future effects of CC in many of the examples we covered here are speculative, and real-world biota may function differently under the predicted CC scenarios. The responses of insects to CC will not necessarily be linear either and even closely related species may show distinctive responses to future stresses. Despite such issues, it is crucial to bear potential processes in mind to consider possible practical conservation efforts to maintain species viability under different CC scenarios. It is also critical to keep thinking of any potential effects of CC on hybridization and speciation to design reactive and proactive actions that could decrease the impact of CC on biodiversity (Sánchez-Guillén et al. 2016).

7.6 Sexual selection and adaptation to change

The effects of sexual selection on population viability can be complex (Candolin and Heuschele 2008), and theory suggests that adverse effects in small populations can become positive in larger ones (Martínez-Ruiz and Knell 2017). On the one hand, sexual selection can decrease the fitness of populations in the short term via sexual conflict. In the broad-horned flour beetle *Gnatocerus cornutus* (Fabricius, 1798), artificial selection for larger mandible horns was shown to lower female fecundity via sexual conflict (Harano et al. 2010). However, at the same time, it is becoming clear that sexual selection can also act as a mechanism that ultimately removes deleterious mutations from populations and expedites their ability to adapt to new environments. Stronger sexual selection pressure in increased competition between males is linked to purging the gene pool of a population of deleterious mutations (Jarzebowska and Radwan 2010; Radwan 2004). This has been shown to protect populations of the flour beetle *Tribolium castaneum* (Herbst, 1797) against extinction (Lumley et al. 2015). Moreover, using an experimental evolution approach, Gómez-Llano et al. (2021) showed that fruit flies *Drosophila melanogaster* presented more thermal adaptation in populations under precopulatory sexual selection than in populations without it. Sexual reproduction is an adaptive response to unpredictable environmental changes in aphids. These animals are famously able to switch between sexual and asexual reproduction (Simon et al. 2010). Environmental stress or instability usually triggers the production of males, which consequently allows sexual selection to start operating.

However, the amount of stress that CC can generate might push populations into a stress zone where natural selection becomes a much stronger force, potentially to the point where sexual selection becomes nearly irrelevant. Anthropogenic CC may suppress the mechanisms of sexual selection in certain conditions, removing the population benefits from these processes. Whether sexual selection can serve as a mechanism that will help a population adapt to the new environmental conditions brought by CC or whether the stress will be strong enough to undermine any potential population benefit from the forces of sexual selection will probably depend on a multitude of factors in each system (Candolin and Heuschele 2008). Nonetheless, compiling a framework (Figure 7.3) that allows us to predict the possible outcomes of this battle between sexual selection and CC will be incredibly important and a powerful tool for informing conservation actions and policies. For this to happen, it seems urgent to gather empirical data on species-specific responses to extreme variation in environmental conditions, especially beyond the range of variation that these populations have historically faced.

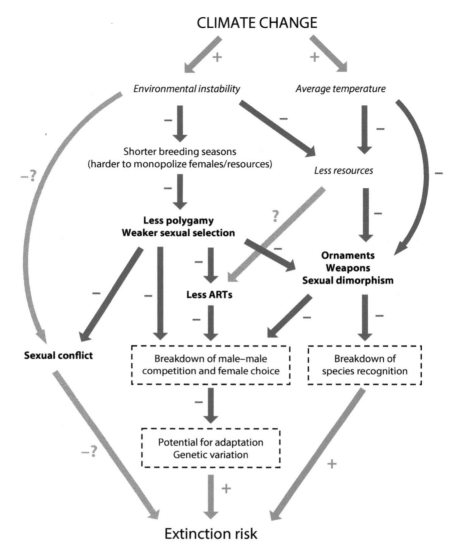

Figure 7.3 Framework for potential effects of climate change (CC) on environmental variables (italics) that secondarily affect sexual selection processes and related traits (bold). "+" represent positive effects (in red as they represent increase in temperature and extinction risk, so "dangerous" effects), whereas "–" represent negative effects (in blue). CC is expected to increase environmental instability and average annual temperatures (in many areas). Environmental instability may shorten breeding seasons (around most parts of the globe), which might then reduce polygamy and the intensity of sexual selection. Increased average temperature may negatively affect sexually selected traits directly or indirectly (by reducing resource availability). Consequences of such effects can disrupt male–male competition and female choice, reducing genetic variation, the potential for adaptation and mechanisms of species recognition, ultimately increasing extinction risk. Alternative reproductive tactics (ARTs) should boost the potential for adaptation, and whether limited resources increase or decrease the expression of ARTs might depend on the system. Weaker sexual selection and less polygamy, however, should decrease ARTs. Finally, environmental instability may also decrease sexual conflict by increasing the selective forces of natural selection (which affects the sexes more similarly than sexual selection), but there is less evidence for this pathway currently).

Key reflections

- Abiotic factors play a role in the evolution of sexually selected traits.
- Climate change can affect the evolution of ornaments used in mate choice and weapons used in male fights.
- Alternative mating tactics and intrasexual dimorphisms are highly influenced by the environment and, therefore, are also subject to the effects of climate change.
- Climate change can severely affect the differences between the sexes (sexual dimorphism) and potentially species-recognition mechanisms.
- The strength of sexual selection itself is influenced by the climate and can affect how a species adapts to future changes.
- We urge future studies to gather empirical data on species responses to extreme variation in environmental conditions, especially beyond the range of variation that populations historically faced

Key further reading

- Baur, J., Jagusch, D., Michalak, P., Koppik, M., and Berger, D. 2022. The mating system affects the temperature sensitivity of male and female fertility. *Functional Ecology* 36(1): 92–106. https://doi.org/10.1111/1365-2435.13952
- Clusella-Trullas, S., and Nielsen, M. 2020. The evolution of insect body coloration under changing climates. *Current Opinion in Insect Science* 41: 25–32. https://doi.org/10.1016/j.cois.2020.05.007
- Dudaniec, R.Y., Carey, A.R., Svensson, E.I., Hansson, B., Yong, C.J., and Lancaster, L.T. 2022. Latitudinal clines in sexual selection, sexual size dimorphism and sex-specific genetic dispersal during a poleward range expansion. *Journal of Animal Ecology* 91(6): 1104–1118.
- Gautam, S., and Kunte, K. 2020. Adaptive plasticity in wing melanisation of a montane butterfly across a Himalayan elevational gradient. *Ecological Entomology* 45(6): 1272–1283.
- Golab, M.J., Johansson, F., and Sniegula, S. 2019. Let's mate here and now—seasonal constraints increase mating efficiency. *Ecological Entomology* 44(5): 623–629.
- Lis, C., Moore, M.P., and Martin, R.A. 2020. Warm developmental temperatures induce non-adaptive plasticity in the intrasexually selected colouration of a dragonfly. *Ecological Entomology* 45(3): 663–70.
- Parrett, J.M., and Knell, R.J. 2018. The effect of sexual selection on adaptation and extinction under increasing temperatures. *Proceedings of the Royal Society of London B-Biological Sciences* 285: 20180303.
- Pato, J., Illera, J.C., Obeso, J.R., and Laiolo, P. 2019. The roles of geography, climate and sexual selection in driving divergence among insect populations on mountaintops. *Journal of Biogeography* 46(4): 784–95.
- Svensson, E.I., Gomez-Llano, M., and Waller, J.T. 2020. Selection on phenotypic plasticity favors thermal canalization. *Proceedings of the National Academy of Sciences* 117(47): 29767–29774.
- Wong, B.B.M., and Candolin, U. 2015. Behavioral responses to changing Environments. *Behavioral Ecology* 26(3): 665–673.

References

AghaKouchak, A., Chiang, F., Huning, L.S., Love, C.A., Mallakpour, I., Mazdiyasni, O., Moftakhari, H., et al. 2020. Climate extremes and compound hazards in a warming world. *Annual Review of Earth and Planetary Sciences* 48: 519–48.

Andersson, M. 1994. *Sexual Selection* (Vol. 72). Princeton: Princeton University Press.

Andersson, M. and Simmons, L.W. 2006. Sexual selection and mate choice. *Trends in Ecology & Evolution* 21 (6): 296–302.

Arnqvist, G. and Nilsson, T. 2000. The evolution of polyandry: Multiple mating and female fitness in insects. *Animal Behaviour* 60 (2): 145–64.

Baroni, D. and Masoero, G. 2018. Complex influence of climate on the distribution and body size of an Alpine species. *Insect Conservation and Diversity* 11 (5): 435–48.

Beckers, O.M. and Schul, J. 2008. Developmental plasticity of mating calls enables acoustic communication in diverse environments. *Proceedings of the Royal Society B: Biological Sciences* 275 (1640): 1243–1248.

Blanckenhorn, W.U. 2015. Investigating yellow dung fly body size evolution in the field: Response to climate change? *Evolution* 69 (8): 2227–2234.

Bonduriansky, R. 2001. The evolution of male mate choice in insects: A synthesis of ideas and evidence. *Biological Reviews* 76 (3): 305–39.

Buzatto, B.A., Tomkins, J.L., and Simmons, L.W. 2014. "Alternative phenotypes within mating systems." In *The Evolution of Insect Mating Systems*, edited by D. M. Shuker and L. W. Simmons. Oxford University Press, 106–28.

Candolin, U. 2019. Mate choice in a changing world. *Biological Reviews of the Cambridge Philosophical Society* 94: 1246–1260.

Candolin, U. and Heuschele, J. 2008. Is sexual selection beneficial during adaptation to environmental change? *Trends in Ecology & Evolution* 23 (8): 446–52.

Capinera, J.L. (ed). 2008. *Encyclopedia of Entomology*. Heidelberg: Springer Science & Business Media.

Chapin, K.J., Peixoto, P.E.C., and Briffa, M. 2019. Further mismeasures of animal contests: A new framework for assessment strategies. *Behavioral Ecology* 30 (5): 1177–1185.

Chapman, R.F., Simpson, S.J., and Douglas, A.E. 2013. *The Insects: Structure and Function*. London: Hodder and Stoughton.

Cocroft, R.B. and Rodríguez, R.L. 2005. The behavioral ecology of insect vibrational communication. *Bioscience* 55 (4): 323–34.

Connallon, T. 2015. The geography of sex-specific selection, local adaptation, and sexual dimorphism. *Evolution* 69 (9): 2333–2344.

Conrad, T., Stöcker, C., and Ayasse, M. 2017. The effect of temperature on male mating signals and female choice in the red mason bee, *Osmia bicornis* (L.). *Ecology and Evolution* 7 (21): 8966–8975.

Corbet, P.S. 1999. *Dragonflies: Behaviour and Ecology of Odonata*. London: Harley Books.

Darwin, C. 1871. *The Descent of Man and Selection in Relation to Sex*. London: John Murray.

David, P., Hingle, A., Greig, D., Rutherford, A., Pomiankowski, A., and Fowler, K. 1998. Male sexual ornament size but not asymmetry reflects condition in stalk–eyed flies. *Proceedings of the Royal Society of London. Series B: Biological Sciences* 265 (1411): 2211–2216.

DeCarvalho, T.N., Watson, P.J., and Field, S.A. 2004. Costs increase as ritualized fighting progresses within and between phases in the sierra dome spider, *Neriene litigiosa*. *Animal Behaviour* 68 (3): 473–82.

Eberhard, W. 1996. *Female Control: Sexual Selection by Cryptic Female Choice* (Vol. 17). Princeton: Princeton University Press.

Eberhard, W.G., Rodríguez, R.L., Huber, B.A., Speck, B., Miller, H., Buzatto, B.A., and Machado, G. 2018. Sexual selection and static allometry: The importance of function. *The Quarterly Review of Biology* 93 (3): 207–50.

Emlen, D.J., Corley Lavine, L., and Ewen-Campen, B. 2007. On the origin and evolutionary diversification of beetle horns. *Proceedings of the National Academy of Sciences* 104 (suppl 1): 8661–8668.

Emlen, S.T. and Oring, L.W. 1977. Ecology, sexual selection, and the evolution of mating systems. *Science* 197 (4300): 215–23.

Fischer, K. and Fiedler, K. 2001. Resource-based territoriality in the butterfly *Lycaena hippothoe* and environmentally induced behavioural shifts. *Animal Behaviour* 61 (4): 723–32.

García-Roa, R., Garcia-Gonzalez, F., Noble, D.W., and Carazo, P. 2020. Temperature as a modulator of sexual selection. *Biological Reviews* 95 (6): 1607–1629.

Gherardi, F., Coignet, A., Souty-Grosset, C., Spigoli, D., and Aquiloni, L. 2013. Climate warming and the agonistic behaviour of invasive crayfishes in Europe. *Freshwater Biology* 58 (9): 1958–1967.

Gómez-Llano, M., Scott, E., and Svensson, E.I. 2021. The importance of pre-and postcopulatory sexual selection promoting adaptation to increasing temperatures. *Current Zoology* 67 (3): 321–27.

Greenfield, M.D. and Medlock, C. 2007. Temperature coupling as an emergent property: Parallel thermal effects on male song and female response do not contribute to species recognition in an acoustic moth. *Evolution* 61 (7): 1590–1599.

Groot, A.T. and Zizzari, Z.V. 2019. Does climate warming influence sexual chemical signaling?. *Animal Biology* 69 (1): 83–93.

Gross, M.R. 1996. Alternative reproductive strategies and tactics: Diversity within sexes. *Trends in Ecology & Evolution* 11 (2): 92–98.

Harano, T., Okada, K., Nakayama, S., Miyatake, T., and Hosken, D.J. 2010. Intralocus sexual conflict unresolved by sex-limited trait expression. *Current Biology* 20 (22): 2036–2039.

Hazel, W., Smock, R., and Lively, C.M. 2004. The ecological genetics of conditional strategies. *The American Naturalist* 163 (6): 888–900.

Huey, R.B. and Kingsolver, J.G. 2019. Climate warming, resource availability, and the metabolic meltdown of ectotherms. *The American Naturalist* 194 (6): E140–E150.

Jaramillo, J., Chabi-Olaye, A., Kamonjo, C., Jaramillo, A., Vega, F.E., Poehling, H.M., and Borgemeister, C. 2009. Thermal tolerance of the coffee berry borer *Hypothenemus hampei*: Predictions of climate change impact on a tropical insect pest. *PloS One* 4 (8): e6487.

Jarzebowska, M. and Radwan, J. 2010. Sexual selection counteracts extinction of small populations of the bulb mites. *Evolution* 64 (5): 1283–1289.

Jocson, D.M.I., Smeester, M.E., Leith, N.T., Macchiano, A., and Fowler-Finn, K.D. 2019. Temperature coupling of mate attraction signals and female mate preferences in four populations of *Enchenopa treehopper* (Hemiptera: Membracidae). *Journal of Evolutionary Biology* 32 (10): 1046–1056.

Johns, A., Gotoh, H., McCullough, E.L., Emlen, D.J. and Lavine, L.C. 2014. Heightened condition-dependent growth of sexually selected weapons in the rhinoceros beetle, *Trypoxylus dichotomus* (Coleoptera: Scarabaeidae). *Integrative and Comparative Biology* 54: 614–21.

Karino, K., Seki, N., and Chiba, M. 2004. Larval nutritional environment determines adult size in Japanese horned beetles *Allomyrina dichotoma*. *Ecological Research* 19: 663–68.

Kelly, C.D. 2008. The interrelationships between resource-holding potential, resource-value and reproductive success in territorial males: How much variation can we explain?. *Behavioral Ecology and Sociobiology* 62 (6): 855–71.

Kokko, H. and Rankin, D.J. 2006. Lonely hearts or sex in the city? Density-dependent effects in mating systems. *Philosophical Transactions of the Royal Society B: Biological Sciences* 361 (1466): 319–34.

Larson, E.L., Tinghitella, R.M., and Taylor, S.A. 2019. Insect hybridization and climate change. *Frontiers in Ecology and Evolution* 7: 348.

Linn, C.E., Campbell, M.G., and Roelofs, W.L. 1988. Temperature modulation of behavioural thresholds controlling male moth sex pheromone response specificity. *Physiological Entomology* 13 (1): 59–67.

López-Sepulcre, A. and Kokko, H. 2005. Territorial defense, territory size, and population regulation. *The American Naturalist* 166 (3): 317–29.

Lumley, A.J., Michalczyk, Ł., Kitson, J.J., Spurgin, L.G., Morrison, C.A., Godwin, J.L., Dickinson, M.E., et al. 2015. Sexual selection protects against extinction. *Nature* 522 (7557): 470–73.

Machado, G., Buzatto, B.A., García-Hernández, S., and Macías-Ordóñez, R. 2016. Macroecology of sexual selection: A predictive conceptual framework for large-scale variation in reproductive traits. *The American Naturalist* 188 (S1): S8–S27.

Macías-Ordóñez, R., Machado, G., and Macedo, R. H. (2013). "Macroecology of sexual selection: Large-scale influence of climate on sexually selected traits." In *Sexual Selection: Perspectives and Models from the Neotropics*, edited by R.H. Macedo and G. Machado. London: Academic Press, 1–32.

Mantyka-pringle, C.S., Martin, T.G., and Rhodes, J.R. 2012. Interactions between climate and habitat loss effects on biodiversity: A systematic review and meta-analysis. *Global Change Biology* 18 (4): 1239–1252.

Martínez-Ruiz, C. and Knell, R.J. 2017. Sexual selection can both increase and decrease extinction probability: Reconciling demographic and evolutionary factors. *Journal of Animal Ecology* 86 (1): 117–27.

Matthews, S.A. and Wong, M.Y. 2015. Temperature-dependent resolution of conflict over rank within a size-based dominance hierarchy. *Behavioral Ecology* 26 (3): 947–58.

Maynard Smith, J. 1982. *Evolution and the Theory of Games.* Cambridge: Cambridge University Press.

McCullough, E.L., Ledger, K.J., O'Brien, D.M., and Emlen, D.J. 2015. Variation in the allometry of exaggerated rhinoceros beetle horns. *Animal Behaviour* 109: 133–40.

McCullough, E.L., Miller, C.W., and Emlen, D.J. 2016. Why sexually selected weapons are not ornaments. *Trends in Ecology & Evolution* 31 (10): 742–51.

Moore, M.P., Hersch, K., Sricharoen, C., Lee, S., Reice, C., Rice, P., Kronick, S., et al. 2021. Sex-specific ornament evolution is a consistent feature of climatic adaptation across space and time in dragonflies. *Proceedings of the National Academy of Sciences* 118 (28): e2101458118.

Moore, M.P., Lis, C., Gherghel, I., and Martin, R.A. 2019. Temperature shapes the costs, benefits and geographic diversification of sexual coloration in a dragonfly. *Ecology letters* 22 (3): 437–46.

Morehouse, N.I. and Rutowski, R.L. 2010. In the eyes of the beholders: Female choice and avian predation risk associated with an exaggerated male butterfly color. *The American Naturalist* 176 (6): 768–84.

Oliveira, R.F., Taborsky, M., and Brockmann, H.J. (eds). 2008. *Alternative reproductive tactics: An integrative approach.* Cambridge: Cambridge University Press.

Ord, T.J. 2021. Costs of territoriality: A review of hypotheses, meta-analysis, and field study. *Oecologia* 197 (3): 615–31.

Ospina-Garces, S.M., Ibarra-Juarez, L.A., Escobar, F., and Lira-Noriega, A. 2021. Growth temperature effect on mandibles' ontogeny and sexual dimorphism in the ambrosia beetle *Xyleborus affinis* (Curculionidae: Scolytinae. *Arthropod Structure and Development* 61: 101029.

Ospina-Garcés, S.M., Ibarra-Juarez, L.A., Escobar, F., and Lira-Noriega, A. 2021. Evaluating sexual dimorphism in the ambrosia beetle *Xyleborus affinis* (Coleoptera: Curculionidae) using geometric morphometrics. *Florida Entomologist* 104 (2): 61–70.

Palaoro, A.V. and Peixoto, P.E.C. 2022. The hidden links between animal weapons, fighting style, and their effect on contest success: A meta-analysis. *Biological Reviews* 97 (5): 1948–1966.

Parker, G.A. 1970. Sperm competition and its evolutionary consequences in insects. *Biological Reviews of the Cambridge Philosophical Society* 45: 525–67.

Parmesan, C. 2006. Ecological and evolutionary responses to recent climate change. *Annual Review of Ecology, Evolution, and Systematics* 37: 637–69.

Peixoto, P.E.C., Medina, A.M., and Mendoza-Cuenca, L. 2014. Do territorial butterflies show a macroecological fighting pattern in response to environmental stability? *Behavioural Processes* 109: 14–20.

Peixoto, P.E.C. and Mendoza-Cuenca, L. 2013. "Territorial mating systems in butterflies: What we know and what neotropical species can show." In *Sexual selection: Perspectives and models from the neotropics*, edited by R.H. Macedo and G. Machado. London: Academic Press, 85–113.

Pinto, N.S., Palaoro, A.V., and Peixoto, P.E. 2019. All by myself? Meta-analysis of animal contests shows stronger support for self than for mutual assessment models. *Biological Reviews* 94 (4): 1430–1442.

Prudic, K.L., Jeon, C., Cao, H., and Monteiro, A. 2011. Developmental plasticity in sexual roles of butterfly species drives mutual sexual ornamentation. *Science* 331 (6013): 73–75.

Punzalan, D., Rodd, F.H., and Rowe, L. 2008. Sexual selection mediated by the thermoregulatory effects of male colour pattern in the ambush bug *Phymata americana*. *Proceedings of the Royal Society B: Biological Sciences* 275 (1634): 483–92.

Radwan, J. 2004. Effectiveness of sexual selection in removing mutations induced with ionizing radiation. *Ecology Letters* 7 (12): 1149–1154.

Rico-Guevara, A. and Hurme, K.J. 2019. Intrasexually selected weapons. *Biological Reviews* 94 (1): 60–101.

Ritchie, M.G., Saarikettu, M., Livingstone, S., and Hoikkala, A. 2001. Characterization of female preference functions for *Drosophila montana* courtship song and a test of the temperature coupling hypothesis. *Evolution* 55 (4): 721–27.

Roeser-Mueller, K., Strohm, E., and Kaltenpoth, M. 2010. Larval rearing temperature influences amount and composition of the marking pheromone of the male beewolf, *Philanthus triangulum*. *Journal of Insect Science* 10 (1): 1–16.

Rosvall, K.A. 2011. Intrasexual competition in females: Evidence for sexual selection? *Behavioral Ecology* 22 (6): 1131–1140.

Sales, K., Vasudeva, R., Dickinson, M.E., Godwin, J.L., Lumley, A.J., Michalczyk, Ł.,... and Gage, M.J. 2018. Experimental heatwaves compromise sperm function and cause transgenerational damage in a model insect. *Nature Communications* 9 (1): 4771.

Samejima, Y. and Tsubaki, Y. 2010. Body temperature and body size affect flight performance in a damselfly. *Behavioral Ecology and Sociobiology* 64: 685–92.

Sánchez-Guillén, R.A., Córdoba-Aguilar, A., Hansson, B., Ott, J., and Wellenreuther, M. 2016. Evolutionary consequences of climate-induced range shifts in insects. *Biological Reviews* 91 (4): 1050–1064.

Shine, R. 1989. Ecological causes for the evolution of sexual dimorphism: A review of the evidence. *The Quarterly Review of Biology* 64 (4): 419–61.

Shuker, D.M. and Kvarnemo, C. 2021. The definition of sexual selection. *Behavioral Ecology* 32 (5): 781–94.

Simon, J.C., Stoeckel, S., and Tagu, D. 2010. Evolutionary and functional insights into reproductive strategies of aphids. *Comptes Rendus Biologies* 333 (6–7): 488–96.

Simpson, S.J., Sword, G.A., and Lo, N. 2011. Polyphenism in insects. *Current Biology* 21 (18): R738–R749.

Sivinski, J. and Wing, S.R. 2008. "Visual mating signals." In *Encyclopedia of Entomology*, edited by J.L. Capinera. Dordrecht: Kluwer Academic Publishers, 4113–4126.

Somjee, U., Woods, H.A., Duell, M., and Miller, C.W. 2018. The hidden cost of sexually selected traits: The metabolic expense of maintaining a sexually selected weapon. *Proceedings of the Royal Society B: Biological Sciences* 285 (1891): 20181685.

Song, H., Béthoux, O., Shin, S., Donath, A., Letsch, H., Liu, S., McKenna, D.D., et al. 2020. Phylogenomic analysis sheds light on the evolutionary pathways towards acoustic communication in Orthoptera. *Nature Communications* 11 (1): 4939.

Spottiswoode, C. and Møller, A. P. 2004. Extrapair paternity, migration, and breeding synchrony in birds. *Behavioral Ecology* 15 (1): 41–57.

Steiger, S. and Stökl, J. 2014. The role of sexual selection in the evolution of chemical signals in insects. *Insects* 5 (2): 423–38.

Stillwell, R.C., Blanckenhorn, W.U., Teder, T., Davidowitz, G., and Fox, C.W. 2010. Sex differences in phenotypic plasticity affect variation in sexual size dimorphism in insects: From physiology to evolution. *Annual Review of Entomology* 55: 227–45.

Svensson, E.I. and Waller, J.T. 2013. Ecology and sexual selection: Evolution of wing pigmentation in calopterygid damselflies in relation to latitude, sexual dimorphism, and speciation. *The American Naturalist* 182 (5): E174–E195.

Thornhill, R. and Alcock, J. 1983. "The evolution of insect mating systems." In *The Evolution of Insect Mating Systems*. Cambridge: Harvard University Press.

Tinghitella, R.M., Broder, E.D., Gallagher, J.H., Wikle, A. W., and Zonana, D.M. 2021. Responses of intended and unintended receivers to a novel sexual signal suggest clandestine communication. *Nature Communications* 12 (1): 797.

Tinghitella, R.M., Wang, J.M., and Zuk, M. 2009. Preexisting behavior renders a mutation adaptive: Flexibility in male phonotaxis behavior and the loss of singing ability in the field cricket *Teleogryllus oceanicus*. *Behavioral Ecology* 20 (4): 722–28.

Tomkins, J.L. and Hazel, W. 2007. The status of the conditional evolutionarily stable strategy. *Trends in Ecology & Evolution* 22 (10): 522–28.

Tsubaki, Y. 2003. The genetic polymorphism linked to mate-securing strategies in the male damselfly *Mnais costalis* Selys (Odonata: Calopterygidae). *Population Ecology* 45: 263–66.

Tsubaki, Y., Samejima, Y., and Siva-Jothy, M.T. 2010. Damselfly females prefer hot males: Higher courtship success in males in sunspots. *Behavioral Ecology and Sociobiology 64*: 1547–1554.

Vande Velde, L. and Van Dyck, H. 2013. Lipid economy, flight activity and reproductive behaviour in the speckled wood butterfly: On the energetic cost of territory holding. *Oikos* 122 (4): 555–62.

Vieira, M.C. and Peixoto, P.E.C. 2013. Winners and losers: A meta-analysis of functional determinants of fighting ability in arthropod contests. *Functional Ecology* 27: 305–13.

Walker Jr, T.J. 1957. Specificity in the response of female tree crickets (Orthoptera, Gryllidae, Oecanthinae) to calling songs of the males. *Annals of the Entomological Society of America 50* (6): 626–36.

Warren, I.A., Gotoh, H., Dworkin, I.M., Emlen, D.J., and Lavine, L.C. 2013. A general mechanism for conditional expression of exaggerated sexually-selected traits. *BioEssays* 35 (10): 889–99.

Weir, L.K., Grant, J.W., and Hutchings, J.A. 2011. The influence of operational sex ratio on the intensity of competition for mates. *The American Naturalist* 177 (2): 167–76.

Wilkinson, G.S., Kahler, H., and Baker, R.H. 1998. Evolution of female mating preferences in stalk-eyed flies. *Behavioral Ecology* 9 (5): 525–33.

Zuk, M., Rotenberry, J.T., and Simmons, L.W. 1998. Calling songs of field crickets (*Teleogryllus oceanicus*) with and without phonotactic parasitoid infection. *Evolution* 52 (1): 166–71.

Zuk, M., Rotenberry, J.T., and Tinghitella, R.M. 2006. Silent night: Adaptive disappearance of a sexual signal in a parasitized population of field crickets. *Biology Letters* 2 (4): 521–24.

CHAPTER 8

Interspecific hybridization in insects in times of climate change

Rosa Ana Sánchez-Guillén, Luis Rodrigo Arce-Valdés, Andrea Viviana Ballén-Guapacha, Jesús Ernesto Ordaz-Morales and Miguel Stand-Pérez

8.1 Introduction

Insects are one of the most abundant groups of living beings, comprising 54 percent of all known species, and they are spread all over the planet, inhabiting terrestrial and aquatic ecosystems (Gullan and Cranston 2010). Their success results from their high adaptability to changing environments (Bale et al. 2002). Components of insects' life cycles, including phenology, voltinism, morphology, physiology and behavior, as well as their geographic ranges, abundance and species interactions are affected by global warming.

Insects respond quickly to global warming in part because their development, reproduction and survival are strongly influenced by temperature but also because they have short generation times, high reproductive rates (Bale et al. 2002) and large effective population sizes (Wright 1938). Insect responses to global warming have been summarized into three main types: adaptation, migration and extinction (Menéndez 2007). Plasticity through short-term adjustments (phenotypic, frequently physiological plasticity) and adaptation through long-term evolutionary responses (such as changes to their life history) allow insect populations to survive in newly altered environments (González-Tokman et al. 2020). Dispersal, by moving to another geographic area to escape unsuitable areas, allows insect populations to follow favorable conditions. However, not all species are able to move or adapt quickly enough, in which case they can become locally or completely extinct. An example of this is the decrease in northern, cold-adapted moth species in Great Britain (Halsch et al. 2021). However, our understanding of how insects respond to climate change is still incomplete and it has many gaps.

Over the past few decades, an increasing number of studies have documented climate-induced changes in insect distributions, ranging from contraction to expansion of their distribution ranges (Parmesan et al. 1999). These responses are species-specific or even population-specific (Musolin and Saulich 2012). Fast-growing, multivoltine and non-diapausing species (e.g., dipterans, hymenopterans, lepidopterans, odonates

Rosa Ana Sánchez-Guillén et al., *Interspecific hybridization in insects in times of climate change*. In: *Effects of Climate Change on Insects*. Edited by: Daniel González-Tokman and Wesley Dáttilo, Oxford University Press. © Oxford University Press (2024). DOI: 10.1093/oso/9780192864161.003.0008

and orthopterans) that can reproduce constantly and rapidly under warmer conditions frequently undergo geographic and altitudinal range expansions (Bale et al. 2002; Sánchez-Guillén et al. 2016). In contrast, slow-growing species that need low temperatures to induce diapause face geographic and altitudinal range contractions (Bale et al. 2002; Menéndez 2007). Climate-induced range shifts can increase the potential for hybridization by increasing the sympatry between parapatric species, altering the established equilibrium of the hybrid zone. Additionally, range expansions can form newly sympatric distributions between species that diverged in allopatry but have not completed their reproductive isolation, in other words create new hybrid zones, introgression and ultimately speciation or extinction (Chunco 2014; Mallet 2018; Sánchez-Guillén et al. 2016).

Hybridization, the phenomenon by which two species produce hybrid offspring, is more frequent than is often recognized (Mallet 2005). For instance, 25–30 percent of plant and 10 percent of animal species are involved in hybridization (Mallet 2005). Hybridization is today distinguished as a widespread phenomenon with significant impact on species evolution (Abbott et al. 2016). The importance of hybridization for the evolutionary diversification of plants and animals has recently been revealed by whole genomic sequencing, which over the last two decades has yielded partial or whole genome sequence data of non-model species (Mallet et al. 2016). Hybridization outcomes depend on two components of reproductive isolation: intrinsic factors, such as genetic incompatibilities, and extrinsic factors, such as environmentally dependent hybrid fitness (Pickup et al. 2019). The degree to which genetic divergence prevents hybridization, and which are the most important reproductive barriers to hybridization, are major research themes in the study of hybridization. Reproductive barriers can be classified according to whether they occur before (pre-mating) or after (post-mating) mating, and whether they occur before (pre-zygotic) or after (post-zygotic) zygote formation (Coyne and Orr 1989).

In many cases hybridization can maintain or increase diversity by introducing new genes or gene combinations or can increase the level of heterozygosity by adding new alleles (Wolfe et al. 2007). There are many examples of this trend, including evolutionary rescue of small, inbred populations, stable hybrid zones, reinforcement of reproductive isolation, the origin of new hybrid lineages and introgression. Introgression refers to the transfer of genetic material from one species into the gene pool of another through hybridization followed by back-crossing (sexual reproduction between a hybrid and an individual from one parental linage). Introgression may include the transfer of alleles that increase fitness and occurs when hybrids can reproduce with one or both parent species (unidirectional and bidirectional introgression, respectively). Introgressed alleles are generally neutral, but they can sometimes result in adaptive variation (i.e., adaptive introgression) (Arnold 2004). In some cases, due to the low fitness of hybrids, natural selection favors adaptations that reduce the formation of hybrids (Butlin and Smadja 2018; Ortiz-Barrientos et al. 2009). This process, known as reinforcement, eliminates from the population alleles from individuals with low heterospecific discrimination ability, leading to a pattern of reproductive character displacement that strengthens premating reproductive isolation in sympatric populations compared to allopatric populations (Ortiz-Barrientos et al. 2009). There is empirical evidence for character displacement in *Calopteryx* and *Heaterina* damselflies from North America (Anderson and Grether 2010; Waage 1975). Reinforcement can also strengthen premating isolation between sympatric and allopatric populations of the same species, known as "cascade reinforcement" (Ortiz-Barrientos et al. 2009). When hybrids are reproductively isolated from the parental species, they can become a new hybrid lineage, which can be homoploid (i.e., the hybrid lineage retains the same number of chromosomes as both parental species) or allopolyploid (a different chromosomal constitution in hybrids (Todesco et al. 2016)).

Alternatively, hybridization can decrease genetic diversity by breaking down reproductive barriers, leading to species fusion and species extinction (Todesco et al. 2016). For example, when hybrids are more viable than the parental species, one or both parental species can become locally or completely extinct or can merge if reproductive isolation between them is incomplete (Seehausen et al. 2008). Additionally, when one of these species is rare, genetic swamping (i.e., when the species is replaced by hybrids) or demographic swamping (i.e., the reduction of population growth rates due to the wasteful production of maladaptive hybrids) can lead to extinction of the rare species (Todesco et al. 2016).

Identifying the insect orders, families and species that are most likely to hybridize and determining the most common outcomes of that hybridization have become relevant questions for understanding the evolutionary consequences of global warming. In this chapter we review short- and long-term consequences of hybridization for species persistence or extinction, with the goal of summarizing the contribution of these studies to our understanding of the outcomes of the climate-induced hybridization. To this end, we focused on five insect orders—dipterans, hymenopterans, lepidopterans, odonates and orthopterans—that have been responsive to climate change in terms of geographic and altitudinal range expansions (Sánchez-Guillén et al. 2016). In the following sections we investigate: (i) the correlation between genetic divergence and reproductive isolation; (ii) the degree of genetic divergence at which different intrinsic reproductive barriers develop; (iii) which are the most important intrinsic reproductive barriers preventing hybridization; and (iv) which are the most frequent evolutionary outcomes of both non-anthropogenic and climate-induced hybridization events.

8.2 Intrinsic factors affecting hybridization

Studies by Coyne and Orr (1989) in dipterans and, a couple of decades later, in butterflies (Presgraves 2002) and in odonates (Sánchez-Guillén et al. 2014) detected a positive relationship between the strength of prezygotic and postzygotic isolation barriers and genetic distance. This trend is referred to as the "speciation clock" by Coyne and Orr (1989) and has been demonstrated in other groups, such as amphibians (Sasa et al. 1998), birds (Price and Bouvier 2002) and angiosperms (Moyle et al. 2004).

We investigated how intrinsic factors such as intrinsic reproductive barriers, and the genetic divergence between hybridizing species, which is positively correlated with the number and strength of genetic incompatibilities that segregate in hybrid offspring, prevent hybridization. We focused our review in five insect orders: dipterans, hymenopterans, lepidopterans, odonates and orthopterans, due to their high responsivity to climate change in terms of range shifts (reviewed in Sánchez-Guillén et al. 2016). Our study was conducted in three-consecutive steps. First, we identified potential reproductive barriers to gene flow. Second, we measured genetic distance between sexually interacting species. Third, we investigated the correlation between the genetic distance and reproductive isolation between hybridizing species and determined the degree of genetic distance between hybridizing species at which the different reproductive barriers develop in different insect orders.

First, to identify potential reproductive barriers and hybridization outcomes, two literature searches were made in the Web of Science Core Collection (Thomson Reuters) from March to April 2022, which includes records starting in 1980. In a first search, we searched for the terms *"reproductive barriers"* or *"reproductive isolation"* in the title, the abstract or as a key word, along with the name of each order (e.g., *"diptera"*) in any part of the document. We found a total of 856 results: Diptera = 261; Hymenoptera = 138; Lepidoptera = 302; Odonata = 27 plus data from previous searches by RAS-G; and Orthoptera = 128. Reproductive barriers can be asymmetric between reciprocal crosses; that is, the strength of the barriers can be different when comparing crosses of males of taxa A with females of taxa B, with crosses of males of taxa B with females of taxa A. When data from both cross directions were available and were asymmetric, we considered only the cross direction with the lowest degree of isolation. Papers reporting natural hybrids or hybrid zones were considered as evidence of interspecific gene flow (adult hybrid formation). After that, we grouped the intrinsic reproductive barriers into a number of categories that vary in the degree of reproductive isolation they produce. Each category was assigned a score based on the completeness of isolation, ranging from zero (no isolation) to 5 (complete isolation) (for similar analyses see Coyne and Orr 1989; Sánchez-Guillén et al. 2014). We used six categories: (i) no reproductive barriers (F_1 hybrids completely fertile, score = 0); (ii) only one sterile sex in F_1 hybrids (partial hybrid sterility, score = 1; (iii) sterile F_1 hybrids (hybrid sterility, score = 2); (iv) non-sexually mature F_1 hybrids (hybrid inviability, score = 3); (v) parental females fail to oviposit following heterospecific crosses (gametic or tactile barriers, score = 4); (vi) assortative mating (preference for conspecifics) and/or mechanical isolation in heterospecific crosses (score = 5). We then divided the score by 5 to yield an index of reproductive isolation ranging from 0 to 1. When species pairs presented reproductive barriers from more than one category, we considered the barriers with the highest score. For instance, the moths *Helicoverpa armigera* and *H. assulta*, despite being completely isolated in the field by assortative mating (pheromones) and mechanical isolation (differences in the aedeagus morphology), they can produce completely fertile F_1 hybrids in the laboratory (no-choice experiments) (Wang 2007). In this case, the reproductive barrier preventing hybridization was prezygotic (assortative mating or mechanical isolation, score = 5). In hymenopterans, non-chromosomal sex-determination systems, such as haplodiploidy and arrhenotokous parthenogenesis, are common (Heimpel and De Boer 2008). For that reason, because hybridization only occurs in the non-hemizygous sex (Koevoets and Beukeboom 2009), hybrids were categorized as fertile or infertile without regard to their sex (i.e., no partial hybrid sterility). Note that in odonates, complete and partial hybrid sterility barriers were not included because of the lack of information (RAS-G personal communication). In the second search, we searched for the terms *"hybrid," "hybridization"* or *"hybridization"* and the name of each order (e.g., *"diptera"*), in the title, abstract or key words. We excluded documents that contained the terms *"fluorescent," "fluorescence"* or *"in situ"* in any part of the document because preliminary searches tended to yield a large volume of irrelevant results due to confusion with the fluorescence *in situ* hybridization (FISH) technique. Additionally, since orthopterans and odonates had fewer results than the other orders, we expanded their search by searching for papers that included *"orthoptera"* and *"odonata"* in any part of the document. We found a total of 1,586 results: Diptera = 465; Hymenoptera = 356; Lepidoptera = 529; Odonata = 46 plus data from previous searches by RAS-G; and Orthoptera = 190. We considered as evidence of hybridization any documented case of live F_1 adult hybrids, confirmed observations of hybrid morphs in nature, or reports of hybrid zones.

A case was considered positive even if hybridization only occurred in one reciprocal cross direction.

Second, to explore genetic divergence between hybridizing species, we used database tools to estimate genetic distance between each species pairs yielded by both literature searches. We measured genetic distance between species pairs as the proportion of divergent sites between species-specific Cytochrome C Oxidase subunit I gene (COI) consensus sequences (Box 8.1). The final data set contained 161 comparisons on phylogenetically corrected reproductive isolation and COI sequences divergence values between pairs of species involved in interspecific interactions recovered for the five insect orders (24 for dipterans, 5 for hymenopterans, 32 for lepidopterans, 94 for odonates and 6 for orthopterans).

Box 8.1 ESTIMATION OF GENETIC DISTANCES BETWEEN SPECIES PAIRS

Cytochrome C Oxidase subunit I gene (COI) sequences were downloaded from BOLD public record database (Ratnasingham and Hebert 2007) using the BOLD-CLI software (<https://github.com/Cnuge/BOLD-CLI>) or GenBank (Benson et al. 2005) when sequences were not available in BOLD.

To reduce a possible bias caused by intraspecific diversity and/or sequencing errors, we downloaded all available sequences from both databases ($n = 60,550$ sequences). Then, we used Muscle v5 (Edgar 2021) to align the sequences from each species using the Super5 algorithm. Finally, we built a species-specific consensus sequence from the resulting alignments using the EMBOSS *cons* tool (Rice et al. 2000). We set plurality to 1 to prevent consensus sequences with non-called nucleotides (N). Using these species-specific COI consensus sequences ($n = 646$) we aligned each species pair's sequences using the Muscle v5 PPP algorithm and estimated the sequence divergence between them using the *ape* 5.0 R package (Paradis and Schliep 2019). Sequence divergences were estimated as the proportion of sites that differed between each sequence pair using the *dist.dna(model = "raw")* command. To account for phylogenetic non-independence of species pairs, we generated a reduced set of phylogenetically "corrected" species pairs using a standardized protocol (Coyne and Orr 1989; Sánchez-Guillén et al. 2014). Briefly, we used nested averages to reduce the comparisons between species pairs at each phylogenetic node to a single comparison. In other words, we averaged sequences divergences and reproductive isolation values in all species pairs involving each species. Raw databases and script files were uploaded at:
   <https://github.com/LuisRodrigoArce-Valdes/Insect_Hybridization_CC>

Third, to investigate the correlation between genetic divergence and reproductive isolation between hybridizing species we used (i) linear regression tests to evaluate the relationship between reproductive isolation and sequence divergence and (ii) Mann–Whitney U tests (corrected for multiple comparisons using the Bonferroni procedure) to evaluate whether the mean sequence divergence for six categories of reproductive isolation was significantly different from each other, and thus evaluate the threshold of genetic divergence preventing hybridization. We discuss these findings in the following sections. Our results are discussed for all five orders together, then separately for each of the five insect orders.

8.3 Reproductive barriers preventing hybridization

To account for a publication bias in the studied insect orders and reproductive barriers (because some taxa or reproductive barriers are easier to study than others), we compared the frequency of species in the surveyed studies on hybridization and reproductive barriers with the frequency of all described insect species representing these insect orders. We also quantified the relative frequency of missing data values between barriers on each insect order. While the first comparison suggested no bias by insect order (GBIF.org 2022; Figure 8.1A), some biases were detected for reproductive barriers (Figure 8.1B). Thus, reported frequencies of different reproductive barriers in the surveyed papers should not be considered as reliable estimates of the real frequencies.

In lepidopterans and odonates, the reproductive barrier that prevented the largest proportion of the hybridization cases (54 percent and 74 percent, respectively) was assortative mating, potentially due to the divergence in secondary sexual characters such as pheromones, songs or colors and/or mechanical isolation due to the incompatibility of the reproductive structures (Figure 8.2). In the case of lepidopterans, this is mostly by pheromone recognition (but there may be some bias because lepidopteran sex pheromones are investigated and used for pest control (Ando et al. 2004)) or by chemical blends (González-Rojas et al. 2020). In the case of odonates, the reproductive barrier that prevented the largest proportion of sexual interactions was mechanical incompatibility to copulate (Figure 8.2). In zygopteran odonates (e.g., *Ischnura*, *Enallagma* and *Argia*) different forces, such as antagonistic mating interactions and sexual conflict (Córdoba-Aguilar and Cordero Rivera 2005) or divergent ecological selection (McPeek et al. 2008), conduct the rapid development of mechanical isolation through the divergence of caudal

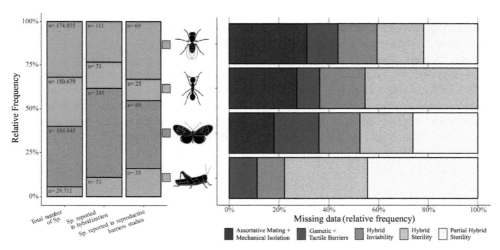

Figure 8.1 Publication bias in insect hybridization studies. A) Relative frequency of known species in each insect order (Diptera = 174,055; Hymenoptera = 150,679; Lepidoptera = 184,845; Orthoptera = 29,711), species involved in hybridization (Diptera = 111; Hymenoptera = 71; Lepidoptera = 241; Orthoptera = 51) and species in reproductive barrier studies (Diptera = 69; Hymenoptera = 25; Lepidoptera = 80; Orthoptera = 33). B) Relative frequency of missing data, i.e., non measured barriers in studies investigating reproductive barriers (assortative mating, mechanical isolation, tactile barriers, gametic barriers, hybrid inviability, hybrid sterility and hybrid partial sterility) mating in each insect order. Odonates were excluded from this analysis since yielded data included additional records compiled by RAS-G besides the literature searches and hybrid sterility was not sampled in these records.

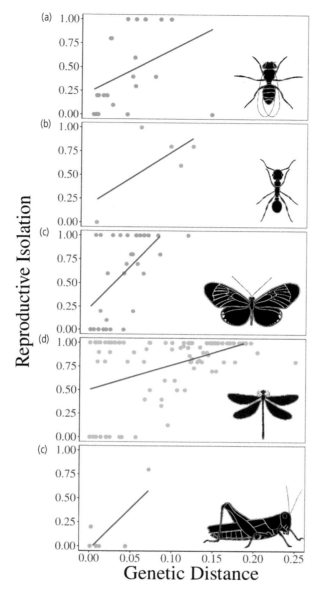

Figure 8.2 Frequency of the reproductive barriers preventing hybridization in the five studied insect orders. Percentage labels show the relative frequency of each barrier, numbers between parenthesis show the absolute number of observations per barrier.

appendages (Paulson 1974; Wellenreuther and Sánchez-Guillén 2016). Assortative mating (visual, acoustical and chemical signals are used to attract females in dipterans) was also the reproductive barrier that prevented the largest proportion (38 percent) of the hybridization cases in dipterans, however partial hybrid sterility was also an important barrier (28 percent). For instance, sex pheromones mediate mate choice in *Drosophila* sp. (González-Rojas et al. 2020), although numerous studies in fruit flies (Coyne and Orr 2004) support also the central role of mechanical compatibility between male and female genitalia in reproductive isolation. Twenty-eight percent of the hybridizing cases in dipterans showed partial hybrid sterility, which may be related to the Haldane's rule, that is, infertility or inviability in the heterogametic sex. Dipterans are in fact models for the study both prezygotic and postzygotic reproductive isolation (Mallet 2006). In hymenopterans, gametic and/or tactile barriers preventing oviposition were the most important reproductive barrier, preventing around 43 percent of hybridization cases. In this order, the presence or absence of the intracellular parasitic *Wolbachia* bacteria determines interspecific mating outcomes, as is the case of interspecific matings in *Nasonia* wasps, for example between *N. vitripennis* and either *N. giraulti* or *N. longicornis* (Beukeboom and Van Den Assem 2001; Mair et al. 2017). Finally, in orthopterans songs are an important component of the specific mate recognition system, playing a role both as a way in which conspecifics can evaluate the "quality" of potential mates during sexual selection and as reproductive barrier between sympatric heterospecifics via the partitioning of communication channels or "acoustic niches" (Robinson and Hall 2002). However, contrary to expectations sex pheromones and acoustic signals that mediate mate choice (González-Rojas et al. 2020) only prevented 20 percent of the hybridization cases. The most frequent barrier preventing the hybridization (40 percent of cases) was the sterility of one sex (Haldane's rule when the sterile sex is the heterogametic). Larger sample sizes are needed in orthopterans to improve the precision of the relative importance of reproductive barriers in this order.

8.4 COI sequence divergence and reproductive isolation between hybridizing insects

We found that, in the studied insects, COI sequence divergence and reproductive isolation between species were correlated (Figure 8.3A). Insect species pairs that form hybrids presented low genetic distances (4.8 percent). However, some outlier species pairs continue to hybridize even at high genetic distances (20 percent) (Figure 8.3B).

When each insect order was analyzed independently, genetic distance and reproductive isolation were correlated (although some correlations were marginal) in dipterans, lepidopterans and odonates (see Figure 8.4) confirming the "speciation clock" previously detected in these orders (Coyne and Orr 1989; Presgraves 2002; Sánchez-Guillén et al. 2014).

The slope of this relationship varied among dipterans, lepidopterans and odonates, suggesting that complete reproductive isolation develops at different rates of genetic distance between orders. Dipterans and lepidopterans completed reproductive isolation at lower genetic distances (i.e., with a higher slope angle) than odonates. In hymenopterans and orthopterans the trend was for increasing reproductive isolation with increasing genetic distance (Figure 8.4). However, it is necessary to increase sample sizes to confirm that trend.

8.5 Climate-induced hybridization

We performed a literature survey to identify published cases of climate-induced hybridization in the five studied insect orders and their outcomes. We found a total of 185 results, of which 26 were relevant to our study, identifying 17 cases of climate induced hybridization, 12 of which were published in the last 5 years (Table 8.1). We also found numerous studies that used potential future distributions to predict the destabilization of current hybrid zones or the formation of new hybrid zones (see Box 8.2). We discuss the findings in each order in the following sections.

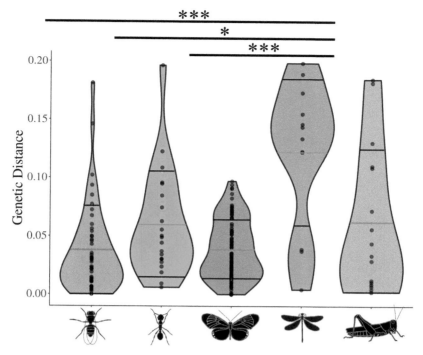

Figure 8.3 Genetic distance and reproductive isolation (mtDNA, COI) in all studied insects. Solid blue line represents the tendency and fussy gray area represents the 95% confidence interval.

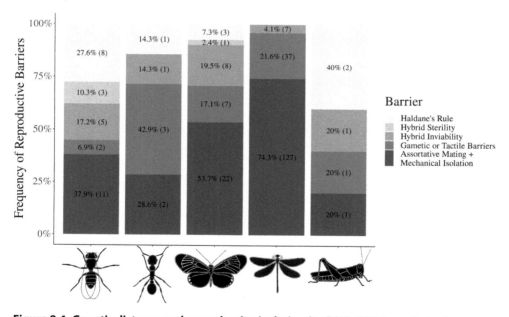

Figure 8.4 Genetic distance and reproductive isolation (mtDNA, COI) in each studied insect order. Solid blue lines represent the tendency line when significant positive correlation between reproductive isolation and COI sequence divergence was found in A) Diptera, B) Hymenoptera, C) Lepidoptera, D) Odonata and E) Orthoptera.

Table 8.1 Summary of study methods on climate-induced hybridization studies in insects. Columns 1–3 indicate order and species involved in hybridization. Column 4 indicates evolutionary consequences. Column 5 indicates genetic distance. Column 6 indicates the geographical region. Column 7 indicates methods used in the hybrid identification. NA indicates no available data.

Order	Sp1	Sp2	Evolutionary consequences	Divergence time	COI sequence divergence	Method	Region	Reference
Diptera	Chironomus riparius	Ch. piger	Postzygotic barriers and reinforcement	Pleistocene (2 mya)	0.1500	Laboratory: Rearing and Backcrossing	Europe	Foucault et al. 2019
Hymenoptera	Formica polyctena	F. aquilonia	Introgression and adaptation to different temperature ranges	Pleistocene	NA	Behavior and genetic data (SSRs)	Europe	Martin-Roy et al. 2021
	Solenopsis invicta	S. richteri	Introgression and adaptation to different temperature ranges	Oligocene (25 mya)	0.0656	Freeze-point exposure	North America	James et al. 2002
Lepidoptera	Hyles euphorbiae	H. tithymali	Introgression and changes in larvae morphology	Pleistocene (1.4–3 mya)	NA	Larvae morphology analysis	Europe	Hundsdoerfer et al. 2011
	Polyommatus agestis	P. artaxerxes	Introgression and voltinism changes	Pleistocene	NA	Genetic data (mtDNA, nDNA)	Europe	Mallet et al. 2011
	Helicoverpa armigera	H. zea	Introgression	Pleistocene (1.5 mya)	0.0293	Genomic (GWAS) and Genetic (SNPs) data	South America	Cordeiro et al. 2020
	Papilo glaucus	P. canadensis	Hybrid speciation and voltinism changes	Pleistocene (0.5–0.6mya)	0.0213	Extensive field work and SNPs	North America	Ryan et al. 2018; Scriber 2011

Order		Species	Processes	Time	Value	Data	Region	References
Odonata	Anax imperator	A. partenophe	Hybridization	NA	0.0136	Genetic data (mtDNA)	Europe, Asia, Africa	Geiger et al. 2021
	Calopteryx splendens	C. virgo	Unidirectional introgression	NA	0.1081	Behavior and Genetic data (mtDNA)	Europe	Geiger et al. 2021; Tynkkynen et al. 2008
	Calopteryx splendens	C. haemorroidalis	Bidirectional introgression	NA	0.1413	Morphological and Genetic data (SSRs)	Europe	Lorenzo-Carballa et al. 2014
	Cordulegaster trinacriae	C. boltonii	Unidirectional introgression, and Reinforcement	NA	NA	Genetic data (mtDNA)	Europe	Geiger et al. 2021; Solano et al. 2018
	Ischnura elegans	I. graellsii	Bidirectional introgression, Local extinction, Reinforcement	Pleistocene (1 mya)	0.0016	Morphological and Genetic data (SSRs)	Europe	Monetti et al. 2002; Sánchez-Guillén et al. 2011; Wellenreuther et al. 2018
	Mnais costalis	M. pruinosa	Introgression, Reinforcement	NA	0.0000	Genetic data (mtDNA)	Asia	Hayashi et al. 2005
Orthoptera	Chorthippus parallelus	C. montanus	Unidirectional hybridization, Local extinction	Pleistocene	0.0385	Genetic data (SSRs)	Europe	Rohde et al. 2015
	Allonemobius socius	A. fasciatus	Unidirectional hybridization, Local extinction	Pleistocene (0.06 mya)	NA	Longitudinal population studies and genetic data (protein electrophoresis)	North America	Britch et al. 2001
	Phaulacridium marginale	P. otagoense	Introgression, Range distribution reduction	NA	NA	Morphometry and phylogenetic reconstruction (nDNA)	Australia	Sivyer et al. 2018
	Trimerotropis agrestis	T. maritima	Breakdown of reproductive isolation, Local extinction	NA	0.0030	Morphology	North America	Brust et al. 2009

8.6 Dipterans

The dipterans, and in particular mosquitoes, are particularly relevant in terms of public health due to their roles as disease vectors and agricultural pests. This order has been frequently investigated in the context of climate change, especially in terms of changes in distribution (e.g., Carvalho et al. 2015; González et al. 2010), because some life-history aspects such as hatching rate, mean development time, body size and blood-feeding rate are affected by rising temperatures (Maier et al. 1990; Muttis et al. 2018). This is the case of the Asian mosquito *Aedes albopictus*, whose range has expanded from its native distribution in southeast Asia to all continents except Antarctica (Rochlin et al. 2013), or *A. aegypti*, *A. atropalpus*, *A. koreicus* and *A. triseriatus*, which have been introduced into Europe over the past forty years (Medlock et al. 2012). The most alarming aspect of the range expansion of these species is their ability to act as vectors for viruses such as Chikunguña, Dengue and Nile Fever (Medlock et al. 2012; Rochlin et al. 2013). Hybridization is relatively common in dipterans and is frequently related to adaptation to pesticides (Sánchez-Guillén et al. 2016). Studies of the inviability and sterility of hybrids and speciation genes in *Drosophila* have been fundamental to our understanding of how reproductive barriers function, an indispensable step for understanding speciation (Mallet 2006). However, cases of hybridization due to distribution changes driven by global climate change are scarce (see Table 8.1 and Figure 8.5). In fact, we found only one example, where Foucault et al. (2019) evaluated reproductive isolation between two harlequin flies, *Chironomus riparius* and *C. piger*, finding that environmental changes can increase the hybridization dynamic between these two species. However, the authors argue that this change in the hybridization dynamic does not necessarily affect the integrity of the two species, since genetic analyses indicated that these two species have probably hybridized over the course of their divergence.

Box 8.2 PREDICTING HYBRIDIZATION RISK UNDER CLIMATE CHANGE

Changes in species distributions have been strongly driven by climate change (Rosenzweig et al. 2008). In order to study how species respond to climate change, researchers have used potential species distribution models, which allow us to address a wide variety of ecological and evolutionary questions (Parmesan 2006), as has been observed in butterflies (Santos et al. 2011), flies (Van Dam et al. 2019), crickets (Borissov et al. 2021) and dragonflies (e.g., Sánchez-Guillén et al. 2013). The possibility that climate change leads to novel interactions between closely related species has generated conservation concerns since the contact between ecologically similar species could induce high levels of interspecific competition in populations that are already stressed by changing climate conditions (Urban et al. 2012). Predicting contact between potentially hybridizing species requires an examination of distribution changes, but also consideration of interspecific interactions and adaptive evolution. This will help generate an integrated understanding of the impacts of climate change on contemporary biodiversity (Taylor et al. 2015). In any case, future studies that combine field and laboratory experiments in an effort to improve our understanding of the physiological, ecological, environmental and adaptive requirements of insects would lay the foundations for more precise niche models of future distribution ranges that could inform conservation decision making to preserve insect biodiversity.

8.7 Hymenopterans

The hymenopterans (ants, bees, ichneumons, chalcids, sawflies and wasps) are another exceptionally diverse group of insects, with around 115,000 described species. Hymenopterans such as ants and bees are relevant in the study of the hybridization outcomes due to their eusocialism and variety of genomic structures (Cohen and Privman 2020; Weyna et al. 2022). This order is usually haplodiploid: males develop from unfertilized eggs and have a haploid genome, while females (including queens) are diploid and are the result of sexual reproduction (Aamidor et al. 2018). Despite the wide framework these groups offer to study hybridization outcomes, we found only two papers where the authors consider hybridization as a consequence of climate change (James et al. 2002; Martin-Roy et al. 2021) (see Table 8.1, Figure 8.5). The first study examined two closely related (genetic distance < 0.06) ant species from South America—*Solenopsis invicta* and *S. richteri*. In this case hybridization had a favorable outcome for the ants. In an area where these species were introduced in North America, they produce hybrids that have lower mortality rates when exposed to low temperatures than the two parental species in their original distribution, suggesting an adaptative process in which introgressive hybridization served as the vehicle for adaptations to thermal conditions (James et al. 2002). The second case of climate-induced hybridization was also in ants, between *Formica polyctena and F. aquilonia* (genetic distance < 0.06). These European ants suffered a negative effect of the hybridization; hybrids are apparently maladaptive, and while each parental species shows thermal tolerances that match their geographical distributions, hybrids are as heat-tolerant as *F. polyctena* but fail to match the cold-tolerance limit of *F. aquilonia* (Martin-Roy et al. 2021).

8.8 Lepidopterans

Lepidopterans have been core species in the study of the impact of climate change on diversity since the 1990s, mainly because their high sensitivity to environmental condition is quickly reflected in their abundance, phenology and/or range distribution (Kocsis and Hufnagel 2011). This sensitivity seems to be given by a phenotypic plasticity in larval life-history traits. Specifically, changes in developmental time and metabolic rates in function of temperature seem to be key for local adaptation in Lepidopterans (Ayres and Scriber 1994; Gomi et al. 2007; Wagner et al. 2011). Hybridization is especially common and well understood in lepidopterans (see Box 8.3).

Box 8.3 LEPIDOPTERA, A HISTORICALLY IMPORTANT SYSTEM FOR THE STUDY OF HYBRIDIZATION

Lepidopterans have proven to be a key group for our understanding of hybridization and its outcomes in nature due to the high rate of naturally occurring interspecific hybridization. For example, Presgraves (2002), based on 212 crosses involving 182 species, reported that nearly 15 percent of species have hybridized. Meanwhile, Mallet et al. (2007) reported that between 26 percent and 29 percent of the species of Heliconiina are involved in interspecific hybridization, which is especially frequent in the genera *Heliconius* and *Eueides*. Furthermore, butterflies have been broadly used to study the strength of the multiple

Continued

Box 8.3 *Continued*

forces underlying interspecific isolation. Mérot et al. (2017), using a comparative analysis in the *Heliconius* subclade, showed two key findings: (i) the wing pattern involved in Müllerian mimicry and mate choice is under natural selection, embodying several isolating barriers, and (ii) even if wing patterns are a main factor early in divergence, it may not be obligatory at later stages. This means that other factors, such as hybrid sterility and chemically mediated mate choice, may complete the speciation process. Interestingly, we found that the principal prezygotic barrier preventing interspecific gene flow among lepidopterans is species-specific pheromone communication, but once this barrier is surpassed, Haldane's rule might be the strongest postzygotic barrier preventing interspecific gene flow (Presgraves 2002).

An exceptional example of a well-understood system of hybridization is the Swallowtail butterflies of North America, where the eastern (*Papilio glaucus*) and Canadian (*Papilio canadensis*) tiger swallowtail butterflies have formed a dynamic hybrid zone. This system has been key for the study of local adaptation, incipient speciation, host-race speciation, multitrait ecological divergence and haploid hybrid speciation constituted by a climate-induced range shift that has resulted in hybridization (details in Scriber 2011).

The dynamic hybrid zone in North America formed by the eastern tiger swallowtail, *Papilio glaucus*, and the Canadian tiger swallowtail, *P. canadensis*, has been studied over the last twenty years, providing a relevant example of hybridization in nature (reviewed by Scriber 2011). Hybridization has been a vehicle for the introgression from *P. glaucus* to *P. canadensis* of the ability to detoxify tulip tree leaves and use this tree as a new host species for oviposition (Scriber 2011). Similarly, Mallet et al. (2011) reported evidence for introgression between two brown argus butterflies in Britain, *Polyommatus agestis* and *P. artaxerxes*. Interestingly, the authors comment that the usually bivoltine *P. agestis* is univoltine in the hybrid zone, probably due to introgressive hybridization. Introgression, although rarely adaptive, is usually considered a positive outcome of hybridization because it might increase heterozygosity (Abbott et al. 2013). However, it could be of concern when it involves pest species, where hybridization could result in the combination of lineage-specific adaptations within a single organism (Cordeiro et al. 2020). This is the concern with *Helicoverpa armigera* (native to the Old World) and *H. zea*. *Helicoverpa armigera* species is considered a threat to crops in North America, where climatic change is facilitating its expansion throughout the range of *H. zea*. Cordeiro et al. (2020), using genome-wide single-nucleotide polymorphisms (SNPs), have detected that each of these species has 10 percent mixed ancestry (i.e., introgression) from the other and have formed a mosaic hybrid zone that could be easily altered by agricultural expansion or climatic fluctuations. The fourth reported case of climate-induced hybridization in lepidopterans has resulted in the formation of a new lineage, *Papilio appalachiensis*, in the *Papilio glaucus* species group (Kunte et al. 2011). Hundsdoerfer et al. (2011) reported several areas of hybridization as a consequence of the northward shift of the North African moth, *Hyles tithymali*, into southern Europe, and the westward expansion of *H. euphorbiae* into Europe from the Iranian Plateau refuge, since larvae in populations in central and southern Italy appear to be hybrids of the two lineages.

8.9 Odonates

Odonates are an especially sensitive order to global warming. Termaat et al. (2019) identified that forty-five odonate species (63 percent of the studied species) have increased their distribution range during last forty years. Of these species, six are already involved in hybridization events. These include five cases of introgression, both unidirectional (*n* = 2) and bidirectional (*n* = 3), one case of simple admixture, two cases of reinforcement of reproductive isolation and one case of the local extinction of one of the involved species (see Table 8.1, Figure 8.5). The expansion of hybrid zones can lead to their destabilization (Buggs 2007). During range expansions, hybridization and introgression can counteract the loss of genetic diversity due to the founder effect and genetic drift (Pfennig et al. 2016). The damselfly *Ischnura elegans*, which has drastically expanded its distribution in Spain over the last twenty years, is an example in damselflies of how genetic diversity is counteracted by introgressive hybridization with its sister species, *Ischnura graellsii*. Both genetic diversity and genetic differentiation in populations of *I. elegans* from Spain were larger than in populations of *I. elegans* where the two species are allopatric (Sánchez-Guillén et al. 2011; Wellenreuther et al. 2018). However, the advantage for the expanding species is disadvantageous for the native species *I. graellsii*, which is endemic to the Iberian Peninsula and is experiencing local extinction in most of the localities occupied by *I. elegans* (Sánchez-Guillén et al. 2011; Wellenreuther et al. 2018). Another case of introgressive hybridization between two sister dragonfly species is that of *Cordulegaster boltonii* and *C. trinacriae*, whose range overlap has increased in central Italy due to a northern shift of *C. trinacriae* in the last few thousand years, probably driven by climate change (Solano et al. 2018). In this study, the authors detected unidirectional introgression of genes from the autochthonous *C. boltonii* into the expanding *C. trinacriae* (Solano et al. 2018).

Another consequence of hybridization is the reinforcement of reproductive isolation, which is usually detected in patchy hybrid zones, as is the case of *Ischnura elegans* and *I. graellsii* in Spain (Sánchez-Guillén et al. 2011; Wellenreuther et al. 2018). Preliminary evidence suggests that pre-mating reproductive isolation has been strengthened (Arce-Valdés et al. 0000) and that there has been reproductive character displacement in female pronotum in *I. elegans* from Spain (personal observation, RAS-G). This pattern has also been detected in the process of introgressive hybridization between *Mnais costalis* and *M. pruinosa* (Hayashi et al. 2005).

8.10 Orthopterans

The orthopterans are another order that is sensitive to changes in temperature since they are distributed mainly in warm, sunny regions (Burton 2003). Their life cycle, specifically the period of growth and emergence, is strongly affected by global warming. Increased rates of growth and development, decreased generation time, and increased fecundity per clutch have been detected, together with changes in their geographic distribution (Porter et al. 1991). In this order, there are at least four published cases of hybridization between species that have come into contact or whose range overlap has increased due to global warming or anthropogenic activity (see Table 8.1 and Figure 8.5). One particularly relevant case is that of *Chorthippus montanus* and *C. parallelus*, two sympatric cricket species whose reproductive barriers are strongly determined by the relative frequency of the two species.

Rohde et al. (2015) detected unidirectional hybridization, with an increase in the genetic diversity of *Chorthippus parallelus* where the two species were in contact. This led to the local extinction of *C. montanus* populations in different parts of Germany (Rohde et al. 2015). The authors point out that other species with similar frequency-based reproductive barriers may be at similar risk, as is the case of *A. socius* and *A. fasciatus* in North America (Britch et al. 2001). In another study, Brust et al. (2009) highlight the vulnerability of parapatric sister species. Using morphological data, the authors detected hybridization between *Trimerotropis maritima* (eastern species) and *T. agrestis* (western species) in a recent contact zone that arose due to the desiccation in the last decade of Lake McConaughy (Nebraska, USA). They showed how habitat alteration can break down reproductive isolation and therefore species integrity. On the other hand, local extinction can be due to the joint action of hybridization and ecological competition. Sivyer et al. (2018) documented a narrow contact zone in New Zealand between *Phaulacridium marginale*, a species with a wide distribution range, and *P. otagoense*, which has a more restricted distribution. Using genetic and morphometric data, they found evidence that the range of *P. marginale* had expanded over that of *P. otagoense*, causing introgressive hybridization and a reduction in the range of *P. otagoense* through ecological competition over the past thirty years where there has been a replacement of native forests by European grasses and herbs.

8.11 Hybridization outcomes: Non-anthropogenic versus climate-induced hybridization

We performed a literature survey to identify published cases of non-anthropogenic hybridization in the five studied insect orders and their outcomes. We found a total of sixty-six results (Diptera = 14; Hymenoptera = 13; Lepidoptera = 22; Odonata = 9 and Orthoptera = 8). Yielded hybridization (non-anthropogenic and climate-induced) cases were categorized in eight categories (Figure 8.5), which were considered as non-mutually exclusive (i.e., a single hybridization case could fit into two or more categories simultaneously). We further classified the eight hybridization categories into positive and negative outcomes based on the effect they have on species and genetic diversity (Adavoudi and Pilot 2022; Runemark et al. 2019). Hybrid speciation, introgression, adaptive introgression and reinforcement can be considered as a beneficial process in some circumstances (Hamilton and Miller 2016; Ortiz-Barrientos et al. 2009). On the other hand, species fusion, local extinction and morphological and phenological changes can be considered as negative processes related with loss of adaptive variation, high mortality rates and increased extinction probability (Adavoudi and Pilot 2022).

The most frequent outcomes of the non-anthropogenic hybridization in dipterans, hymenopterans, lepidopterans and odonates were introgression and hybrid speciation. In hymenopterans, introgression, hybrid speciation, local adaptation and phenological changes were similarly frequent (Figure 8.5). Unfavorable outcomes, such as local extinction, phenological changes and morphological changes have low prevalence in non-anthropogenic hybridization events (Figure 8.5). However, outcomes of climate-induced hybridization in lepidopterans were mainly positive, including three cases of introgression and one case of hybrid speciation (see Table 8.1 and Figure 8.5). It is worth mentioning that the majority of the cases of hybridization in orthopterans had negative effects, such as reduction of genetic diversity, local extinction, morphological anomalies

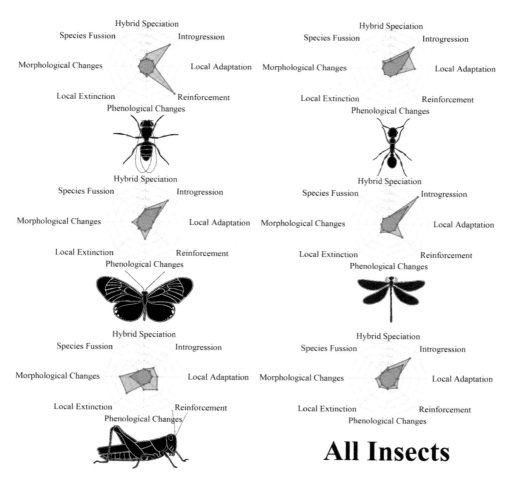

Figure 8.5 Spider plot comparison of hybridization outcomes in climate-induced and non-anthropogenic hybrid zones. Comparison was done for all five focal orders together and for each order separately. Dots in the center of each plot depicts absence of each consequence; points are placed increasingly towards the outer areas of the plot as the frequency of each consequence increases (0%, representing "none"; to 100% representing "all"). Red was used for the outcomes of climate-induced hybridization cases and green for the outcomes of non-climate-induced hybridization cases.

and breakdown of reproductive barriers. Notice that due to bias towards evidence in some particular outcomes, reported frequencies of different hybridization outcomes in the surveyed papers should not be considered as reliable estimates of the real frequencies.

Acknowledgments

We thank Lynna Kiere for her English revision and constructive comments. We also thank Janet Nolasco for her technical support. We also thank the editors for the invitation to contribute with this chapter and an anonymous reviewer whose comments greatly helped improving the content of this manuscript. The research was funded by the CONACYT grant to RAS-G (no. 282922).

Key reflections

- Although genetic distances and reproductive isolation are correlated in insects, the slope of these relationships varies among insects evidencing that complete reproductive isolation develops at different genetic divergence values in orders such as dipterans, butterflies and hymenopterans than in odonates or orthopterans. Therefore, odonates and orthopterans might be especially vulnerable to hybridization. In odonates, mechanical isolation (prezygotic barrier) prevents the largest proportion of the hybridization events, while in orthopterans partial and total hybrid sterility (postzygotic barrier) are the barriers that prevent the largest proportion of hybridization events, making the hybridization in orthopterans more expensive, in terms of fitness reduction, than in odonates.
- The effects of hybridization on genetic diversity and biodiversity during climate-induced range shifts are of major interest in conservation biology. However, in the surveyed papers, climate-induced hybridization outcomes that were unfavorable for the involved studied insect species were much less frequent than those that were favorable. Moreover, we detected a shift in the frequency of the outcomes in non-anthropogenic climate-induced versus hybridization events in the surveyed papers: (i) in non-anthropogenic hybridization events, introgression and hybrid speciation were the most frequent; while (ii) in climate-induced hybridization events, introgression, reinforcement and local extinction were the most frequent outcomes. In orthopterans, negative outcomes of climate-induced hybridization, such as local extinction and phenological changes, were the most frequent.
- The fact that most of the consequences of hybridization that occurred following range shifts (in the surveyed papers) were positive (e.g., introgression, adaptive introgression, reinforcement of reproductive isolation and hybrid speciation) suggests that on the whole, hybridization is not the most serious threat that global climate change poses to insects. However, in the case of endemic, threatened and ecologically or economically important species, the possibility of local extinction or species fusion due to hybridization, or hybrids becoming significant invasive pests is cause for some concern for local economies and biodiversity (Sakai et al. 2001). Note that reported frequencies of different hybridization outcomes in the surveyed papers should not be considered as reliable estimates of the real frequencies.
- The increased prevalence of hybridization due to climate-induced range shifts provides a unique opportunity to improve our understanding of the processes and the short- and long-term consequences of hybridization (Abbott et al. 2013). Future studies should combine different approaches, such as next-generation sequencing techniques to investigate the genetic consequences of genome mixing, climate modelling to predict the extent of range overlap following range expansions, and ecological experiments, to investigate the integrity of prezygotic and postzygotic barriers.

Key further reading

- For understanding global warming on insects we recommend following papers: González-Tokman et al. (2020) review the physiological mechanisms that allow insects tolerating high temperatures. Garnas (2018) compiles the drivers and limitations of the rapid evolution in insects due to climate change. Musolin and Saulich (2012) discuss the effect of global warming on insects and their effects on distribution, abundance, phenology, morphology, physiology, behavior, etc.
- To deepen the understanding of the outcomes climate-induced hybridization in insects, we recommend following papers: Sánchez-Guillén et al. (2016) review the evolutionary consequences of distribution changes in the short and long term, with a special focus on genomics. Sánchez-Guillén et al. (2014) evaluate how climate change affects hybridization in endangered species.

- For a focused approach on hybridization and speciation, as well as their evolutionary consequences, we recommend the books by Dobzhansky (1937), who addresses topics such as diversity, laws of inheritance, mutations in populations, natural selection, mechanisms of reproductive isolation and speciation; and Coyne and Orr (2004) who elucidate much on topics related to speciation, such as sympatric speciation, reinforcement, the role of hybridization in speciation, the search for genes that cause reproductive isolation and the growing evidence of the role of natural and sexual selection in the origin of species. In addition, we recommend the works of Abbott et al. (2016), Harrison and Larson (2016), Mallet (2005), Martin and Jiggins (2017), Noor (1999) and Runemark et al. (2019).

References

Aamidor, S.E., Yagound, B., Ronai, I., and Oldroyd, B.P. 2018. Sex mosaics in the honeybee: How haplodiploidy makes possible the evolution of novel forms of reproduction in social Hymenoptera. *Biology Letters* 14 (11): 670–76.

Abbott, R., Albach, D., Ansell, S., Arntzen, J.W., Baird, S.J., Bierne, N., Boughman, J., et al. 2013. Hybridization and speciation. *Journal of Evolutionary Biology* 26 (2): 229–46.

Abbott RJ, Barton NH and Good JM. 2016. Genomics of hybridization and its evolutionary consequences. *Molecular Ecology* 25: 2325–2332.

Adavoudi, R. and Pilot, M. 2022. Consequences of hybridization in mammals: A systematic review. *Genes* 13 (1): 50.

Anderson, C.N. and Grether, G.F. 2010. Interspecific aggression and character displacement of competitor recognition in Hetaerina damselflies. *Proceedings of the Royal Society B: Biological Sciences* 277 (1681): 549–55.

Ando T., Inomata S., and Yamamoto M. 2004. "Lepidopteran sex pheromones." In *The Chemistry of Pheromones and Other Semiochemicals I. Topics in Current Chemistry, Vol 1*, edited by S. Schulz. Berlin: Springer Science & Business Media, 51–96.

Arce-Valdés, L.R., Ballén-Guapacha, A.V., Rivas-Torres, A., Chávez-Ríos, J.R., Wellenreuther, M., Hansson, B., and Sánchez-Guillén, R.A. Testing the predictions of reinforcement: long-term empirical data from a damselfly mosaic hybrid zone. *Under review.*

Arnold, M.L. 2004. Transfer and origin of adaptations through natural hybridization: were Anderson and Stebbins right? *The Plant Cell* 16 (3): 562–70.

Ayres, M.P., and Scriber, J.M. 1994. Local adaptation to regional climates in Papilio canadensis (Lepidoptera: Papilionidae). *Ecological Monographs* 64 (4): 465–82.

Bale, J.S., Masters, G.J., Hodkinson, I.D., Awmack, C., Bezemer, T.M., Brown, V. K., Butterfield, J. et al. 2002. Herbivory in global climate change research: Direct effects of rising temperature on insect herbivores. *Global Change Biology* 8 (1): 1–16.

Benson, D.A., Karsch-Mizrachi, I., Lipman, D.J., Ostell, J., and Wheeler, D.L. 2005. GenBank. *Nucleic Acids Research* 33 (suppl-1): D34–D38.

Beukeboom, L.W. and Van den Assem, J. 2001. Courtship and mating behaviour of interspecific Nasonia hybrids (Hymenoptera, Pteromalidae): A grandfather effect. *Behavior Genetics* 31: 167–77.

Borissov, S.B., Hristov, G.H., and Chobanov, D.P. 2021. Phylogeography of the Poecilimon ampliatus species group (Orthoptera: Tettigoniidae) in the context of the Pleistocene glacial cycles and the origin of the only thelytokous parthenogenetic phaneropterine bush-cricket. *Arthropod Systematics and Phylogeny* 79: 401–18.

Britch, S.C., Cain, M.L., and Howard, D.J. 2001. Spatio-temporal dynamics of the Allonemobius fasciatus–A. socius mosaic hybrid zone: A 14-year perspective. *Molecular Ecology* 10 (3): 627–38.

Brust, M.L., Hoback, W.W., and Wright, R.J. 2009. Apparent hybridization between Trimerotropis agrestis and Trimerotropis maritima (Orthoptera: Acrididae) in a recently disturbed habitat. *Journal of the Kansas Entomological Society* 82 (3): 215–22.

Buggs, R.J.A. 2007. Empirical study of hybrid zone movement. *Heredity* 99 (3): 301–12.

Burton, J.F. 2003. The apparent influence of climatic change on recent changes of range by European insects (Lepidoptera, Orthoptera). Changes in ranges: invertebrates on the move. *Processings 13th International Colloquium European Invertebrate Survey, Leiden*, September 2–5, 2001: 13–21.

Butlin, R.K. and Smadja, C.M. 2018. Coupling, reinforcement, and speciation. *The American Naturalist* 191 (2): 155–72.

Carvalho, B.M., Rangel, E.F., Ready, P.D., and Vale, M.M. 2015. Ecological niche modelling predicts southward expansion of *Lutzomyia (Nyssomyia) flaviscutellata* (Diptera: Psychodidae: Phlebotominae), vector of *Leishmania (Leishmania) amazonensis* in South America, under climate change. *PLoS One* 10: 1–21.

Carvalho, B.M., Rangel, E.F., Ready, P.D., and Vale, M.M. 2015. Ecological niche modelling predicts southward expansion of Lutzomyia (Nyssomyia) flaviscutellata (Diptera: Psychodidae: Phlebotominae), vector of Leishmania (Leishmania) amazonensis in South America, under climate change. *PLoS One* 10 (11): e0143282.

Chunco, A.J. 2014. Hybridization in a warmer world. *Ecology and Evolution* 4 (10): 2019–2031.

Cohen, P. and Privman, E. 2020. The social supergene dates back to the speciation time of two Solenopsis fire ant species. *Scientific Reports* 10 (1): 11538.

Cordeiro, E.M., Pantoja-Gomez, L.M., de Paiva, J.B., Nascimento, A.R., Omoto, C., Michel, A.P., and Correa, A.S. 2020. Hybridization and introgression between Helicoverpa armigera and H. zea: An adaptational bridge. *BMC Evolutionary Biology* 20 (1): 1–12.

Córdoba-Aguilar, A. and Cordero-Rivera, A. 2005. Evolution and ecology of Calopterygidae (Zygoptera: Odonata): Status of knowledge and research perspectives. *Neotropical Entomology* 34: 861–79.

Coyne, J.A. and Orr, H.A. 1989. Patterns of speciation in Drosophila. *Evolution 43* (2): 362–81.

Coyne, J.A. and Orr, H.A. 2004. *Speciation*. Sunderland: Sinauer Associates.

Dobzhansky, T. 1937. *Genetics and the Origin of Species*. New York: Columbia University Press.

Edgar, R. C. 2021. MUSCLE v5 enables improved estimates of phylogenetic tree confidence by ensemble bootstrapping. *BioRxiv* 2021–2026. doi: 10.1101/2021.06.20.449169.

Foucault, Q., Wieser, A., Heumann-Kiesler, C., Diogo, J., Cocchiararo, B., Nowak, C., . . . and Pfenninger, M. 2019. An experimental assessment of reproductive isolation and its consequences for seasonal hybridization dynamics. *Biological Journal of the Linnean Society* 126 (2): 327–37.

Garnas, J.R. 2018. Rapid evolution of insects to global environmental change: conceptual issues and empirical gaps. *Current Opinion in Insect Science* 29: 93–101.

GBIF.org. 2022. GBIF website, accessed August 16, 2022, <https://www.gbif.org>.

Geiger, M., Koblmüller, S., Assandri, G., Chovanec, A., Ekrem, T., Fischer, I., Galimberti, A. et al. 2021. Coverage and quality of DNA barcode references for Central and Northern European Odonata. *PeerJ* 9: e11192.

Gomi, T., Nagasaka, M., Fukuda, T., and Hagihara, H. 2007. Shifting of the life cycle and life-history traits of the fall webworm in relation to climate change. *Entomologia Experimentalis et Applicata* 125 (2): 179–84.

González, C., Wang, O., Strutz, S.E., González-Salazar, C., Sánchez-Cordero, V., and Sarkar, S. 2010. Climate change and risk of leishmaniasis in North America: predictions from ecological niche models of vector and reservoir species. *PLoS Neglected Tropical Diseases* 4 (1): e585.

González-Rojas, M.F., Darragh, K., Robles, J., Linares, M., Schulz, S., McMillan, W. O., Jiggins, C.D. et al. 2020. Chemical signals act as the main reproductive barrier between sister and mimetic Heliconius butterflies. *Proceedings of the Royal Society B: Biological Sciences* 287 (1926): 20200587.

González-Tokman, D., Córdoba-Aguilar, A., Dáttilo, W., Lira-Noriega, A., Sánchez-Guillén, R.A., and Villalobos, F. 2020. Insect responses to heat: Physiological mechanisms, evolution and ecological implications in a warming world. *Biological Reviews* 95 (3): 802–21.

Gullan, P.J. and Cranston, P.S. 2010. *The insects: An outline of entomology*. Oxford: Blackwell Publishing.

Halsch, C.A., Shapiro, A.M., Fordyce, J.A., Nice, C.C., Thorne, J.H., Waetjen, D.P., and Forister, M.L. 2021. Insects and recent climate change. *Proceedings of the National Academy of Sciences* 118(2): e2002543117.

Hamilton, J.A. and Miller, J.M. 2016. Adaptive introgression as a resource for management and genetic conservation in a changing climate. *Conservation Biology 30* (1): 33–41.

Harrison, R.G., and Larson, E.L. 2016. Heterogeneous genome divergence, differential introgression, and the origin and structure of hybrid zones. *Molecular Ecology 25* (11): 2454–2466.

Hayashi, F., Dobata, S., and Futahashi, R. 2005. Disturbed population genetics: suspected introgressive hybridization between two Mnais damselfly species (Odonata). *Zoological Science 22* (8): 869–81.

Heimpel, G.E. and De Boer, J.G. 2008. Sex determination in the Hymenoptera. *Annual Reviews in Entomology* 53: 209–30.

Huber, J. 2017. "Biodiversity of Hymenoptera." In *Insect Biodiversity: Science and Society, Vol 1,* edited by R.G. Footit and P.H. Adler. 2nd edn, Sussex, United Kingdom, John Wiley and Sons Ltd, 419–61.

Hundsdoerfer, A.K., Mende, M.B., Harbich, H., Pittaway, A.R., and Kitching, I.J. 2011. Larval pattern morphotypes in the Western Palaearctic Hyles euphorbiae complex (Lepidoptera: Sphingidae: Macroglossinae). Insect Systematics & Evolution 42: 41–86.

James, S.S., Pereira, R.M., Vail, K.M., and Ownley, B.H. 2002. Survival of imported fire ant (Hymenoptera: Formicidae) species subjected to freezing and near-freezing temperatures. *Environmental Entomology* 31 (1): 127–33.

Kocsis, M. and Hufnagel, L. 2011. Impacts of climate change on Lepidoptera species and communities. *Applied Ecology and Environmental Research* 9 (1): 43–72.

Koevoets, T. and Beukeboom, L.W. 2009. Genetics of postzygotic isolation and Haldane's rule in haplodiploids. *Heredity* 102 (1): 16–23.

Kunte, K., Shea, C., Aardema, M.L., Scriber, J.M., Juenger, T.E., Gilbert, L.E., and Kronforst, M.R. 2011. Sex chromosome mosaicism and hybrid speciation among tiger swallowtail butterflies. *PLoS Genetics* 7 (9): e1002274.

Lorenzo-Carballa, M.O., Watts, P.C., and Cordero-Rivera, A. 2014. Hybridization between Calopteryx splendens and C. haemorrhoidalis confirmed by morphological and genetic analyses. *International Journal of Odonatology* 17 (2–3): 149–60.

Maier, K.J., Kosalwat, P., and Knight, A.W. 1990. Culture of Chironomus decorus (Diptera: Chironomidae) and the effect of temperature on its life history. *Environmental Entomology* 19 (6): 1681–1688.

Mair, M.M., Kmezic, V., Huber, S., Pannebakker, B.A., and Ruther, J. 2017. The chemical basis of mate recognition in two parasitoid wasp species of the genus n asonia. *Entomologia Experimentalis et Applicata,* 164 (1): 1–15.

Mallet, J. 2005. Hybridization as an invasion of the genome. *Trends in Ecology & Evolution* 20 (5): 229–37.

Mallet, J. 2006. What does Drosophila genetics tell us about speciation? *Trends in Ecology & Evolution* 21 (7): 386–93.

Mallet, J. 2018. Invasive insect hybridizes with local pests. *Proceedings of the National Academy of Sciences* 115 (19): 4819–4821.

Mallet, J., Beltrán, M., Neukirchen, W., and Linares, M. 2007. Natural hybridization in heliconiine butterflies: The species boundary as a continuum. *BMC Evolutionary Biology* 7 (1): 1–16.

Mallet, J., Besansky, N., and Hahn, M.W. 2016. How reticulated are species? *BioEssays* 38 (2): 140–49.

Mallet, J., McMillan, W.O., and Jiggins, C.D. 1998. Mimicry and warning color at the boundary between races and species. In *Endless Forms: Species and Speciation*, edited by D.J. Howard and S.H. Berlocher. Oxford, Oxford University Press. 390–403.

Mallet, J., Wynne, I.R., and Thomas, C.D. 2011. Hybridisation and climate change: Brown argus butterflies in Britain (Polyommatus subgenus Aricia). *Insect Conservation and Diversity* 4 (3): 192–99.

Martin, S.H. and Jiggins, C.D. 2017. Interpreting the genomic landscape of introgression. *Current Opinion in Genetics and Development* 47, 69–74.

Martin-Roy, R., Nygård, E., Nouhaud, P., and Kulmuni, J. 2021. Differences in thermal tolerance between parental species could fuel thermal adaptation in hybrid wood ants. *The American Naturalist* 198 (2): 278–94.

McPeek, M.A., Shen, L., Torrey, J.Z., and Farid, H. 2008. The tempo and mode of three-dimensional morphological evolution in male reproductive structures. *The American Naturalist* 171 (5): E158–E178.

Medlock, J.M., Hansford, K.M., Schaffner, F., Versteirt, V., Hendrickx, G., Zeller, H., and Bortel, W.V. 2012. A review of the invasive mosquitoes in Europe: Ecology, public health risks, and control options. *Vector-borne and Zoonotic Diseases 12* (6): 435–47.

Mende, M., Kitching, I., Hundsdoerfer, A., Harbich, H., and Pittaway, A. 2011. Larval pattern morphotypes in the Western Palaearctic Hyles euphorbiae complex (Lepidoptera: Sphingidae: Macroglossinae). *Insect Systematics and Evolution* 42 (1): 41–86.

Menéndez, R. 2007. How are insects responding to global warming? *Tijdschrift voor Entomologie* 150 (2): 355.

Mérot, C., Salazar, C., Merrill, R.M., Jiggins, C.D., and Joron, M. 2017. What shapes the continuum of reproductive isolation? Lessons from Heliconius butterflies. *Proceedings of the Royal Society B: Biological Sciences* 284 (1856): 20170335.

Monetti, L., Sánchez-Guillén, R.A., and Rivera, A.C. 2002. Hybridization between Ischnura graellsii (Vander Linder) and 2. elegans (Rambur)(Odonata: Coenagrionidae): Are they different species? *Biological Journal of the Linnean Society 76* (2): 225–35.

Moyle, L.C., Olson, M.S., and Tiffin, P. 2004. Patterns of reproductive isolation in three angiosperm genera. *Evolution* 58 (6): 1195–1208.

Musolin, D.L., and Saulich, A.K. 2012. Responses of insects to the current climate changes: From physiology and behavior to range shifts. *Entomological Review* 92: 715–40.

Muttis, E., Balsalobre, A., Chuchuy, A., Mangudo, C., Ciota, A.T., Kramer, L.D., and Micieli, M. 2018. Factors related to Aedes aegypti (Diptera: Culicidae) populations and temperature determine differences on life-history traits with regional implications in disease transmission. *Journal of Medical Entomology* 55 (5): 1105–1112.

Noor, M.A. 1999. Reinforcement and other consequences of sympatry. *Heredity 83*(5): 503–08.

Ortiz-Barrientos, D., Grealy, A., and Nosil, P. 2009. The genetics and ecology of reinforcement implications for the evolution of prezygotic isolation in sympatry and beyond. *The Year in Evolutionary Biology: Annals of the New York Academic of Scieince* 1168: 156–82.

Paradis, E. and Schliep, K. 2019. ape 5.0: An environment for modern phylogenetics and evolutionary analyses in R. *Bioinformatics* 35 (3): 526–28.

Parmesan, C. 2006. Ecological and evolutionary responses to recent climate change. *Annual Review of Ecology & Environmental Systems* 37: 637–69.

Parmesan, C., Ryrholm, N., Stefanescu, C., Hill, J.K., Thomas, C.D., Descimon, H., Huntley, B. et al. 1999. Poleward shifts in geographical ranges of butterfly species associated with regional warming. *Nature* 399 (6736): 579–83.

Paulson, D.R. 1974. Reproductive isolation in damselflies. *Systematic Biology 23* (1): 40–49.

Pfennig, K.S., Kelly, A.L., and Pierce, A.A. 2016. Hybridization as a facilitator of species range expansion. *Proceedings of the Royal Society B: Biological Sciences* 283 (1839): 20161329.

Pickup, M., Brandvain, Y., Fraïsse, C., Yakimowski, S., Barton, N.H., Dixit, T., Lexer, C., et al. 2019. Mating system variation in hybrid zones: Facilitation, barriers and asymmetries to gene flow. *New Phytologist* 224 (3): 1035–1047.

Porter, J.H., Parry, M.L., and Carter, T.R. 1991. The potential effects of climatic change on agricultural insect pests. *Agricultural and Forest Meteorology* 57 (1–3): 221–40.

Presgraves, D. C. 2002. Patterns of postzygotic isolation in Lepidoptera. *Evolution 56*(6): 1168–1183.

Price, T.D. and Bouvier, M.M. 2002. The evolution of F1 postzygotic incompatibilities in birds. *Evolution* 56 (10): 2083–2089.

Ratnasingham, S. and Hebert, P.D. 2007. BOLD: The Barcode of Life Data System (http://www.barcodinglife.org). *Molecular Ecology Notes* 7 (3): 355–64.

Rice, P., Longden, I., and Bleasby, A. 2000. EMBOSS: The European molecular biology open software suite. *Trends in Genetics* 16 (6): 276–77.

Robinson, D.J. and Hall, M.J. 2002. Sound signalling in orthoptera. *Advances in Insect Physiology* 29: 151–278.

Rochlin, I., Ninivaggi, D.V., Hutchinson, M.L., and Farajollahi, A. 2013. Climate change and range expansion of the Asian tiger mosquito (Aedes albopictus) in Northeastern USA: Implications for public health practitioners. *PloS One* 8 (4): e60874.

Rohde, K., Hau, Y., Weyer, J., and Hochkirch, A. 2015. Wide prevalence of hybridization in two sympatric grasshopper species may be shaped by their relative abundances. *BMC Evolutionary Biology* 15: 1–14.

Rosenzweig, C., Karoly, D., Vicarelli, M., Neofotis, P., Wu, Q., Casassa, G., Menzel A. et al. 2008. Attributing physical and biological impacts to anthropogenic climate change. *Nature* 453 (7193): 353–57.

Runemark, A., Vallejo-Marin, M., and Meier, J.I. 2019. Eukaryote hybrid genomes. *PLoS Genetics* 15 (11): e1008404.

Ryan, S.F., Deines, J.M., Scriber, J.M., Pfrender, M.E., Jones, S.E., Emrich, S.J., and Hellmann, J.J. 2018. Climate-mediated hybrid zone movement revealed with genomics, museum collection, and simulation modeling. *Proceedings of the National Academy of Sciences* 115 (10): E2284–E2291.

Sakai, A.K., Allendorf, F.W., Holt, J.S., Lodge, D.M., Molofsky, J., With, K.A., Baughman, S., et al. 2001. The population biology of invasive species. *Annual Review of Ecology and Systematics* 32 (1): 305–32.

Sánchez-Guillén, R.A., Córdoba-Aguilar, A., Cordero-Rivera, A., and Wellenreuther, M. 2014. Genetic divergence predicts reproductive isolation in damselflies. *Journal of Evolutionary Biology* 27 (1): 76–87.

Sánchez-Guillén, R. A., Córdoba-Aguilar, A., Hansson, B., Ott, J., and Wellenreuther, M. 2016. Evolutionary consequences of climate-induced range shifts in insects. *Biological Reviews* 91 (4): 1050–1064.

Sánchez-Guillén, R.A., Muñoz, J., Rodríguez-Tapia, G., Feria Arroyo, T.P., and Córdoba-Aguilar, A. 2013. Climate-induced range shifts and possible hybridisation consequences in insects. *PloS One* 8 (11): e80531.

Sánchez-Guillén, R.A., Wellenreuther, M., Cordero-Rivera, A., and Hansson, B. 2011. Introgression and rapid species turnover in sympatric damselflies. *BMC Evolutionary Biology* 11 (1): 1–17.

Santos, H., Paiva, M.R., Tavares, C., Kerdelhue, C., and Branco, M. 2011. Temperature niche shift observed in a Lepidoptera population under allochronic divergence. *Journal of Evolutionary Biology* 24 (9): 1897–1905.

Sasa, M.M., Chippindale, P.T., and Johnson, N.A. 1998. Patterns of postzygotic isolation in frogs. *Evolution* 52 (6): 1811–1820.

Scriber, J.M. 2011. Impacts of climate warming on hybrid zone movement: Geographically diffuse and biologically porous "species borders." *Insect Science* 18 (2): 121–59.

Seehausen, O.L.E., Takimoto, G., Roy, D., and Jokela, J. 2008. Speciation reversal and biodiversity dynamics with hybridization in changing environments. *Molecular Ecology* 17 (1): 30–44.

Sivyer, L., Morgan-Richards, M., Koot, E., and Trewick, S.A. 2018. Anthropogenic cause of range shifts and gene flow between two grasshopper species revealed by environmental modelling, geometric morphometrics and population genetics. *Insect Conservation and Diversity* 11 (5): 415–34.

Solano, E., Hardersen, S., Audisio, P., Amorosi, V., Senczuk, G., and Antonini, G. 2018. Asymmetric hybridization in Cordulegaster (Odonata: Cordulegastridae): Secondary postglacial contact and the possible role of mechanical constraints. *Ecology and Evolution* 8 (19): 9657–9671.

Taylor, S.A., Larson, E.L., and Harrison, R.G. 2015. Hybrid zones: Windows on climate change. *Trends in Ecology & Evolution* 30 (7): 398–406.

Termaat, T., van Strien, A. J., van Grunsven, R.H., De Knijf, G., Bjelke, U., Burbach, K., Conze, K.J., et al. 2019. Distribution trends of European dragonflies under climate change. *Diversity and Distributions* 25 (6): 936–50.

Todesco, M., Pascual, M.A., Owens, G.L., Ostevik, K.L., Moyers, B.T., Hübner, S., Heredia, S.M., et al. 2016. Hybridization and extinction. *Evolutionary Applications* 9 (7): 892–908.

Tynkkynen, K., Grapputo, A., Kotiaho, J.S., Rantala, M.J., Väänänen, S., and Suhonen, J. 2008. Hybridization in Calopteryx damselflies: The role of males. *Animal Behaviour* 75 (4): 1431–1439.

Urban, M.C., Tewksbury, J.J., and Sheldon, K.S. 2012. On a collision course: Competition and dispersal differences create no-analogue communities and cause extinctions during climate change. *Proceedings of the Royal Society B: Biological Sciences* 279 (1735): 2072–2080.

Van Dam, M.H., Rominger, A.J., and Brewer, M.S. 2019. Environmental niche adaptation revealed through fine scale phenological niche modelling. *Journal of Biogeography* 46 (10): 2275–2288.

Waage, J.K. 1975. Reproductive isolation and the potential for character displacement in the damselflies, Calopteryx maculata and C. aequabilis (Odonata: Calopterygidae. *Systematic Biology* 24 (1): 24–36.

Wagner, K.D., Krauss, J., and Steffan-Dewenter, I. 2011. Changes in the life history traits of the European Map butterfly, Araschnia levana (Lepidoptera: Nymphalidae) with increase in altitude. *European Journal of Entomology* 108 (3): 447.

Wang, C. 2007. Interpretation of the biological species concept from interspecific hybridization of two Helicoverpa species. *Chinese Science Bulletin* 52 (2): 284–86.

Wellenreuther, M., Muñoz, J., Chávez-Ríos, J.R., Hansson, B., Cordero-Rivera, A., and Sánchez-Guillén, R.A. 2018. Molecular and ecological signatures of an expanding hybrid zone. *Ecology and Evolution* 8 (10): 4793–4806.

Wellenreuther, M. and Sánchez-Guillén, R.A. 2016. Nonadaptive radiation in damselflies. *Evolutionary Applications* 9(1): 103–18.

Weyna, A., Bourouina, L., Galtier, N., and Romiguier, J. 2022. Detection of F1 hybrids from single-genome data reveals frequent hybridization in Hymenoptera and particularly ants. *Molecular Biology and Evolution* 39 (4): msac071.

Wolfe, L.M., Blair, A.C., and Penna, B.M. 2007. Does intraspecific hybridization contribute to the evolution of invasiveness? An experimental test. *Biological Invasions* 9: 515–21.

Wright, S. 1938. Size of population and breeding structure in relation to evolution. *Science* 87: 430–31.

CHAPTER 9

Changes in insect population dynamics due to climate change

Carol L. Boggs

9.1 Introduction

Climate change at local to global scales alters the environment that organisms inhabit. Effects of climate change propagate back and forth across levels of biological organization, from metabolic, physiological and behavioral levels, to phenology, life history traits and fitness, to community and ecosystem structure and function. A full integration of climate change effects across these levels would of necessity also include all interacting organisms. An integrated understanding of the effects of climate change on any set of organisms is therefore a daunting enterprise. Nonetheless, we can obtain some general insight from a focus on one set of taxa and one organizational level, including other actors as needed. Here I focus on the effects of climate change on insect population dynamics and underlying life history traits, emphasizing examples from temperate and arctic taxa. Insects occupy nearly all trophic levels and most terrestrial and freshwater habitats. Due to the abundance and diversity of insects, they are fundamental to life as we know it (Wilson 1987). Insects are thus a critical group that is inclusive of many lifestyles. I begin this section with a short review of climate parameters that are relevant to insect life history and population dynamics, next briefly examine how these are changing on different scales and then outline the scope of this chapter.

9.2 Relevant climate and weather parameters

Individual insects experience weather, or the specific expression of a particular climate regime at a particular point in space and time. As discussed further below, which climate and weather parameters are critical depends on whether the insect's habitat is terrestrial or aquatic; arctic/antarctic, temperate or tropical; and sheltered or not from weather extremes. Not only are weather values for a given climate parameter important but we must also be cognizant of the moderating effects of insect behavior, physiology, morphology and access to microhabitats.

Carol L. Boggs, *Changes in insect population dynamics due to climate change*. In: *Effects of Climate Change on Insects*. Edited by: Daniel González-Tokman and Wesley Dáttilo, Oxford University Press. © Oxford University Press (2024). DOI: 10.1093/oso/9780192864161.003.0009

9.2.1 *Climate and weather parameters important to terrestrial insects*

The temporal scale studied determines in part which climate variables have important effects on individuals and populations. Air and substrate temperature, solar flux and windspeed are important modulators of body temperature and hence activity of terrestrial insects on minute-to-minute timescales, taking into account mechanisms used to buffer weather events (e.g., Kingsolver and Watt 1983). At the scale of parts of a lifetime, growing degree days, or the sum of time above a minimum temperature, is also critical for various life-history traits, including development rate, fecundity, and survival and resulting phenology. Precipitation (either as rain or snow) comes into play as a modulator of air temperature and solar flux, especially for diapausing insects.

These climate parameters also have indirect effects on insect life history and population dynamics through effects on the plants on which the insects depend, as well as on their competitors, predators and parasites. For example, drought and heat can alter plant traits ranging from nectar quantity and quality (Descamps et al. 2021) to secondary defense compounds (Kuczyk et al. 2021), which will affect the quantity and quality of both preferred and less-preferred resources available to insects. In an extension of this idea, Foote et al. (2017) documented asymmetrical competition between two aphid species (*Metopolophium festucae cerealium* and *Rhopalosiphum padi*, both Hemiptera: Aphidae) on wheat (*Triticum aestivum* Poaceae). Competitive effects suppressing population growth of *M. festucae cerealium* vanished under drought conditions.

9.2.2 *Climate and weather parameters important to aquatic life stages*

Insect orders with aquatic life stages include Collembola, Ephemeroptera, Odonata, Plecoptera, Megaloptera, Neuroptera, Coleoptera, Hemiptera, Hymenoptera, Diptera, Mecoptera, Lepidoptera and Trichoptera—in short, the major insect orders. As for terrestrial insects discussed above, successful survival and development of aquatic life stages of insects depend on ambient temperature and precipitation patterns. However, oxygenation, turbidity and solute concentrations are uniquely critical to the niche structure of aquatic life forms. In turn, water level and flow rates are critical determinants of water temperature, oxygenation, turbidity and solute concentrations (Knouft and Ficklin 2017; Showalter 2011). Water level and flow rates are determined by precipitation patterns, air temperature, solar flux and groundwater and soil water flow. Precipitation patterns include rainfall amount, intensity and duration; snow cover depth and duration; and seasonal timing of both rain and snow. Similarly to terrestrial systems, aquatic systems can be affected by human land- and water-use patterns as well as pollution.

9.2.3 *Buffering of climate phenology, means and variances in the absence of climate change*

The statistical description of a given climate parameter includes its mean and variance, which may change across seasons. However, individual insects generally experience climate as shorter-term weather phenomena, drawn from the distribution that makes up climate. Where an individual's performance optimum falls within the range of weather values that comprise a given climate parameter will determine how sensitive that individual is to changes in that climate parameter. Performance optima have been described by measuring fitness components such as survival or fecundity across a range of, for example,

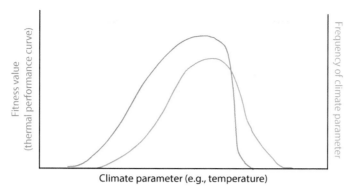

Figure 9.1 Example of a hypothetical relationship between the performance curve of a species (blue line) and the distribution of values (e.g., temperature) of a climate parameter (brown line). Fitness is 0 outside the performance curve envelop, which will result in a population decline.

temperatures. The shape of the thermal performance curve and the lower and upper critical temperatures at which the organism loses the ability to move defines the organism's response to temperature. The performance curve is often skewed, with the optimal performance closer to the upper critical point. The superimposition of the performance curve on weather distributions underlying a given climate parameter will determine the performance of the individual in a given climate scenario (Figure 9.1). However, Kingsolver and Woods (2016) showed that the details of shape of the performance curve depend on duration of temperature exposure, which complicates interpretation in the context of climate change.

Some insects can acclimatize to a set of weather conditions that change on a short timescale relative to the individual's lifespan. Such plasticity results in shifts in the performance curve with respect to the specific weather condition. For example, populations of the emerald ash borer, *Agrilus planipennis* (Coleoptera: Buprestidae), in Winnipeg, Canada exhibit freeze tolerance plasticity that allows them to survive extreme cold temperature events (Duell et al. 2022). In contrast, *Nysius groenlandicus* (Hemiptera: Lygaeidae) exhibits a rapid heat-hardening response in the arctic (Sørensen et al. 2019). Hence, temporal and spatial autocorrelation in weather parameters or in extreme events also affect performance. Correlations resulting in seasonal weather shifts are predictable and may be accommodated via selection for plasticity. However, not all weather patterns are so straightforward. For example, sea surface temperature and air pressure anomalies such as the El Niño-Southern Oscillation (ENSO) or the Arctic Oscillation (AO) create patterns of drought or excess rainfall regionally for several months at a time. Such patterns differ from seasonality in their lack of predictability. As a result, individual life-history traits and resulting population dynamics may not be able to be buffered against such weather patterns by selection for plasticity since selection fluctuates through time.

In addition, insects experience weather on a micro-scale, or below 10 centimeters (cm) in horizontal distance (Gols et al. 2021; reviewed in Harvey et al. 2023; Pincebourde and Woods 2020; Storlie et al. 2014). Climate heterogeneity goes up as the scale on which climate is measured goes down, due to increasing architectural heterogeneity of the habitat. As a result, insects may buffer exposure to climate variance by habitat choice (e.g., Gols et al. 2021; Pincebourde and Woods 2020).

9.3 Current and projected patterns of climate change across space and time

In order to understand how climate change has and will affect insects, we must also understand both current and projected changes in climate. With respect to temperature, the global temperature was 1°C warmer in 2010–19 relative to 1850–1900 (IPCC 2021). The incidence of hot extremes increased on every land mass except for central and eastern North America (southern South America and central Africa had insufficient data) (IPCC 2021). Over the next century, Arctic and Antarctic regions, the South American monsoon region and some mid-latitude and semi-arid regions are projected to experience warming at a rate that is much higher than the global rate (IPCC 2021). In general, land masses are projected to warm more rapidly than the oceans (IPCC 2021), such that "average" global warming will be greater over land areas.

With respect to precipitation, average global precipitation has increased relative to 1950, and storm tracks at mid-latitudes have shifted towards the poles (IPCC 2021). Changes in precipitation are somewhat consistent across the globe, where long-term data exist. The IPPCC Working Group 1 divided global land areas minus Antarctica into forty-five regions of equal land mass (IPCC 2021). The incidence of heavy precipitation has increased since 1950, with low to high certainty, in nine of those forty-five regions across the globe, with no decreases in any of the forty-five regions (IPCC 2021). There is low agreement in the type of change in eight regions, and insufficient data from the remaining eighteen regions, mostly in the Global South, to determine precipitation changes. Drought has decreased in northern Australia, but has increased in twelve to forty-five global regions since 2010 (IPCC 2021). Taken together, this reflects an intensification of the global water cycle (IPCC 2021). Projections call for a continuation of this pattern. They include an increase in precipitation in the Arctic and Antarctic, the equatorial Pacific and some other regions, and a decrease in some areas of the subtropics and tropics (IPCC 2021). The intensity of precipitation is expected to increase in general, leading to more variable surface water flows in most areas within and between years (IPCC 2021). This variation in water flow will be exacerbated in temperate and montane areas, as snow melt is expected to occur earlier, leading to less water storage as snow and higher earlier run-off (IPCC 2021).

9.4 Chapter overview

I synthesize the recent literature concerning documented insect responses to climate change. I focus on local and regional changes in population dynamics, including population state changes, or stable shifts in mean abundances. The recent focus on whether or not insects are declining globally has led to a number of reviews that summarize the available literature for changes in species richness, abundance or total biomass (Halsch et al. 2021; Hill et al. 2021; Høye 2020; Sánchez-Bayo and Wyckhuys 2019; van Klink et al. 2020a, 2020b; Wagner 2020; Wagner et al. 2021a, 2021b), as well as the mechanisms by which insect respond to climate change (Harvey, et al. 2022). I rely on these reviews and integrate additional studies appearing in the last two years, using key words based on each section below. Note that this was not meant to be an exhaustive literature search as would be done for a meta-analysis. I then examine climate-driven shifts in population vital rates at the local scale. These changes in vital rates are the mechanism underlying changes in population size. I end with an agenda for further understanding the impact of climate change on insects.

9.5 Is climate change driving an "Insect Apocalypse"?

Early reports of declines in insect abundance and biomass at specific locations led to a *New York Times* article in 2018, titled "The Insect Apocalypse is Here" (Jarvis 2018). That title notwithstanding, most questions remained unresolved. These questions are critical for understanding first, whether insect abundance and biomass is declining globally and catastrophically, and second, whether the primary driver of such declines is climate change. The first set of questions includes the geographical extent of declines; whether species' declines globally outweigh expansions; whether biomass, species richness and population numbers are all declining; what taxonomic groups are declining; and what species' traits are associated with declines. The second set of questions includes determination of the factors that are driving observed changes in insect mass and numbers. To grasp the limits of our current understanding, I next examine basic extinction dynamics, the types of evidence available, indirect effects of climate change and interactions with other types of global change, including land-use change.

9.6 Scales and trajectories of population declines and species extinctions

Extinction events can happen on several organizational, spatial and temporal scales. Populations may go extinct. Populations may go extinct on a regional scale, associated with changes in species' ranges. And populations may go extinct globally, such that the species is extinct. Population extinction may also be permanent. Alternatively, extinction may be an element of metapopulation dynamics, with repeated local extinction and colonization. Depending on the turnover rate for demes within metapopulations, and the spatial and temporal scale of monitoring, data from a metapopulation may look like a simple population extinction. This confounds efforts to document extinctions due to climate or other environmental changes.

Population extinction also may follow different temporal trajectories. These include a gradual decline in population size, extinctions due to large population fluctuations around a given mean, extinction vortices in which feedback loops hasten extinction (e.g., Fagan and Holmes 2006) and extinction debts in which extinctions are delayed beyond that expected (e.g., Figueiredo et al. 2019; Hylander and Ehrlén 2013;). We have mathematical theories supporting these trajectories (e.g., Fagan and Holmes 2006; Nieddu et al. 2014; Lande 1993; Williams et al. 2021). However, there is almost no experimental or observational data in wild insect populations connecting climate change to these specific dynamics. An example not connected with climate change is the work of Estay et al. (2012), who examined changes in the variance in population size in eight populations of each of two species of aphids in Britain. They showed that most changes in population size variance in *Elatobium abietinum* (Hemiptera: Aphididae) were associated with changes in the carrying capacity, but most observed changes in variance in *Drepanosiphum platanoidis* (Hemiptera: Aphididae) were associated with changes in the growth rate, hence reproductive and/or mortality rates. While none of these shifts resulted in population extinctions, they did alter the structure of the population dynamics. As we accumulate extinctions of populations that have long-term monitoring data, the opportunity exists to develop a better understanding of the trajectory leading to extinction itself, rather than just population dynamic state changes. This is important, because which trajectory a given population follows has implications for rates of decline and extinction.

9.7 Sources of evidence of population decline and species extinction

The first step in linking population declines and species extinctions to climate change is to demonstrate that in fact populations are declining and species are going extinct. Several approaches have been used to examine possible declines or extinctions of insect populations or species. First, long-term trapping sites have used malaise traps, pitfall traps, pan traps, light traps, emergence traps, fruit traps, etc. Standardized transect sighting surveys (summarized in Halsch et al. 2021) have also been done, usually for Lepidoptera with some version of a Pollard walk (Pollard and Yates 1993). These all generate data on species richness and often abundance and/or biomass (summarized in Halsch et al. 2021; see also Burner et al. 2021). Increasingly, DNA metabarcoding has been used to document trapped samples (e.g., Villalta et al. 2021). A twist on the trapping approach is Janzen and Wallach's (2021) surveys of tri-trophic interactions among plants, Lepidoptera and their parasitoids in Costa Rica. With colleagues, they have collected and reared lepidopteran larvae, and simultaneously passively sampled their parasitoids. These data yield information on changes in abundance of the parasitoid species through time. Perhaps the most exciting developing approach to trapping integrates technology with deep learning to fully automate monitoring (Besson et al. 2022).

A second approach relies on databases accumulated by citizen scientists. The most famous of these are the UK Butterfly Monitoring Scheme (UKMBS 2022), along with the North American Butterfly Association's (NABA) 4th of July butterfly counts (North American Butterfly Association 2017), PollardBase (PollardBase 2022) and Monarch Watch (MonarchWatch 2022). Some of these surveys occur sporadically at any given locale. This can make it difficult to detect trends, since the timing of the survey may occur at different phenological points across years. Correcting for accumulated degree days can help solve this problem (Clarridge 2016). The more robust of these datasets have yielded valuable insights (e.g., Wepprich et al. 2019). iNaturalist (2022) also provides a growing repository of community-vetted information on more diverse species occurrences, which may prove productive in the future.

Third, either museum records or earlier surveys combined with modern-day resampling can generate insights into population declines (e.g., Grixti et al. 2009; Sheard et al. 2021; Theng et al. 2020). Museum records may be biased by either geography or preference for rare species or phenotypes. For example, readily accessible locales, such as near campgrounds, may be more heavily sampled than areas far from roads. This can be an advantage, as more time depth is often available for these sites. Preference for rare species means that museum records provide a better approximation to species richness than to abundance.

All these methods have limitations for use in determining whether insect populations are declining. The first limitation derives from geographic coverage. Many long-term trap sites are in Europe, and to a lesser extent in North America and Costa Rica. Only six of seventy-three sites recorded by Sánchez-Bayo and Wyckhuys (2019) are located in the Global South. Citizen science efforts have historically been similarly biased, but a search on iNaturalist indicates that this is changing (<http://www.iNaturalist.org>). The second limitation derives from temporal depth of the data. Insect populations include classic examples of time-lagged population dynamics, population cycling, chaos and outbreak dynamics (e.g., Kozlov et al. 2010; Turchin and Taylor 1992). Avoiding confounding such dynamics with population declines is best done using data sets with multi-generational time depth.

Once we understand which populations and species are declining, determining whether climate change is a driver is the next step. Are the dynamics altered, compared to dynamics prior to an identifiable change in a relevant climate variable? This requires first separating effects of climate from land-use changes. This is not always easy for European long-term trapping sites, which are generally located at the interface between agricultural and non-agricultural land (e.g., Halsch et al. 2021). Long-Term Ecological Research sites (LTER) in the United States are generally located in areas where land-use change is not as prevalent, allowing a focus on effects of climate change (e.g., Crossley et al. 2020). Second, trap, transect and citizen science data generally date from no earlier than the 1960s (e.g., Halsch et al. 2021; Hellmann et al. 2003), whereas climate variables have been changing since the beginning of the Industrial Revolution in the 1800s (Abram et al. 2016). Finally, assigning temporal variation to climate change drivers is particularly problematic for community samples, for which some species may be increasing while others are declining due to differently lagged responses.

9.8 Spatial distribution and timelines of declines and extinctions due to climate change

Determining the extent of insect declines and extinctions that are due to climate change has proven to be a difficult undertaking. Early synthetic studies were critiqued as over-reach for global extrapolation from geographically limited data sets (e.g., by Wagner 2020), for an inability to partition land-use change from other factors driving declines (e.g., by Wagner 2020) or in some cases on grounds of faulty statistical methodology (by Daskalova et al. 2021). Wepprich et al. (2019) and Høye et al. (2021) also note that inclusion of even a few outlier data points, capturing sample variability, can affect the statistical analysis. Nonetheless, here I make some predictions and compare those to available data to draw some conclusions. I also discuss factors that will confound measurement of species' declines and extinctions.

9.8.1 *Where a given species resides within its climate envelope determines its fate*

As noted earlier (section 9.2.1.3; Figure 9.1), insects experience weather on spatial scales approximating the insect's size (e.g., Kingsolver and Watt 1983), and insect performance is dependent in part on where a given weather value falls within the performance–weather envelope. This leads to two predictions. First, declines should be greater for insects in architecturally or topographically simpler environments, where they are less likely to be able to buffer weather through behavior. Second, declines should be greater for populations at the edge of their climate envelope, in the context of their microclimate (e.g., Storlie et al. 2014).

Available studies support both predictions. However, data are scarce in support of the first prediction that lack of microclimate diversity derived from spatial or topographic heterogeneity should be associated with population declines (Pincebourde and Woods 2020). In one example, populations of *Euphydryas editha bayensis* (Lepidoptera: Nymphalidae) in topographically flatter habitats went extinct more rapidly than those in topographically heterogeneous habitats (Hellmann et al. 2003). More extensive support is accruing for the influence of a species' location within its climate envelope on population expansions versus extinctions. For example, Antão et al. (2022) documented a shift in location within the climate envelope for moths and butterflies across a latitudinal gradient in Finland. Such shifts led to both increases and decreases in species' presence, depending on the extent of climate change and the initial location of the species in its climate envelope. Likewise, ant communities at thirty-three sites across the United States showed a significant nonlinear relationship between average upper critical temperature for species within a genus and the proportion of sites with higher incidence for that genus over a twenty-year period (Roeder et al. 2021). That is, the incidence of genera increased under climate change with increasing average upper critical temperature to a maximum, and then fell slightly. This would be expected if the sampling sites span species with a range of locations within the thermal range, and suggests that the upper value of the insects' climate envelope affected their fate. No such relationship was found for the average lower critical temperature, again as expected under climate change.

9.8.2 *Rates of climate change are correlated with rates of decline and extinction*

Geographic areas with a greater change in climate means, variances or extremes should exhibit a faster rate of decline and more extinctions. However, we still expect considerable variation in responses due to the relationship between the baseline weather relative to the species' climate envelope (see Figure 9.1).

Areas with greater rates of temperature increase include northern latitudes, while extremes in hot temperature were less pronounced in central and eastern North America (IPCC 2021). The work of Antão et al. (2022) across a latitudinal gradient in Finland (discussed above) yielded differences in population response rates as a function of rates of climate change. The authors noted that the initial position within the species' climate envelope was critical to the type of response detected. This further highlights the need for more fine-grained studies.

However, not all high latitude studies provide support for the hypothesis. In the Arctic, using twenty-four years of data from Zackenberg, Greenland Høye et al. (2021) found, first, that temperature and winter duration did not change linearly with time, and summer precipitation showed no trend with time. Thus, local change did not follow regional patterns. Second, as expected based on local change, they found either nonlinear changes in abundance and diversity, or no detectable trend, depending on habitat. Third, trends differed among families, as well as by position in the food web. Amalgamating across species with diverse positions in their climate envelope, diverse microhabitats and diverse relations to other species muddies the analysis.

9.8.3 *Extreme weather events have significant effects on insects*

A shift in mean climate and/or in climate variance results in an increase in the frequency and/or severity in extreme weather events (Figure 9.2). Even if the new mean climate is included within a species' climate envelope, there may be more extreme events that fall outside the climate envelope. The effects on population dynamics of these extreme weather events will of course depend in part on the duration of the event relative to the tolerance of the insect. Ma et al. (2021) and Filazzola et al. (2021) review the effects of more extreme events on insects. Such effects range from shifts in ion balance in response to heat

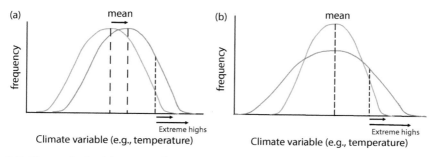

Figure 9.2 Change in the (a) mean or (b) variance of a climate parameter result in a change in the frequency or severity of extreme events. The blue curve represents the original distribution of values for the climate parameter. The brown curve represents the new distribution.

extremes, which lead to reduced mobility in *Locusta migratoria* (Orthoptera: Acrididae) (O'Sullivan et al. 2017); to changes in thermoregulatory morphology as a plastic response to heat stress in the dragonfly *Erythemis collocata* (Odonata: Libellulidae), which leads to reduced flight ability (McCauley et al. 2018); to desiccation in response to drought stress in collembolans (Elnitsky et al. 2008).

9.8.4 *Confounding factors: Metapopulation structure*

Detection of declines and extinctions will be particularly fraught for species with a metapopulation structure. Local demes normally go extinct and are recolonized, independent of climate change. Detecting extinctions due to metapopulation dynamics versus true extinctions, or changes in rates of extinctions will necessitate regional sampling. This dynamic is so far not addressed in analyses of general insect trap data.

Rates of colonization–extinction in a metapopulation may also affect the response to climate change. Johansson et al. (2022) showed that *Euphydryas aurinia*'s (Lepidoptera: Nymphalidae) survival of an extreme drought in Gotland was due in part to high connectivity among metapopulation patches resulting in favorable colonization–extinction dynamics. This metapopulation also has relatively large patches, which may insulate it from climate extremes due to both increased colonization associated with large patches and increased availability of diverse microclimatic refuges. The increase in overall butterfly population size and patch occupancy occurred in spite of a lack of full recovery of the larval host plant. The authors acknowledge that factors such as drought-induced changes in host plant quality could have played a role in the rapid recovery of the butterfly, but they point to documented effects of patch connectivity and size as critical factors. Acevedo et al. (2015) found that asymmetric differences in effective distances between patches affected colonization–extinction dynamics in a wind dispersed orchid. That is, prevailing winds make distances effectively shorter for downwind patches, and longer for upwind patches. Since climate change could affect wind intensity and direction, this needs greater attention in insects.

Climate change may also alter the spatial synchrony of weather drivers of metapopulation dynamics. The classic metapopulation system, *Melitaea cinxia* (Lepidoptera: Nymphalidae) in the Åland Islands provides an example. Increasing frequency of drought (a spatially synchronized weather driver) has increased the synchrony of regional metapopulation dynamics. This has resulted in an increased risk of extinction of the system (Tack et al. 2015).

9.8.5 *Confounding factors: Direct effects of phenology and voltinism*

Climate change can lead to changes in insect phenology, and in particular to changes in voltinism. As the climate warms in temperate and arctic areas, a longer growing season can lead to opportunities for an increase in the number of generations per year (reviewed in Forrest 2016). Three patterns may result from the longer growing season, depending on the relative importance of temperature and photoperiod in controlling diapause. First, if photoperiod plays no role, an increase in growing season length may lead to population growth and outbreaks (see section 3). Second, a longer growing season for these same species can lead to a final partial generation, which results in a decline in the first generation the following year. If this pattern repeats over several years, it may manifest as a population decline, as in *Lasiommata megera* (Lepidoptera: Nymphalidae) (Van Dyck et al. 2015) in Europe. The decline may be temporary in the face of continued climate change. As the growing season continues to expand, the last generation becomes a complete generation, and the population may resume increasing in size. This will be true as long as the insects' thermal climate envelope is not exceeded. Third, heat accumulation may result in a cue for diapause induction before the solstice such that the insect interprets the short daylength as fall and goes into diapause early (Grevstad and Coop 2015). This reduces voltinism and may result in a decline in abundance, since sequential generations cannot build population numbers through the growing season.

These diverse responses to changes in temperature and growing season highlight underlying causes of diverse changes in species richness, abundance or mass over time at sampling stations. This draws attention again to the need to analyze such data at finer taxonomic scales.

9.8.6 *Confounding factors: Declines in abundance vs biomass vs species richness*

Depending on the monitoring scheme, researchers record some combination of species richness, biomass and abundance. These can show different patterns of change (Vereecken et al. 2021). For example, species may be going extinct over time at a particular location but new species may be recorded as species' ranges change, such that species richness remains constant. Likewise, the relationship between biomass and abundance will depend on the relative numbers of small- and large-bodied insects. For example, Homburg et al. (2019) found no change in carabid beetle biomass in a twenty-four-year trapping study in a nature reserve in northern Germany, but significant declines in species richness and phylogenetic diversity. In northwestern China, a carabid beetle assemblage showed different rates of decline in species abundance and richness (Liu et al. 2021). In Greenland, Høye et al. (2021) documented opposing trends in abundance and diversity for total arthropod trap data across three habitat types, but with large variation at the family level.

Declines in species richness, biomass and abundance, along with functional and phylogenetic diversity, have different implications for ecosystem health, including food web structure. Discussion of insect declines needs to be broadened to include and separate all of these metrics at various taxonomic and functional group resolutions if we are to be able to assess the full impact of climate change. This discussion also necessitates data collection across an increased geographic range, with varying land uses, at a finer grained scale than currently occurs.

9.9 Indirect effects of climate on insect declines and extinctions

Climate change directly affects species with which insects interact, leading to indirect effects on the focal insect's population dynamics. Affected species can be food resources, predators, parasites and pathogens, or competitors. Mechanisms include phenological mismatches (reviewed in Renner and Zohner 2018), changes in host plant quality (e.g., Kuczyk et al. 2021) or floral resources (e.g., Descamps et al. 2021) and changes in abundance of one or both interacting species.

Evidence for phenological mismatches between insects and flowers or between insects and their herbaceous larval hosts in temperate areas is scant (reviewed in Kharouba and Wolkovich 2020; Renner and Zohner 2018). Kudo and Ida (2013) present one of the most thorough studies, showing that in early spring, onset of flowering of *Corydalis ambigua* (Papaveraceae) was earlier than first emergence of its queen bee pollinators. They demonstrate negative effects on seed set under such circumstances, but do not address effects on abundances of the insects.

Selection or plasticity that allows populations to buffer their phenology will be critical to whether a mismatch occurs. Examining a set of Mediterranean butterflies, Colom et al. (2022) found that the sensitivity of phenology to climate change, and the ability to match changes in climate variability across years, is critical to avoiding a phenological mismatch.

The question also arises as to at what point the phenologies of two interacting species no longer overlap, even under extreme selection to do so. Simmonds et al. (2020) used structural equation modeling to explore the point at which the great tit-caterpillar system becomes phenologically mismatched. Even under directional selection, if the bird phenology lagged the caterpillar phenology by more than twenty-four days, the birds went extinct. Potentially, the result for the insects is new predator-free space, and a population outbreak.

Currently available studies of phenological mismatch have several limitations. First, they have generally neglected consideration of the distribution of phenological events (Inouye et al. 2019). Using data from a long-term monitoring project in Colorado, USA, Stemkovski et al. (2022) show that bees and flowers differ in the extent of skew in their phenology curves for emergence. Such differences, if common, would have significant implications for resilience of interactions to climate change. Second, studies have ignored community context—that is, the broader set of competitors, predators and mutualists within which the two focal species are embedded. Pardikes et al. (2022) demonstrated in a laboratory experimental setting using *Drosophila* and its parasitoids that community context could alter the effects of phenological mismatch.

Changes in host plant quality in temperate regions as a function of heat and drought are relatively well studied. For example, Kuczyk et al. (2021) showed that *Sinapis alba* (Brassicaceae), the host plant for larval *Pieris napi* (Lepidoptera: Pieridae) in Europe, had greater C:N ratios and glucosinolate concentrations in response to high temperatures but lower glucosinolate concentrations in response to drought stress. Female butterflies preferred to lay eggs on plants with high glucosinolate concentrations but larvae performed better on plants with low glucosinolate concentrations. These results reflect species interaction responses that will alter the population dynamics of both participants in the interaction, under scenarios of changes in the means and extremes of temperature and precipitation.

Finally, changes in the abundance of one or more competitors, predators or mutualists may affect the probability of a given insect species' decline. For example, the population dynamics of *Speyeria mormonia* (Lepidoptera: Nymphalidae) in Colorado, USA, are partially determined by floral nectar availability, which in turn is affected by early season frosts in years with early snow melt (Boggs and Inouye 2012). Top-down changes in insect population size can also occur via indirect effects of climate change on predator behavior. Barton (2014) showed that reduced wind results in more efficient foraging by lady bugs (*Harmonia axyridis* (Coleoptera: Coccinellidae)) on aphids (*Aphis glycines* (Hemiptera: Aphididae)), reducing aphid population sizes. One observed change in climate is reduced average wind speeds (e.g., Vautard et al. 2010). While not yet the case for the butterfly or the aphid, such indirect effects of climate change may result in extinction cascades, as reviewed for insects by Kehoe et al. (2021). McPeek (2022) provides a template for using food web dynamics to understand the effects of changes in abundance on other members of a community. Such an approach will be necessary to tease apart the complex effects of climate change via species interactions.

9.10 Interaction between climate and other types of global change

One of the difficulties in detecting effects of climate change on insect dynamics is that ongoing global change includes not just climate but also land use, pollution, etc. Interactions among climate and other global change parameters may be non-additive and/or indirect, complicating analyses at the species level, let alone the community level (see Yang et al. 2021 and Outhwaite et al. 2022 for recent reviews). For example, Forister et al. (2010) documented greater declines through time in butterfly species richness at lower elevations in California's Sierra Nevada as compared to higher elevations. They attribute this difference to more extensive habitat loss at lower elevations. Likewise, Assandri (2021) argued that changes in land use (including drainage and reclamation, intensification or abandonment) and change in pollutant levels have interacted with climate change to result in significant extinctions of dragonfly communities in low elevations of the European Alps. For freshwater systems, van Klink et al. (2020a, 2020b) suggested that increases in species abundance in freshwater systems was due to cleaning up pollution, which outweighed effects of climate change. At the most basic level, understanding whether and how the effects of climate change on insect populations are amplified or lessened by other aspects of global change is critical in order to mitigate insect declines.

9.11 So, is there an insect apocalypse?

Despite concerns mentioned above about geographic, habitat and taxonomic coverage, effects of land use change and pollution, and statistical analysis, overall current mean rates of local extinction for insect species as a result of climate change are estimated in the vicinity of 1 percent per year (summarized in Pinedo-Escatel et al. 2021; Theng et al. 2020; van Klink et al. 2020b; Wagner 2020; Wagner et al. 2021; Warren et al. 2021;). Rates vary across habitats, location and taxa (reviewed in Wagner et al. 2021), and depend on the time depth of the study (Bell et al. 2020). Not all groups in all locations are declining; there is much heterogeneity in the record. Intriguingly, insect cyclic outbreak dynamics are also damped in some species (Esper et al. 2007). In contrast, Harvey et al. (2020) argued that the interplay between climate-driven change and trophic disruption may lead to increased incidence of outbreaks. We do not yet have the data or the insights to understand the mechanistic drivers underlying this heterogeneity.

The data thus suggest that insects are participating in the global Sixth Extinction. But are they? Schachat and Labandeira (2021) argue that the insect fossil record shows faunal turnover, rather than mass extinction during previous global mass extinction events. If extinctions progress to the loss of families or orders, they argue that insects will actually be entering their first mass extinction. That would indeed be an insect apocalypse.

9.12 Is climate change driving insect population increases and explosions?

Not all insect populations are declining. We see a mix of "winner and loser species" (*sensu* Filgueiras et al. 2021). Especially in terrestrial temperate and arctic areas, species at the cooler edge of their thermal climate envelope are increasing (e.g., Antão et al. 2022). In areas where winter temperatures are warming, species that used to suffer high mortality from winter cold stress are generally increasing in abundance, although species subjected to changes in snow cover leading to intermittent freeze–thaw cycles or that lose their required chilling period during diapause may not show this pattern (reviewed in Bale and Hayward 2010; Simler-Williamson et al. 2019). Species that can now add an entire additional generation during the growing season are also increasing in abundance, as exemplified by European Lepidoptera (Altermatt 2009). However, as for population declines, responses to climate change are dependent on species, community and habitat characteristics. This mix of winners and losers has implications beyond the idea of an insect apocalypse, encompassing changes in community structure that will propagate through the ecosystem.

9.13 Underlying vital rates

Vital rates, or age-specific survival and fecundity, are the mechanism that underlie changes in population abundances, hence species richness, abundance and biomass. Development rates control generation time, which also influences population dynamics. Dispersal is also important at a regional scale. To build a predictive understanding of insect declines and outbreaks, we must understand (i) how vital rates respond to climate change, (ii) which vital rates control rates of population growth, and (iii) whether the identity of vital rates controlling population growth changes with changes in climate. Changes in climate means and variances, including extreme events, are important considerations.

9.14 Effects of climate change on vital rates

9.14.1 *Survival, fecundity and virility: Responses to change*

In general, survival and fecundity in insects follow a hump-shaped curve in response to temperature. There is a peak at some intermediate temperature, with declines around that peak (reviewed in Kingsolver and Buckley 2020). The response can be plastic. For example, heat-hardening improves survival under temperature extremes in *Drosophila virilis* (Diptera: Drosophilidae) (Walsh et al. 2021).

Virility, including sperm production and viability, is reduced under temperature extremes in those insects that have been studied (e.g., in *D. virilis*: Walsh et al. 2021; in *Tribolium castaneum* (Coleoptera: Tenebrionidae): Sales et al. 2018). Heat extremes can sterilize males. Sales et al. (2018) further describe a transgenerational effect of heat extremes on male sperm viability. Offspring fathered by males subjected to heat stress, or by sperm stored in females subjected to heat stress, had reduced survival and fecundity. In *T. castaneum*, the same heat extreme had no effect on female fecundity, indicating that the effects of heat waves on each sex's reproductive vital rates are independent (Sales et al. 2018).

Complicating matters, survival and reproduction may have different thermal optima. Survival was better at low temperatures in the field for two damselfly species (*Calopteryx splendens* and *C. virgo* (Odonata: Calopterygidae)), while male mating success and hence female fecundity was better at higher temperatures (Svensson et al. 2020). Integrating these different optima and responses to extreme events into a population dynamics model would allow us to explore the population dynamic response to changing temperatures (see Amarasekare and Savage 2012 for a review and interesting approach). Exploring population dynamic outcomes for an array of temperature responses would give insight into both the characteristics of species that are likely to decline in response to temperature change and to the breadth of declines across types of species in response to give climate scenarios.

Understanding the interaction between precipitation (including snow) and temperature on vital rates in the field has proven elusive. Diverse studies have documented effects or lack thereof on both, on population growth patterns or on over-winter survival (e.g., Convey et al. 2015; Mitrus 2016; Roland et al. 2021), but studies across the life cycle are missing. One exception is work by Radchuk et al. (2013), who document differences in the effects of warming on different life stages of *Boloria eunomia* (Lepidoptera: Nymphalidae). Warming had a positive impact on nearly all life stages, except for over-winter larval survival, for which the effect was negative. The net result of warming for population dynamics was also negative.

9.14.2 *Development rate and generation time: Responses to change*

Thermal sensitivity in insects varies through ontogeny (reviewed in Kingsolver and Buckley 2020). Eggs, larvae and pupae differ in their minimum temperature threshold for development, and in the total number of degree-days above that minimum required to move to the next life stage. The minimum temperature threshold was on average greater for pupae than for eggs and larvae (Kingsolver and Buckley 2020). Insect order also was a significant effect. Although the differences among life stages were small, they have a significant impact on time spent in each life stage, and on overall generation time. Such differences in thermal sensitivity will also influence the effect on development rate of differences among seasons in thermal climate change. For example, if the minimum temperature for development is high for larvae compared to current temperatures during the spring, increasing spring temperatures should result in faster larval development, shorter generation times, alterations in phenology and, possibly, population growth. This area cries out for more work to enhance our understanding.

9.14.3 *Buffering, lability and changes in which vital rate(s) drives the population size, and implications for unexpected state changes in population dynamics*

The relationship between age-specific vital rates and population dynamics in the context of climate change will be affected by two factors. First, what is the elasticity of population growth in response to changes in means and variance in each vital rate (e.g., Boyce at al. 2006)? This reflects the extent to which the vital rate drives the population dynamics, with greater elasticity reflecting a greater importance of the vital rate. Second, how buffered or labile is each vital rate to changes in the mean and variance of climate parameters (e.g., McDonald et al. 2017)? I know of no case studies in insects exploring this question. It is particularly important because of the possibility that the degree of buffering or lability of vital rates to changing climates might alter the relative importance of each vital rate as drivers of population dynamics. It is not inconceivable that such an event could result in a permanent state change (extinction or explosion) in population size. Exploration of this possibility is needed, through some combination of modeling, experimentation or observation.

9.15 Conclusions

The available evidence indicates that insects are generally declining, although the decline is far from uniform across space and taxa, or with respect to abundance, species richness and biomass. Separating the effects of climate change from other global changes is challenging. Much of the available evidence is correlative rather than either experimental or based on a solid understanding of the mechanisms driving insect population and community dynamics. Gaining that insight, and moving beyond correlations, will be crucial for advancing our understanding of the effects of climate change.

In addition, much of our present understanding focuses on the effects of temperature. Precipitation, be it as rain or snow, is also changing and projected to change further. More attention is needed to the effects of this climate variable, including its interaction with temperature, humidity, and soil moisture.

I have focused here primarily on terrestrial insects, but insects with aquatic life stages are a significant component of insect biomass. Some of the conclusions drawn here apply to aquatic insects as well. Nonetheless, a deeper understanding of the trends in abundance, richness and biomass separated from changes in land use and pollution is warranted.

Of necessity, ecologists tend to focus on one or a few levels of organization when understanding challenges posed by climate change. How is species richness changing? How is the food web changing? How is abundance changing? But nature doesn't have such a restricted focus. To understand the effects that we are having on the insect world via climate change fully we need to integrate across levels of organization, from cellular to community, and then return to our key questions about insect decline (Harvey et al. 2020). One way to accomplish this is through integrative studies at "model" locations, using information from basic and applied studies, to generate and test models, using techniques such as piece-wise structural equation modeling. An alternative is a similar focus on "model" communities or habitats. Either or both will help provide the answers that we need to understand the effects of climate change on insects, as important parts of terrestrial and freshwater ecosystems.

Key reflections

- The response of insect diversity, abundance and biomass to climate change is heterogeneous with respect to geography, habitat and taxon. However, the data we have indicate that a number of groups are in significant decline in various locations.
- Our understanding of insects' population dynamics in response to climate change is hampered by confounding effects of other types of global change, insufficient baseline data across diverse geographies and habitats and inadequate understanding of indirect effects due to changes in predator, competitor and prey populations.

- We also lack an understanding of the responses of insect vital rates to climate change under field conditions. Are vital rates that control population dynamics buffered against climate change or are they labile in situations where weather fluctuates on various time scales? Does climate change result in a shift in which vital rates control population dynamics? If so, what are the resulting effects on insect abundance?
- Answers to these questions will increase our ability to predict which species are most at risk of decline in response to climate change, and which are most likely to exhibit population outbreaks.

Key further reading

- Forrest, J.R.K. 2016. Complex responses of insect phenology to climate change. *Current Opinion in Insect Science* 17: 49–54. doi:10.1016/j.cois.2016.07.002
- Halsch, C.A., Shapiro, A.M., Fordyce, J.A., Nice, C.C., Thorne, J.H., Waetjen, D.P., and Forister, M.L. 2021. Insects and recent climate change. *Proceedings of the National Academy of Sciences* 118: e2002543117. doi:10.1073/pnas.2002343117
- Harvey, J.A., Tougeron, K., Gols, R., Heinen, R., Abarca, M., Abram, P.K., Bassett, Y., *et al.* 2023. Scientists' warning on climate change and insects. *Ecological Monographs.* doi:10.1002/ecm.1553
- Pincebourde, S., and Woods, H.A. 2020. There is plenty of room at the bottom: microclimates drive insect vulnerability to climate change. *Current Opinion in Insect Science* 41: 63–70. doi:10.1016/j.cois.2020.07.001
- Schachat, S.R., and Labandeira, C.C. 2021. Are insects heading toward their first mass extinction? Distinguishing turnover from crises in their fossil record. *Annals of the Entomological Society of America* 114: 99–118. doi:10.1093/aesa/saaa042
- Yang, L.H., Postema, E.G., Hayes, T.E., Lippey, M.K., and MacArthur-Waltz, D.J. 2021. The complexity of global change and its effects on insects. *Current Opinion in Insect Science* 47: 90–102. doi:10.1016/j.cois.2021.05.001

References

Abrams, N.J., McGregor, H.V., Tierney, J.E., Evans, M.N., McKay, N.P., Kaufman, D.S., and the PAGES 2K Consortium. 2016. Early onset of industrial-era warming across oceans and continents. *Nature* 536: 411–18.

Acevedo, M.A., Fletcher, R.J., Tremblay, R.L., and Melendez-Ackerman, E.J. 2015. Spatial asymmetries in connectivity influence colonization-extinction dynamics. *Oecologia* 179: 415–24.

Altermatt, F. 2009. Climatic warming increases voltinism in European butterflies and moths. *Proceedings of the Royal Society B: Biological Sciences* 277: 1281–1287. doi:10.1098/rspb.2009.1910.

Amarasekare, P. and Savage, V. 2012. A framework for elucidating the temperature dependence of fitness. *The American Naturalist* 179: 178–91. doi:10.1086/663677.

Antão, L.H., Weigel, B., Strona, G., Hällfors, M., Kaarlejärvi, E., Dallas, T., Opedal, O., et al. 2022. Climate change reshuffles northern species within their niches. *Nature Climate Change.* doi:10.1038/s41558-022-01381-x.

Assandri, G. 2021. Anthropogenic-driven transformations of dragonfly (Insecta: Odonata) communities of low elevation mountain wetlands during the last century. *Insect Conservation and Diversity* 14 (1): 26–39.

Bale, J.S. and Hayward, S. 2010. Insect overwintering in a changing climate. *Journal of Experimental Biology* 213: 980–94.

Barton, B.T. 2014. Reduced wind strengthens top-down control of an insect herbivore. *Ecology* 95 (9): 2375–2381.

Bell, J. R., Blumgart, D., and Shortall, C.R. 2020. Are insects declining and at what rate? An analysis of standardised, systematic catches of aphid and moth abundances across Great Britain. *Insect Conservation and Diversity* 13 (2): 115–26.

Besson, M., Alison, J., Bjerge, K., Gorochowski, T.E., Høye, T., Jucker, T., Mann, H.M.R., et al.2022. Towards the fully automated monitoring of ecological communities. *Ecology Letters* 25 (12): 2753–2775.

Boggs, C.L. and Inouye, D.W. 2012 A single climate driver has direct and indirect effects on insect population dynamics. *Ecology Letters* 15 (5): 502–08.

Boyce, M.S., Haridas, C.V., Lee, C.T., and the NCEAS Stochastic Demography Working Group. 2006. Demography in an increasingly variable world. *Trends in Ecology and Evolution* 21: 141–48. doi:10.1016/j.tree.2005.11.018.

Burner, R.C., Selås, V., Kobro, S., Jacobsen, R.M., and Sverdrup-Thygeson, A. 2021. Moth species richness and diversity decline in a 30-year time series in Norway, irrespective of species' latitudinal range extent and habitat. *Journal of Insect Conservation* 25: 887–96.

Clarridge, A. 2016. *North American Butterfly Association counts at Congaree National Park: A case study for connecting citizen science to management.* MEERM thesis. University of South Carolina.

Colom, P., Ninyerola, M., Pons, X., Traveset, A., and Stefanescu, C. 2022. Phenological sensitivity and seasonal variability explain climate-driven trends in Mediterranean butterflies. *Proceedings of the Royal Society B: Biological Sciences* 289 (1973): 20220251.

Convey, P., Abbandonato, H., Bergan, F., Beumer, L.T., Biersma, E.M., Bråthen, V.S., D'Imperio, L. et al. 2015. Survival of rapidly fluctuating natural low winter temperatures by High Arctic soil invertebrates. *Journal of Thermal Biology* 54: 111–17.

Crossley, M.S., Meier, A.R., Baldwin, E.M., Berry, L.L., Crenshaw, L.C., Hartman, G.L., and Moran, M.D. 2020. No net insect abundance and diversity declines across US Long Term Ecological Research sites. *Nature Ecology and Evolution* 4 (10): 1368–1376.

Daskalova, G.N., Phillimore, A.B., and Myers-Smith, I.H. 2021. Accounting for year effects and sampling error in temporal analyses of invertebrate population and biodiversity change: A comment on Seibold et al. 2019. *Insect Conservation and Diversity* 14 (1): 149–54.

Descamps, C., Quinet, M. and Jacquemart, A.-L. 2021. Climate change-induced stress reduce quantity and alter composition of nectar and pollen from a bee-pollinated species (*Borago officinalis*, Boraginaceae). *Frontiers in Plant Science* 12: 755843. doi:10.3389/fpls.2021.755843.

Duell, M.E., Gray, M.T., Roe, A.D., MacQuarrie, C.J., and Sinclair, B.J. 2022. Plasticity drives extreme cold tolerance of emerald ash borer (Agrilus planipennis) during a polar vortex. *Current Research in Insect Science* 2: 100031.

Elnitsky, M.A., Benoit, J.B., Denlinger, D.L., and Lee Jr., R.E. 2008. Desiccation tolerance and drought acclimation in the Antarctic collembolan Cryptopygus antarcticus. *Journal of Insect Physiology* 54 (10–11): 1432–1439.

Esper, J., Büntgen, U., Frank, D.C., Nievergelt, D., and Liebhold, A. 2007. 1200 years of regular outbreaks in alpine insects. *Proceedings of the Royal Society B: Biological Sciences* 274 (1610): 671–79.

Estay, S.A., Lima, M., Labra, F.A., and Harrington, R. 2012. Increased outbreak frequency associated with changes in the dynamic behaviour of populations of two aphid species. *Oikos* 121(4): 614–22.

Fagan, W.F. and Holmes, E. 2006. Quantifying the extinction vortex. *Ecology Letters* 9 (1): 51–60.

Figueiredo, L., Krauss, J., Steffan-Dewenter, I., and Cabral, J.S. 2019. Understanding extinction debts: Spatio-temporal scales, mechanisms and a roadmap for future research. *Ecography* 42: 1973–1990. doi:10.1111/ecog.04740.

Filazzola, A., Matter, S.F., and MacIvor, J.S. 2021. The direct and indirect effects of extreme climate events on insects. *Science of the Total Environment* 769: 145161.

Filgueiras, B.K., Peres, C.A., Melo, F.P., Leal, I.R., and Tabarelli, M. 2021. Winner–loser species replacements in human-modified landscapes. *Trends in Ecology and Evolution* 36 (6): 545–55.

Foote, N.E., Davis, T.S., Crowder, D.W., Bosque-Pérez, N.A., and Eigenbrode, S.D. 2017. Plant water stress affects interactions between an invasive and a naturalized aphid species on cereal crops. *Environmental Entomology* 46 (3): 609–16.

Forister, M.L., McCall, A.C., Sanders, N.J., Fordyce, J.A., Thorne, J.H., O'Brien, J., Waetjen, D.P., et al. 2010. Compounded effects of climate change and habitat alteration shift patterns of butterfly diversity. *Proceedings of the National Academy of Sciences* 107 (5): 2088–2092.

Forrest, J. R. 2016. Complex responses of insect phenology to climate change. *Current Opinion in Insect Science* 17: 49–54.

Gols, R., Ojeda-Prieto, L.M., Li, K., van Der Putten, W.H., and Harvey, J.A. 2021. Within-patch and edge microclimates vary over a growing season and are amplified during a heatwave: Consequences for ectothermic insects. *Journal of Thermal Biology* 99: 103006.

Grevstad, F.S. and Coop, L.B. 2015. The consequences of photoperiodism for organisms in new climates. *Ecological Applications* 25 (6): 1506–1517.

Grixti, J.C., Wong, L.T., Cameron, S., and Favret, C. 2009. Decline of bumble bees (*Bombus*) in the North American midwest. *Biological Conservation* 142: 75–84. doi:10.1016/j.biocon.2008.09.027.

Halsch, C.A., Shapiro, A.M., Fordyce, J.A., Nice, C.C., Thorne, J.H., Waetjen, D.P., and Forister, M.L. 2021. Insects and recent climate change. *Proceedings of the National Academy of Sciences* 118 (2): e2002543117.

Harvey, J.A., Heinen, R., Gols, R., and Thakur, M.P. 2020. Climate change-mediated temperature extremes and insects: From outbreaks to breakdowns. *Global Change Biology* 26 (12): 6685–6701.

Harvey, J.A., Tougeron, K., Gols, R., Heinen, R., Abarca, M., Abram, P.K., Bassett, Y., et al. 2023. Scientists' warning on climate change and insects. *Ecological Monographs* 93 (1): e1553.

Hellmann, J.J., Weiss, S.B., Mclaughlin, J.F., Boggs, C.L., Ehrlich, P.R., Launer, A.E., and Murphy, D.D. 2003. Do hypotheses from short-term studies hold in the long-term? An empirical test. *Ecological Entomology* 28 (1): 74–84.

Hill, G.M., Kawahara, A.Y., Daniels, J.C., Bateman, C.C., and Scheffers, B.R. 2021. Climate change effects on animal ecology: Butterflies and moths as a case study. *Biological Reviews* 96 (5): 2113–2126.

Homburg, K., Drees, C., Boutaud, E., Nolte, D., Schuett, W., Zumstein, P., von Ruschkowski, E. et al. 2019. Where have all the beetles gone? Long-term study reveals carabid species decline in a nature reserve in Northern Germany. *Insect Conservation and Diversity* 12: 268–77. doi:10.1111/icad.12348.

Høye, T.T. 2020. Arthropods and climate change—Arctic challenges and opportunities. *Current Opinion in Insect Science* 41: 40–45.

Høye, T.T., Loboda, S., Koltz, A.M., Gillespie, M.A., Bowden, J.J., and Schmidt, N.M. 2021. Nonlinear trends in abundance and diversity and complex responses to climate change in Arctic arthropods. *Proceedings of the National Academy of Sciences* 118 (2): e2002557117.

Hylander, K. and Ehrlén, J. 2013. The mechanisms causing extinction debts. *Trends in Ecology and Evolution* 28 (6): 341–46.

iNaturalist. 2022. <https://www.inaturalist.org/> accessed June 21, 2022.

Inouye, B.D., Ehrlén, J., and Underwood, N. 2019. Phenology as a process rather than an event: From individual reaction norms to community metrics. *Ecological Monographs* 89 (2): e01352.

IPCC. 2021. "Summary for policymakers." In *Climate Change 2021: The Physical Science Basis. Contribution of Working Group I to the Sixth Assessment Report of the Intergovernmental Panel on Climate Change*, edited by V. Masson-Delmotte, P. Zhai, A. Pirani, S.L. Connors, C. Péan, S. Berger, N. Caud, Y. Chen, L. Goldfarb, M.I. Gomis, M. Huang, K. Leitzell, E. Lonnoy, J.B.R. Matthews, T.K. Maycock, T. Waterfield, O. Yelekçi, R. Yu, and B. Zhou. Cambridge: Cambridge University Press, 3–32. doi:10.1017/9781009157896.001.

Janzen, D.H. and Hallwachs, W. 2021. To us insectometers, it is clear that insect decline in our Costa Rican tropics is real, so let's be kind to the survivors. *Proceedings of the National Academy of Sciences* 118 (2): e2002546117.

Jarvis, B. 2018. The insect apocalypse is here. *New York Times Magazine*, 27 (2018). https://www.nytimes.com/2018/11/27/magazine/insect-apocalypse.html.

Johansson, V., Kindvall, O., Askling, J., Säwenfalk, D.S., Norman, H., and Franzén, M. 2022. Quick recovery of a threatened butterfly in well-connected patches following an extreme drought. *Insect Conservation and Diversity* 15 (5): 572–82.

Kehoe, R., Frago, E., and Sanders, D. 2021. Cascading extinctions as a hidden driver of insect decline. *Ecological Entomology* 46 (4): 743–56.

Kharouba, H.M. and Wolkovich, E.M. 2020. Disconnects between ecological theory and data in phenological mismatch research. *Nature Climate Change* 10 (5): 406–15.

Kingsolver, J.G. and Buckley, L.B. 2020. Ontogenetic variation in thermal sensitivity shapes insect ecological responses to climate change. *Current Opinion in Insect Science* 41: 17–24.

Kingsolver, J.G. and Watt, W.B. 1983. Thermoregulatory strategies in Colias butterflies: Thermal stress and the limits to adaptation in temporally varying environments. *The American Naturalist* 121 (1): 32–55.

Kingsolver, J.G. and Woods, H.A. 2016. Beyond thermal performance curves: Modeling time-dependent effects of thermal stress on ectotherm growth rates. *The American Naturalist* 187 (3): 283–94.

Knouft, J.H. and Ficklin, D.L. 2017. The potential impacts of climate change on biodiversity in flowing freshwater systems. *Annual Review of Ecology, Evolution, and Systematics* 48: 111–33.

Kozlov, M.V., Hunter, M.D., Koponen, S., Kouki, J., Niemelä, P., and Price, P.W. 2010. Diverse population trajectories among coexisting species of subarctic forest moths. *Population Ecology* 52: 295–305.

Kuczyk, J., Müller, C., and Fischer, K. 2021. Plant-mediated indirect effects of climate change on an insect herbivore. *Basic and Applied Ecology* 53: 100–13.

Kudo, G. and Ida, T.Y. 2013. Early onset of spring increases the phenological mismatch between plants and pollinators. *Ecology* 94 (10): 2311–2320.

Lande, R. 1993. Risks of population extinction from demographic and environmental stochasticity and random catastrophes. *The American Naturalist* 142 (6): 911–27.

Liu, X., Wang, X., Bai, M., and Shaw, J.J. 2021. Decrease in carabid beetles in grasslands of northwestern China: Further evidence of insect biodiversity loss. *Insects* 13: 35. doi:10.3390/insects13010035.

Ma, C.-S., Ma, G., and Pincebourde, S. 2021. Survive a warming climate: Insect responses to extreme high temperatures. *Annual Review of Entomology* 66: 163–84. doi:10.1146/annurev-ento-041520-074454.

Masson-Delmotte, V., Zhai, P., Pirani, A., Connors, S.L., Péan, C., Berger, S., Caud, N., et al. (eds). 2021. *Climate Change 2021: The Physical Science Basis. Contribution of Working Group I to the Sixth Assessment Report of the Intergovernmental Panel on Climate Change.* Cambridge : Cambridge University Press.

McCauley, S.J., Hammond, J.I., and Mabry, K.E. 2018. Simulated climate change increases larval mortality, alters phenology, and affects flight morphology of a dragonfly. *Ecosphere* 9: e02151.

McDonald, J.L., Franco, M., Townley, S., Ezard, T.H., Jelbert, K., and Hodgson, D.J. 2017. Divergent demographic strategies of plants in variable environments. *Nature Ecology and Evolution* 1 (2): 0029.

McPeek, M.A. 2022. *Coexistence in Ecology: A Mechanistic Perspective.* Princeton: Princeton University Press.

Mitrus, S. 2016. Size-related mortality during overwintering in cavity-nesting ant colonies (Hymenoptera: Formicidae). *European Journal of Entomology* 113: 524.

MonarchWatch. 2022. <https://www.monarchwatch.org> accessed on June 21, 2022.

Nieddu, G., Billings, L., and Forgoston, E. 2014. Analysis and control of pre-extinction dynamics in stochastic populations. *Bulletin of Mathematical Biology* 76: 3122–3137. doi:10.1038/s41586-022-04644-x.

North American Butterfly Association. 2017. <https://www.naba.org/counts.html> accessed on June 21, 2022.

O'Sullivan, J.D., MacMillan, H.A., and Overgaard, J. 2017. Heat stress is associated with disruption of ion balance in the migratory locust, Locusta migratoria. *Journal of Thermal Biology* 68: 177–85.

Outhwaite, C.L., McCann, P., and Newbold, T. 2022. Agriculture and climate change are reshaping insect biodiversity worldwide. *Nature* 605 (7908): 97–102.

Pardikes, N.A., Revilla, T.A., Lue, C.H., Thierry, M., Souto-Vilarós, D., and Hrcek, J. 2022. Effects of phenological mismatch under warming are modified by community context. *Global Change Biology* 28 (13): 4013–4026.

Pincebourde, S. and Woods, H.A. 2020. There is plenty of room at the bottom: Microclimates drive insect vulnerability to climate change. *Current Opinion in Insect Science* 41: 63–70.

Pinedo-Escatel, J.A., Moya-Raygoza, G., Dietrich, C.H., Zahniser, J.N., and Portillo, L. 2021. Threat-ened Neotropical seasonally dry tropical forest: Evidence of biodiversity loss in sap-sucking herbivores over 75 years. *Royal Society Open Science* 8 (3): 201370.

Pollard, E. and Yates, T.J. 1993. *Monitoring Butterflies for Ecology and Conservation: The British Butterfly Monitoring Scheme.* Springer Science and Business Media.

PollardBase. 2022. <https://pollardbase.org/> accessed on June 21, 2022.

Radchuk, V., Turlure, C., and Schtickzelle, N. 2013. Each life stage matters: The importance of assessing the response to climate change over the complete life cycle in butterflies. *Journal of Animal Ecology* 82 (1): 275–85.

Renner, S.S. and Zohner, C.M. 2018. Climate change and phenological mismatch in trophic interactions among plants, insects, and vertebrates. *Annual Review of Ecology, Evolution, and Systematics* 49: 165–82.

Roeder, K.A., Bujan, J., de Beurs, K.M., Weiser, M.D., and Kaspari, M. 2021. Thermal traits pre-dict the winners and losers under climate change: An example from North American ant communities. *Ecosphere* 12 (7): e03645.

Roland, J., Filazzola, A., and Matter, S.F. 2021. Spatial variation in early-winter snow cover determines local dynamics in a network of alpine butterfly populations. *Ecography* 44 (2): 334–43.

Sales, K., Vasudeva, R., Dickinson, M.E., Godwin, J.L., Lumley, A.J., Michalczyk, Ł., Hebberecht, L., et al. 2018. Experimental heatwaves compromise sperm function and cause transgenerational damage in a model insect. *Nature Communications* 9: 4771. doi:10.1038/s41467-018-07273-z.

Sánchez-Bayo, F. and Wyckhuys, K.A.G. 2019. Worldwide decline of the entomofauna: A review of its drivers. *Biological Conservation* 232: 8–27. doi:10.1016/j.biocon.2019.01.020.

Schachat, S.R. and Labandeira, C.C. 2021. Are insects heading toward their first mass extinction? Distinguishing turnover from crises in their fossil record. *Annals of the Entomological Society of America* 114 (2): 99–118.

Schneider, L., Rebetez, M., and Rasmann, S. 2022. The effect of climate change on invasive crop pests across biomes. *Current Opinion in Insect Science* 100895. doi:10.1016/j.cois.2022.100895.

Sheard, J.K., Rahbek, C., Dunn, R.R., Sanders, N.J., and Isaac, N.J. 2021. Long-term trends in the occupancy of ants revealed through use of multi-sourced datasets. *Biology Letters* 17 (10): 20210240.

Showalter, T.D. 2011. *Insect Ecology.* 3rd edn, London: Academic Press.

Simler-Williamson, A.B., Rizzo, D.M., and Cobb, R.C. 2019. Interacting effects of global change on forest pest and pathogen dynamics. *Annual Review of Ecology, Evolution, and Systematics* 50: 381–403.

Simmonds, E.G., Cole, E.F., Sheldon, B.C., and Coulson, T. 2020. Phenological asynchrony: A ticking time-bomb for seemingly stable populations? *Ecology Letters* 23 (12): 1766–1775.

Sørensen, M.H., Kristensen, T.N., Lauritzen, J.M.S., Noer, N.K., Høye, T.T., and Bahrndorff, S. 2019. Rapid induction of the heat hardening response in an Arctic insect. *Biology Letters* 15 (10): 20190613.

Stemkovski, M., Dickson, R.G., Griffin, S.R., Inouye, B.D., Inouye, D.I., Pardee, G.L., Underwood, N., et al. 2022. Skewness in bee and flower phenological distributions. *bioRxiv*, [Preprint]. doi:10.1101/2022.05.17.492369.

Storlie, C., Merino-Viteri, A., Phillips, B., VanDerWal, J., Welbergen, J., and Williams, S. 2014. Stepping inside the niche: Microclimate data are critical for accurate assessment of species' vulnerability to climate change. *Biology Letters* 10 (9): 20140576.

Svensson, E.I., Gomez-Llano, M., and Waller, J.T. 2020. Selection on phenotypic plasticity favors thermal canalization. *Proceedings of the National Academy of Sciences* 117 (47): 129767–29774.

Tack, A.J., Mononen, T., and Hanski, I. 2015. Increasing frequency of low summer precipitation synchronizes dynamics and compromises metapopulation stability in the Glanville fritillary butterfly. *Proceedings of the Royal Society B: Biological Sciences*, 282 (1806): 20150173.

Theng, M., Jusoh, W.F., Jain, A., Huertas, B., Tan, D.J., Tan, H.Z., Kristensen, N.P., et al. 2020. A comprehensive assessment of diversity loss in a well-documented tropical insect fauna: Almost half of Singapore's butterfly species extirpated in 160 years. *Biological Conservation* 242: 108401.

Turchin, P. and Taylor, A.D. 1992. Complex dynamics in ecological time series. *Ecology* 73 (1): 289–305.

UKMBS. 2022. <https://ukbms.org/> accessed on June 21, 2022.

Van Dyck, H., Bonte, D., Puls, R., Gotthard, K., and Maes, D. 2015. The lost generation hypothesis: Could climate change drive ectotherms into a developmental trap? *Oikos* 124: 54–61. doi:10.1111/oik.02066.

van Klink, R., Bowler, D.E., Gongalsky, K.B., Swengel, A.B., Gentile, A., and Chase, J.M. 2020a. Meta-analysis reveals declines in terrestrial but increases in freshwater insect abundances. *Science* 368 (6489): 417–20.

van Klink, R., Bowler, D.E., Gongalsky, K.B., Swengel, A.B., Gentile, A., and Chase, J.M. 2020b. Erratum for the report "Meta-analysis reveals declines in terrestrial but increases in freshwater insect abundances by R. van Klink, D.E. Bowler, K.B. Gongalsky, A.B. Swengel, A. Gentile, J.M. Chase. *Science* 370: 6515.

Vautard, R., Cattiaux, J., Yiou, P., Thépaut, J.N., and Ciais, P. 2010. Northern Hemisphere atmospheric stilling partly attributed to an increase in surface roughness. *Nature Geoscience* 3 (11): 756–61.

Vereecken, N.J., Weekers, T., Leclercq, N., De Greef, S., Hainaut, H., Molenberg, J.M., Martin, Y., et al. 2021. Insect biomass is not a consistent proxy for biodiversity metrics in wild bees. *Ecological Indicators* 121: 107132.

Villalta, I., Ledet, R., Baude, M., Genoud, D., Bouget, C., Cornillon, M., Moreau, S., et al. 2021. A DNA barcode-based survey of wild urban bees in the Loire Valley, France. *Scientific Reports* 11 (1): 4770.

Wagner, D.L. 2020. Insect declines in the Anthropocene. *Annual Review of Entomology* 65: 457–480. doi:10.1146/annur ev-ento-01101 9–025151.

Wagner, D.L., Fox, R., Salcido, D.M., and Dyer, L.A. 2021. A window to the world of global insect declines: Moth biodiversity trends are complex and heterogeneous. *Proceedings of the National Academy of Sciences* 118 (2): e2002549117.

Wagner, D.L., Grames, E.M., Forister, M.L., and Stopak, D. 2021b. Insect decline in the Anthropocene: Death by a thousand cuts. *Proceedings of the National Academy of Sciences* 117: e2023989118. doi:10.1073/pnas. 2023989118.

Walsh, B.S., Parratt, S.R., Mannion, N.L., Snook, R.R., Bretman, A., and Price, T.A. 2021. Plastic responses of survival and fertility following heat stress in pupal and adult Drosophila virilis. *Ecology and Evolution* 11 (24): 18238–18247.

Warren, M.S., Maes, D., van Swaay, C.A., Goffart, P., Van Dyck, H., Bourn, N.A., Wynhoff, I., et al. 2021. The decline of butterflies in Europe: Problems, significance, and possible solutions. *Proceedings of the National Academy of Sciences* 118 (2): e2002551117.

Wepprich, T., Adrion, J.R., Ries, L., Wiedmann, J., and Haddad, N.M. 2019. Butterfly abundance declines over 20 years of systematic monitoring in Ohio, USA. *PLoS One* 14 (7): e0216270.

Williams, N.F., McRae, L., Freeman, R., Capdevila, P., and Clements, C.F. 2021. Scaling the extinction vortex: Body size as a predictor of population dynamics close to extinction events. *Ecology and Evolution* 11 (11): 7069–7079.

Wilson, E.O. 1987. The little things that run the world (the importance and conservation of invertebrates). *Conservation Biology* 1: 344–46.

Yang, L.H., Postema, E.G., Hayes, T.E., Lippey, M.K., and MacArthur-Waltz, D.J. 2021. The complexity of global change and its effects on insects. *Current Opinion in Insect Science* 47: 90–102.

CHAPTER 10

Evidence of climate change effects on insect diversity

The wind and the pinwheel

Kleber Del-Claro, Vitor Miguel da Costa Silva, Eduardo S. Calixto, Elliot Centeno de Oliveira, Iasmim Pereira, Diego Anjos, Helena Maura Torezan-Silingardi and Renan Fernandes Moura

10.1 Introduction

In the last 200–250 years, humankind has lived through profound changes, where human welfare has improved immeasurably due to a combination of economic, social and political forces that have given us longer life spans, better education, better health care, increasing volumes of information and consumer goods. However, the cost of all these improvements to human life has been the over-exploitation of everything the Earth has to offer, like water, soils, plants and animals. This new geological epoch, named the Anthropocene (Crutzen 2002), has provoked a huge transformation of the Earth's ecosystems with cascading effects in aquatic, terrestrial and marine communities (Lewis and Maslin 2020;). Historical demographers estimate that around the year 1800 the world population was around 1 billion people, but by 2023 we expect it to exceed 8 billion people (Worldometer 2019), an immense increase in such a short period. According to Sage (2020), "if human population and resource demands continue to grow unabated, there would come a time when the use of energy, fresh water, and land by the human population would leave little for species other than humanity's domesticates, leading to widespread biodiversity loss and potential collapse of the life support system." This could result in the sixth mass extinction.

The humanity success is based in the exploration of natural resources, air, water, oil and minerals, but fundamentally from the net primary productivity (NPP) green plants and their interactions with animals and microbes (Reverchon and Méndez-Bravo 2021). The loss of biodiversity which directly impacts NPP is not represented only by extinctions but a reduction in the size of the populations can make a species nonviable in a mid- to long term. Reduced populations may break links and nodes in biotic interaction networks and this has the potential to completely stop ecological services provided by animals such

Kleber Del-Claro et al., *Evidence of climate change effects on insect diversity*. In: *Effects of Climate Change on Insects.*
Edited by: Daniel González-Tokman and Wesley Dáttilo, Oxford University Press. © Oxford University Press (2024).
DOI: 10.1093/oso/9780192864161.003.0010

as pollinator insects (Del-Claro and Dirzo 2021; Torezan-Silingardi et al. 2021). Furthermore, if reduced populations reach levels below their minimum viable size this might lead to an irreversible extinction vortex (Thomas 1990). Everything is connected. Herbivores can control the size and reproduction of plant populations, predators can control herbivore populations and both may influence the presence and efficiency of pollinators. The microbiome also directly impacts the quality of resources that plants offer to herbivores and protective mutualists (Moura et al. 2021). Thus, the extinction or extensive reduction in the population of key species will have dramatic, cascading effects throughout the whole trophic chain.

In this chapter we discuss the main drivers of global change that can disrupt trophic chains and ecological services and degrade the environment—true threats to life on Earth—using insects as a focal group.

10.2 Why insects?

At the beginning of the Devonian, more than 450 million years ago, the class Insecta emerged on Earth. They were mainly composed of small creatures with brief life cycles, able to reproduce several times a year. First came flying animals, then came those that walk, run, jump, climb, dig or swim. For 300 million years, insects occupied almost all terrestrial ecosystems of the planet, making very successful use of new habitats and microhabitats. In the Mesozoic, between Jurassic and Cretaceous, around 135 million years ago, with the emergence of Angiosperms and their associations with insects, every aspect of the insect life was amplified resulting in a huge diversification of traits that led to rapid adaptive radiations (e.g., Abrahamson 1989; Del-Claro and Torezan-Silingardi 2021; Labandeira 1998). Then, insect–plant interactions, either antagonistic like herbivory interactions or mutualistic like pollination, stood in the spotlight of terrestrial ecological networks (Luna and Dátillo 2021). There are more than 1 million described insect species but we estimate that living species of insects easily surpass 5.5 million (Stork 2018). As Paul Eggleton (2020) recently told us:

> Insects appear to be everywhere, and they seem eternal. They are a normal part of routine life: whether it is ants raiding your pantry, bees pollinating your apple tree, moths eating your clothes, or wasps spoiling a summer picnic. Insects have been thought to be an unchanging part of our world, and so it is with extreme urgency that recent reports have raised the alarm that everywhere insects may be in decline.

Insects are extremely diverse and numerous, occupying all terrestrial environments such as air, ground and underground, lakes, ponds and rivers. Immature insects and adults are at the base of all trophic chains and structure most ecological networks (Del-Claro and Torezan-Silingardi 2021). They can be herbivores, predators, detritivores, parasites. Coleoptera and Diptera together account for more than 50 percent of all insect species, presenting almost all known feeding habits (Figure 10.1). The six most numerous and diverse orders (Coleoptera, Diptera, Lepidoptera, Hymenoptera, Hemiptera and Orthoptera) comprise most herbivore species—the basis of trophic chains—in terrestrial and aquatic environments (Figure 10.1; Del-Claro 2019).

Insects, the main herbivores, pollinators and controllers of insect pests, fungi and weeds, have a huge impact on natural systems and the entire human food supply chain (Gullan and Cranston 2014). An example of immediate economic concern is that more

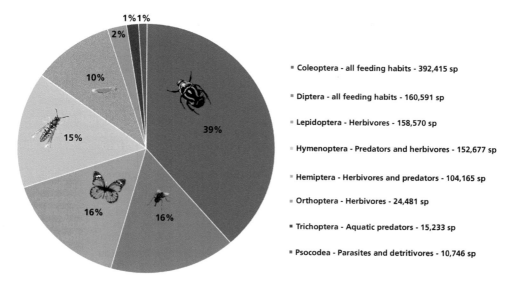

■ Coleoptera - all feeding habits - 392,415 sp

■ Diptera - all feeding habits - 160,591 sp

■ Lepidoptera - Herbivores - 158,570 sp

■ Hymenoptera - Predators and herbivores - 152,677 sp

■ Hemiptera - Herbivores and predators - 104,165 sp

■ Orthoptera - Herbivores - 24,481 sp

■ Trichoptera - Aquatic predators - 15,233 sp

■ Psocodea - Parasites and detritivores - 10,746 sp

Figure 10.1 The eight most numerous insect orders and their predominant feeding habits (% of species), adapted from Eggleton (2020). The predominant eating habit is presented first; numbers represent the known species in the order.

than 80 percent of the world's flowering plants, including approximately three-quarters of all crop species, are thought to be dependent on insects for pollination (Torezan-Silingadi et al. 2021; Vanbergen et al. 2013). The decline or extinction of insects threatens to cause a collapse not only of natural ecological networks but our whole food supply system (Wagner 2020). Recent reviews lit the warning sign: insect declines are being reported worldwide in flying, ground and aquatic lineages. There is an urgent need for global monitoring efforts which will help to identify important causal factors in population declines (Wagner 2020; Wilson and Fox 2021).

10.3 Drivers of global change impacting insect diversity

The same factors that threaten plants, animals and other life forms on our planet threaten insects. Here, we will center on four main drivers and one fuel of global change and their effects on insect diversity (Table 10.1). The first driver is habitat change: the structural alteration of landscapes, lakes, rivers, including fragmentation due to establishment of pastures and cities, and biotic homogenization due to urbanization, for example, all comprise a clear cause of loss in insect diversity and abundance. The second driver is over-exploitation; excessive logging, the establishment of monocultures across huge areas, the use of fire in managing specific environments, excessive use of water in irrigation and industrial process, all are clear examples of over-exploitation of natural resources that threaten insect life. Third, environmental pollution, including industrial, urban and agricultural, and light and noise as a result of human activities reduce insect species in numbers and diversity. Fourth, invasive species; human activities amplify the effect of competitors, parasites, pathogens, ecosystem engineers and invasive plants reduce native diversity by direct or indirect competition, predation or by altering habitat structure (see Table 10.1).

Table 10.1 Four drivers and one fuel of global change and their effects on insect diversity*.

Driver	Examples of impacts causing insect biodiversity loss
1. Habitat change	Unrestricted and broad habitat loss due to direct conversion of land to cities, farms, pastures, roads and other human utilities. Structural alteration of rivers, lakes and springs in order to irrigate pastures and crops or attend a city's needs. The main effects are reduction in total natural area provoking fragmentation and loss of connectivity. Consequent disruption in gene flow and interacting ecological networks, which amplifies risk of local extinctions. Urbanization may cause biotic homogenization, resulting in a negative large-scale impact on the diversity of specialist insect species.
2. Over-exploitation	Overgrazing, firewood collection, excessive logging, monoculture in huge areas, excessive use of water in irrigation and industrial processes. These actions cause widespread reduction in the population size of insect species, many of them key species in trophic chains and providers of ecological services. On the other hand, some species may benefit from monocultures expanding its populations and competitively excluding many others.
3. Pesticide use and light	Environmental pollution, including industrial, urban and agricultural pollution, and light and noise pollution from human activities reduce insect species in number and diversity. Insecticides, for example, have direct lethal and sub-lethal effects on mortality and fecundity. Fertilizers can cause habitat loss and degradation. Light can cause disruption in plant phenology with direct effects on its interactions with insects (synchrony). Population reduction due to pesticides directly impacts ecological networks reducing food availability.
4. Invasive species	Human activities like transportation of food, agricultural products, and livestock represent a huge facility to invasive species go through geographic barriers. Invasive competitors, parasites, pathogens, ecosystem engineers and invasive plants all reduce native diversity by direct competition, predation or by altering habitat structure, microclimate or exposure to fire.
Fuel	Examples of impacts on drivers and direct on insect biodiversity loss
1. Climate change	Climate warming and altered precipitation act as accelerants on the drivers of impact biodiversity (see Figure 10.1). Climatic change has the potential to intensify and be intensified by human activities (habitat change, over-exploitation, pollution) that directly impact Earth's biodiversity. By altering the synchrony in plant–animal interactions, climatic changes can disrupt ecological interaction networks. In particular, mutualisms (pollination, seed dispersal, biotic defenses) that are sources of biodiversity may be strongly affected. Climate change can facilitate invasive species dispersion in new areas, intensify the effects of pesticides due to excessive dry periods, facilitate fire in areas of overgrazing and logging and change habitats irreversibly. Some species will benefit from climatic change but most species will suffer drastic population reduction.

* See also Sage (2019) and Wilson and Fox (2021) for other perspectives of global change drivers.

These are the main drivers we consider that impact insect diversity and survival (e.g., Eggleton 2020; Forister et al. 2019; Knop 2016; Wagner 2020; Wilson and Fox 2021). However, there is a factor that enhances all drivers that threatens insects and that is climate change. Imagine that the drivers of global change impacting insect diversity are pinwheel shovels. The climate change is the fuel, the wind, that will blow the pinwheel shovels and drive their movement. Climate change has been accelerated by human activities (habitat change, over-exploitation, pollution), amplifying global changes that directly impact Earth's biodiversity (Figure 10.2).

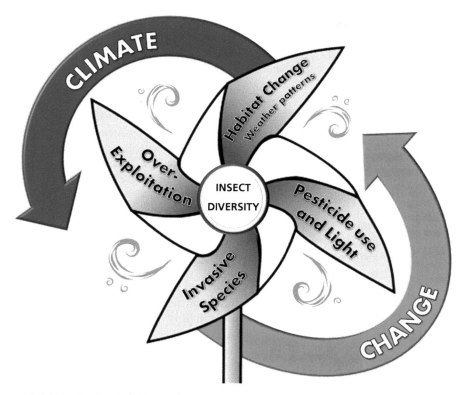

Figure 10.2 The pinwheel of climate change impacting insect diversity. Stressors of global change have been linked to population declines and insect extinctions, with varying importance depending on regional factors. Among these stressors, habitat change (including change in weather patterns) and over-exploitation have had the most severe and wide-ranging global impacts on the whole terrestrial biodiversity over the past hundred years. Environmental pollutants including pesticides (also light) reduce suitable habitat to insects particularly in regions of intense human interest like agricultural areas. Some specific insect declines have been related to direct invasive alien species. Climate change, with direct effects on weather patterns (precipitation and temperature), emerges as the "wind" most geographically pervasive that drives the pinwheel with even greater force, with huge impacts on insect diversity and abundance.

10.4 Habitat change and over-exploitation

Human activity has negatively impacted different insect groups over time, causing sequential declines in the diversity and abundance of these organisms in recent decades and increasing the numbers of extinct or endangered species (Dirzo et al. 2014; Sánchez-Bayo and Wyckhuys 2019; Wagner 2020). Recent studies have pointed to a 45 percent decline in the largest groups of insects (Coleoptera, Lepidoptera and Hymenoptera) (Dirzo et al. 2014). Although this decline can be consistently attributed to the direct effects of pollution, climate change and the expansion of human activities (which can be interconnected), the loss of habitat promoted by these factors has been the main responsible for insect decline (Sánchez-Bayo and Wyckhuys 2019; Figure 10.2; Table 10.1). More than 50 percent of Earth's terrestrial environment has already undergone some type of manmade change (Hooke et al. 2012). Deforestation, modification of natural areas into housing and commercial establishments, industrialization, conversion of natural forests on land for agriculture and silviculture and recurrent fires are all factors that have altered the conditions necessary for the survival and reproduction of various species (Lewis et al. 2015; Maxwell et al. 2016; Sánchez-Bayo and Wyckhuys 2019; Wilson and Fox 2021).

Recurrent and non-accidental fire is an important contributor to habitat loss and is often related to the exploitation of biomes. Due to the constant expansion of agriculture recently, tropical environments have been seriously affected by increasingly devastating fires (Del-Claro and Dirzo 2021; Maxwell et al. 2016; Regan et al. 2010; Watson et al. 2018). Furthermore, climate change has also had a major influence on fire regimes, facilitating the ignition and spread of fires in different regions of the planet (Bowman 2017; Regan et al. 2010; Stephens et al. 2013). Thus, since fire can radically modify the environment, it seems certain that it is fostering the decline and extinction rates of insects (Sage 2020). Furthermore, recent events such as the intense fires between 2019 and 2021 in the different Brazilian biomes (Amazonia, Cerrado and Pantanal), in Australia and in California, have drawn even more attention to the devasting effects of fire (Del-Claro and Dirzo 2021).

Lepidoptera is the most threated insect order due to human activities (Box 10.1) (Dirzo et al. 2014; Wagner 2020). Several studies have pointed to its sharp decline in several European countries and in North America, driven mainly by the expansion of agricultural activities (Conrad et al. 2004, 2006; Fox 2013; Fox et al. 2015; Habel et al. 2019; Mattila et al. 2006; Wagner 2020). Although insect responses to climate change are biased due to an overrepresentation of studies involving lepidopterans (Dirzo et al. 2014; Wagner 2020), similar patterns of decline have been observed in other orders such as Coleoptera, Hymenoptera, Orthoptera and Odonata, with habitat loss also being the main reason for the decline (Dirzo et al. 2014; Sánchez-Bayo and Wyckhuys 2019).

Although there is no doubt about the decline of insects in the Anthropocene due to habitat loss (Dirzo et al. 2014; Del-Claro and Dirzo 2021; Wagner 2020), one factor that has worried many researchers is the fact that several studies are also showing a decrease in insect abundance and diversity even in fully preserved areas, which suggests that factors other than habitat loss are already influencing insect decline (Franzén and Johannesson 2007; Hallmann et al. 2017; Janzen and Hallwachs 2019; Lister and Garcia 2018; Rada et al. 2019; Salcido et al. 2020; Schuch et al. 2012). According to Sánchez-Bayo and Wyckhuys (2019), habitat loss accounts for 49.7 percent of the decline of insects, while pollution accounts for 25.8 percent. Pesticides, fertilizers, pollution and climate change account for 30.8 percent of insect decline. Deforestation, fire and agriculture together account for 34.6 percent of the group's decline. Thus, even if insects are protected from habitat loss, the action of other factors is already causing relevant declines (Habel et al. 2019).

Factors that prompt habitat loss (deforestation, fire, urbanization) are followed by a series of events that directly and/or indirectly affect the survival of insects (climate change, pollution, extinction of other groups of animals, collapse of ecological interactions, pesticides) (Habel et al. 2019; Sage 2020; Wagner 2020). Aquatic or semiaquatic insects, for example, represent a good modelling group for these interacting factors. Although they suffer from habitat loss, pollution also has a devastating effect on these insects (Sánchez-Bayo and Wyckhuys 2019; Van Dijk et al. 2013;). The same is also true for pollinators, which are affected by habitat loss, but who are also severely punished by the pesticides used on crops (Goulson et al. 2015; Torezan-Silingardi et al. 2021).

The accumulation of harmful factors can generate catastrophic events for insect biodiversity across the spectrum of interactions (Del-Claro and Dirzo 2021) since habitat loss also affects plants and vertebrates (Del-Claro and Dirzo 2021; Dirzo et al. 2014). Many insects, such as plant pollinators and parasites, are completely dependent on other species to survive. Thus, factors such as habitat loss which negatively affect key plant or vertebrate species can put these insects at risk (Del-Claro and Dirzo 2021; Halsch et al. 2021; Powney et al. 2019). Despite a general decline in insects (Dirzo et al. 2014), it is suggested that specialist species are suffering from higher rates of extinction than generalists (Box 10.2; Biesmeijer et al. 2006; Carvalheiro et al. 2013; Halsch et al. 2021; Newbold et al. 2018; Powney et al. 2019). Therefore, the loss of specialist species can generate cascading effects that affect an entire group of related species (Carvalheiro et al. 2013; Kehoe et al. 2021).

Box 10.1 THE HISTORY OF STUDIES OF THE ORDER LEPIDOPTERA

Studies of the order Lepidoptera began more than 200 years ago, and over time they have offered a series of important information about the biology, ecology and distribution of the group, such as herbivore habit, pollination, mutualistic interactions and antagonistic with other animals and wide distribution across the planet (Thomas et al. 2004; Wagner et al. 2021). This ecological diversity made the group an important model for understanding the effects caused by human activity on insect over time (Dirzo et al. 2014). The most complete studies on the decline of insects due to habit loss and climate change are those carried out with lepidopterans and offer the richness and depth of studies over time. Among these, several point to a severe decline of the group due to habitat loss, climate change and pollution (Conrad et al. 2004, 2006; Fox 2013; Fox et al. 2015; Habel et al. 2019; Mattila et al. 2006; Wagner 2020). However, some have also pointed out that some lepidopterans may not be affected by these effects or even benefit from these changes (Halsch et al. 2021; Lamarre et al. 2022; Wagner et al. 2021).

Species of Nymphalidae (Danainae) as this *Episcada* are seriously endangered in Brazil (see <https://www2.ib.unicamp.br/labor/site/?page_id=1345>).

Box 10.2 THE DECLINE OF INSECTS IS HAPPENING BUT ARE THEY RESPONDING EQUALLY?

Several studies in different regions of the planet show that insects have been in steady decline for some years (Dirzo et al. 2014; Wagner 2020). However, the responses of insects to human activities have been shown to vary depending on the context in which insects are found; there is no standardized response for all insects (van Klink et al. 2020). Thus, while van Klink et al. (2020) showed that terrestrial insects are declining and aquatic insects are increasing in number, Sánchez-Bayo and Wyckhuys (2019) point to a general decline in insect population overall. Several studies point to a high decline in the order Lepidoptera (Conrad et al. 2004, 2006; Fox 2013; Fox et al. 2015; Mattila et al. 2006), but in some regions populations have remained constant (Wagner et al. 2021). Furthermore, species living in different environmental gradients show different responses to changes in the environment (Halsch et al. 2021), and something similar has been observed in both generalist and specialist species (Powney et al. 2019). Thus although the decline of insects is a fact of global influence, it should not be extrapolated to all insects (Watson et al. 2018).

Most studies showing that insects are in decline have been carried out in temperate regions (Sánchez-Bayo and Wyckhuys 2019; Wagner 2020), but we are in urgent need of knowing how insects are currently behaving in tropical areas, a place that harbors the greatest diversity of insects on the planet (Stork 2018). Furthermore, the fact that insects from tropical zones are much more sensitive to small disturbances than those living in temperate zones (Deutsch et al. 2008; García-Robledo et al. 2016) has further worried the scientific community since the loss of native vegetation for crops, logging and cattle has reached alarming levels in recent years (Watson et al. 2018). For example, in the Brazilian Amazon rainforest, rates of logging, burning and agricultural occupation have increased in recent years to the point of endangering the ability of this vital forest to regenerate itself (Box 10.3; Leite-Filho et al. 2021; Lovejoy and Nobre 2018; Marengo et al. 2022; Matricardi et al. 2020; Nobre et al. 2016). Thus, as several tropical biomes have already lost and continue to lose a large part of their original area (Watson et al. 2018), this pessimism regarding the real state of insect biodiversity on the planet becomes inevitable.

Box 10.3 THE UNPROFITABLE RISKS OF EXPLOITING THE AMAZON RAINFOREST

It is estimated that if the Amazon Forest loses more than 40 percent of its original area or has its average temperature changed by more than 4 °C, the forest will reach a point where it can no longer regenerate itself (Nobre et al. 2016). Given this, several studies have sought to show the importance of conserving the Amazon and preventing the collapse of the forest. Among these studies, several point to weak performance of the Amazon biome as a source of income for agriculture. This is because in 2016, the Amazon corresponded to 14.5 percent of the gross domestic product (GDP) of Brazilian agriculture and that in order to account for this, it was necessary to deforest 750,000 km^2 of the forest. In the state of São Paulo, an area of 193,000 km^2 corresponds to 11.3 percent of the agricultural sector (GDP) (Nobre et al. 2016). Furthermore, as a consequence of totally unmanaged exploration and a poor understanding of the forest's ecological functions, some studies have pointed out that the loss the Amazon forest will cause the collapse of the rainfall cycle in Brazil, affecting the southeastern region in particular which is key for the country's economy (Marengo et al. 2022).

Amazonian forest, Carajás' region, PA, Brazil 1988: today this area is deforested.

10.5 Pesticides, pollution and alien species

The constant increase in the human population intensifies the demand for food around the globe. The current response to this crisis is a rampant increase in food production, commonly through extensive areas of monoculture (Tudi et al. 2021). To sustain areas devoted to monocultures and keep them healthy, free from disease, weeds and pests, pesticides are used. Around the world, the production of pesticides increased at a rate of about 11 percent per year (Hayes et al. 2017), but careless use of these chemicals can have disastrous consequences (Temreshev et al. 2018). In addition to the growing human population, over-exploitation of soils via agriculture and livestock and the changes in climate modifying the dynamics of insect and plant pests all lead to an increased use of pesticides (Tirado et al. 2010). Although these chemicals are necessary for the food supply (a very controversial issue, see Lykogianni et al. 2021), they may harm the environment (Gouin et al. 2008), human health and other non-target organisms such plants and insects (Aktar et al. 2009; Elgueta et al. 2017; Mingo et al. 2017). In addition, the impact of these substances is not limited to the location where they were applied and they can contaminate distant places through spread by water, wind currents, run-off and animals (Fang et al. 2017).

Pesticides are one of the main drivers leading to insect decline (Goulson et al. 2015). Bees are emblematic of this as they are in constant contact with pesticides applied to crops when they seek food such as pollen, nectar or honeydew (Dicks et al. 2021; Torezan-Silingardi et al. 2021; Box 10.4). The effects of these pollutants vary depending on species and could affect the life cycle, body size and weight and foraging period of insects (Arena and Sgolastra 2014). Toxic substances collected by bees can be carried over long distances and contaminate whole colonies. Researchers have mainly focused on acute toxicity tests in pollinator insects (Bargańska et al. 2018), however pesticides cause harm not only in lethal levels; the constant exposure of sub-lethal levels of these pollutants can lead to altered behaviors and increased susceptibility to diseases (Brittain and Potts 2011; Dicks et al. 2021). Since climate change can alter several environmental factors, pesticide use in food production may be on the rise. This increase in pesticide use may well have the knock-on effect of increasing the abundance of those insect pests that may require ever-greater doses of pesticides for their control (Tudi et al. 2021).

Therefore, tighter control over pesticide use is urgently required in order to mitigate environmental pollution. For this, new studies must be carried out to improve our understanding of the risks of pesticides to local non-target species, in addition to new frameworks aiming to develop less toxic and contaminant products.

Box 10.4 POLLINATORS ENDANGERED

Studies in different biomes of the planet show that insects and mainly pollinators have been in steady decline due to over-exploitation and reduction and pollution of natural habitats (Dirzo et al. 2014; Torezan-Silingardi et al. 2021; Wagner 2020). According to Pots et al. (2016) "there are well-documented declines in some wild and managed pollinators in several regions of the world. However, many effective policy and management responses can be implemented to safeguard pollinators and sustain pollination services." All over the world researchers and some governments are leading initiatives for studies aiming to revert the causes in the reduction of pollinators. Some examples are:

1) Brazil—ICMBio (Chico Mendes Institute) coordinates PAN (national plan) for the conservation of pollinating insects <https://www.icmbio.gov.br/portal/ultimas-noticias/20-geral/11328-icmbio-coordena-pan-para-a-conservacao-de-insetos-polinizadores>
2) USA-NAPPC (North American Pollinator Protection Campaign) <https://www.pollinator.org/about#about-team>
3) The Intergovernmental Science-Policy Platform on Biodiversity and Ecosystem Services (IPBES) is an independent intergovernmental body seeking to strengthen the interface between science and policy for biodiversity and ecosystem services for the conservation and sustainable use of biodiversity, long-term human wellbeing, and sustainable development. It was established in Panama City on April 21, 2012, by ninety-four governments <https://ipbes.net/about>

In the Brazilian tropical savanna native bees are endangered due to the introduction of invasive species, habitat reduction, indiscriminate use of pesticides and climate change.

Anthropogenic noise is a human disturbance that is expanding nowadays, present not only in cities but in natural environments (Barber et al. 2010). This kind of pollution can alter the behavior of animals that communicate through sound in many ways, interfering with hunting, mate attraction and territory protection (Morley et al. 2014). Some insects have adaptations that allow them to communicate even in noisy environments, thus minimizing risks to reproduction and survival. Some grasshoppers inhabiting areas near roads change the frequency of their songs to improve communication (Lampe et al. 2012). Other insects, such as crickets, can stop singing for short periods of time when exposed to car noise (Gallego-Abenza et al. 2020). Bowen et al. (2020) showed that crickets raised in quiet environments produced larger spermatophores when compared to crickets raised in places with intense traffic sounds. Females of *Teleogryllus oceanicus* raised in noise pollution conditions have more difficulty in locating conspecific males, taking 80 percent more time to locate singing males than females reared in silent conditions (Gurule-Small and Tinghitella 2018). Noise pollution leads to negative effects on the reproductive success of these animals, and consequently the decline of their populations, especially insects that use airborne sounds as their main channel of communication.

Studies focusing on target species in determined regions are needed to give us a better understanding of how noise pollution affects the local biodiversity of insects. The creation of green spaces in urban areas, even on the side of the roads, is a possible way of mitigating this problem, since these would act as a barrier so that noise pollution does not spread through the landscape, minimizing the impact of this type of pollution on the communication of animals (Levenhagen et al. 2020).

Light pollution is another modern phenomenon that causes environmental disturbance, particularly to insect populations (Owens et al. 2020). The effect of this pollution on insects depends upon the species. While some species retreat from light, as occurs in tree weta (*Hemideina thoracica*) and cave weta (Rhaphidophoridae), other insects, such as moths, could be attracted to illuminated spots (Boyes et al. 2021). Artificial light at night can disorient sensitive insects, making them highly susceptible to predators (Owens et al. 2020) such as bats, rats, shorebirds, geckos and toads (Owens et al. 2020; Russo et al. 2019). Artificial light is especially harmful to fireflies as they use bioluminescent flashes as signals to their conspecifics in courtship to find and attract mates. Firebaugh et al. (2016) showed that the firefly *Photuris versicolor* reduces its flash activity by about 70 percent when exposed to light pollution conditions.

The direct impact of a lamp's light can affect insects but also skyglow—diffuse luminance of night sky caused by artificial light, which reduces the contrast of stars and the moon—also contributes to impacts in insect behavior. Natural nocturnal light signals are important elements in orientation in nocturnal dung beetles (*Scarabaeus satyrus*), and Dacke et al. (2021) showed that skyglow disrupts the orientation elements of this insect by obscuring natural celestial cues. As result, *S. satyrus* was attracted toward artificial lights, leading to increased interspecific competition and reduced dispersal capacity. Further studies are necessary to mitigate the impacts of light pollution on insects since each insect species has a specific response to light and its different wavelengths. However, taking simple measures such as installing motion sensors for lighting can reduce light pollution and ease its negative impacts on insects (Stewart 2021).

Alien (invasive) species are the second most important cause of extinction of the five major taxa (plants, amphibians, reptiles, birds and mammals) (Bellard et al. 2016). Invasive species also drive biodiversity loss among insects (Table 10.1). Invasive and non-native species, be these pathogens, plants or animals, pose an ever-increasing threat to global biodiversity (Wagner 2020). According to Wagner (2020), "invasive and ornamental plants pose a combined and growing threat, especially in areas or systems of high human occupation and disturbance; exotic plants can diminish insect herbivore loads by more than 90%, which, in turn, negatively affects birds and other insectivores." Some iconic invasive species have strong economic and natural impacts. The Argentine ant (*Linepithema humile*), for instance, is native to northern Argentina and is a globally distributed invasive pest in urban, natural and agricultural habitats. This species facilitates the occurrence of plant-eating insect pests (e.g., honeydew-producing insects), disturbing native ants, pollinators and even vertebrates. The problems caused by invasive species run so deep that governments have created complete centers dedicated to their study like the Center for Invasive Species Research in the USA-University of California Riverside Campus (<https://cisr.ucr.edu/invasive-species/argentine-ant>). Honeybees (*Apis mellifera*) are another example. Since their introduction in the Americas, they have been causing huge impacts on natural systems. Torezan-Silingardi and Del-Claro (1998) showed honeybees can disrupt the entire pollination system of plants in a native tropical savanna, significantly reducing fruit production and possibly affecting feeding and survival of native bees.

10.6 Climate change, the wind in the pinwheel

Insects are ectodermal animals so they are highly susceptible to the microclimatic conditions (Boggs 2016). Changes in weather patterns are affecting insect biodiversity on a global scale (Eggleton 2020). Anthropogenic changes encompass multiple components of stressors for insects such as increased average temperature, increased greenhouse gases (CO_2), water stress and changes in rainfall patterns (Halsch et al. 2021; Wagner 2020). In general, all these factors are expected to have a negative and significant impact on insects, especially in tropical regions (Baranov et al. 2020; Lamarre et al. 2020; Román-Palacios and Wiens 2020). Climate change, with direct effects on weather patterns (precipitation and temperature), emerges as a strong and pervasive "wind" that drives the pinwheel faster, having huge impacts on insect diversity and abundance (Figure 10.2). To understand how these multiple factors can impact insect diversities, a broad approach is needed. We will highlight some of these stressor components in order to facilitate better comprehension of this approach.

10.7 Physiology

The effects of climate change on the physiology of tropical insects are especially concerning (García-Robledo et al. 2016). Tropical insects have less thermal tolerance than temperate species (Pincebourde and Suppo 2016; Sunday et al. 2014). Among the physiological stresses we should highlight protein denaturation (Farahani et al. 2020; King and Macrae 2015; Martín-Folgar et al. 2015), changes in the fluidity of phospholipid membranes and cell homeostasis (Bowler 2018), disturbance of ionic balance (O'Sullivan et al. 2017), desiccation (Chown et al. 2011; Woon et al. 2019) and changes in the immune response (Gherlenda et al. 2016; Wojda 2017). The increase in temperature will influence the hatching and incubation periods of eggs and the sex of aquatic insects (Del-Claro 2019; Strachan et al. 2015; Terblanche et al. 2005).

10.8 Phenology

Insects have short life cycles, making them intensely influenced by changing climatic factors. The shortening of seasons and the increase in temperatures in winter cause changes in species voltinism, such as the increase in the number of annual life cycles (Altermatt 2010; Bell et al. 2019; Mitton and Ferrenberg 2012). The increase in precipitation can also influence the phenology of some species, causing delay or early emergence (Chown et al. 2011; Cohen et al. 2018; Vilela et al. 2014). Changes in the phenology of species will not only influence their population but also their interactions, a topic discussed below. All these changes will affect the survival and fitness of populations (Bonebrake and Mastrandrea 2010; Forrest 2016; Kingsolver et al. 2011). Vilela et al. (2018) showed that climatic parameters can influence the timing of phenological events affecting the degree of synchrony among plant species, their interactions with insects and the outcomes in plant reproduction.

10.9 Other stressors

Temperature alone is often used to analyze future climate change scenarios but other abiotic factors can influence insects. One of the main stressors for insects will be the increase of dry periods and low levels of precipitation and humidity, which will bring challenges to the maintenance of water balance (Addo-Bediako et al. 2001; Chown et al. 2011; Fisher 1978). For terrestrial insects, drought causes death by desiccation, affecting not only the individual but entire populations (Herrando et al. 2019). Aquatic insects, on the other hand, will have to face water loss concomitantly with changes in water quality (e.g., changes in salinity and in oxygen and carbon dioxide levels) (Davies 2010; Strachan et al. 2015).

CO_2 is an important sensory signal for insects (Jones 2013). Increasing CO_2 levels can lead to behavioral changes (e.g., changes in oviposition behavior) (Stange 1997) and physiological changes (Guerenstein and Hildebrand 2008). The increase in CO_2 has the potential to change the structure of entire insect communities (Hillstrom and Lindroth 2008). Indirect effects through insect–plant interactions can also be observed on parasitism (Klaiber et al. 2013) and plant immune response (Gherlenda et al. 2016).

10.10 Climate change impacts on ecological interactions

As we move further into the Anthropocene, climate change may prove to be the greatest threat to biodiversity, rivaled only by habitat destruction in the tropical regions of the planet (Wagner 2020). The effects of climate change have largely been directed at the loss of biodiversity. However, a neglected component of biodiversity is ecological interactions, responsible for ecosystem functions such as pollination, seed dispersal and decomposition (Del-Claro and Torezan-Silingardi 2021). The loss of interactions between species can anticipate and accompany the loss of biodiversity. Increases or decreases in regional precipitation rates, for example, will trigger changes in plant abundance and distribution. As plant communities across the planet increase, decrease, change, adapt and disappear, so too will their associated insect faunas. Changing floras will have cascading consequences for specialized herbivores (Casas-Pinilla et al. 2022; Del-Claro and Torezan-Silingardi 2021). Despite the imminent scenario of climate change, studies considering the loss of ecological interactions involving insects are incipient (e.g., Wagner et al. 2021). Recently, Bellaver et al. (2022) proposed that the extinction risk of insect species "should be adjusted to critically endangered and point that species interactions and climate change must be accounted for in conservation planning."

The increase in CO_2 and temperature, both drivers of climate change, have altered the complex interactions between insects and other groups, whether plants or animals (e.g., Vilela et al. 2014, 2018; Wagner et al. 2021). In addition to these climatic factors, nitrogen deposition affects plant–pollinator interactions, directly interfering with plant traits such as floral morphology, phenology and nectar chemical composition (Hoover et al. 2012). These climate changes can also directly impact the insects involved in these interactions. In plantations, high temperatures have increased the number of generations of some insects that can damage crops. In general, the increase in temperature has led to an increase in the intensity and outbreaks of diseases in parasite populations (Thierry et al. 2019). Thus, climate changes may decrease the populations of most insect species, but it could be beneficial to some particular species.

10.11 Synchronism (phenology of interactions)

The main effect of climate change on ecological interactions is the direct impact on populations, which generates temporal and spatial changes in interactions and cascading effects (Tylianakis et al. 2008; Vilela et al. 2014). For example, asynchronism between plants and pollinators due to temperature rise has been widely studied (Memmott et al. 2007). This rupture compromises food production and the conservation of natural ecosystems (Settele et al. 2016; Torezan-Silingardi et al. 2021). Asynchronism can also have consequences on herbivory insects (Warren and Bradford 2014), considering the interactions between caterpillars and host plants. Many herbivorous insects increase plant consumption in high CO_2 settings as plants tend to increase carbon concentrations and decrease nitrogen concentrations in leaves. Asynchronism is also evident in the interactions between ants and plants, directly affecting the dispersal of myrmecochorous seeds (Anjos et al. 2020; Warren and Bradford 2014). The increase in aridity, mainly due to the increase in temperature, has decreased seed dispersal rates by ants as well as dispersal distances (Oliveira et al. 2021). In this scenario of more arid environments, plant protection by ants has also been compromised (Oliveira et al. 2021). Some indirect interactions, involving decomposing insects (e.g., termites, beetles), have also been affected by climate change and may be beneficial in future scenarios. For example, termites are important in mitigating the effects of increasingly frequent droughts in tropical forests (Ashton et al. 2019) and dung beetles are responsible for reducing the greenhouse gas emissions (e.g., methane) (Slade et al. 2016). Therefore, although most interactions are harmed by climate change, some interacting species can mitigate such effects.

10.12 Insects will resist climate change

Defaunation (*sensu* Dirzo et al. 2014) is a useful term to examine the "terrestrial biodegradation" variable of Steffan's (2015) great acceleration of the Anthropocene. Recent reviews (e.g., Del-Claro and Dirzo 2021) indicate that the main direct driver of defaunation is over-exploitation (poaching and illegal trade). Land-use change is a critical driver of indirect defaunation, followed by invasive taxa, intoxication by pesticides and pollution and, finally, climate change. According to Dirzo et al. (2014, and sources therein) 67 percent of monitored populations of invertebrates (mainly insects) show a 45 percent mean decline in abundance. Focusing on insects, van Klink et al. (2020) report an average decline of 9 percent in abundance, much lower than that of the previous studies predicting imminent Armageddon (e.g., Sánchez-Bayo and Wyckhuys 2019). Indeed, insects are resilient species and evolutionary biology shows that the natural dynamics of speciation and extinction maintain, over time, a huge diversity of insect species interacting with other animals and plants. However, the main problem highlighted by Del-Claro and Dirzo (2021) is that the Anthropocene defaunation is occurring at such a rapid pace and globally that the question of whether natural systems will have time to restore the constellations of species interaction networks that sustain Earth biomes and the humanity remains open.

It is already possible to observe populations adapting to climate change. Rising temperatures are raising increased thermal tolerance (Buckley and Huey 2016; González-Tokman et al. 2020), dietary breadth (Lancaster 2020; Singer and Parmesan 2020), higher feeding rates and niche shifting (González-Tokman et al. 2020; Hill et al. 2011; Martin et al. 2020). Pest insects are favored as their survival increases with higher temperature rates and higher CO_2 concentration (Schneider et al. 2020). Accelerated development, reduced body size and greater heat tolerance, along with melanization of wings, were adaptations observed in lepidopterans (Günter et al. 2020). For aquatic insects, some beetle species were analyzed for cross-adaptability to two stressors, water salinity and desiccation (Pallarés et al. 2017). These adaptations favor saline species, indicating that freshwater species may be more sensitive to desiccation (Pallarés et al. 2016). On the other hand, in species that live in lotic environments the flow of water favors them, making it easier for them to tolerate environments with higher temperatures and lower rates of oxygen in the water (Frakes et al. 2021).

As discussed throughout this chapter, insect biodiversity and the direction, frequency and intensity of ecological interactions can be altered by climate change. Therefore understanding the complex relationship between biodiversity/ecological interactions and climate change is critical to predicting how ecosystems will respond to these inevitable changes. Robust predictions about the future are an immense challenge as they require multi-species models combined with large databases of complex systems (e.g., long-term monitoring), which must include multiple variables both biotic and abiotic (e.g., distribution models of species) (Araújo et al. 2011). However, we can infer future scenarios based on past and current climate change. An expected consequence is the favoring of species that are more tolerant to higher temperatures and more unstable climatic conditions, increasing the performance of these species and the amplitude of occurrence. For example, moth species living in the north of the United Kingdom (adapted to cold climates) have decreased their populations because of climate change, while species in the south (adapted to milder climates) have seen population increases (Fox et al. 2014). Insect decline has been a recurring and contrasting issue in recent years (Hallmann et al. 2017; Janzen and Hallwachs 2021; Wagner et al. 2021). Another predicted pattern is that in regions with rapid environmental changes, often regions with high latitudes and altitudes, generalist interactions should predominate (Dynesius and Jansson 2000), and mutualistic interactions and parasitism may be strongly threatened (Dunn et al. 2009). In addition, new biotic interactions may emerge (Lurgi et al. 2012; Urban et al. 2012) with several unknown consequences. Our knowledge is still very limited considering the effects of climate change on insect biodiversity and their ecological interactions (Del-Claro and Torezan-Silingardi 2021; Moura et al. 2021). However, despite rapid changes in ecosystems, promising tools have emerged (e.g., species distribution models, paleoclimatic and ecological databases) along with efforts focused on this issue (e.g., increased funding, training of manpower and research in syntheses), contributing to a better understanding of future scenarios and mitigation of climate change impacts on insect biodiversity.

Key reflections

- The Anthropocene has caused major transformations in Earth's ecosystems with cascading effects on aquatic, terrestrial and marine communities.
- The loss of biodiversity that directly impacts net primary productivity is not only represented by extinctions but by a reduction in population size which can make a species unviable in the medium-long term. Biodiversity loss can break links and nodes in biotic interaction networks, involving, for example, herbivorous insects, pollinators and seed dispersers.
- We have discussed the main drivers of global changes that are fueled by climate change and that can break trophic chains, ecological services, degrade the environment, and threats the life on Earth, with insects as a focus group.

Key further reading

- Del-Claro, K., Torezan-Silingardi, H.M. 2021. *Plant–animal interactions*. Cham: Springer.
- Dirzo, R., Young, H.S., Galetti, M., Ceballos, G., Isaac, N.J., and Collen, B. 2014. Defaunation in the Anthropocene. *Science 345* (6195): 401–06.
- Sage, R.F. 2020. Global change biology: A primer. *Global Changes in Biology 26* (1): 3–30.
- Wagner, D.L. 2020. Insect declines in the anthropocene. *Annual Review of Entomology* 65: 457–80.

References

Abrahamson, W.G. 1989. "Plant–animal interactions: An overview." In *Plant-Animal Interactions*, edited by W.G. Abrahamson. New York: McGraw-Hill Publishing, 1–22.

Addo-Bediako, A., Chown, S.L., and Gaston, K.J. 2001. Revisiting water loss in insects: a large scale view. *Journal of Insect Physiology* 47 (12): 1377–1388.

Aktar, W., Sengupta, D., and Chowdhury, A. 2009. Impact of pesticides use in agriculture: their benefits and hazards. *Interdisciplinary Toxicology* 2 (1): 1.

Altermatt, F. 2010. Climatic warming increases voltinism in European butterflies and moths. *Proceedings of the Royal Society B: Biological Sciences* 277 (1685): 1281–1287.

Anjos, D.V., Leal, L.C., Jordano, P., and Del-Claro, K. 2020. Ants as diaspore removers of non-myrmecochorous plants: a meta-analysis. *Oikos* 129 (6): 775–86.

Araújo, M.B., Rozenfeld, A., Rahbek, C., and Marquet, P.A. 2011. Using species co-occurrence networks to assess the impacts of climate change. *Ecography* 34 (6): 897–908.

Arena, M. and Sgolastra, F. 2014. A meta-analysis comparing the sensitivity of bees to pesticides. *Ecotoxicology* 23: 324–34.

Ashton, L.A., Griffitthis, H.M., Parr, C.L., Evans, T.A., Didham, R.K., Hasan, F., Teh, Y.A. et al. 2019.Termites mitigate the effects of drought in tropical rainforest. *Science* 363: 174–77. doi:10.1126/science.aau9565.

Baranov, V., Jourdan, J., Pilotto, F., Wagner, R., and Haase, P. 2020. Complex and nonlinear climate-driven changes in freshwater insect communities over 42 years. *Conservation Biology* 34 (5): 1241–1251.

Barber, J.R., Crooks, K.R., and Fristrup, K.M. 2010. The costs of chronic noise exposure for terrestrial organisms. *Trends in Ecology and Evolution* 25 (3): 180–89.

Bargańska, Ż., Lambropoulou, D., and Namieśnik, J. 2018. Problems and challenges to determine pesticide residues in bumblebees. *Critical Reviews in Analytical Chemistry* 48 (6): 447–58.

Bell, J.R., Botham, M.S., Henrys, P.A., Leech, D.I., Pearce-Higgins, J.W., Shortall, C.R., Brereton, T., Pickup J., and Thackeray, S.J. 2019. Spatial and habitat variation in aphid, butterfly, moth and bird phenologies over the last half century. *Global Change Biology* 25 (6): 1982–1994.

Bellard, C., Cassey, P., and Blackburn, T.M. 2016. Alien species as a driver of recent extinctions. *Biology Letters* 12 (2): 20150623.

Bellaver, J.M., Lima-Ribeiro, M.D.S., Hoffmann, D., and Romanowski, H.P. 2022. Rare and common species are doomed by climate change? A case study with neotropical butterflies and their host plants. *Journal of Insect Conservation* 26 (4): 651–61.

Biesmeijer, J.C., Roberts, S.P.M., Reemer, M., Ohlemueller, R., Edwards, M., Peeters, T., Schaffers. A.P., et al. 2006. Parallel declines in pollinators and insect-pollinated plants in Britain and the Netherlands. *Science* 313 (1): 351–54.

Boggs, C.L. 2016. The fingerprints of global climate change on insect populations. *Current Opinion in Insect Science* 17: 69–73.

Bonebrake, T.C. and Mastrandrea, M.D. 2010. Tolerance adaptation and precipitation changes complicate latitudinal patterns of climate change impacts. *Proceedings of the National Academy of Sciences* 107 (28): 12581–12586.

Bowen, A.E., Gurule-Small, G.A., and Tinghitella, R.M. 2020. Anthropogenic noise reduces male reproductive investment in an acoustically signaling insect. *Behavioral Ecology and Sociobiology* 74: 1–8.

Bowler, K. 2018. Heat death in poikilotherms: Is there a common cause? *Journal of Thermal Biology* 76: 77–79.

Bowman, D. 2017. When will the jungle burn? *Nature Climate Change* 7 (6): 390–91.

Boyes, D.H., Evans, D.M., Fox, R., Parsons, M.S., and Pocock, M.J. 2021. Is light pollution driving moth population declines? A review of causal mechanisms across the life cycle. *Insect Conservation and Diversity* 14 (2): 167–87.

Brittain, C. and Potts, S.G. 2011. The potential impacts of insecticides on the life-history traits of bees and the consequences for pollination. *Basic and Applied Ecology* 12 (4): 321–31.

Buckley, L.B. and Huey, R.B. 2016. How extreme temperatures impact organisms and the evolution of their thermal tolerance. *Integrative and Comparative Biology* 56 (1): 98–109.

Carvalheiro, L.G., Kunin, W.E., Keil, P., Aguirre-Gutierrez, J., Ellis, W.N., Fox, R., Groom, Q., et al. 2013. Species richness declines and biotic homogenisation have slowed down for NW-European pollinators and plants. *Ecology Letters* 16 (7): 870–78.

Casas-Pinilla, L.C., Iserhard, C.A., Richter, A., Gawlinski, K., Cavalheiro, L.B., Romanowski, H.P., and Kaminski, L.A. 2022. Different-aged Pinus afforestation does not support typical Atlantic Forest fruit-feeding butterfly assemblages. *Forest Ecology and Management* 518: 120279.

Chown, S.L., Sørensen, J.G., and Terblanche, J.S. 2011. Water loss in insects: an environmental change perspective. *Journal of Insect Physiology* 57 (8): 1070–1084.

Cohen, J.M., Lajeunesse, M.J. and Rohr, J.R. 2018 A global synthesis of animal phenological responses to climate change. *Nature Climate Change* 8 (3): 224–28.

Conrad, K.F., Warren, M.S., Fox, R., Parsons, M.S., and Woiwod, I.P. 2006. Rapid declines of common, widespread British moths provide evidence of an insect biodiversity crisis. *Biological Conservation* 132 (3): 279–91.

Conrad, K.F., Woiwod, I.P., Parsons, M., Fox, R., and Warren, M.S. 2004. Long-term population trends in widespread British moths. *Journal of Insect Conservation* 8 (2–3): 119–36.

Crutzen, P.J. 2002. Geology of mankind. *Nature* 415 (6867): 23–23.

Dacke, M., Baird, E., El Jundi, B., Warrant, E.J., and Byrne, M. 2021. How dung beetles steer straight. *Annual Review of Entomology* 66: 243–56.

Davies, P.M. 2010. Climate change implications for river restoration in global biodiversity hotspots. *Restoration Ecology,* 18(3): 261–68.

Del-Claro, K. 2019. Aquatic insects: Why it is important to dedicate our time on their study? In *Aquatic Insects: Behavior and Ecology,* edited by Del-Claro, K. and Guillermo, R. 1–9.

Del-Claro, K. and Dirzo, R. 2021. Impacts of Anthropocene defaunation on plant–animal interactions. In *Plant–Animal Interactions*, edited by K. Del-Claro and H.M. Torezan-Silingardi. Cham: Springer, 333–45.

Del-Claro, K. and Torezan-Silingardi, H.M. 2021. "An evolutionary perspective on plant-animal interactions." In *Plant–Animal Interactions*, edited by K. Del-Claro and H.M. Torezan-Silingardi. Cham: Springer, 1–15. doi:10.1007/978-3-030-66877-8_1.

Deutsch C.A., Tewksbury, J.J., Huey, R.B., Sheldom, K.S., Galambour, C.K., Haak, D.C., and Martim, P.R. 2008. Impacts of climate warming on terrestrial ectotherms across latitude. *Proceedings of the National Academy of Sciences* 105 (18): 6668–6672. doi:10.1073/pnas.0709472105.

Dicks, L.V., Breeze, T.D., Ngo, H.T., Senapathi, D., An, J., Aizen, M. A., Potts, S.G., et al. 2021. A global-scale expert assessment of drivers and risks associated with pollinator decline. *Nature Ecology and Evolution* 5 (10): 1453–1461.

Dirzo, R., Young, H.S., Galetti, M., Ceballos, G., Isaac, N.J., and Collen, B. 2014. Defaunation in the Anthropocene. *Science* 345 (6195): 401–06.

Dunn, R.R., Harris, N.C., Colwell, R.K., Koh, L.P., and Sodhi, N.S. 2009. The sixth mass coextinction: are most endangered species parasites and mutualists? *Proceedings of the Royal Society B: Biological Sciences* 276 (1670): 3037–3045.

Dynesius, M. and Jansson, R. 2000. Evolutionary consequences of changes in species' geographical distributions driven by Milankovitch climate oscillations. *Proceedings of the National Academy of Sciences* 97 (16): 9115–9120.

Eggleton, P. 2020 The state of the world's insects. *Annual Review of Environment and Resources 45*: 61–82. doi:10.1146/annurev-environ-012420-050035.

Elgueta, S., Moyano, S., Sepúlveda, P., Quiroz, C., and Correa, A. 2017. Pesticide residues in leafy vegetables and human health risk assessment in North Central agricultural areas of Chile. *Food Additives and Contaminants: Part B* 10 (2): 105–12.

Fang, Y., Nie, Z., Die, Q., Tian, Y., Liu, F., He, J., and Huang, Q. 2017. Organochlorine pesticides in soil, air, and vegetation at and around a contaminated site in southwestern China: concentration, transmission, and risk evaluation. *Chemosphere* 178: 340–49.

Farahani, S., Bandani, A.R., Alizadeh, H., Goldansaz, S.H., and Whyard, S. 2020. Differential expression of heat shock proteins and antioxidant enzymes in response to temperature, starvation, and parasitism in the Carob moth larvae, Ectomyelois ceratoniae (Lepidoptera: Pyralidae). *PloS One* 15 (1): e0228104.

Firebaugh, A. and Haynes, K.J. 2016. Experimental tests of light-pollution impacts on nocturnal insect courtship and dispersal. *Oecologia* 182 (4): 1203–1211.

Fisher, R.C. 1978. Water balance in land arthropods. *Journal of Arid Environments* 1 (2): 196–97.

Forister, M.L., Pelton, E.M., and Black, S.H. 2019. Declines in insect abundance and diversity: We know enough to act now. *Conservation Science and Practice* 1 (8): e80.

Forrest, J.R. 2016. Complex responses of insect phenology to climate change. *Current Opinion in Insect Science* 17: 49–54.

Fox, R. 2013. The decline of moths in Great Britain: a review of possible causes. *Insect Conservation and Diversity* 6 (1): 5–19.

Fox, R., Brereton, T.M., Asher, J., August, T.A., Botham, M.S., Bourn, N.A.D., Chruickshank K.L., et al. 2015. *The State of the UK's Butterflies 2015*. Wareham, UK: Butterfly Conserv/Cent Ecol Hydrol.

Fox, R., Oliver, T.H., Harrower, C., Parsons, M.S., Thomas, C.D., and Roy, D.B. 2014. Long-term changes to the frequency of occurrence of British moths are consistent with opposing and synergistic effects of climate and land-use changes. *Journal of Applied Ecology* 51 (4): 949–57.

Frakes, J.I., Birrell, J.H., Shah, A.A., and Woods, H.A. 2021. Flow increases tolerance of heat and hypoxia of an aquatic insect. *Biology Letters* 17 (5): 20210004.

Franzén, M. and Johannesson, M. 2007. Predicting extinction risk of butterflies and moths (Macrolepidoptera) from distribution patterns and species characteristics. *Journal of Insect Conservation* 11: 367–90.

Gallego-Abenza, M., Mathevon, N., and Wheatcroft, D. 2020. Experience modulates an insect's response to anthropogenic noise. *Behavioral Ecology* 31 (1): 90–96.

García-Robledo, C., Kuprewicz, E.K., Staines, C.L., Erwin, T.L., and Kress, W.J. 2016. Limited tolerance by insects to high temperatures across tropical elevational gradients and the implications of global warming for extinction. *Proceedings of the National Academy of Sciences* 113 (3): 680–85.

Gherlenda, A.N., Haigh, A.M., Moore, B.D., Johnson, S.N., and Riegler, M. 2016. Climate change, nutrition and immunity: effects of elevated CO2 and temperature on the immune function of an insect herbivore. *Journal of Insect Physiology* 85: 57–64.

González-Tokman, D., Córdoba-Aguilar, A., Dáttilo, W., Lira-Noriega, A., Sánchez-Guillén, R.A., and Villalobos, F. 2020. Insect responses to heat: Physiological mechanisms, evolution and ecological implications in a warming world. *Biological Reviews* 95 (3): 802–21.

Gouin, T., Shoeib, M., and Harner, T. 2008. Atmospheric concentrations of current-use pesticides across south-central Ontario using monthly-resolved passive air samplers. *Atmospheric Environment* 42 (34): 8096–8104.

Goulson, D., Nicholls, E., Botías, C., and Rotheray, E.L. 2015. Bee declines driven by combined stress from parasites, pesticides, and lack of flowers. *Science* 347 (6229): 1255957.

Guerenstein, P.G. and Hildebrand, J.G. 2008 Roles and effects of environmental carbon dioxide in insect life. *Annual Review of Entomology* 53: 161–78.

Gullan, P.J. and Cranston, P.S. 2014. *The Insects: An Outline of Entomology*. Malden, US: John Wiley and Sons.

Günter, F., Beaulieu, M., Freiberg, K.F., Welzel, I., Toshkova, N., Žagar, A. Fischer, K., et al. 2020. Genotype-environment interactions rule the response of a widespread butterfly to temperature variation. *Journal of Evolutionary Biology* 33 (7): 920–29.

Gurule-Small, G.A., and Tinghitella, R.M. 2018. Developmental experience with anthropogenic noise hinders adult mate location in an acoustically signalling invertebrate. *Biology Letters* 14 (2): 20170714.

Habel, J.C., Samways, M.J. and Schmitt, T. 2019 Mitigating the precipitous decline of terrestrial European insects: Requirements for a new strategy. *Biodiversity and Conservation* 28 (6): 1343–1360.

Hallmann, C.A., Sorg, M., Jongejans, E., Siepel, H., Hofland, N., Schawan, H., Stenmans, W. et al. 2017. More than 75 percent decline over 27 years in total flying insect biomass in protected areas. *PLoS One* 12 (10): e0185809.

Hallwachs, W. and Janzen, D.H. 2019. Perspective: Where might be many tropical insects?

Halsch, C.A., Shapiro, A.M., Fordyce, J.A., Nice, C.C., Thorne, J.H., Waetjen, D.P., and Forister, M.L. 2021. Insects and recent climate change. *Proceedings of the National Academy of Sciences* 118 (2): e2002543117.

Hayes, T.B. and Hansen, M. 2017. From silent spring to silent night: Agrochemicals and the anthropocene. *Elementa: Science of the Anthropocene* 5: 57.

Herrando, S., Titeux, N., Brotons, L., Anton, M., Ubach, A., Villero Pi D., Garcia-Barros E., Munguira, M.L., Godinho C., and Stefanescu, C. 2019. Contrasting impacts of precipitation on Mediterranean birds and butterflies. *Scientific Reports* 9 (1): 5680.

Higgins, J.K., MacLean, H.J., Buckley, L.B., and Kingsolver, J.G. 2014. Geographic differences and microevolutionary changes in thermal sensitivity of butterfly larvae in response to climate. *Functional Ecology* 28: 57.

Hill, J.K., Griffiths, H.M., and Thomas, C.D. 2011. Climate change and evolutionary adaptations at species' range margins. *Annual Review of Entomology* 56: 143–59.

Hillstrom, M.L. and Lindroth, R.L. 2008. Elevated atmospheric carbon dioxide and ozone alter forest insect abundance and community composition. *Insect Conservation and Diversity* 1 (4): 233–41.

Hooke, R.L.B., Martín-Duque, J.F. and Pedraza, J. 2012 Land transformation by humans: A review. *GSA Today* 22 (12): 4–10. doi:10.1130/GSAT151A.1.

Hoover, S.E., Ladley, J.J., Shchepetkina, A.A., Tisch, M., Gieseg, S.P., and Tylianakis, J.M. 2012. Warming, CO2, and nitrogen deposition interactively affect a plant-pollinator mutualism. *Ecology Letters* 15 (3): 227–34.

Janzen D. H. and Hallwachs W. 2019. Perspective: Where might be many tropical insects? *Biology and Conservation* 233: 102–08. doi:10.1016/j.biocon.2019.02.030.

Janzen, D.H. and Hallwachs, W. 2021. To us insectometers, it is clear that insect decline in our Costa Rican tropics is real, so let's be kind to the survivors. *Proceedings of the National Academy of Sciences* 118 (2): e2002546117.

Jones, W. 2013. Olfactory carbon dioxide detection by insects and other animals. *Molecules and Cells* 35: 87–92.

Kehoe, R., Frago, E., and Sanders, D. 2021. Cascading extinctions as a hidden driver of insect decline. *Ecological Entomology* 46 (4): 743–56.

King, A.M. and MacRae, T.H. 2015. Insect heat shock proteins during stress and diapause. *Annual Review of Entomology* 60: 59–75.

Kingsolver, J.G., Arthur Woods, H., Buckley, L.B., Potter, K.A., MacLean, H.J. and Higgins, J.K. 2011 Complex life cycles and the responses of insects to climate change. *Integrative and Comparative Biology* 51 (5): 719–32. doi:10.1093/icb/icr015.

Klaiber, J., Najar-Rodriguez, A.J., Dialer, E., and Dorn, S. 2013. Elevated carbon dioxide impairs the performance of a specialized parasitoid of an aphid host feeding on Brassica plants. *Biological Control* 66 (1): 49–55.

Knop, E. 2016. Biotic homogenization of three insect groups due to urbanization. *Global Change Biology* 22 (1): 228–36.

Labandeira, C.C. 1998. Plant-insect associations from the fossil record. *Geotimes* 43 (9): 18–24.

Lamarre, G.P., Fayle, T.M., Segar, S.T., Laird-Hopkins, B.C., Nakamura, A., Souto-Vilaros, D., and Basset, Y. 2020. "Monitoring tropical insects in the 21st century". *Advances in Ecological Research* 62: 295–330.

Lamarre, G.P., Pardikes, N.A., Segar, S., Hackforth, C.N., Laguerre, M., Vincent, B., and Basset, Y. 2022. More winners than losers over 12 years of monitoring tiger moths (Erebidae: Arctiinae) on Barro Colorado Island, Panama. *Biology Letters* 18 (4): 20210519.

Lampe, U., Schmoll, T., Franzke, A., and Reinhold, K. 2012. Staying tuned: Grasshoppers from noisy roadside habitats produce courtship signals with elevated frequency components. *Functional Ecology* 26 (6): 1348–1354.

Lancaster, L.T. 2020. Host use diversification during range shifts shapes global variation in Lepidopteran dietary breadth. *Nature Ecology and Evolution* 4 (7): 963–69.

Leite-Filho, A.T., Soares-Filho, B.S., Davis, J.L., Abrahão, G.M., and Borner, J. 2021 Deforestation reduces rainfall and agricultural revenues in the Brazilian Amazon. *Nature Communications* 12 (1): 1–7. doi:10.1038/s41467-021-22840-7.

Levenhagen, M.J., Miller, Z.D., Petrelli, A.R., Ferguson, L.A., Shr, Y., Gomes, D.G., Taff, B.D., White, C., Fristrup, K., Monz, C. Ecosystem services enhanced through soundscape management link people and wildlife. *People Nat.* (00):1–14.

Lewis, S.L., Edwards, D.P., and Galbraith, D. 2015. Increasing human dominance of tropical forests. *Science* 349 (6250): 827–32.

Lewis, S.L. and Maslin, M.A. 2020. The human planet: How we created the Anthropocene. *Global Environment* 13 (3): 674–80.

Lister, B.C. and Garcia, A. 2018 Climate-driven declines in arthropod abundance restructure a rainforest food web. *Proceedings of the National Academy of Sciences* 115 (44): E10397–E10406. doi:10.1073/pnas.1722477115.

Lovejoy, T.E., and Nobre. C. 2018. Amazon tipping point. *Science Advances 4* (2): eaat2340. doi:10.1126/sciadv.aat2340.

Luna, P. and Dáttilo, W. 2021. "Disentangling plant–animal interactions into complex networks: A multi-view approach and perspectives." In *Plant–Animal Interactions*, edited by K. Del-Claro and H.M. Torezan-Silingardi. Cham: Springer, 261–81. doi:10.1007/978-3-030-66877-8_10.

Lurgi, M., López, B.C., and Montoya, J.M. 2012. Novel communities from climate change. *Philosophical Transactions of the Royal Society B: Biological Sciences* 367 (1605): 2913–2922.

Lykogianni, M., Bempelou, E., Karamaouna, F., and Aliferis, K.A. 2021. Do pesticides promote or hinder sustainability in agriculture? The challenge of sustainable use of pesticides in modern agriculture. *Science of the Total Environment* 795: 148625.

Marengo, J.A., Jimenez, J.C., Espinoza, J.C., Cunha, A.P., and Aragão, L.E. 2022. Increased climate pressure on the agricultural frontier in the Eastern Amazonia–Cerrado transition zone. *Scientific Reports* 12 (1): 457.

Martin, Y., Van Dyck, H., Legendre, P., Settele, J., Schweiger, O., Harpke, A., Wiemers, M., et al. 2020. A novel tool to assess the effect of intraspecific spatial niche variation on species distribution shifts under climate change. *Global Ecology and Biogeography* 29 (3): 590–602.

Martín-Folgar, R., de la Fuente, M., Morcillo, G., and Martínez-Guitarte, J.L. 2015. Characterization of six small HSP genes from Chironomus riparius (Diptera, Chironomidae): Differential expression under conditions of normal growth and heat-induced stress. *Comparative Biochemistry and Physiology Part A: Molecular and Integrative Physiology* 188: 76–86.

Matricardi, E.A.T., Skole, D.L., Costa, O.B., Pedlowski, M.A., Samek, J.H., and Miguel, E.P. 2020. Long-term forest degradation surpasses deforestation in the Brazilian Amazon. *Science* 369 (6509): 1378–1382.

Mattila, N., Kaitala, V., Komonen, A., Kotiaho, J.S., and Päivinen, J. 2006. Ecological determinants of distribution decline and risk of extinction in moths. *Conservation Biology* 20 (4): 1161–1168.

Maxwell, S.L., Fuller, R.A., Brooks, T.M., and Watson, J.E. 2016. Biodiversity: The ravages of guns, nets and bulldozers. *Nature* 536 (7615): 143–45.

Memmott, J., Craze, P.G., Waser, N. ., and Price, M.V. 2007. Global warming and the disruption of plant–pollinator interactions. *Ecology Letters* 10 (8): 710–17.

Mingo, V., Lötters, S., and Wagner, N. 2017. The impact of land use intensity and associated pesticide applications on fitness and enzymatic activity in reptiles—a field study. *Science of the Total Environment* 590: 114–24.

Mitton, J.B. and Ferrenberg, S.M. 2012. Mountain pine beetle develops an unprecedented summer generation in response to climate warming. *The American Naturalist* 179 (5): E163–E171.

Morley, E.L., Jones, G., and Radford, A.N. 2014. The importance of invertebrates when considering the impacts of anthropogenic noise. *Proceedings of the Royal Society B: Biological Sciences* 281 (1776): 20132683.

Moura, R.F., Colberg, E., Alves-Silva, E., Mendes-Silva, I., Fagundes, R., Stefani, V., and Del-Claro, K. 2021. "Biotic defenses against herbivory." In *Plant–Animal Interactions*, edited by K. Del-Claro and H.M. Torezan-Silingardi. Cham: Springer International Publishing, 93–118.

Newbold, T., Hudson, L.N., Contu, S., Hill, S.L., Beck, J., Liu, Y., Meyer C., et al. 2018. Widespread winners and narrow-ranged losers: Land use homogenizes biodiversity in local assemblages worldwide. *PLoS Biology* 16 (12): e2006841.

Nobre, C.A., Sampaio, G., Borma, L.S., Castilla-Rubio, J.C., Silva, J.S., and Cardoso, M. 2016. Land-use and climate change risks in the amazon and the need of a novel sustainable development paradigm. *Proceedings of the National Academy of Sciences* 3 (39): 10759–10768. https://doi.org/10.1073/pnas.1605516113.

Oliveira, F.M., Câmara, T., Durval, J.I., Oliveira, C.L., Arnan, X., Andersen, A.N., Ribeiro E. 2021. Plant protection services mediated by extrafloral nectaries decline with aridity but are not influenced by chronic anthropogenic disturbance in Brazilian Caatinga. *Journal of Ecology* 109 (1): 260–72.

Orci, K.M., Petróczki, K., and Barta, Z. 2016. Instantaneous song modification in response to fluctuating traffic noise in the tree cricket Oecanthus pellucens. *Animal Behaviour* 112: 187–94.

O'Sullivan, J.D., MacMillan, H.A., and Overgaard, J. 2017. Heat stress is associated with disruption of ion balance in the migratory locust, Locusta migratoria. *Journal of Thermal Biology* 68: 177–85.

Owens, A.C., Cochard, P., Durrant, J., Farnworth, B., Perkin, E.K., and Seymoure, B. 2020. Light pollution is a driver of insect declines. *Biological Conservation 241*: 108259.

Pallarés, S., Botella-Cruz, M., Arribas, P., Millán, A., and Velasco, J. 2017. Aquatic insects in a multistress environment: cross-tolerance to salinity and desiccation. *Journal of Experimental Biology* 220 (7): 1277–1286.

Pallarés, S., Velasco, J., Millán, A., Bilton, D.T., and Arribas, P. 2016. Aquatic insects dealing with dehydration: Do desiccation resistance traits differ in species with contrasting habitat preferences? *PeerJ* 4: e2382.

Pincebourde, S. and Suppo, C. 2016. The vulnerability of tropical ectotherms to warming is modulated by the microclimatic heterogeneity. *Integrative and Comparative Biology* 56 (1): 85–97.

Portner, H.O. and Farrell, A.P. 2008. Physiology and climate change. *Science 322*: 690–92.

Potts, S., Imperatriz-Fonseca, V., Ngo, H., Aizen, J.C., Bieismeijer, T.D., Breeze, L.V., Dicks, L.A. et al. 2016. Safeguarding pollinators and their values to human well-being. *Nature* 540 (7632): 220–29.

Powney, G.D., Carvell, C., Edwards, M., et al. 2019. Widespread losses of pollinating insects in Britain. *Nature Communications* 10 (1): 1–6.

Rada S., Schweiger O., Harpke A., Kühn, E., Kuras, T., Settele, T., Musche, M. 2019. Protected areas do not mitigate biodiversity declines: A case study on butterflies. *Diversity and Distribution* 25 (2): 217–24.

Regan, H.M., Crookston, J.B., Swab, R., Franklin, J., and Lawson, D.M. 2010. Habitat fragmentation and altered fire regime create trade-offs for an obligate seeding shrub. *Ecology*, 91(4): 1114–1123.

Reverchon, F. and Méndez-Bravo, A. 2021. "Plant-mediated above-belowground interactions: A phytobiome story." In *Plant–Animal Interactions*, edited by K. Del-Claro and H.M. Torezan-Silingardi. Springer. Cham: Springer International Publishing, 205–31.

Román-Palacios, C., and Wiens, J.J. 2020. Recent responses to climate change reveal the drivers of species extinction and survival. *Proceedings of the National Academy of Sciences* 117 (8): 4211–4217.

Russo, D., Cosentino, F., Festa, F., De Benedetta, F., Pejic, B., Cerretti, P., and Ancillotto, L. 2019. Artificial illumination near rivers may alter bat-insect trophic interactions. *Environmental Pollution* 252: 1671–1677.

Sage, R.F. 2020 Global change biology: A primer. *Global Changes in Biology* 26 (1): 3–30.

Salcido, D.M., Forister, M.L., Garcia Lopez, H., and Dyer, L.A. 2020. Loss of dominant caterpillar genera in a protected tropical forest. *Scientific Reports* 10 (1): 1–10.

Sánchez-Bayo, F. and Wyckhuys, K.A.G. 2019 Worldwide decline of the entomofauna: A review of its drivers. *Biology and Conservation* 232: 8–27.

Schneider, D., Ramos, A.G., and Córdoba-Aguilar, A. 2020. Multigenerational experimental simulation of climate change on an economically important insect pest. *Ecology and Evolution* 10 (23): 12893–12909.

Schuch, S., Wesche, K., and Schaefer, M. 2012. Long-term decline in the abundance of leafhoppers and planthoppers (Auchenorrhyncha) in Central European protected dry grasslands. *Biological Conservation* 149 (1): 75–83.

Singer, M.C. and Parmesan, C. 2020. Colonizations drive host shifts, diversification of preferences and expansion of herbivore diet breadth. *BioRxiv* 2020–2023.

Slade, E., Riutta, T., Roslin, T., Tuomitsuo, H.L. 2016. The role of dung beetles in reducing greenhouse gas emissions from cattle farming. *Sci Rep* 6: 18140. doi:10.1038/srep18140.

Stange, G. 1997. Effects of changes in atmospheric carbon dioxide on the location of hosts by the moth, Cactoblastis cactorum. *Oecologia* 110: 539–45.

Steffen, W., Broadgate, W., Deutsch, L., Gaffney, O., and Ludwig, C. 2015. The trajectory of the Anthropocene: the great acceleration. *The Anthropocene Review* 2 (1): 81–98.

Steffen, W., Persson, S.A., Deutsch, L., Zalasiewicz, J., Williams, M., Richardson, K., Crumley, C., et al. 2011. The anthropocene: From global change to planetary stewardship. *Ambio* 40: 73961.

Stephens, S.L., Agee, J.K., Fulé, P.Z., Nort, M.P., Romme, W.H., Swetnam, T.W., Turner, M.G. 2013. Managing forests and fire in changing climates. *Science* 342 (6154): 41–42. doi:10.1126/science.1240294.

Stewart, A.J. 2021. Impacts of artificial lighting at night on insect conservation. *Insect Conservation and Diversity* 14 (2): 163–166.

Stork, N.E. 1997. Measuring global biodiversity and its decline. *Biodiversity II: Understanding and Protecting Our Biological Resources* 41: 41–68.

Stork, N.E. 2018. How Many Species of Insects and Other Terrestrial Arthropods Are There on Earth? *Annual Review of Entomology* 63: 31–45. doi:10.1146/annurev-ento-020117-043348.

Strachan, S.R., Chester, E.T., and Robson, B.J. 2015. Freshwater invertebrate life history strategies for surviving desiccation. *Springer Science Reviews* 3: 57–75.

Sunday, J.M., Bates, A.E., Kearney, M.R., Colwell, R.K., Dulvy, N.K., Longino, J.T., and Huey, R.B. 2014. Thermal-safety margins and the necessity of thermoregulatory behavior across latitude and elevation. *Proceedings of the National Academy of Sciences* 111 (15): 5610–5615.

Terblanche, J.S., Sinclair, B.J., Klok, C.J., McFarlane, M.L., and Chown, S.L. 2005. The effects of acclimation on thermal tolerance, desiccation resistance and metabolic rate in Chirodica chalcoptera (Coleoptera: Chrysomelidae). *Journal of Insect Physiology* 51 (9): 1013–1023.

Temreshev, I.I., Esenbekova, P.A., Sagitov, A.O., Mukhamadiev, N.S., Sarsenbaeva, G.B., Ageenko, A.V., Homziak, J. 2018. Evaluation of the effect of locally produced biological pesticide (AKKөbelek ™) on biodiversity and abundance of beneficial insects in four forage crops in the Almaty region of Kazakhstan. *International Journal of Environment Agriculture and Biotechnology* 3 (1): 072–091.

Thierry, M., Hrček, J., and Lewis, O.T. 2019. Mechanisms structuring host–parasitoid networks in a global warming context: a review. *Ecological Entomology* 44 (5): 581–92.

Thomas, C. D. 1990. What do real population dynamics tell us about minimum viable population sizes? *Conservation Biology* 4 (3): 324–27.

Thomas, J.A., Telfer, M.G., Roy, D.B., Preston, C.D., Greenwood, J.J.D., Asher, J., Fox, R., et al. 2004. Comparative losses of British butterflies, birds, and plants and the global extinction crisis. *Science* 303 (5665): 1879–1881. doi:10.1126/science.1095046.

Tirado, M.C., Clarke, R., Jaykus, L.A., McQuatters-Gollop, A., and Frank, J.M. 2010. Climate change and food safety: A review. *Food Research International* 43 (7): 1745–1765.

Torezan-Silingardi, H.M. and Del-Claro, K.L.E.B.E.R. 1998. Behavior of visitors and reproductive biology of Campomanesia pubescens (Myrtaceae) in cerrado vegetation. *Ciencia e Cultura (Sao Paulo)* 50 (4): 281–83.

Torezan-Silingardi, H.M., Silberbauer-Gottsberger, I., and Gottsberger, G. 2021. "Pollination ecology: natural history, perspectives and future directions". In Plant–Animal Interactions, edited by K. Del-Claro and H.M. Torezan-Silingardi. Cham: Springer International Publishing 119–74.

Tudi, M., Daniel Ruan, H., Wang, L., Lyu, J., Sadler, R., Connell, D., Phung, D. T., et al. 2021. Agriculture development, pesticide application and its impact on the environment. *International Journal of Environmental Research and Public Health* 18 (3): 1112.

Tylianakis, J.M., Didham, R.K., Bascompte, J., and Wardle, D.A. 2008. Global change and species interactions in terrestrial ecosystems. *Ecology Letters* 11 (12): 1351–1363.

Urban, M.C., Tewksbury, J.J., and Sheldon, K.S. 2012. On a collision course: competition and dispersal differences create no-analogue communities and cause extinctions during climate change. *Proceedings of the Royal Society B: Biological Sciences* 279 (1735): 2072–2080.

Vanbergen, A.J., Baude, M., Biesmeijer, J.C., and Insect Pollinator Initiative, et al. 2013. Threats to an ecosystem service: pressures on pollinators. *Frontiers in Ecology and the Environment* 11 (5): 251–59.

Van Dijk, T.C., Van Staalduinen, M.A., and Van der Sluijs, J.P. 2013. Macro-Invertebrate decline in surface water polluted with imidacloprid. *PLoS One* 8 (4): e61336. doi:10.1371/journal.pone.0062374.

van Klink, R., Bowler, D.E., Gongalsky, K.B., Swengel, A.B., Gentile, A., and Chase, J.M. 2020. Meta-analysis reveals declines in terrestrial but increases in freshwater insect abundances. *Science* 368 (6489): 417–20.

Vilela, A.A., Del Claro, V.T.S., Torezan-Silingardi, H.M., and Del-Claro, K. 2018. Climate changes affecting biotic interactions, phenology, and reproductive success in a savanna community over a 10-year period. *Arthropod–Plant Interactions* 12: 215–27.

Vilela, A.A., Torezan-Silingardi, H.M., and Del-Claro, K. 2014. Conditional outcomes in ant–plant–herbivore interactions influenced by sequential flowering. *Flora-Morphology, Distribution, Functional Ecology of Plants* 209 (7): 359–66.

Wagner, D.L. 2020. Insect declines in the Anthropocene. *Annual Review of Entomology* 65: 457–80.

Wagner, D.L., Fox, R., Salcido, D.M., and Dyer, L.A. 2021. A window to the world of global insect declines: Moth biodiversity trends are complex and heterogeneous. *Proceedings of the National Academy of Sciences* 118 (2): e2002549117.

Wagner, D.L., Grames, E.M., Forister, M.L., Berenbaum, M.R., and Stopak, D. 2021. Insect decline in the Anthropocene: Death by a thousand cuts. *Proceedings of the National Academy of Sciences* 118 (2): e2023989118.

Warren, R.J., and Bradford, M.A. 2014. Mutualism fails when climate response differs between interacting species. *Global Change Biology* 20 (2): 466–74.

Watson, J.E.M., Evans, T., Venter, O., Willians, B., Tulloc, A., Stewart, C., Thompson, I. et al. 2018. The exceptional value of intact forest ecosystems. *Nature Ecology and Evolution* 2 (4): 599–610. doi:10.1038/s41559-018-0490-x.

Wilson, R.J., and Fox, R. 2021. Insect responses to global change offer signposts for biodiversity and conservation. *Ecological Entomology* 46 (4): 699–717.

Wojda, I. 2017. Temperature stress and insect immunity. *Journal of Thermal Biology* 68, 96–103.

Woon, J.S., Boyle, M.J.W., Ewers, R.M., Chung, A. and Eggleton, P. 2019. Termite environmental tolerances are more linked to desiccation than temperature in modified tropical forests. *Insectes Sociaux* 66 (1): 57–64. doi:10.1007/s00040-018-0664-1.

Worldometer. 2019. Elaboration of Data by United Nations, Department of Economic and Social Affairs, Population Division. World Population Prospects: The 2017 Revision. <https://www.worldometers.info>.

CHAPTER 11

Effects of climate change on insect distributions and invasions

Lucie Aulus-Giacosa*, Olivia K. Bates*, Aymeric Bonnamour*, Jelena Bujan*, Jérôme M. W. Gippet*, Gyda Fenn-Moltu*, Tristan Klaftenberger* and Cleo Bertelsmeier*

11.1 Introduction

Since the early twentieth century, the global surface temperature (land and ocean) has increased by approximately 1°C (Brönnimann 2018), and models predict continued warming of 2–6°C by 2100 (Christensen et al. 2007). Climate change during the past fifty years has mainly been caused by human activities (Cooper et al. 2002; IPCC 2012) and temperatures are rising ten to 10,000 times faster than during the last deglaciation. Climate change also involves cascading effects such as rising sea level, and the increasing frequency of extreme weather events such as floods, storms and droughts (Bale et al. 2002). Several metrics can be used to describe the multiple dimensions of climate change: the magnitude (difference in climate parameters and probability of extremes), the timing of climatic events (e.g., change in seasonality) and the availability (area of analogous climates and emergence of novel climates) or position (change in distance to analogous climate and climate change velocity) of climates (for more detail see Garcia et al. 2014). The combination of the several dimensions of climate change will not only affect species specific responses to climate change but also the pattern of population dynamics and global biodiversity of insects (Kiritani 2013), thus providing local and regional opportunities for some species to maintain or expand their range while others will be threatened (Figure 11.1).

11.1.1 *Niches and distributions*

In order to understand how climate change will affect future species distributions, we will first discuss the determinants of historical species distributions and in particular the

---

* All authors contributed equally.

Lucie Aulus-Giacosa et al., *Effects of climate change on insect distributions and invasions*. In: *Effects of Climate Change on Insects.*
Edited by: Daniel González-Tokman and Wesley Dáttilo, Oxford University Press. © Oxford University Press (2024).
DOI: 10.1093/oso/9780192864161.003.0011

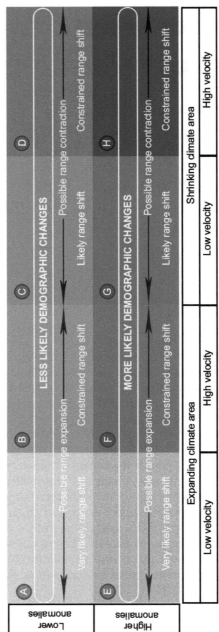

Figure 11.1 Spatial overlap of climate change metric and its effect on species range shift. The interplay of three metrics is displayed in this illustration adapted from Garcia et al. 2014 (see their paper for more details). Climatic metrics are displayed in black. One metric is at the local scale: the standardized local climate anomalies such as change in climate extremes and seasonality. The two other metrics are at the regional scale: change in area-baseline analogous climate and velocity of climate change. Consequences of the interplay of climate change metrics are displayed in white. Lower (A–B-C-D) and higher (E-F-G-H) local anomalies refer to values below and above the median, reflecting lower and higher chance of demographic changes. Expansion of analogous climatic area (A–B-E-F) is expected to increase the probability of species range expansion, whereas shrinking analogous climatic area may favor species range contraction (C-D-G-H). At low velocity changes (A-E-C-G), species may be able to track suitable climate over the region's topography and habitat structure, while this task may become more difficult at high velocity (B-F-D-H).

link between a species' climatic niche and its spatial distribution. All species have range limits beyond which they do not occur. It has been a central mystery in biogeography and evolutionary biology to understand what determines range limits (Bridle and Vines 2007). Across space, many environmental factors change including temperature, precipitation, solar radiation, wind speed or snow cover (Spence and Tingley 2020). Therefore, range limits might simply correspond to the most extreme conditions that a particular species can tolerate. This simple link between the species' biological requirements and its spatial distribution has been formalized in 1917 by Grinnell as the species' "niche" (Grinnell 1917). The environmental niche encompasses the range of conditions under which a species can thrive. As numerous different factors may be important for a species, the niche is generally viewed today as a space within an n-dimensional hypervolume of environmental variables defining the full set of conditions enabling a species to survive and reproduce (Blonder et al. 2014).

However, it is only rarely true that a species' range limits correspond perfectly to its niche limits (Gaston 2009; Sexton et al. 2009). Transplant experiments have shown that there are often large areas with suitable environmental conditions that are not fully occupied (Hargreaves, Samis and Eckert 2014). To account for the frequent failure of species to establish under all suitable environments, Hutchinson has introduced the concept of the "realized niche" corresponding to the set of conditions where a species actually lives and not to where it could potentially live (i.e., its fundamental niche) (Hutchinson 1957). The constraints of realized niches can be summarized by the BAM (Biotic, Abiotic, Movement) model (Figure 11.2) (Soberón and Peterson 2005). Species distributions are defined by biotic interactions (B), eco-physiological adaptations that determine the range of abiotic conditions they can tolerate (A) and the ability of the species to disperse and move (M) across geographic barriers such as mountain ranges or oceans. A species can survive where all three factors are met (Soberón and Peterson 2005). Most current research focuses on the realized niche (which is easy to infer from the species' current distribution) and not the fundamental niche (which is hard to measure) (Figure 11.2a).

Occasionally, a population may be found *outside* its fundamental niche. This may be because of source-sink dynamics at the metapopulation scale (Anderson et al. 2009; Watts et al. 2013) and this population would not be able to persist under unfavorable conditions without the constant arrival of immigrants. A population outside of its fundamental niche may also subsist after the environment has changed and go extinct with a certain time delay. Ongoing research is attempting to capture these types of complex time-delayed range dynamics (Fordham et al. 2018; Lurgi et al. 2015; Zurell et al. 2016).

11.1.2 *Possible species responses to novel climates*

The distinction between realized and fundamental niches becomes crucial for understanding species' responses to climate change. Following climate change, what does it mean for a species to experience abiotic conditions outside the hypervolume of conditions previously experienced by the species? A large part of biodiversity might be confronted to novel climatic conditions in the future and there is currently no clear answer to that question (Bellard et al. 2012). If the "novel" climatic conditions are outside of the fundamental niche of a species, it will no longer be able to survive there. However, the novel climatic conditions may as well be outside the species' previous realized niche (Figure 11.2, the intersection of B, A and M), but still within its fundamental niche (Figure 11.2, the intersection between B and A). In that case, the species will be able to persist. Without knowing how much of its fundamental niche a species currently occupies it is difficult to predict how it will fare under "novel" climatic conditions.

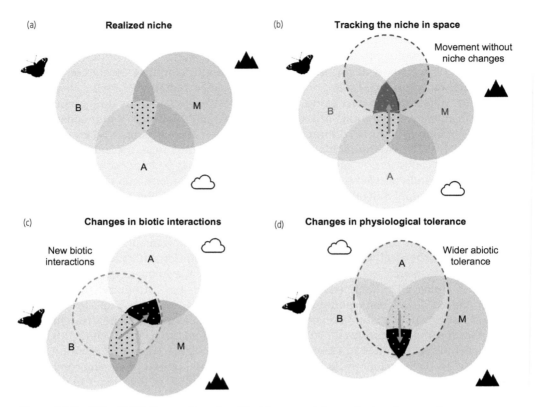

Figure 11.2 a) The "BAM diagram" is a simplified framework for understanding species distributions. Three sets of factors determine where a species occurs: the abiotic niche (A) and biotic interactions (B), and the possibility to access the area (through species "movement" (M)). The realized niche corresponds to the area where the three factors intersect (light gray area). Areas at the intersection of A and B are areas that correspond to the species' fundamental niche (biotic and abiotic) but are currently not colonized due to geographic constraints. b) Following climate change, the location where the abiotic requirements (A) are met may shift and the species could survive by tracking its niche in space through movement, resulting in range shifts (to the black area). c) Climate change-induced range shifts may require modifications in species interactions. D) Species may persist at their current location when the conditions have become unsuitable (black area) by changing their physiological requirements through adaptive evolution or acclimation (Adapted from Peterson 2011; Bates and Bertelsmeier 2021; phylopic by Andy Wilson).

If a species is exposed to a climate outside of the range of conditions that it can tolerate, there are three possible ways to respond to avoid extinction (Bellard et al. 2012).

First, a species may change its spatial distribution to remain in equilibrium with suitable climatic conditions (through movement—M, Figure 11.2b). This may occur at large spatial scales in the form of range shifts to higher latitudes or elevations. But it can also happen at a local scale, within habitats presenting heterogeneous microclimatic conditions. Especially small-sized and mobile organisms may be able to select more favorable microclimatic conditions, buffering against unfavorable macroclimatic changes (Montejo-Kovacevich et al. 2020; Pincebourde and Woods 2020; Suggitt et al. 2011). Elevational range shifts are predicted to be more likely than latitudinal range shifts (Colwell et al. 2008) as elevational temperature gradients are steeper. Elevational range shifts are also a complex interplay between land-use constraints and topography (Elsen et al. 2020). Here, species at the highest altitudes are the most vulnerable to climate changes (Urbani et al. 2017) because they cannot move further up in elevation (Colares et al. 2021). On the other hand, species living at high latitudes might be more robust to climate change than tropical species because they evolved under large daily and seasonal variations in temperature.

Second, a species may subsist through trait changes altering its biotic interactions (B) (Figure 11.2c) or abiotic requirements (A) (Figure 11.2d) that limit its range, either via plastic changes (such as acclimation) or via adaptive evolution (see section 11.5). It is a hotly debated question to what extent species will be able to evolve in response to novel climates. Indeed, the time scale at which current changes are happening appears too short for many long-lived organisms to adapt.

A final possibility (where B, A and M remain stable) for the species is to change the timing of life-cycle events to match the new conditions, for example by changing phenology (flowering, reproduction) or diurnal rhythms (shifting activity patterns to cooler hours of the day). Most of these species' responses are expected to occur in the future as the climate will dramatically change over the next few decades. Much research effort has concentrated on predicting species distribution. To achieve that, increasingly complex statistical tools and methods to work with large-scale data sets are being developed. However, empirical observations and tests of species responses *in situ* are comparatively rare, as they rely on long-term datasets which are difficult to collate (except for some taxonomic groups such as butterflies, see section 2).

To observe species responses to novel climates, one possibility is to use introduced species as model system (Moran and Alexander 2014). These are species that have been moved outside their native area and frequently encounter novel conditions (Mack et al. 2000). Studying introduced species allows us to address questions about the expected frequency of niche shifts, the importance of life-history and eco-physiological traits in enabling such shifts and the role of biotic interactions. It may also allow testing for adaptative evolution potentially underlying these shifts using common garden experiments or reciprocal transplantation (Bertelsmeier and Keller 2018). Moreover, introduced species are among the greatest threats to biodiversity besides climate change (Bellard et al. 2021). Introduced species are thought to be successful at expanding geographically because they are assumed to be rather highly competitive generalist species with high adaptability (van Kleunen et al. 2010; Weis 2010). Multiple drivers of global change including the globalization of trade, habitat loss and land-use change may increase the risks of biological invasions in the future (Bertelsmeier 2021; Gippet et al. 2019). Throughout this chapter, we will therefore consider effects of climate change on the distribution of native and introduced species.

11.1.3 *Current and future distributions of insects*

Knowledge of species distributions arise from observational records across the world. Despite representing the largest group of animals on Earth with more than 1 million described species, insects are in fact the group of animals with the greatest gaps in knowledge of taxonomy (Linnean shortfall) and distribution (Wallacean shortfall) (Diniz-Filho et al. 2010). While more than 70 percent of all animals are insects (Lobo 2016), insects represent only 10 percent of species occurrences in the widely used Global Biodiversity Information Facility database (GABI). However, insects are a particularly interesting group to study the impact of climate change on biodiversity. As ectotherms, insects depend on the thermal conditions of their environment for activities such as flight, reproduction, and foraging (Cox and Dolder 1995; Kenna et al. 2021; Régnière et al. 2012). Thus, climate change is expected to impact insects' spatial distribution to a large extent. Understanding how insect distributions will respond to contemporary climate change is urgent in light of recent population declines (Didham et al. 2020; Vogel 2017).

The number of studies on impacts of climate change on spatial distributions of insects has nearly doubled between 2000 and 2010 (Halsch et al. 2021). Here we consider how studies on insects' distributions and climate change are gaining interest while summarizing the available literature on Web of Science by performing a qualitative metadata analysis of 4,195 articles from 1990 until present (Figure 11.3a). The literature search was performed on Web of Science in December 2021 using the topic search terms (title, abstract, keywords): (insects* OR PROTURA* OR COLLEMBOLA* OR DIPLURA* OR MICROCORYPHIA* OR THYSANURA* OR EPHEMEROPTERA* OR ODONATA* OR ORTHOPTERA* OR PHASMOTODEA* OR GRYLLOBLATTODEA* OR MANTOPHASMA-TODEA* OR DERMAPTERA* OR PLECOPTERA* OR EMBIIDINA* OR ZORAPTERA* OR ISOPTERA* OR MANTODEA* OR BLATTODEA* OR HEMIPTERA* OR THYSANOPTERA* OR PSOCOPTERA* OR PHTHIRAPTERA* OR COLEOPTERA* OR NEUROPTERA* OR HYMENOPTERA* OR TRICHOPTERA* OR LEPIDOPTERA* OR SIPHONAPTERA* OR MECOPTERA* OR STREPSIPTERA* OR DIPTERA*) AND ((climate OR weather) AND change*) AND (range*OR migration* OR distribution*), which identified 4,615 studies. To be included in our metadata-analysis, we performed several filters. We removed 166 articles and book belonging to categories recording fewer than twenty articles. We restricted the language to English (forty-three additional articles were removed). We removed books and duplicated articles using the Cadima website[1] which led to ninety-three additional disregarded articles. Finally, we removed conference papers and journal reviews (118 articles), yielding 4,195 articles. To study detailed information on taxonomy and geographical study range, we randomly selected 210 articles for which title and abstract specifically referred to insects, range shifts and climate change among the filtered 4,195 articles. Using these 210 articles we extracted information on species name, species order, geographical scale of study and continent of the study.

The number of published papers on the topic has multiplied by ten during the past decade compared to the 1990s. However, it can also be noted that these studies are taxonomically (Figure 11.3b) and geographically restricted (Figure 11.3c), with most studies considering Lepidoptera in Europe.

This chapter will focus on possible and observed spatial changes, at large scales (section 2) and small scales (section 3) and discuss the role of natural and human-mediated dispersal (section 4), species traits (section 5) and biotic interactions (section 6) in enabling range shifts. We will also review novel modeling techniques (section 7) that are attempting to characterize the fundamental niche and dispersal processes more mechanistically.

11.2 Large-scale range shifts of insects

11.2.1 *Latitudinal range shifts*

Large-scale patterns in climate and biodiversity on Earth are strongly linked to latitude, with an increase in biological diversity from the poles to the equator (for details see Willig

---

[1] <https://www.cadima.info/index.php>.

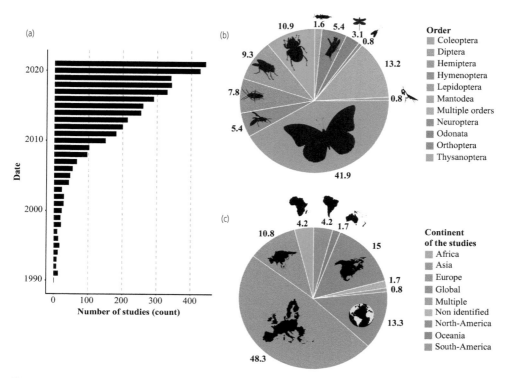

Figure 11.3 Trends in studies on insects' distribution responses to climate change from 1990 until present. a) General trends in publications over time (4195 studies); b) insect orders studied. Lepidoptera were 2.4 times overrepresented compared to the total number of species described (~160,000 species, 17% of described insects, while Coleoptera (~400,000 species, 40% of described insects) were four times underrepresented; c) Study locations. Most studies on insects' distribution responses to climate change were done in Europe (48.3%), followed by North America (15%).

et al. 2003). Research has focused on range expansions at the cold range margin (Lenoir and Svenning 2015), yet range shift may involve more than one geographical limit at the same time, including for example southern and northern limits (Figure 11.4). Depending on how the different limits are changing, the original range size may increase or decrease. Changes in range size depend on the differential change between leading and trailing range limits (Figure 11.4).

11.2.2 *Empirical observations of latitudinal range shifts*

Latitudinal range shifts have been reported in many different taxa (Figure 11.5). Studies generally indicate a trend of poleward range shifts in insects as expected (Hickling et al. 2006; Lowe et al. 2011; Mason et al. 2015; Parmesan et al. 1999; Parmesan and Yohe 2003; Poniatowski et al. 2020; Warren et al. 2001). To study latitudinal range shifts due to climate change, historical occurrence data collected at either the northern or southern range limit, but seldom both, are compared to current occurrence data and correlated with past climate change. Therefore, few studies on range shifts can conclude on the impact of climate change on total range size.

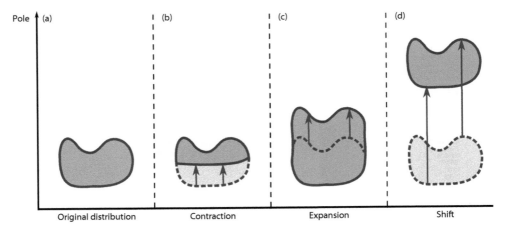

Figure 11.4 The main types of distribution changes and their consequences on range size. a) the original distribution area, b) a poleward contraction, c) an expansion of the original distribution, d) a complete poleward shift of the original distribution with an equal shift of both lower and upper limit. This figure is not intended to be exhaustive. More complex spatial changes can occur resulting in differences between the lower and upper limits of the original range distribution, impacting the original range size. Fragmentation or changes in abundance can also occur (for more details see Lenoir and Svenning 2015; Yang et al. 2021).

The first major large-scale study of latitudinal range shifts focused on thirty-five non-migratory European butterflies (Parmesan et al. 1999). Within one century, 63 percent of species moved poleward (with shifts ranging from 35 kilometers (km) to 240 km), while 34 percent maintained their original distribution and 3 percent shifted southward. Many subsequent studies have focused on Lepidopteran species in Europe and North America (Lenoir and Svenning 2015) because historical data are more easily available (Andrew et al. 2013). But Southern Hemisphere insects, for which there are far fewer data and studies, seem to show a similar pattern. For example, three Coleoptera and five Lepidoptera species in South Africa showed a poleward range shift over the last two decades (Perissinotto et al. 2011), with an extension of the southern limit (ranging from 90 to 830 km). Two distinct metanalyses confirmed a general trend of a poleward range shift (Chen et al. 2011; Parmesan and Yohe 2003). The first meta-analysis found an average poleward shift of 6.1 km per decade (Parmesan and Yohe 2003), while the second study found an even higher median rate of 16.9 km per decade with a higher rate in places that experiment higher warming levels (Chen et al. 2011). Both studies included only multi-species studies to limit the impact of publication biases towards reports of shifting compared to non-shifting species. However, average range shifts may not be relevant at a finer taxonomic scale because species may differ greatly in terms of physiology and dispersal ability. Here, we calculated the mean median response for vertebrates, non-insect invertebrates, and insects based on data from Chen et al. (2011) (Figure 11.5). Compared to the average range shift of 16.9 km across all taxa reported by the Chen et al. (2011), vertebrates respond on average slower (7.6 km per decade), non-insect invertebrates faster (22.9 km per decade) and insects even faster (26.8 km per decade) (Figure 11.5). Perhaps insects show larger range shifts in response to climate because as ectotherms they are more sensitive to external temperature changes (Sheldon et al. 2011). Higher temperatures increase development and survival rates in insects in temperate areas, enhancing establishment likelihood at higher latitudes (Strange and Ayres 2010).

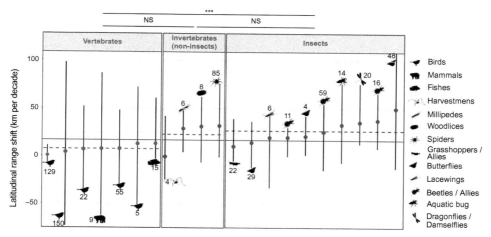

Figure 11.5 Latitudinal shift per decade for three main groups, based on data available from a previous meta-analysis (Chen et al. 2011). Each circle represents the median latitudinal shift per decade for a taxonomic group in a specific study region as described in the metanalysis. For two reasons, we may find the same taxonomic group represented several times because the data may come from different studies (e.g., butterflies) or from the same study with more precise taxonomic groups than the one represented on the figure (e.g., beetles/allies). The error bars represent the 90th and 10th percentile. Note that the last point on the right has a 90th percentile equal to 268.75. The size of each circle is proportional to the number of species studied. The red solid line represents the median latitudinal shift across all species, reported by the authors. The dotted lines represent the mean median latitudinal shift per decade for the three main taxonomic groups. For each pair of main groups, a Man–Whitney test was calculated. The mean median latitudinal range shift (dotted lines) of vertebrates and insects is different (p value < 0.001). In total, we used seven different studies which are referenced from 1 to 7 in brackets. (1): (Zuckerberg et al. 2009); (2): (Brommer 2004); (3): (Hickling et al. 2006); (4): (Hitch and Leberg 2007); (5): (Peterson 2003); (6): (Franco et al. 2006); (7): (Pöyry et al. 2009).

11.2.3 *Limitations of studies on large-scale range shifts*

Large-scale range shift studies suffer from several limitations. First, there is a geographical bias towards the northern hemisphere (Lenoir and Svenning 2015). In particular, studies in tropical areas have focused on elevation rather than latitudinal range shifts (Lenoir and Svenning 2015; Thomas 2010). Although, the climate is expected to change more slowly in tropical than in temperate areas (Loarie et al. 2009), limiting the necessity for a latitudinal shift. However, tropical insect species live closer to their thermal tolerance limits than temperate species (Khaliq et al. 2014), increasing potential impacts on insects (Deutsch et al. 2008) and their need to shift poleward. To gain a better understanding of range shifts in insects, it would be interesting to study their link with variation in thermal tolerance and the complex geography of climate change.

Second, scientists focus mainly on the north-south gradient with five times as many studies on latitudinal range shift as on longitudinal range shift (Lenoir and Svenning 2015). However, focusing on a unidirectional distribution shift neglects the complex interactions between temperature and precipitation (VanDerWal et al. 2013). For example, it is commonly accepted that continentality (distance from the sea) can have a significant impact on temperature and precipitation (Brunt 1924; Makarieva et al. 2009). Consequently, large-scale range shift studies should include both latitude and longitude to capture the complexity of climate change.

Finally, research on large-scale range shifts focuses mainly on one distribution limit (Lenoir and Svenning 2015). However, changes in a single distribution limit (without considering changes in total range size) are a poor proxy of the real species' response to climate change and can lead to a misleading estimate of a species' local extinction risk (McCain and Garfinkel 2021).

11.3 Fine-scale range shifts of insects

Biotic and abiotic conditions such as air and ground temperature, precipitation, soil moisture and vegetation type can strongly vary across short distances because of fine-scale environmental heterogeneity linked to topography, habitats, land use and microclimates. These fine-scale environmental variations might affect species response to climate change, especially small organisms like insects (Pincebourde and Woods 2020). They can either exacerbate the risks of local extinction associated with climate change (Raven and Wagner 2021) or, on the opposite, act as microrefugia (by buffering the effects of climate change (Suggitt et al. 2018) or help species track their niche locally (McCain and Garfinkel 2021) (Figure 11.6).

Figure 11.6 Landscapes are heterogenous. They vary in topography, comprise different types of habitats, and are subject to diverse anthropogenic modifications such as agriculture, urbanization and pollutions. Therefore, within landscapes, environmental conditions can greatly vary over short distances and modulate the effects of climate change on insects' distribution (Artwork: P. Gippet-Vinard, based on vector images from <http://www.Freepik.com> authors: macrovector, pch.vector, uplyak, vectorpocket, vectorpouch, freepik and all-free-download.com).

11.3.1 *Topography*

Changes in elevational ranges are the most studied fine-scale range shifts associated with climate change (McCain and Garfinkel 2021). It is expected that insect species shift, or at least contract, their range upward as climate warms (see McCain and Garfinkel 2021) for a detailed review) and multiple studies have already confirmed this trend over the last decades (Chen et al. 2009; Dolson et al. 2021; Halsch et al. 2021). However, upward shifts might not be a general pattern as many species do not show any elevational shift and some even shift unexpectedly downward (Halsch et al. 2021; McCain and Garfinkel 2021). For example, among 102 moth species sampled along an elevation gradient in Borneo, only 57 percent moved uphill over 42 years, with an average elevation increase of approximately 100 meters (m) (Chen et al. 2009).

In addition to elevation, slope and aspect (i.e., the compass direction that a terrain faces) can also affect insects' spatial distribution (Bennie et al. 2008; Buse et al. 2015; Vessby and Wiktelius 2003) as they strongly impact air and ground temperature by creating variations in solar radiation or wind regimes (Ebel 2012; Liu and Luo 2011; Oorthuis et al. 2021) as well as soil properties (by affecting water movement, nutrient content) (Oorthuis et al. 2021). North-exposed slopes can be up to 10 °C colder than adjacent south-exposed slopes (Rita et al. 2021) and could thus offer opportunities for horizontal spatial shifts that might have equivalent outcomes than vertical (elevational) shifts. However, the influence of slope and aspect on insects' response to climate change remains mostly unexplored so far (but see Suggitt et al. 2018).

11.3.2 *Habitats and land use*

Depending on the habitat they occupy, species might not experience climate change in the same way or at the same pace (Uhler et al. 2021). Therefore, insects' response to climate change (including latitudinal and elevational shifts and contractions) might vary across habitats (Guo et al. 2018; Stralberg et al. 2020). For example, among animals, plants and fungi, terrestrial insects have the greatest average upward shift (i.e., + 36.2 m per decade), while semi-aquatic insects do not show any elevational shift, perhaps because freshwater habitats are heavily fragmented (e.g. dams) and thus difficult to colonize or because water provides microclimate buffering on warming rate, especially within mountain ranges where streams are fed by snow and glacier melt (Vitasse et al. 2021). In terrestrial environments, trees strongly buffer ground level temperatures by preventing solar radiation to reach the ground and by limiting air fluxes (De Frenne et al. 2021). Thus, species living in forests' understory might experience climate change to a lesser extent than species inhabiting less buffered environments such as drylands, shrublands or grasslands (Rita et al. 2021; Wagner et al. 2021).

The size, number and distance between patches of habitats affects insects' ability to track climate change (Platts et al. 2019; Wagner et al. 2021) because small and isolated patches are difficult to colonize or to recolonize after local extinction associated with extreme climatic events (Oliver et al. 2015). The quality of habitat patches also varies within landscapes due to natural features (e.g., soil properties) and anthropogenic pollutants (e.g., heavy metals, plastics, pesticides, artificial light) that can affect insects' survival, reproduction and dispersal (Wagner et al. 2021). For example, cultivated areas are frequently sprayed with insecticides (Colin et al. 2020; Raven and Wagner 2021) and, with the combined effect of rising temperature, could become even more inhospitable to many insect species as the climate warms (Raven and Wagner 2021; but see Maino et al. 2018).

Among many land-use changes, urbanization is perhaps the closest to climate change in terms of environmental modifications because urban areas experience higher temperatures than adjacent rural or semi-natural areas. This is due to the urban heat island effect, a phenomenon caused by the thermal properties of artificial materials such as concrete and asphalt (e.g., low albedo, high emissivity) and by the lack of evaporative cooling associated with sparse vegetation, elevating the temperature in urban areas by up to 5 °C relative to surrounding natural areas (Chapman et al. 2017). Urbanization is thus regarded as a potential unintentional global experiment to study and predict the effects of climate change on ecological and evolutionary dynamics (Lahr et al. 2018). But the effect of urbanization on insects' distribution is more complex than a simple "fine-scale replicate" of climate change. The interaction of climatic conditions and urbanization is known to affect insects' spatial distribution at local to continental scale (Cordonnier et al. 2020; Gippet et al. 2017; Polidori et al. 2021). For example, some species colonize northern locations that are outside of their climatic niche by exploiting the warmer urban microclimate (e.g., the invasive mud-dauber wasp (Polidori et al. 2021). Also, as drought events increase in frequency, some species might find refugia in urban areas by exploiting irrigated areas such as public parks and private gardens (e.g., ants in Arizona; Miguelena and Baker 2019). Finally, as urban conditions favor heat-tolerant species, it is expected that urban specialist species will expand their distribution outside cities as the climate warms (Menke et al. 2011). The effects of urbanization and climate change on insects' range shifts (at local to continental scales) might also depend on the background climatic conditions, as arid, temperate or tropical areas will not experience the same relative climatic changes along urbanization gradients (Diamond et al. 2015).

11.3.3 *Microclimates*

Because insects are small, they experience environmental conditions at very fine spatial scale (i.e., ~10 cm around them (Pincebourde and Woods 2020). Thus, insects could, in theory, exploit microscale variations in temperature and humidity by, for example, moving around tree trunks or going deeper in the ground (Pincebourde and Woods 2020). Air and ground temperature can vary greatly over a few centimeters (up to ~15°C; Pincebourde et al. 2016) because of differences in the amount of direct solar radiation that are mainly due to natural and artificial vertical features such as trees and buildings (Gippet et al. 2022; Napoli et al. 2016). Very few studies have tested the effect of microscale shading conditions on insects' spatial distribution. Shades created by human buildings or experimentally have been shown to affect the foraging patterns of native and invasive ant species (Gippet et al. 2022; Wittman et al. 2010). However, to our knowledge, it is still unknown if insects can exploit shading conditions as microrefugia in response to climate change. Microclimatic conditions (e.g., temperature, water content, nutrients) can also vary depending on the depth of soil (Duffy et al. 2015; Krab et al. 2010). With climate warming, deep soil layers will heat less than upper layers and might thus offer microrefugia for many insect species (Duffy et al. 2015).

11.4 The role of dispersal

11.4.1 *Natural dispersal*

Dispersal ability is key to determining how insect species will cope with climate change. Highly mobile species might be more successful in shifting their ranges (Pöyry et al. 2009; Figure 11.7). In Europe for example, highly dispersive insects might be able to naturally spread from the continent to Great Britain as the climate becomes suitable (Hulme 2017). Conversely, species with limited dispersal might not be able to track the shifting climate and are therefore more likely to be limited by suitable niche space and risk extinction. Among butterflies, for instance, species with low dispersal capacities showed smaller altitudinal shift than more mobile species (Rödder et al. 2021).

In response to climate change, insect populations at expanding range boundaries might evolve greater dispersal capacities because dispersive individuals are more likely to establish new populations beyond their current range limits and transmit their genes. Descendants of these individuals will be more likely to found populations at range margins and therefore transmit traits favoring high dispersal rates (Hill et al. 2011; Parmesan 2006). This phenomenon has already been described in several insect species. For example, two bush cricket species show a higher proportion of long-winged (dispersive) individuals in populations at range margins in the United Kingdom (Thomas et al. 2001). Similar observations were made in Germany, where Roesel's bush-cricket (*Metrioptera roeseli*) has increased proportions of long-winged individuals in populations at the expanding range margin (Poniatowski et al. 2012). Likewise, populations of the European map butterfly (*Araschnia levana*) at the expansion front in Finland show higher frequency of the *Pgi-1* allele, associated with superior flight metabolic rate, compared to historical Estonian populations (Mitikka and Hanski 2010).

Rising temperatures may also affect insect dispersal directly, by increasing activity levels (Lantschner et al. 2014). Insects may be able to disperse over longer distances in regions with higher temperatures because of increased metabolic rate and extended flying period (Robinet and Roques 2010). For example, females of the winter pine processionary moth (*Thaumetopoea pityocampa*) showed increased flight activity with higher temperatures, allowing them to disperse over longer distances (Battisti et al. 2006). This is likely to have facilitated the rapid increase in the altitudinal range limit of this species during the record hot summer of 2003 in southern Europe. But temperature is not the only climatic factor affecting insect dispersal. For example, increased precipitation was shown to facilitate the spread of the invasive Argentine ant (*Linepithema humile*) in California (Heller et al. 2008). Moreover, the increased prevalence, intensity and duration of extreme climatic events (IPCC 2012) could facilitate species dispersal to new regions (Diez et al. 2012; Hellmann et al. 2008). Many insect species depend on wind currents for natural long-distance dispersal (Chapman et al. 2015; Leitch et al. 2021). More frequent storms could therefore increase the probability of insects moving over long distances and across physical barriers. It is likely that the cactus moth (*Cactoblastis cactorum*) benefited from the 2005 hurricane season to travel from the Caribbean to Mexico where it now has important ecological and economic impacts (Burgiel and Muir 2010). Similarly, the red palm mite (*Raoiella indica*) is thought to have spread throughout the Caribbean due to storms and hurricanes (Burgiel and Muir 2010).

11.4.2 *Human-mediated dispersal*

Climate change will not only affect natural dispersal. Many insect species are transported and introduced accidentally to new regions as contaminants or stowaways on traded commodities (Gippet et al. 2019; Meurisse et al. 2019). Consequently, more than 5,000 insect species have established outside of their native range (Seebens et al. 2017). Climate change is likely to alter patterns of trade and transport, and thus change the dispersal dynamics of introduced insects (Hellmann et al. 2008; Figure 11.7). For instance, the opening of Arctic shipping routes due to the loss of sea ice might considerably reshape trade flows (Bekkers et al. 2018). Trade between Europe and Eastern Asia is expected to grow, which could increase introduction opportunities. The opening of the Arctic shipping routes will

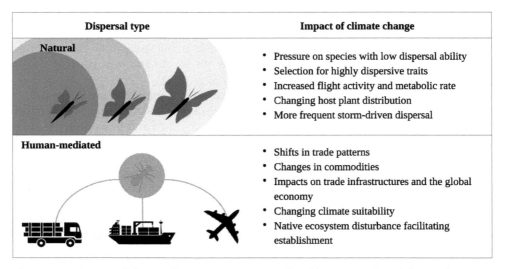

Dispersal type	Impact of climate change
Natural	• Pressure on species with low dispersal ability • Selection for highly dispersive traits • Increased flight activity and metabolic rate • Changing host plant distribution • More frequent storm-driven dispersal
Human-mediated	• Shifts in trade patterns • Changes in commodities • Impacts on trade infrastructures and the global economy • Changing climate suitability • Native ecosystem disturbance facilitating establishment

Figure 11.7 Impact of climate change on insect natural and human-mediated dispersal.

also greatly reduce sailing distances and travel time (Bekkers et al. 2018; Dellink et al. 2017), which could enhance the survival of insects during transport. Climate change will also affect the production of many commodities (Dellink et al. 2017) which could change global trade flows and thus the dynamics of insect dispersal. In particular, climate change will impact agricultural production. Changes in temperatures, precipitations and more frequent heat extremes may lead to crop yield loss in most parts of the world (Dellink et al. 2017). Some productions might also be relocated as new regions become suitable for certain crops. International trade could therefore play a key role in compensating for these shifts in production potential (Huang et al. 2011), which may enhance unintentional insect introductions.

Climate change will also affect the establishment probability of introduced insects (Hulme 2017; Robinet and Roques 2010). For example, it is predicted that climate suitability for pest arthropod species will increase in northeastern European countries but decrease in central European countries (Bacon et al. 2014). The Argentine ant (*Linepithema humile*) and the yellow crazy ant (*Anoplolepis gracilipes*) are two highly invasive species that are regularly intercepted at the British border. They have so far failed to establish there (at outdoor locations, but see Charrier et al. 2020), but this may change with warming climate (Hulme 2017). In Antarctica, insect establishment probability is low due to the harsh climatic conditions, but this region might become more susceptible to insect invasions in the future (Bergstrom 2022). More frequent extreme climatic events might also strongly disturb ecosystems, which could facilitate the establishment and spread of introduced species as they may have broader physiological tolerance than native species (Diez et al. 2012). Extreme climatic events can create "invasion windows," for instance by generating resource pulses that introduced species can exploit more quickly than native species (e.g., thanks to more rapid growth and colonization). Heat waves and droughts can also cause significant stress to native communities and increase the mortality of native species, thus reducing biotic resistance (e.g., competition) against introduced species.

Finally, climate change might impact global trade as damage to trade infrastructures will increase with more frequent extreme climatic events (Dellink et al. 2017). Storms, extreme precipitations and sea level rise may affect operations of airports, cause more frequent port closure, require the use of alternative shipping routes, affect sailing time and increase port and ship maintenance costs. Moreover, climate change will affect the global economy and potentially lead to countries having lower gross domestic products (GDPs), which is also likely to impact global trade (Dellink et al. 2017). These negative impacts of climate change on international trade and the world economy overall could also reduce the rate of insect invasions in the future as it is tightly linked to the level of trade globalization (Bertelsmeier et al. 2017; Bonnamour et al. 2021).

11.5 Eco-physiological and life history traits

With ongoing climate change, species traits will be beneficial or limiting, creating both "winners" and "losers." Some traits relevant to responses to climate change are difficult to measure in the field and therefore current knowledge predominately relies on lab experiments, with sufficient data in some cases only for *Drosophila* species. Yet some of these traits might hold the key to understanding persistence of insect populations in a changing environment. Given that traits are not as fixed as often assumed when predicting species distributions, we also address insects' potential for adaptation and plasticity.

11.5.1 *Thermal traits*

Critical thermal limits (CTs)—temperatures at which insects lose voluntary muscle control—have gained popularity in recent years for predicting species distributions (Lutterschmidt and Hutchison 1997; Rezende et al. 2011). However, the data on insect CTs are biased towards three insect groups, Drosophila, Coleoptera and Formicidae, which comprise 95 percent of data on heat tolerance (Calosi et al. 2010; Diamond and Chick 2018a; Kellermann and van Heerwaarden 2019). Because of lower geographic variability and stronger phylogenetic signal in insect heat tolerance compared to cold tolerance (Addo-Bediako et al. 2000; Bujan et al. 2020; Hoffmann et al. 2013), insect heat tolerance is expected to have low adaptive potential, which could be detrimental in a warming world. Tropical insects have narrower thermal ranges (CTmax–CTmin), and they are expected be under higher extinction risk compared to insects from more variable, temperate climates (Diamond and Chick 2018b; Diamond et al. 2012; Sunday et al. 2014). However, in bumblebees, cold-adapted species are more sensitive to extreme heat events (Martinet et al. 2021). Thus, tropical species and cold-specialist species are predicted to be most threatened by climate change.

Thermal sensitivity of insect reproductive organs is also crucial to determine fitness under climate change. Temperatures that sterilize (Parratt et al. 2021) *Drosophila* males predicted global species distributions better than critical thermal limits (van Heerwaarden and Sgrò 2021). Additionally, temperature can differentially impact insect ovarian development across their geographic range (Everman et al. 2018). To assess which insect taxa are reproductively challenged by temperature, it would be necessary to measure optimal reproductive temperatures for a wide range of species and populations, in turn improving our ability to understand species distributions.

11.5.2 *Desiccation resistance*

Climate change increases the frequency of severe droughts, particularly in the tropics (Dai 2011). Therefore, desiccation resistance—the ability to withstand water stress—is especially important for small insects. Generally, larger species of fruit flies (Gibbs and Matzkin 2001), tiger beetles (Schultz and Hadley 1987) and ants (Hood and Tschinkel 1990) are more resistant to desiccation compared to smaller species. Insects from dry areas adapted to withstand desiccation lose water slower than their mesic counterparts (Gibbs and Matzkin 2001). Given that tropical insects are not exposed to desiccation stress, they have low desiccation resistance and low adaptive potential for this trait, as shown in *Drosophila* (Hoffmann et al. 2003; Rajpurohit et al. 2013). However, the tropics are not uniform and in the same tropical forest, the drier tropical canopy holds species able to withstand drought stress longer than their understory counterparts (Bujan et al. 2016). Increase in drought frequency in the future is expected to limit the spread of some introduced insects, such as Argentine ants whose spread depends on soil moisture (Couper et al. 2021; Holway et al. 2002). Other introduced species like pollinator bees in Fiji show higher desiccation resistance than native pollinators (da Silva et al. 2021), and the larger grain borer (*Prostephanus truncates*) can acclimate its desiccation resistance (Mutamiswa et al. 2021).

11.5.3 *Dispersal ability*

Dispersal ability is expected to be key to withstanding climate change (Berg et al. 2010). Generally, insect populations at the range edges have traits that favor dispersal, such as higher proportion of long-winged morphs linked to increased flight ability (Simmons and Thomas 2004). However, there is mixed evidence that warming increases wing size and consequently dispersal ability in flying insects. Wing sizes of a social wasp decreased in response to warming (Polidori et al. 2020) but also increased with elevation in introduced *Drosophila suzuki* (Jardeleza et al. 2022). While stable environmental conditions are assumed to be one reason for evolution of flightlessness in insects (Wagner and Liebherr 1992), this trait might now be disadvantageous.

11.5.4 *Voltinism*

Number of generations produced per year—voltinism—is negatively correlated with latitude in many insect taxa (Musolin 2007; Zeuss et al. 2017). Thus, warmer areas promote multiple insect generations. Climatic warming increased voltinism in European butterflies and moths which now reproduce more frequently, giving them a potential evolutionary advantage, as insects with faster reproductions cycles have higher chances of adaptation to novel conditions (Altermatt 2010). But this is risky as some species, like the European wall brown butterfly, have been known to fall into "developmental traps" risking extinction if they fail to predict climates of the upcoming season (van Dyck et al. 2015). Some species might have an advantage by producing more generations per year, which is best demonstrated in introduced insects in which multi-voltinism enables faster spread. For example, introduced populations of gypsy moths develop faster due to temperatures and length of the growing season (Faske et al. 2019). Multivoltine introduced insects spread on average 72.9 km/year, and univoltine insects only 16.9 km/year (Fahrner and Aukema 2018), highlighting a worrying competitive advantage in the face of warmer climates in introduced species that can increase the number of generations per year (Tobin et al. 2008; Ziter et al. 2012).

11.5.5 *Adaptation vs plasticity*

Species can alter their traits in two ways to persist under novel environmental conditions: through plasticity and evolutionary adaptation. Adaptations to novel environmental conditions are beneficial trait changes that are underpinned by genetic changes resulting in increased fitness (Bertelsmeier and Keller 2018). However, direct experimental tests of adaptive potential of introduced and native populations are rare (Chevin and Lande 2011; Colautti and Lau 2015), and the presence of adaptation is usually inferred (Bertelsmeier and Keller 2018). Adaptive responses of insects to climate change are not limited to thermal tolerances but can involve changes in melanism (Brakefield and de Jong 2011), voltinism (Altermatt 2010), morphology (Huey et al. 2000), desiccation resistance (Tejeda et al. 2016) and dispersal ability (Hill et al. 2011).

In the absence of adaptation, species may benefit from phenotypic plasticity—the potential of one genotype to express multiple phenotypes (Agrawal 2001). For example, phenotypic plasticity can include changes in morphology, diet and physiology under new environmental conditions. When insects are faced with extreme heat, thermal plasticity enables them to withstand temperature changes and provides a competitive advantage over insects with static thermal tolerances (Rodrigues and Beldade 2020). Some studies suggest there is a trade-off between basal thermal tolerance and thermal plasticity (Esperk et al. 2016). However, a recent metanalysis shows this evidence to be equivocal (van Heerwaarden and Kellermann 2020). A plastic phenotype can be costly, either because of timing costs associated with developmental stages or costly production of heat shock proteins. Therefore, phenotypic plasticity is assumed to be lost in a stable environment (Sgrò et al. 2016).

Introduced insects may benefit from short-term phenotypic changes when they arrive in novel environments. In plants, introduced species showed higher phenotypic plasticity than native species (Davidson et al. 2011), but in springtails there were little differences between introduced and native populations (Janion-Scheepers et al. 2018). A frequently studied type of phenotypic plasticity is acclimatization—a reversible physiological change which enhances performance (Angilletta 2009). It has been shown that some introduced insects are able to acclimate (Bujan et al. 2021; Coulin et al. 2019; Nyamukondiwa et al. 2010) but it remains unknown if acclimation is a major factor contributing to invasion success.

11.6 Species interactions and responses to climate change

11.6.1 *Interactions regulate ability to track climate change*

Climate change impacts virtually every type of biotic interaction among species in a community (Table 11.1) by altering the conditions that species experience (Tylianakis et al. 2008). In turn, these interactions determine how species respond to changing environmental conditions (e.g., Davis et al. 1998a, 1998b). Moreover, community composition is not static. Climate change and human activities are reshuffling species distributions, creating new communities worldwide (e.g., Alexander et al. 2016; Schweiger et al. 2010).

Table 11.1 Main types of interactions between species, and their impact on each partner. Impacts are either positive (+), neutral (0) or negative (−).

Type of interaction	Response	
	Species 1	Species 2
Mutualism	+	+
Commensalism	+	0
Neutral	0	0
Predation/parasitism	+	−
Amensalism	−	0
Competition	−	−

New interactions emerge in these novel communities while others no longer take place. Interactions at all trophic levels are implicated: herbivores, predators, parasitoids and pathogens, hyperparasitoids and tertiary predators and their prey or host species. Insect endosymbiont functions are also potentially altered by rising temperatures (van Baaren et al. 2010). The nature and relative importance of interactions depend on the climatic conditions, which has important implications for determining species' current and future distributions (Wisz et al. 2013). Warmer range-edge responses in particular depend strongly on biotic interactions (Paquette and Hargreaves 2021), potentially due to stronger negative interactions in warmer and more productive ecosystems (e.g., Hargreaves et al. 2019; Roslin et al. 2017; Vamosi et al. 2006), or interactions being relatively more important under benign climatic conditions (Dobzhansky 1950).

11.6.2 *Range shifts disrupt interactions*

Species' range shifts driven by climate change are often not synchronized within a community (Schweiger et al. 2010; Urban et al. 2012). Biotic interactions are thus disrupted by shifts either in space or time, which can have considerable impacts on species' fitness. These impacts can be positive, offering a release from negative interactions in the new range (Figure 11.8a). The enemy release hypothesis, which was originally formulated in the context of biological invasions, predicts that introduced species are more successful in new areas where their native enemies are absent (Keane and Crawley 2002). So far, support for this is mixed (Heger and Jeschke 2014; Mlynarek 2015) and the hypothesis may be more relevant for species under strong enemy effects in their historic range (Prior et al. 2015). Similarly, range shifts of native species may disrupt enemy interactions. Subsequent fitness increases in the new range could facilitate tracking suitable climatic conditions (Figure 11.8a). This has been illustrated in grassland communities, where less spatial overlap between predatory spiders and their grasshopper prey due to differential responses to warming allows the grasshoppers to increase their feeding (Barton 2010).

Disrupting interactions can also have negative impacts. Species may be prevented from colonizing climatically suitable areas if their interaction partners are absent (Figure 11.8b). For example, eggs of the butterfly *Aporia crataegi* survive at higher elevations with increased warming, but upward colonization is restricted by the lack of host plants (Merrill et al. 2008). Future projections indicate increasing range mismatch between host-plant-limited European butterflies and their hosts, restricting their climate-tracking ability (Schweiger et al. 2010).

11.6.3 *Range shifts create new interactions*

If interacting species' distributions change in synchrony, or if they find new partners, interactions can facilitate spread (Figure 11.8c). For example, *Polygonia c-album* has undergone the fastest range expansion of any resident butterfly species in Britain (Warren et al. 2001), most likely due to a shift in larval host preferences (Braschler and Hill 2007). The effects of global change on insect mutualisms are highly variable however, making general patterns difficult to detect (Vidal et al. 2021). Range shifts can also be restricted by novel interactions. Competition may slow species spread into suitable habitats (Urban et al. 2012; Figure 11.8d) and can negatively impact species in the recipient community. For example, numbers of migratory butterfly species in southern Britain are increasing with rising temperatures, posing a competitive threat to less mobile and more specialized native insects (Sparks et al. 2007).

Insects may join existing species interaction networks in their new range, for example as pollinators or seed dispersers. This has been observed for introduced species (e.g., Aizen et al. 2008; Traveset et al. 2013) and may also be the case with naturally dispersing species. New species can fundamentally alter network structure, transferring links from generalist native to super-generalist introduced species (Aizen et al. 2008). Novel networks with many links may be more stable, making them more resistant to certain types of disturbance, and more vulnerable to others (Aizen et al. 2008; Traveset et al. 2013). Novel species can also, to some extent, replace lost or declining native pollinators (Dick et al. 2003; Gross 2001), and can either partly compensate for negative impacts of climate change on pollination networks, or intensify them (Schweiger et al. 2010).

11.6.4 *Hitchhiking pathogens and interactions across trophic levels*

The effects of climate change are likely to be stronger at higher trophic levels (van Baaren et al. 2010). Predators, parasitoids and hyper-parasitoids must locate and exploit their prey or hosts. Many of these interactions are temperature-dependent, and vulnerable to environmental change (Hance et al. 2007). Additionally, as insect distributions change, parasites and pathogens may hitch a ride, forming their own novel interactions. Introduced bees and bumblebees have been shown to transmit pathogens to less-resistant native species, and even less virulent ones can be lethal to new hosts when combined with environmental stressors (Arbetman et al. 2013; Vilcinskas 2019). The co-introduction of parasites and pathogens with their hosts can also impact human health. For example, introduced mosquito species pose significant threats to public health due to the diseases they transmit (e.g., Schaffner et al. 2013), and climate change may increase their invasion potential (Iwamura et al. 2020).

11.6.5 *Novel interactions can facilitate additional species establishing*

Species may also form interactions or create environmental conditions that promote the establishment of additional species as they spread. For instance, introduced insects are often key pollinators of introduced plants (Goulson 2003; Olesen et al. 2002; Simberloff and von Holle 1999; Stout, et al. 2002), potentially facilitating their spread (Morales and Aizen 2002). The hemlock woolly adelgid, *Adelges tsugae*, reduces light interception by the forest canopy, indirectly creating conditions favoring introduced plants (Eschtruth et al. 2006). Such interactions can have knock-on effects on community composition (e.g., Brightwell and Silverman 2010), potentially also in response to species shifting their range due to climate change.

At the community level, facilitation between several introduced species may lead to increasing establishment rates or accelerating impacts, termed "invasional meltdown" (Simberloff 2006; Simberloff and von Holle 1999). While this concept is based on species invasions, the same processes could occur following natural dispersal induced by climate change. It is challenging to show experimentally that mutualisms increase the populations of both partners at a regional scale, due to ethical considerations and the complexity of factors involved, but the circumstantial evidence is often strong (Simberloff 2006). On Christmas Island, populations of the introduced ant *Anoplolepis gracilipes* increased dramatically after the introduction of scale insects. The ants protect the scales from predators and parasites and also devastate populations of the native land crab *Gecarcoidea natalis*. *G. natalis* no longer controls ground cover plants, and sooty mold growing on honeydew causes canopy dieback, altering forest community composition (Abbott 2004; O'Dowd et al. 2003). In turn, the absence of *G. natalis* has facilitated the establishment of Giant African Land Snails, *Achatina fulica* (Green et al. 2011). In the face of global change, restoring "pristine" interaction networks is likely impossible. The question is whether these novel communities can absorb new species, while simultaneously sustaining complex interactions between native species (Roubik 2000; Traveset and Richardson 2006).

11.7 Predicting the distribution of species under climate change

Different approaches can be used to make predictions about future species distributions under climate change. Most use species occurrences in combination with climate data to model species' climate niches and project potential future distribution under future climatic conditions (Gallien et al. 2010; Guisan and Thuiller 2005; Mammola et al. 2021). But the complexity of models can range from simple correlational models to process-based mechanistic models. Here, we discuss the use of correlational models (section 7.1), hybrid models and semi-mechanistic models (section 7.2), and mechanistic models (section 7.3) in predicting the future distributions of insect species under climate change.

11.7.1 *Correlational models*

The simplest and most widely used model type are species distribution models (SDMs; also called ecological niche models (ENMs) (Evans et al. 2015). These correlative models use the present-day occurrence point locations of a species to determine the current environmental conditions experienced by the species, and then map areas where these climatic conditions are expected to occur in the future (Dormann 2007; Evans et al. 2015; Hill et al. 2015). An underlying assumption of these models is that future distributions will reflect the current realized niches of species. Although all SDMs link species occurrences with climatic data, there are many different algorithms that can be used to model this species-environment relationship. Options range from statistical models such as general linear models (GLMs), for example in recent predictions of future distributions of hoverflies (*Syrphidae*) in the Balkans (Radenković et al. 2017), to other machine learning models such as neural networks (e.g., boosted regression trees, for example seen in a study of Odonata species (Jaeschke et al. 2013)). One of the most commonly used SDM method is "Maxent," an easy-to-use machine learning approach (Phillips et al. 2006). For example, a recent study on eighteen meso-American bumblebee (*Bombus*) species used a Maxent

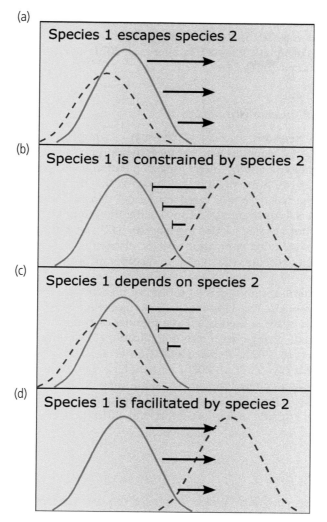

Figure 11.8 Interactions determine species' climate-tracking ability, impacting the distributions of focal species (blue line) and their interaction partners (dashed green line). The gradient indicates changing climate and habitat suitability. a) species track climate change through enemy release, b) species lag behind climate change due to enemy interactions in their new range, c) species lag behind climate change due to mismatched range shifts between interaction partners, or d) species track climate change through facilitation by existing or new partners (Adapted from HilleRisLambers et al. 2013).

SDM to predict area losses of up to 67 percent by 2050 of their current range along with altitudinal shifts upwards (Martínez-López et al. 2021). However, there is no single "best" model type (Carvalho et al. 2017). Multiple model outputs may have statistically good predictive outputs (usually evaluated using true skill statistics (TSS) or the area under the receiver operating characteristic curve (AUC) scores), while predicting significantly different distributions. A widely used solution to deal with this variation in individual model outputs is to combine them by averaging across models. The contribution of individual models to a final "ensemble model" can be weighted based on evaluation statistics (Hao et al. 2019; Thuiller et al. 2016). Recent examples of ensemble models include studies on future distributions of aquatic insects (Timoner et al. 2021) and introduced bee species in Hawaii (Tabor and Koch 2021).

11.7.2 *Hybrid/Semi-mechanistic models*

There are several extensions to classic SDMs that can inform models with additional biologically relevant information. For instance, hybrid or niche-population models can predict distributional changes by integrating population-level responses (Aragón et al. 2010; Zurell et al. 2016). A recent study on Japanese stag beetles (*Allomyrina dichotoma*) used estimated dispersal distances based on kernel densities and found that models incorporating dispersal constraints performed better than simpler models, ultimately predicting large reductions of stag beetle ranges by 2070 (Zhang and Kubota 2021). Another study on butterflies included population-specific distributions, site-specific species richness, as well as species-specific growth rates to incorporate dispersal into their predictions (Isaac et al. 2011).

Other traits are more commonly used to inform predictions of insect distributions by semi-mechanistic models. The semi-mechanistic modeling tool "CLIMEX" (Jung et al. 2016; Kriticos et al. 2016) is used in a large number of studies to predict future distributions under climate change, with recent applications including studies on the orders Coleoptera (Zhou et al. 2021), Lepidoptera (Guo et al. 2021), Hemiptera (Falla et al. 2021), Hymenoptera (Byeon et al. 2020) and Diptera (Kim et al. 2020). CLIMEX uses laboratory-based growth, phenology, and stress information on a specific species to create an ecoclimatic index (EI) which is used to quantify habitat suitability in different areas and under different climatic scenarios (Byeon et al. 2018; Jung et al. 2016; Kriticos et al. 2016).

11.7.3 *Mechanistic models*

Currently, the most "complex" models employed to predict insect species distributions under climate change are mechanistic models (Evans et al. 2015; Maino et al. 2016). These models can predict which regions will remain suitable under climate change by modeling species-specific responses to climate-based eco-physiological data and vital rates, sometimes even including life-stage specific growth and death rates in response to climate (Kearney 2006; Kearney and Porter 2009).

Physiological based demography models (PBDMs) base predictions on physiological, phenological and demographic responses in space and time. For example, the range expansion of the pest species *Bemisia tabaci* (Hemiptera) has been predicted under climate change with a PBDM using estimates of developmental rates, temperature-dependent morality rates, age-specific fecundity (Gilioli et al. 2014). PBDMs can ultimately incorporate mechanistically both demographic and physiological responses to climate (Gutierrez and Ponti 2014). It is important to note, however, that eco-physiological limits and growth rates measured under laboratory conditions may not represent the realized limits of a species in the field. Moreover, physiologically based models do not account for biotic and dispersal constraints (Soberón and Peterson 2005). But other mechanistic models can explicitly incorporate dispersal into estimates of distribution and abundance, for example done in a study that used a random-walk model to predict distributional changes of the European grasshopper *Chorthippus albomarginatus* (Walters et al. 2006). However, detailed models incorporating empirical data on dispersal are rare for insects.

Another type of mechanistic model estimates "degree-days," that is, the number of days within a certain climate range needed for a species development. This phenological model type is particularly interesting for insects because many species go through multiple life stages with different climatic optima, and potentially impacted differently by climate change. For example, brood and early life cycles are known to be particularly sensitive to climate. Degree-day models are calibrated using data from physiological experiments measuring developmental response curves under different temperatures (Lemoine 2021), to estimate "degree days" across a season. Degree-day models have been used to predict the spread of the introduced potato beetle (*Leptinotarsa decemlineata*) in Scandinavia (Pulatov et al. 2014). This approach allows researchers to delimit when daily temperature is above the species' threshold for survival or development, and therefore using a combination of timing of first- to second-generation development as well as generational numbers per years. Degree-day models have also been used to predict brood survival and oviposition rates in the introduced Argentine ant *Linepithema humile* (Abril et al. 2009). Such phenological models have in some cases even been shown to outperform SDMs, for example as was seen in a study on in UK butterfly species (Buckley et al. 2011), and even more complex physiological-based mechanistic models (Buffo et al. 2007; Bryant et al. 1997), as was exemplified in a study of the pine processionary moth in Italy (Buffo et al. 2007).

11.8 Future of modeling

Overall, a combination of data availability and expert knowledge can help determine which approaches are best for a particular species. Employing multiple model types, from correlative to mechanistic, across various future environmental predictions will allow us to predict future species distributions with increased confidence (Benito Garzón 2019; Mammola et al. 2019, 2021; Violle et al. 2014).

Key reflections

Important open research questions include:

- What are the similarities and differences in how alien and range-shifting species interact with biological communities in their new range?
- What is the importance of thermal plasticity for the spread of invasive insects under novel climates?
- What is the role of microhabitats (and their spatial and temporal heterogeneity) in shaping insect distributions and responses to climate change?
- Can we anticipate future trade network and shifts in traded commodities due to climate change to predict future insect invasions?
- How much of their fundamental niche do species currently occupy?
- How important and frequent is evolutionary adaptation in response to climate change?
- What is the expected frequency of niche shifts?
- What are the links between range dynamics and species' life histories, ecophysiological traits and biotic interactions?

Key further reading

- Bates, O.K., and Bertelsmeier, C. 2021. Climatic niche shifts in introduced species. *Current Biology 31*: 1252–1266.
- Bekkers, E., Francois, J.F., and Rojas-Romagosa, H. 2018. Melting ice caps and the economic impact of opening the northern sea route. *Economic Journal 128*: 1095–1127.
- Bertelsmeier, C. 2021. Globalization and the anthropogenic spread of invasive social insects. *Current Opinion in Insect Science 46*: 16–23.
- Bertelsmeier, C., and Keller, L. 2018. Bridgehead effects and role of adaptive evolution in invasive populations. *Trends in Ecology and Evolution 33*: 527–34.
- Bridle, J.R., and Vines, T.H. 2007. Limits to evolution at range margins: when and why does adaptation fail? *Trends in Ecology and Evolution 22*: 140–7.
- Diez, J.M., D'Antonio, C.M., Dukes, J.S., Grosholz, E.D., Olden, J.D., Sorte, C.J., Blumenthal, D.M., *et al.* 2012. Will extreme climatic events facilitate biological invasions? *Frontiers in Ecology and the Environment 10*: 249–57.
- Fahrner, S., and Aukema, B.H. 2018. Correlates of spread rates for introduced insects. *Global Ecology and Biogeography 27*: 734–43.
- Hill, J.K., Griffiths, H.M., and Thomas, C.D. 2011. Climate change and evolutionary adaptations at species' range margins. *Annual Review of Entomology 56*: 143–59.
- HilleRisLambers, J., Harsch, M.A., Ettinger, A.K., Ford, K.R., and Theobald, E.J. 2013. How will biotic interactions influence climate change-induced range shifts? *Annals of the New York Academy of Science 1297*: 112–25.
- Hulme, P.E. 2017. Climate change and biological invasions: evidence, expectations, and response options. *Biological Reviews 92*: 1297–1313.
- Kellermann, V., and van Heerwaarden, B. 2019. Terrestrial insects and climate change: adaptive responses in key traits. *Physiological Entomology 44*: 99–115.
- McCain, C.M., and Garfinkel, C.F. 2021. Climate change and elevational range shifts in insects. *Current Opinion in Insect Science 47*: 111–18.
- Moran, E. v., and Alexander, J.M. 2014. Evolutionary responses to global change: Lessons from invasive species. *Ecology Letters 17*: 637–49.
- Pincebourde, S., and Woods, H.A. 2020. There is plenty of room at the bottom: microclimates drive insect vulnerability to climate change. *Current Opinion in Insect Science 41*: 63–70.
- Spence, A.R., and Tingley, M.W. 2020. The challenge of novel abiotic conditions for species undergoing climate-induced range shifts. *Ecography 43*: 1571–1590.
- Suggitt, A.J., Wilson, R.J., Isaac, N.J.B., Beale, C.M., Auffret, A.G., August, T., Bennie, J.M., *et al.* 2018. Extinction risk from climate change is reduced by microclimatic buffering. *Nature Climate Change 8*: 713–17.
- Vidal, M.C., Anneberg, T.J., Curé, A.E., Althoff, D.M., and Segraves, K.A. 2021. The variable effects of global change on insect mutualisms. *Current Opinion in Insect Science 47*: 46–52.
- Wisz, M.S., Pottier, J., Kissling, W.D., Pellissier, L., Lenoir, J., Damgaard, C.F., Dormann, C.F., *et al.* 2013. The role of biotic interactions in shaping distributions and realised assemblages of species: Implications for species distribution modelling. *Biological Reviews 88*: 15–30.

References

Abbott, K.L. 2004. *Alien Ant Invasion on Christmas Island, Indian Ocean: The Role of Ant-Scale Associations in the Dynamics of Supercolonies of the Yellow Crazy Ant, Anoplolepis Gracilipes*. Melbourne: Monash University.

Abril, S., Roura-Pascual, N., Oliveras, J., and Gomez, C. 2009. Assessing the distribution of the Argentine ant using physiological data. *Acta Oecologica* 35 (5): 739–45.

Addo-Bediako, A., Chown, S.L., and Gaston, K.J. 2000. Thermal tolerance, climatic variability and latitude. *Proceedings of the Royal Society of London. Series B: Biological Sciences* 267 (1445): 739–45.

Agrawal, A.A. 2001. Phenotypic plasticity in the interactions and evolution of species. *Science* 294 (5541): 321–26.

Aizen, M.A., Morales, C.L., and Morales, J.M. 2008. Invasive mutualists erode native pollination webs. *PLoS Biology* 6 (2): e31.

Alexander, J.M., Diez, J.M., Hart, S.P., and Levine, J.M. 2016. When climate reshuffles competitors: A call for experimental macroecology. *Trends in Ecology and Evolution* 31 (11): 831–841.

Altermatt, F. 2010. Climatic warming increases voltinism in European butterflies and moths. *Proceedings of the Royal Society B: Biological Science* 277 (1685): 1281–1287.

Anderson, B.J., Akçakaya, H.R., Araújo, M.B., Fordham, D.A., Martinez-Meyer, E., Thuiller, W., and Brook, B.W. 2009. Dynamics of range margins for metapopulations under climate change. *Proceedings of the Royal Society B: Biological Sciences* 276 (1661): 1415–1420.

Andrew, N.R., Hill, S.J., Binns, M., Bahar, M.H., Ridley, E.V., Jung, M.P., Fyfe, C., et al., M. 2013. Assessing insect responses to climate change: What are we testing for? Where should we be heading? *PeerJ* 1: e11.

Angilletta, M.J. 2009. *Thermal Adaptation: A Theoretical and Empirical Synthesis*. Oxford: Oxford University Press.

Aragón, P., Baselga, A., and Lobo, J.M. 2010. Global estimation of invasion risk zones for the western corn rootworm Diabrotica virgifera virgifera: Integrating distribution models and physiological thresholds to assess climatic favourability. *Journal of Applied Ecology* 47 (5): 1026–1035.

Arbetman, M.P., Meeus, I., Morales, C.L., Aizen, M.A., and Smagghe, G. 2013. Alien parasite hitchhikes to Patagonia on invasive bumblebee. *Biological Invasions* 15: 489–94.

van Baaren, J., Le Lann, C., and van Alphen, J. 2010. "Consequences of climate change for aphid-based multi-trophic systems." In *Aphid Biodiversity under Environmental Change: Patterns and Processes*, edited by Kindlmann, P., Dixon, A., Michaud, J. Dordrecht: Springer, 55–68. doi:10.1007/978-90-481-8601-3_4.

Bacon, S.J., Aebi, A., Calanca, P., and Bacher, S. 2014. Quarantine arthropod invasions in Europe: The role of climate, hosts and propagule pressure. *Diversity and Distributions* 20 (1): 84–94.

Bale, J.S., Masters, G.J., Hodkinson, I.D., Awmack, C., Bezemer, T.M., Brown, V.K., Butterfield, J., et al. 2002. Herbivory in global climate change research: Direct effects of rising temperature on insect herbivores. *Global Change Biology* 8 (1): 1–16.

Barton, B.T. 2010. Climate warming and predation risk during herbivore ontogeny. *Ecology* 91 (10): 2811–2818.

Bates, O.K., and Bertelsmeier, C. 2021. Climatic niche shifts in introduced species. *Current Biology* 31 (19): R1252–R1266.

Battisti, A., Stastny, M., Buffo, E., and Larsson, S. 2006. A rapid altitudinal range expansion in the pine processionary moth produced by the 2003 climatic anomaly. *Global Change Biology* 12 (4): 662–71.

Bekkers, E., Francois, J.F., and Rojas-Romagosa, H. 2018. Melting ice caps and the economic impact of opening the northern sea route. *Economic Journal* 128: 1095–1127.

Bellard, C., Bernery, C., and Leclerc, C. 2021. Looming extinctions due to invasive species: Irreversible loss of ecological strategy and evolutionary history. *Global Change Biology* 27 (20): 4967–4979.

Bellard, C., Bertelsmeier, C., Leadley, P., Thuiller, W., and Courchamp, F. 2012. Impacts of climate change on the future of biodiversity. *Ecology Letters* 15: 365–77.

Benito Garzón, M., Robson, T.M., and Hampe, A. 2019. ΔTrait SDMs: Species distribution models that account for local adaptation and phenotypic plasticity. *New Phytologist* 222 (4): 1757–1765.

Bennie, J., Huntley, B., Wiltshire, A., Hill, M.O., and Baxter, R. 2008. Slope, aspect and climate: Spatially explicit and implicit models of topographic microclimate in chalk grassland. *Ecological Modelling* 216 (1): 47–59.

Berg, M.P., Kiers, E.T., Driessen, G., Van Der Heijden, M., Kooi, B.W., Kuenen, F., Liefting, M., et al. 2010. Adapt or disperse: Understanding species persistence in a changing world. *Global Change Biology* 16 (2): 587–98.

Bergstrom, D.M. 2022. Maintaining Antarctica's isolation from non-native species. *Trends in Ecology and Evolution* 37 (1): 5–9.

Bertelsmeier, C. 2021. Globalization and the anthropogenic spread of invasive social insects. *Current Opinion in Insect Science* 46: 16–23.

Bertelsmeier, C. and Keller, L. 2018. Bridgehead effects and role of adaptive evolution in invasive populations. *Trends in Ecology and Evolution* 33 (7): 527–34.

Bertelsmeier, C., Ollier, S., Liebhold, A., and Keller, L. 2017. Recent human history governs global ant invasion dynamics. *Nature Ecology and Evolution* 1 (7): 0184.

Blonder, B., Lamanna, C., Violle, C., and Enquist, B.J. 2014. The n-dimensional hypervolume. *Global Ecology and Biogeography* 23 (5): 595–609.

Bonnamour, A., Gippet, J.M., and Bertelsmeier, C. 2021. Insect and plant invasions follow two waves of globalisation. *Ecology Letters* 24 (11): 2418–2426.

Brakefield, P.M., and De Jong, P.W. 2011. A steep cline in ladybird melanism has decayed over 25 years: A genetic response to climate change? *Heredity* 107 (6): 574–78.

Braschler, B., and Hill, J.K. 2007. Role of larval host plants in the climate-driven range expansion of the butterfly Polygonia c-album. *Journal of Animal Ecology* 76 (3): 415–23.

Bridle, J.R., and Vines, T.H. 2007. Limits to evolution at range margins: When and why does adaptation fail? *Trends in Ecology and Evolution* 22 (3): 140–47.

Brightwell, R.J., and Silverman, J. 2010. Invasive Argentine ants reduce fitness of red maple via a mutualism with an endemic coccid. *Biological Invasions* 12: 2051–2057.

Brommer, J.E. 2004. The range margins of northern birds shift polewards. *Annales Zoologici Fennici* 41: 391–97.

Brönnimann, S. 2018. "Global warming (1970–present)". In *The Palgrave Handbook of Climate History*. Palgrave Macmillan, 321–28.

Brunt, D. 1924. Climatic Continentality and Oceanity. *The Geographical Journal* 64 (1): 43–49.

Bryant, S., Thomas, C., and Bale, J. 1997. Nettle-feeding nymphalid butterflies: Temperature, development and distribution. *Ecological Entomology* 22 (4): 390–98.

Buckley, L.B., Waaser, S.A., MacLean, H.J., and Fox, R. 2011. Does including physiology improve species distribution model predictions of responses to recent climate change? *Ecology* 92 (12): 2214–2221.

Buffo, E., Battisti, A., Stastny, M., and Larsson, S. 2007. Temperature as a predictor of survival of the pine processionary moth in the Italian Alps. *Agricultural and Forest Entomology* 9 (1): 65–72.

Bujan, J., Charavel, E., Bates, O.K., Gippet, J.M., Darras, H., Lebas, C., and Bertelsmeier, C. 2021. Increased acclimation ability accompanies a thermal niche shift of a recent invasion. *Journal of Animal Ecology* 90 (2): 483–91.

Bujan, J., Roeder, K.A., de Beurs, K., Weiser, M.D., and Kaspari, M. 2020. Thermal diversity of North American ant communities: Cold tolerance but not heat tolerance tracks ecosystem temperature. *Global Ecology and Biogeography* 29 (9): 1486–1494.

Bujan, J., Yanoviak, S.P., and Kaspari, M. 2016. Desiccation resistance in tropical insects: Causes and mechanisms underlying variability in a Panama ant community. *Ecology and Evolution* 6 (17): 6282–6291.

Burgiel, S.W. and Muir, A.A. 2010. *Invasive Species, Climate Change and Ecosystem-based Adaptation: Addressing Multiple Drivers of Global Change*. Global Invasive Species Programme (GISP), ZA.

Buse, J., Fassbender, S., Entling, M.H., and Pavlicek, T. 2015. Microclimatic divergence in a Mediterranean canyon affects richness, composition, and body size in saproxylic beetle assemblages. *PloS One* 10 (6): e0129323.

Byeon, D.H., Jung, J.M., Park, Y., Lee, H.S., Lee, J.H., Jung, S., and Lee, W.H. 2020. Model-based assessment of changes in the potential distribution of Solenopsis geminata (Hymenoptera: Formicidae) according to climate change scenarios. *Journal of Asia-Pacific Biodiversity* 13 (3): 331–38.

Byeon, D.H., Jung, S., and Lee, W.H. 2018. Review of CLIMEX and MaxEnt for studying species distribution in South Korea. *Journal of Asia-Pacific Biodiversity* 11 (3): 325–33.

Calosi, P., Bilton, D.T., Spicer, J.I., Votier, S.C., and Atfield, A. 2010. What determines a species' geographical range? Thermal biology and latitudinal range size relationships in European diving beetles (Coleoptera: Dytiscidae). *Journal of Animal Ecology* 79 (1): 194–204.

Carvalho, B.M., Rangel, E.F., and Vale, M.M. 2017. Evaluation of the impacts of climate change on disease vectors through ecological niche modelling. *Bulletin of Entomological Research* 107 (4): 419–30.

Chapman, J.W., Reynolds, D.R., and Wilson, K. 2015. Long-range seasonal migration in insects: Mechanisms, evolutionary drivers and ecological consequences. *Ecology Letters* 18 (3): 287–302.

Chapman, S., Watson, J.E., Salazar, A., Thatcher, M., and McAlpine, C.A. 2017. The impact of urbanization and climate change on urban temperatures: A systematic review. *Landscape Ecology* 32, 1921–1935.

Charrier, N. P., Hervet, C., Bonsergent, C., Charrier, M., Malandrin, L., Kaufmann, B., and Gippet, J.M. 2020. Invasive in the North: New latitudinal record for Argentine ants in Europe. *Insectes Sociaux* 67 (2): 331–35.

Chen, I.C., Hill, J.K., Ohlemüller, R., Roy, D.B., and Thomas, C.D. 2011. Rapid range shifts of species associated with high levels of climate warming. *Science* 333 (6045): 1024–1026.

Chen, I.C., Shiu, H.J., Benedick, S., Holloway, J.D., Chey, V.K., Barlow, H.S., Hill, J.K., et al. 2009. Elevation increases in moth assemblages over 42 years on a tropical mountain. *Proceedings of the National Academy of Sciences* 106 (5): 1479–1483.

Chevin, L.M., and Lande, R. 2011. Adaptation to marginal habitats by evolution of increased phenotypic plasticity. *Journal of Evolutionary Biology* 24 (7): 1462–1476.

Christensen, J.H., Hewitson, B., Busuioc, A., Chen, A., Gao, X., Held, I., Jones, R., et al. 2007. "Regional Climate Projections. Chapter 11. IPCC Working Group I". In *Climate Change 2007: The Physical Science Basis. Contribution of Working Group I to the Fourth Assessment Report of the Intergovernmental Panel on Climate Change*, edited by Solomon S., Qin, D., Manning, M., Chen, Z., Marquis, M., Averyt, K.B., Tignor, M., and Miller, H.L. Cambridge, United Kingdom and New York, NY, USA: Cambridge University Press, 847–940.

Colares, C., Roza, A.S., Mermudes, J.R., Silveira, L.F., Khattar, G., Mayhew, P.J., Monteiro, R.F., et al. 2021. Elevational specialization and the monitoring of the effects of climate change in insects: Beetles in a Brazilian rainforest mountain. *Ecological Indicators* 120, 106888.

Colautti, R.I. and Lau, J.A. 2015. Contemporary evolution during invasion: Evidence for differentiation, natural selection, and local adaptation. *Molecular Ecology* 24: 1999–2017.

Colin, T., Monchanin, C., Lihoreau, M., and Barron, A.B. 2020. Pesticide dosing must be guided by ecological principles. *Nature Ecology and Evolution* 4 (12): 1575–1577.

Colwell, R.K., Brehm, G., Cardelús, C.L., Gilman, A.C., and Longino, J.T. 2008. Global warming, elevational range shifts, and lowland biotic attrition in the wet tropics. *Science* 322 (5899): 258–61.

Cooper, R.N., Houghton, J.T., McCarthy, J.J., and Metz, B. 2002. Climate change 2001: The scientific basis. *Foreign Affairs* 81: 208.

Cordonnier, M., Bellec, A., Escarguel, G., and Kaufmann, B. 2020. Effects of urbanization–climate interactions on range expansion in the invasive European pavement ant. *Basic and Applied Ecology* 44, 46–54.

Coulin, C., de la Vega, G. J., Chifflet, L., Calcaterra, L.A., and Schilman, P.E. 2019. Linking thermotolerances of the highly invasive ant, Wasmannia auropunctata, to its current and potential distribution. *Biological Invasions* 21, 3491–3504.

Couper, L.I., Sanders, N.J., Heller, N.E., Gordon, D.M. 2021. Multiyear drought exacerbates long-term effects of climate on an invasive ant species. *Ecology* 102: e03476.

Cox, P.D. and Dolder, H.S. 1995. A simple flight chamber to determine flight activity in small insects. *Journal of Stored Products Research* 31 (4): 311–16.

Dai, A. 2011. Drought under global warming: A review. *Wiley Interdisciplinary Reviews: Climate Change* 2 (1): 45–65.

Da Silva, C.R., Beaman, J.E., Dorey, J.B., Barker, S.J., Congedi, N.C., Elmer, M.C., Galvin, S. et al. 2021. Climate change and invasive species: A physiological performance comparison of invasive and endemic bees in Fiji. *Journal of Experimental Biology* 224 (1): jeb230326.

Davidson, A.M., Jennions, M., and Nicotra, A.B. 2011. Do invasive species show higher phenotypic plasticity than native species and, if so, is it adaptive? A meta-analysis. *Ecology Letters* 14 (4): 419–31.

Davis, A.J., Jenkinson, L.S., Lawton, J.H., Shorrocks, B., and Wood, S. 1998a. Making mistakes when predicting shifts in species range in response to global warming. *Nature* 391 (6669): 783–86.

Davis, A.J., Lawton, J.H., Shorrocks, B., and Jenkinson, L.S. 1998b. Individualistic species responses invalidate simple physiological models of community dynamics under global environmental change. *Journal of Animal Ecology* 67 (4): 600–12.

Dellink, R., Hwang, H., Lanzi, E., and Chateau, J. 2017. International trade consequences of climate change. OECD Trade and Environment Working Papers 2017/1, OECD Publishing.

Deutsch, C.A., Tewksbury, J.J., Huey, R.B., Sheldon, K.S., Ghalambor, C.K., Haak, D.C., and Martin, P.R. 2008. Impacts of climate warming on terrestrial ectotherms across latitude. *Proceedings of the National Academy of Sciences* 105 (18): 6668–6672.

Diamond, S.E. and Chick, L.D. 2018a. The Janus of macrophysiology: Stronger effects of evolutionary history, but weaker effects of climate on upper thermal limits are reversed for lower thermal limits in ants. *Current Zoology* 64 (2): 223–30.

Diamond, S.E. and Chick, L.D. 2018b. Thermal specialist ant species have restricted, equatorial geographic ranges: Implications for climate change vulnerability and risk of extinction. *Ecography* 41 (9): 1507–1509.

Diamond, S.E., Dunn, R.R., Frank, S.D., Haddad, N.M., and Martin, R.A. 2015. Shared and unique responses of insects to the interaction of urbanization and background climate. *Current Opinion in Insect Science* 11: 71–77.

Diamond, S.E., Sorger, D.M., Hulcr, J., Pelini, S.L., Toro, I.D., Hirsch, C., Oberg, E. et al. 2012. Who likes it hot? A global analysis of the climatic, ecological, and evolutionary determinants of warming tolerance in ants. *Global Change Biology* 18 (2): 448–56.

Dick, C.W., Etchelecu, G., and Austerlitz, F. 2003. Pollen dispersal of tropical trees (Dinizia excelsa: Fabaceae) by native insects and African honeybees in pristine and fragmented Amazonian rainforest. *Molecular Ecology* 12 (3): 753–64.

Didham, R.K., Barbero, F., Collins, C.M., Forister, M.L., Hassall, C., Leather, S.R., Packer, L. et al. 2020. Spotlight on insects: Trends, threats and conservation challenges. *Insect Conservation and Diversity* 13 (2): 99–102.

Diez, J.M., D'Antonio, C.M., Dukes, J.S., Grosholz, E.D., Olden, J.D., Sorte, C.J., Blumenthal, D.M. et al. 2012. Will extreme climatic events facilitate biological invasions? *Frontiers in Ecology and the Environment* 10 (5): 249–57.

Diniz-Filho, J.A.F., De Marco Jr., P., and Hawkins, B.A. 2010. Defying the curse of ignorance: Perspectives in insect macroecology and conservation biogeography. *Insect Conservation and Diversity* 3 (3): 172–79.

Dobzhansky T. 1950. Evolution in the tropics. *American Scientist* 38: 209–21.

Dolson, S.J., Loewen, E., Jones, K., Jacobs, S.R., Solis, A., Hallwachs, W., Brunke, A.J., et al. 2021. Diversity and phylogenetic community structure across elevation during climate change in a family of hyperdiverse neotropical beetles (Staphylinidae). *Ecography* 44 (5): 740–52.

Dormann, C.F. 2007. Promising the future? Global change projections of species distributions. *Basic and Applied Ecology* 8 (5): 387–97.

Duffy, G.A., Coetzee, B.W., Janion-Scheepers, C., and Chown, S.L. 2015. Microclimate-based macrophysiology: Implications for insects in a warming world. *Current Opinion in Insect Science* 11: 84–89.

Ebel, B.A. 2012. Impacts of wildfire and slope aspect on soil temperature in a mountainous environment. *Vadose Zone Journal* 11(3): vzj2012.0017.

Elsen, P.R., Monahan, W.B., and Merenlender, A.M. 2020. Topography and human pressure in mountain ranges alter expected species responses to climate change. *Nature Communications* 11 (1): 1974.

Eschtruth, A.K., Cleavitt, N.L., Battles, J.J., Evans, R.A., and Fahey, T.J. 2006. Vegetation dynamics in declining eastern hemlock stands: 9 years of forest response to hemlock woolly adelgid infestation. *Canadian Journal of Forest Research* 36 (6): 1435–1450.

Esperk, T., Kjaersgaard, A., Walters, R. J., Berger, D., and Blanckenhorn, W.U. 2016. Plastic and evolutionary responses to heat stress in a temperate dung fly: Negative correlation between basal and induced heat tolerance? *Journal of Evolutionary Biology* 29 (5): 900–15.

Evans, T.G., Diamond, S.E., and Kelly, M.W. 2015. Mechanistic species distribution modelling as a link between physiology and conservation. *Conservation Physiology* 3 (1): cov056.

Everman, E.R., Freda, P.J., Brown, M., Schieferecke, A.J., Ragland, G.J., and Morgan, T.J. 2018. Ovary development and cold tolerance of the invasive pest Drosophila suzukii (Matsumura) in the central plains of Kansas, United States. *Environmental Entomology* 47 (4): 1013–1023.

Fahrner, S., and Aukema, B.H. 2018. Correlates of spread rates for introduced insects. *Global Ecology and Biogeography* 27 (6): 734–43.

Falla, C.M., Avila, G.A., McColl, S.T., Minor, M., and Najar-Rodríguez, A.J. 2021. The current and future potential distribution of Gargaphia decoris: A biological control agent for Solanum mauritianum (Solanaceae). *Biological Control* 160: 104637.

Faske, T.M., Thompson, L.M., Banahene, N., Levorse, A., Quiroga Herrera, M., Sherman, K., Timko, S.E., et al. 2019. Can gypsy moth stand the heat? A reciprocal transplant experiment with an invasive forest pest across its southern range margin. *Biological Invasions* 21: 1365–1378.

Fordham, D.A., Bertelsmeier, C., Brook, B.W. et al. 2018. How complex should models be? Comparing correlative and mechanistic range dynamics models. *Global Change Biology* 24: 1357–1370.

Franco, A.M., Hill, J.K., Kitschke, C., Collingham, Y.C., Roy, D.B., Fox, R., Huntley, B. et al. 2006. Impacts of climate warming and habitat loss on extinctions at species' low-latitude range boundaries. *Global Change Biology* 12 (8): 1545–1553.

De Frenne, P., Lenoir, J., Luoto, M., Scheffers, B.R., Zellweger, F., Aalto, J., Ashcroft, M.B. et al. 2021. Forest microclimates and climate change: Importance, drivers and future research agenda. *Global Change Biology* 27 (11): 2279–2297.

Gallien, L., Münkemüller, T., Albert, C.H., Boulangeat, I., and Thuiller, W. 2010. Predicting potential distributions of invasive species: Where to go from here? *Diversity and Distributions* 16 (3): 331–42.

Garcia, R.A., Cabeza, M., Rahbek, C., and Araújo, M.B. 2014. Multiple dimensions of climate change and their implications for biodiversity. *Science* 344 (6183): 1247579.

Gaston, K.J. Geographic range limits of species. 2009. *Proceedings of the Royal Society B: Biological Sciences* 276: 1391–1393.

Gibbs, A.G. and Matzkin, L.M. 2001. Evolution of water balance in the genus Drosophila. *Journal of Experimental Biology* 204 (13): 2331–2338.

Gilioli, G., Pasquali, S., Parisi, S., and Winter, S. 2014. Modelling the potential distribution of Bemisia tabaci in Europe in light of the climate change scenario. *Pest Management Science* 70 (10): 1611–1623.

Gippet, J. M., George, L., and Bertelsmeier, C. 2022. Local coexistence of native and invasive ant species is associated with micro-spatial shifts in foraging activity. *Biological Invasions* 24 (3): 761–73.

Gippet, J.M., Liebhold, A.M., Fenn-Moltu, G., and Bertelsmeier, C. 2019. Human-mediated dispersal in insects. *Current Opinion in Insect Science* 35: 96–102.

Gippet, J.M., Mondy, N., Diallo-Dudek, J., Bellec, A., Dumet, A., Mistler, L., and Kaufmann, B. 2017. I'm not like everybody else: Urbanization factors shaping spatial distribution of native and invasive ants are species-specific. *Urban Ecosystems* 20: 157–69.

Goulson, D. 2003. Effects of introduced bees on native ecosystems. *Annual Review of Ecology, Evolution, and Systematics* 34 (1): 1–26.

Green, P.T., O'Dowd, D.J., Abbott, K.L., Jeffery, M., Retallick, K., and Mac Nally, R. 2011. Invasional meltdown: Invader–invader mutualism facilitates a secondary invasion. *Ecology*, 92(9): 1758–1768.

Grinnell, J. 1917. The niche-relationships of the California Thrasher. *The Auk* 34 (4): 427–33.

Gross, C.L. 2001. The effect of introduced honeybees on native bee visitation and fruit-set in Dillwynia juniperina (Fabaceae) in a fragmented ecosystem. *Biological Conservation* 102 (1): 89–95.

Guisan, A., and Thuiller, W. 2005. Predicting species distribution: Offering more than simple habitat models. *Ecology Letters* 8 (9): 993–1009.

Guo, F., Lenoir, J., and Bonebrake, T.C. 2018. Land-use change interacts with climate to determine elevational species redistribution. *Nature Communications* 9 (1): 1315.

Guo, S., Ge, X., Zou, Y., Zhou, Y., Wang, T., and Zong, S. 2021. Projecting the global potential distribution of Cydia pomonella (Lepidoptera: Tortricidae) under historical and RCP4. 5 climate scenarios. *Journal of Insect Science* 21 (2): 15.

Gutierrez, A.P., and Ponti, L. 2014. Analysis of invasive insects: Links to climate change. *Invasive Species and Global Climate Change* 4: 45–61.

Halsch, C.A., Shapiro, A.M., Fordyce, J.A., Nice, C.C., Thorne, J.H., Waetjen, D.P., and Forister, M.L. 2021. Insects and recent climate change. *Proceedings of the National Academy of Sciences* 118 (2): e2002543117.

Hance, T., van Baaren, J., Vernon, P., and Boivin, G. 2007. Impact of extreme temperatures on parasitoids in a climate change perspective. *Annual Review of Entomology* 52: 107–26.

Hao, T., Elith, J., Guillera-Arroita, G., and Lahoz-Monfort, J.J. 2019. A review of evidence about use and performance of species distribution modelling ensembles like BIOMOD. *Diversity and Distributions* 25 (5): 839–52.

Hargreaves, A.L., Samis, K.E., and Eckert, C.G. 2014. Are species' range limits simply niche limits writ large? A review of transplant experiments beyond the range. *The American Naturalist* 183(2): 157–73.

Hargreaves, A.L., Suárez, E., Mehltreter, K., Myers-Smith, I., Vanderplank, S.E., Slinn, H.L., Vargas-Rodriguez, Y.L. et al. 2019. Seed predation increases from the Arctic to the Equator and from high to low elevations. *Science Advance* 5 (2): eaau4403.

Heger, T., and Jeschke, J.M. 2014. The enemy release hypothesis as a hierarchy of hypotheses. *Oikos* 123 (6): 741–50.

Heller, N.E., Sanders, N.J., Shors, J.W., and Gordon, D.M. 2008. Rainfall facilitates the spread, and time alters the impact, of the invasive Argentine ant. *Oecologi* 155: 385–95.

Hellmann, J.J., Byers, J.E., Bierwagen, B.G., and Dukes, J.S. 2008. Five potential consequences of climate change for invasive species. *Conservation Biology* 22 (3): 534–43.

Hickling, R., Roy, D.B., Hill, J.K., Fox, R., and Thomas, C.D. 2006. The distributions of a wide range of taxonomic groups are expanding polewards. *Global Change Biology* 12 (3): 450–55.

Hill, J.K., Griffiths, H.M., and Thomas, C.D. 2011. Climate change and evolutionary adaptations at species' range margins. *Annual Review of Entomology* 56: 143–59.

Hill, M.P., Thomson, L.J., Björkman, C., and Niemelä, P. 2015. Species distribution modelling in predicting response to climate change. *Climate Change and Insect Pests* 16.

HilleRisLambers, J., Harsch, M.A., Ettinger, A.K., Ford, K.R., and Theobald, E.J. 2013. How will biotic interactions influence climate change–induced range shifts? *Annals of the New York Academy of Sciences* 1297 (1): 112–25.

Hitch, A.T., and Leberg, P.L. 2007. Breeding distributions of North American bird species moving north as a result of climate change. *Conservation Biology* 21 (2): 534–39.

Hoffmann, A.A., Chown, S.L., and Clusella-Trullas, S. 2013. Upper thermal limits in terrestrial ectotherms: How constrained are they?. *Functional Ecology* 27 (4): 934–49.

Hoffmann, A.A., Hallas, R.J., Dean, J.A., and Schiffer, M. 2003. Low potential for climatic stress adaptation in a rainforest Drosophila species. *Science* 301 (5629): 100–02.

Holway, D.A., Suarez, A.V., and Case, T.J. 2002. Role of abiotic factors in governing susceptibility to invasion: A test with Argentine ants. *Ecology* 83 (6): 1610–1619.

Hood, W.G., and Tschinkel, W.R. 1990. Desiccation resistance in arboreal and terrestrial ants. *Physiological Entomology* 15 (1): 23–35.

Huang, H., von Lampe, M., and van Tongeren, F. 2011. Climate change and trade in agriculture. *Food Policy* 36: S9–S13.

Huey, R.B., Gilchrist, G.W., Carlson, M.L., Berrigan, D., and Serra, L. 2000. Rapid evolution of a geographic cline in size in an introduced fly. *Science* 287 (5451): 308–09.

Hulme, P.E. 2017. Climate change and biological invasions: Evidence, expectations, and response options. *Biological Reviews* 92 (3): 1297–1313.

Hutchinson, G.E. 1957. Concluding remarks. *Cold Spring Harbor Symposium on Quantitative Biology* 22: 415–27.

IPCC. 2012. *Managing the risks of extreme events and disasters to advance climate change adaptation: Special Report of the Intergovernmental Panel on Climate Change.* Field, C.B., V. Barros, T.F. Stocker, D. Qin, D.J., Dokken, K.L., Ebi, M.D., Mastrandrea, K.J., Mach, G.-K., Plattner, S.K., Allen, M., Tignor, and P.M., Midgley (Eds.) Available from Cambridge University Press, The Edinburgh Building, Shaftesbury Road, Cambridge CB2 8RU ENGLAND, 582. Available from June 2012.

Isaac, N.J., Girardello, M., Brereton, T.M., and Roy, D.B. 2011. Butterfly abundance in a warming climate: Patterns in space and time are not congruent. *Journal of Insect Conservation* 15: 233–40.

Iwamura, T., Guzman-Holst, A., and Murray, K.A. 2020. Accelerating invasion potential of disease vector Aedes aegypti under climate change. *Nature Communications* 11 (1): 2130.

Jaeschke, A., Bittner, T., Reineking, B., and Beierkuhnlein, C. 2013. Can they keep up with climate change?–Integrating specific dispersal abilities of protected Odonata in species distribution modelling. *Insect Conservation and Diversity* 6 (1): 93–103.

Janion-Scheepers, C., Phillips, L., Sgrò, C.M., Duffy, G.A., Hallas, R., and Chown, S.L. 2018. Basal resistance enhances warming tolerance of alien over indigenous species across latitude. *Proceedings of the National Academy of Sciences* 115 (1): 145–50.

Jardeleza, M.K.G., Koch, J.B., Pearse, I.S., Ghalambor, C.K., and Hufbauer, R.A. 2022. The roles of phenotypic plasticity and adaptation in morphology and performance of an invasive species in a novel environment. *Ecological Entomology* 47 (1): 25–37.

Jung, J.M., Lee, W.H., and Jung, S. 2016. Insect distribution in response to climate change based on a model: Review of function and use of CLIMEX. *Entomological Research* 46 (4): 223–35.

Keane, R.M. and Crawley, M.J. 2002. Exotic plant invasions and the enemy release hypothesis. *Trends in Ecology and Evolution* 17 (4): 164–70.

Kearney, M. 2006. Habitat, environment and niche: What are we modelling? *Oikos* 115 (1): 186–91.

Kearney, M. and Porter, W. 2009. Mechanistic niche modelling: Combining physiological and spatial data to predict species' ranges. *Ecology Letters* 12 (4): 334–50.

Kellermann, V. and van Heerwaarden, B. 2019. Terrestrial insects and climate change: Adaptive responses in key traits. *Physiological Entomology* 44 (2): 99–115.

Kenna, D., Pawar, S., and Gill, R.J. 2021. Thermal flight performance reveals impact of warming on bumblebee foraging potential. *Functional Ecology* 35 (11): 2508–2522.

Khaliq, I., Hof, C., Prinzinger, R., Böhning-Gaese, K., and Pfenninger, M. 2014. Global variation in thermal tolerances and vulnerability of endotherms to climate change. *Proceedings of the Royal Society B: Biological Sciences* 281 (1789): 20141097.

Kim, S.B., Park, J.J., and Kim, D.S. 2020. CLIMEX simulated predictions of the potential distribution of Bactrocera dorsalis (Hendel) (Diptera: Tephritidae) considering the northern boundary: With special emphasis on Jeju, Korea. *Journal of Asia-Pacific Entomology* 23 (3): 797–808.

Kiritani, K. 2013. Different effects of climate change on the population dynamics of insects. *Applied Entomology and Zoology* 48 (2): 97–104.

Krab, E.J., Oorsprong, H., Berg, M.P., and Cornelissen, J.H. 2010. Turning northern peatlands upside down: Disentangling microclimate and substrate quality effects on vertical distribution of Collembola. *Functional Ecology* 24 (6): 1362–1369.

Kriticos, D., Maywald, G., Yonow, T., Zurcher, E.J., Herrmann, N.I., and Sutherst, R.W. 2016. *CLIMEX Version 4: Exploring the Effects of Climate on Plants, Animals and Diseases.* Canberra: CSIRO.

Lahr, E.C., Dunn, R.R., and Frank, S.D. 2018. Getting ahead of the curve: Cities as surrogates for global change. *Proceedings of the Royal Society B: Biological Sciences* 285 (1882): 20180643.

Lantschner, M.V., Villacide, J.M., Garnas, J.R., Croft, P., Carnegie, A.J., Liebhold, A.M., and Corley, J.C. 2014. Temperature explains variable spread rates of the invasive woodwasp Sirex noctilio in the Southern Hemisphere. *Biological Invasions* 16: 329–39.

Leitch, K.J., Ponce, F.V., Dickson, W.B., van Breugel, F., and Dickinson, M.H. 2021. The long-distance flight behavior of Drosophila supports an agent-based model for wind-assisted dispersal in insects. *Proceedings of the National Academy of Sciences* 118 (17): e2013342118.

Lemoine, N.P. 2021. Phenology dictates the impact of climate change on geographic distributions of six co-occurring North American grasshoppers. *Ecology and Evolution* 11 (24): 18575–18590.

Lenoir, J., and Svenning, J.C. 2015. Climate-related range shifts—A global multidimensional synthesis and new research directions. *Ecography* 38 (1): 15–28.

Liu, X., and Luo, T. 2011. Spatiotemporal variability of soil temperature and moisture across two contrasting timberline ecotones in the Sergyemla Mountains, Southeast Tibet. *Arctic, Antarctic, and Alpine Research* 43 (2): 229–38.

Loarie, S.R., Duffy, P.B., Hamilton, H., Asner, G.P., Field, C.B., and Ackerly, D.D. 2009. The velocity of climate change. *Nature* 462 (7276): 1052–1055.

Lobo, J.M. 2016. The use of occurrence data to predict the effects of climate change on insects. *Current Opinion in Insect Science* 17: 62–68.

Lowe, C.B., Kellis, M., Siepel, A., Raney, B.J., Clamp, M., Salama, S.R., Kingsley, D.M., et al. 2011. Three periods of regulatory innovation during vertebrate evolution. *Science* 333: 1019–1024.

Lurgi, M., Brook, B.W., Saltre, F., and Fordham, D.A. 2015. Modelling range dynamics under global change: Which framework and why? *Methods in Ecology and Evolution* 6 (3): 247–56.

Lutterschmidt, W.I. and Hutchison, V.H. 1997. The critical thermal maximum: History and critique. *Canadian Journal of Zoology* 75 (10): 1561–1574.

Mack, R., Simberloff, D., Lonsdale, W. et al. 2000. Biotic invasions: Causes, epidemiology, global consequences, and control. *Ecological Applications* 86: 249–50.

Maino, J.L., Kong, J.D., Hoffmann, A.A., Barton, M.G., and Kearney, M.R. 2016. Mechanistic models for predicting insect responses to climate change. *Current Opinion in Insect Science* 17: 81–86.

Maino, J.L., Umina, P.A., and Hoffmann, A.A. 2018. Climate contributes to the evolution of pesticide resistance. *Global Ecology and Biogeography* 27(2): 223–32.

Makarieva, A.M., Gorshkov, V.G., and Li, B.L. 2009. Precipitation on land versus distance from the ocean: Evidence for a forest pump of atmospheric moisture. *Ecological Complexity* 6 (3): 302–07.

Mammola, S., Milano, F., Vignal, M., Andrieu, J., and Isaia, M. 2019. Associations between habitat quality, body size and reproductive fitness in the alpine endemic spider Vesubia jugorum. *Global Ecology and Biogeography* 28 (9): 1325–1335.

Mammola, S., Pétillon, J., Hacala, A., Monsimet, J., Marti, S.L., Cardoso, P., and Lafage, D. 2021. Challenges and opportunities of species distribution modelling of terrestrial arthropod predators. *Diversity and Distributions* 27 (12): 2596–2614.

Martinet, B., Dellicour, S., Ghisbain, G., Przybyla, K., Zambra, E., Lecocq, T., Boustani, M., Baghirov, R. et al. 2021. Global effects of extreme temperatures on wild bumblebees. *Conservation Biology* 35 (5): 1507–1518.

Martínez-López, O., Koch, J.B., Martínez-Morales, M.A., Navarrete-Gutiérrez, D., Enríquez, E., and Vandame, R. 2021. Reduction in the potential distribution of bumble bees (Apidae: Bombus) in Mesoamerica under different climate change scenarios: Conservation implications. *Global Change Biology* 27 (9): 1772–1787.

Mason, S.C., Palmer, G., Fox, R., Gillings, S., Hill, J.K., Thomas, C.D., and Oliver, T.H. 2015. Geographical range margins of many taxonomic groups continue to shift polewards. *Biological Journal of the Linnean Society* 115: 586–97.

McCain, C.M. and Garfinkel, C.F. 2021. Climate change and elevational range shifts in insects. *Current Opinion in Insect Science* 47: 111–18.

Menke, S.B., Guénard, B., Sexton, J.O., Weiser, M.D., Dunn, R.R., and Silverman, J. 2011. Urban areas may serve as habitat and corridors for dry-adapted, heat tolerant species; an example from ants. *Urban Ecosystems* 14: 135–63.

Merrill, R.M., Gutiérrez, D., Lewis, O.T., Gutiérrez, J., Díez, S.B., and Wilson, R.J. 2008. Combined effects of climate and biotic interactions on the elevational range of a phytophagous insect. *Journal of Animal Ecology* 77 (1): 145–55.

Meurisse, N., Rassati, D., Hurley, B.P., Brockerhoff, E.G., and Haack, R.A. 2019. Common pathways by which non-native forest insects move internationally and domestically. *Journal of Pest Science* 92: 13–27.

Miguelena, J.G., and Baker, P.B. 2019. Effects of urbanization on the diversity, abundance, and composition of ant assemblages in an arid city. *Environmental Entomology* 48 (4): 836–46.

Mitikka, V., and Hanski, I. 2010. Pgi genotype influences flight metabolism at the expanding range margin of the European map butterfly. *Annales Zoologici Fennici* 47 (1): 1–14.

Mlynarek, J.J. 2015. Testing the enemy release hypothesis in a native insect species with an expanding range. *PeerJ* 3: e1415.

Montejo-Kovacevich, G., Martin, S.H., Meier, J.I., Bacquet, C.N., Monllor, M., Jiggins, C.D., and Nadeau, N.J. 2020. Microclimate buffering and thermal tolerance across elevations in a tropical butterfly. *Journal of Experimental Biology* 223 (8): jeb220426.

Morales, C.L., and Aizen, M.A. 2002. Does invasion of exotic plants promote invasion of exotic flower visitors? A case study from the temperate forests of the southern Andes. *Biological Invasions* 4: 87–100.

Moran, E.V., and Alexander, J.M. 2014. Evolutionary responses to global change: Lessons from invasive species. *Ecology Letters* 17 (5): 637–49.

Musolin, D.L. 2007. Insects in a warmer world: Ecological, physiological and life-history responses of true bugs (Heteroptera) to climate change. *Global Change Biology* 13 (8): 1565–1585.

Mutamiswa, R., Machekano, H., Singano, C., Joseph, V., Chidawanyika, F., and Nyamukondiwa, C. 2021. Desiccation and temperature resistance of the larger grain borer, Prostephanus truncatus (Horn) (Coleoptera: Bostrichidae): Pedestals for invasion success? *Physiological Entomology* 46 (2): 157–66.

Napoli, M., Massetti, L., Brandani, G., Petralli, M., and Orlandini, S. 2016. Modeling tree shade effect on urban ground surface temperature. *Journal of Environmental Quality* 45 (1): 146–56.

Nyamukondiwa, C., Kleynhans, E., and Terblanche, J.S. 2010. Phenotypic plasticity of thermal tolerance contributes to the invasion potential of Mediterranean fruit flies (Ceratitis capitata). *Ecological Entomology* 35 (5): 565–75.

O'Dowd, D.J., Green, P.T., and Lake, P.S. 2003. Invasional "meltdown" on an oceanic island. *Ecology Letters* 6 (9): 812–17.

Olesen, J.M., Eskildsen, L.I., and Venkatasamy, S. 2002. Invasion of pollination networks on oceanic islands: Importance of invader complexes and endemic super generalists. *Diversity and Distributions* 8(3): 181–92.

Oliver, T.H., Marshall, H.H., Morecroft, M.D., Brereton, T., Prudhomme, C., and Huntingford, C. 2015. Interacting effects of climate change and habitat fragmentation on drought-sensitive butterflies. *Nature Climate Change* 5 (10): 941–45.

Oorthuis, R., Vaunat, J., Hürlimann, M., Lloret, A., Moya, J., Puig-Polo, C., and Fraccica, A. 2021. Slope orientation and vegetation effects on soil thermo-hydraulic behavior. An experimental study. *Sustainability (Switzerland)* 13: 1–13.

Paquette, A., and Hargreaves, A.L. 2021. Biotic interactions are more often important at species' warm versus cool range edges. *Ecology Letters* 24 (11): 2427–2438.

Parmesan, C. 2006. Ecological and evolutionary responses to recent climate change. *Annual Review of Ecology, Evolution, and Systematics* 37: 637–69.

Parmesan, C. and Yohe, G. 2003. A globally coherent fingerprint of climate change impacts across natural systems. *Nature* 421 (6918): 37–42.

Parmesan, C., Ryrholm, N., Stefanescu, C., Hill, J.K., Thomas, C.D., Descimon, H., Huntley, B., et al. 1999. Poleward shifts in geographical ranges of butterfly species associated with regional warming. *Nature* 399 (6736): 579–83.

Parratt, S.R., Walsh, B.S., Metelmann, S., White, N., Manser, A., Bretman, A.J., Hoffman, A.A. et al. 2021. Temperatures that sterilize males better match global species distributions than lethal temperatures. *Nature Climate Change* 11 (6): 481–84.

Perissinotto, R., Pringle, E.L., and Giliomee, J.H. 2011. Southward expansion in beetle and butterfly ranges in South Africa. *African Entomology* 19 (1): 61–69.

Peterson, A.T. 2003. Subtle recent distributional shifts in Great Plains bird species. *The Southwestern Naturalist* 48 (2): 289–92.

Peterson, A.T. 2011. Ecological niche conservatism: A time-structured review of evidence. *Journal of Biogeography* 38 (5): 817–27.

Phillips, S.J., Anderson, R.P., Schapire, R.E. 2006. Maximum entropy modeling of species geographic distributions. *Ecological Modelling* 190: 231–59.

Pincebourde, S., Murdock, C.C., Vickers, M., and Sears, M.W. 2016. Fine-scale microclimatic variation can shape the responses of organisms to global change in both natural and urban environments. *Integrative and Comparative Biology* 56 (1): 45–61.

Pincebourde, S. and Woods, H.A. 2020. There is plenty of room at the bottom: Microclimates drive insect vulnerability to climate change. *Current Opinion in Insect Science* 41: 63–70.

Platts, P.J., Mason, S.C., Palmer, G., Hill, J.K., Oliver, T.H., Powney, G.D., Fox, R., and Thomas, C.D. 2019. Habitat availability explains variation in climate-driven range shifts across multiple taxonomic groups. *Scientific Reports* 9: 1–10.

Polidori, C., García-Gila, J., Blasco-Aróstegui, J., and Gil-Tapetado, D. 2021. Urban areas are favouring the spread of an alien mud-dauber wasp into climatically non-optimal latitudes. *Acta Oecologica* 110: 103678.

Polidori, C., Gutiérrez-Cánovas, C., Sánchez, E., Tormos, J., Castro, L., and Sánchez-Fernández, D. 2020. Climate change-driven body size shrinking in a social wasp. *Ecological Entomology* 45 (1): 130–41.

Poniatowski, D., Beckmann, C., Löffler, F., Münsch, T., Helbing, F., Samways, M.J., and Fartmann, T. 2020. Relative impacts of land-use and climate change on grasshopper range shifts have changed over time. *Global Ecology and Biogeography* 29 (12): 2190–2202.

Poniatowski, D., Heinze, S., and Fartmann, T. 2012. The role of macropters during range expansion of a wing-dimorphic insect species. *Evolutionary Ecology* 26: 759–70.

Pöyry, J., Luoto, M., Heikkinen, R. K., Kuussaari, M., and Saarinen, K. 2009. Species traits explain recent range shifts of Finnish butterflies. *Global Change Biology*, 15(3): 732–43.

Prior, K.M., Powell, T.H., Joseph, A.L., and Hellmann, J.J. 2015. Insights from community ecology into the role of enemy release in causing invasion success: The importance of native enemy effects. *Biological Invasions* 17: 1283–1297.

Pulatov, B., Hall, K., Linderson, M.L., and Jönsson, A.M. 2014. Effect of climate change on the potential spread of the Colorado potato beetle in Scandinavia: An ensemble approach. *Climate Research* 62 (1): 15–24.

Radenković, S., Schweiger, O., Milić, D., Harpke, A., and Vujić, A. 2017. Living on the edge: Forecasting the trends in abundance and distribution of the largest hoverfly genus (Diptera: Syrphidae) on the Balkan Peninsula under future climate change. *Biological Conservation*, 212, 216–29.

Rajpurohit, S., Nedved, O., and Gibbs, A.G. 2013. Meta-analysis of geographical clines in desiccation tolerance of Indian drosophilids. *Comparative Biochemistry and Physiology Part A: Molecular and Integrative Physiology* 164 (2): 391–98.

Raven, P.H., and Wagner, D.L. 2021. Agricultural intensification and climate change are rapidly decreasing insect biodiversity. *Proceedings of the National Academy of Sciences* 118 (2): e2002548117.

Régnière, J., Powell, J., Bentz, B., and Nealis, V. 2012. Effects of temperature on development, survival and reproduction of insects: Experimental design, data analysis and modeling. *Journal of Insect Physiology* 58 (5): 634–47.

Rezende, E.L., Tejedo, M., and Santos, M. 2011. Estimating the adaptive potential of critical thermal limits: Methodological problems and evolutionary implications. *Functional Ecology* 25 (1): 111–21.

Rita, A., Bonanomi, G., and Allevato, E. et al. 2021. Topography modulates near-ground microclimate in the Mediterranean Fagus sylvatica treeline. *Scientific Reports* 11: 1–14.

Robinet, C., and Roques, A. 2010. Direct impacts of recent climate warming on insect populations. *Integrative Zoology* 5 (2): 132–42.

Rödder, D., Schmitt, T., Gros, P., Ulrich, W., and Hable, J.C. 2021. Climate change drives mountain butterflies towards the summits. *Scientific Reports* 11: 1–12.

Rodrigues, Y.K., and Beldade, P. 2020. Thermal plasticity in insects' response to climate change and to multifactorial environments. *Frontiers in Ecology and Evolution* 8: 271.

Roslin, T., Hardwick, B., Novotny, V., Petry, W.K., Andrew, N.R., Asmus, A., Barrio, I.C. et al. 2017. Higher predation risk for insect prey at low latitudes and elevations. *Science* 356 (6339): 742–44.

Roubik, D.W. 2000. Pollination system stability in tropical America. *Conservation Biology* 14: 1235–1236.

Schaffner, F., Bellini, R., Petrić, D., Scholte, E.J., Zeller, H., and Marrama Rakotoarivony, L. 2013. Development of guidelines for the surveillance of invasive mosquitoes in Europe. *Parasites and Vectors* 6: 1–10.

Schultz, T.D., and Hadley, N.F. 1987. Microhabitat segregation and physiological differences in co-occurring tiger beetle species, Cicindela oregona and Cicindela tranquebarica. *Oecologia* 73: 363–70.

Schweiger, O., Biesmeijer, J. C., Bommarco, R., Hickler, T., Hulme, P.E., Klotz, S., Künh, I. et al. 2010. Multiple stressors on biotic interactions: How climate change and alien species interact to affect pollination. *Biological Reviews* 85 (4): 777–95.

Seebens, H., Blackburn, T.M., Dyer, E.E., Genovesi, P., Hulme, P.E., Jeschke, J.M., Pagad, S. et al. 2017. No saturation in the accumulation of alien species worldwide. *Nature Communications* 8 (1): 14435.

Sexton, J.P., McIntyre, P.J., Angert, A.L., and Rice, K.J. 2009. Evolution and ecology of species range limits. *Annual Review of Ecology, Evolution, and Systematics* 40: 415–36.

Sgrò, C.M., Terblanche, J.S., and Hoffmann, A.A. 2016. What can plasticity contribute to insect responses to climate change? *Annual Review of Entomology* 61: 433–51.

Sheldon, K.S., Yang, S., and Tewksbury, J.J. 2011. Climate change and community disassembly: Impacts of warming on tropical and temperate montane community structure. *Ecology Letters* 14 (12): 1191–1200.

Simberloff, D. 2006. Invasional meltdown 6 years later: Important phenomenon, unfortunate metaphor, or both? *Ecology Letters* 9 (8): 912–19.

Simberloff, D. and Von Holle, B. 1999. Positive interactions of nonindigenous species: Invasional meltdown? *Biological Invasions* 1: 21–32.

Simmons, A.D. and Thomas, C.D. 2004. Changes in dispersal during species' range expansions. *The American Naturalist* 164 (3): 378–95.

Soberón, J. and Peterson, A.T. 2005. Interpretation of models of fundamental ecological niches and species' distributional areas. *Biodiversity Informatics* 2: 1–10.

Sparks, T.H., Dennis, R.L., Croxton, P.J., and Cade, M. 2007. Increased migration of Lepidoptera linked to climate change. *European Journal of Entomology* 104 (1): 139–43.

Spence, A.R., and Tingley, M.W. 2020. The challenge of novel abiotic conditions for species undergoing climate-induced range shifts. *Ecography* 43 (11): 1571–1590.

Stout, J.C., Kells, A.R., and Goulson, D. 2002. Pollination of the invasive exotic shrub Lupinus arboreus (Fabaceae) by introduced bees in Tasmania. *Biological Conservation* 106 (3): 425–34.

Stralberg, D., Arseneault, D., Baltzer, J.L., Barber, Q.E., Bayne, E.M., Boulanger, Y., Brown, C.D. et al. 2020. Climate-change refugia in boreal North America: What, where, and for how long? *Frontiers in Ecology and the Environment* 18 (5): 261–70.

Strange, E.E., and Ayres, M.P. 2010. Climate change impacts: Insects. In *Encyclopedia of Life Sciences*, (Ed.). doi:10.1002/9780470015902.a0022555.

Suggitt, A.J., Gillingham, P.K., Hill, J.K., Huntley, B., Kunin, W.E., Roy, D.B., and Thomas, C.D. 2011. Habitat microclimates drive fine-scale variation in extreme temperatures. *Oikos* 120 (1): 1–8.

Suggitt, A.J., Wilson, R.J., Isaac, N.J., Beale, C.M., Auffret, A.G., August, T., Bennie, J.J. et al. 2018. Extinction risk from climate change is reduced by microclimatic buffering. *Nature Climate Change* 8 (8): 713–17.

Sunday, J.M., Bates, A.E., Kearney, M.R., Colwell, R.K., Dulvy, N.K., Longino, J.T., and Huey, R.B. 2014. Thermal-safety margins and the necessity of thermoregulatory behavior across latitude and elevation. *Proceedings of the National Academy of Sciences* 111 (15): 5610–5615.

Tabor, J.A., and Koch, J.B. 2021. Ensemble models predict invasive bee habitat suitability will expand under future climate scenarios in Hawai'i. *Insects* 12 (5): 443.

Tejeda, M.T., Arredondo, J., Liedo, P., Pérez-Staples, D., Ramos-Morales, P., and Díaz-Fleischer, F. 2016. Reasons for success: Rapid evolution for desiccation resistance and life-history changes in the polyphagous fly Anastrepha ludens. *Evolution* 70 (11): 2583–2594.

Thomas, C.D. 2010. Climate, climate change and range boundaries. *Diversity and Distributions* 16 (3): 488–95.

Thomas, C.D., Bodsworth, E.J., Wilson, R.J., Simmons, A.D., Davies, Z.G., Musche, M., and Conradt, L. 2001. Ecological and evolutionary processes at expanding range margins. *Nature* 411 (6837): 577–81.

Thuiller, A.W., Georges, D., Engler, R., Georges, M.D., and Thuiller, C.W. 2016. Package "biomod2". Species distribution modeling within an ensemble forecasting framework 600 https://CRAN.R-project.org/package=biomod2.

Timoner, P., Fasel, M., Ashraf Vaghefi, S.S., Marle, P., Castella, E., Moser, F., and Lehmann, A. 2021. Impacts of climate change on aquatic insects in temperate alpine regions: Complementary modeling approaches applied to Swiss rivers. *Global Change Biology* 27 (15): 3565–3581.

Tobin, P.C., Nagarkatti, S., Loeb, G., and Saunders, M.C. 2008. Historical and projected interactions between climate change and insect voltinism in a multivoltine species. *Global Change Biology* 14 (5): 951–57.

Traveset, A., Heleno, R., Chamorro, S., Vargas, P., McMullen, C.K., Castro-Urgal, R., Nogales, M. et al. 2013. Invaders of pollination networks in the Galápagos Islands: Emergence of novel communities. *Proceedings of the Royal Society B: Biological Sciences* 280 (1758): 20123040.

Traveset, A. and Richardson, D.M. 2006. Biological invasions as disruptors of plant reproductive mutualisms. *Trends in Ecology and Evolution* 21 (4): 208–16.

Tylianakis, J.M., Didham, R.K., Bascompte, J. and Wardle, D.A. 2008. Global change and species interactions in terrestrial ecosystems. *Ecology Letters* 11: 1351–1363.

Uhler, J., Redlich, S., Zhang, J., Hothorn, T., Tobisch, C., Ewald, J., Thorn, S. *et al.* 2021. Relationship of insect biomass and richness with land use along a climate gradient. *Nature Communications* 12 (1): 5946.

Urban, M.C., Tewksbury, J.J., and Sheldon, K.S. 2012. On a collision course: Competition and dispersal differences create no-analogue communities and cause extinctions during climate change. *Proceedings of the Royal Society B: Biological Sciences* 279 (1735): 2072–2080.

Urbani, F., D'Alessandro, P., and Biondi, M. 2017. Using Maximum Entropy Modeling (MaxEnt) to predict future trends in the distribution of high altitude endemic insects in response to climate change. *Bulletin of Insectology* 70 (2): 189–200.

Vamosi, J.C., Knight, T.M., Steets, J.A., Mazer, S.J., Burd, M., and Ashman, T.L. 2006. Pollination decays in biodiversity hotspots. *Proceedings of the National Academy of Sciences* 103 (4): 956–61.

VanDerWal, J., Murphy, H.T., Kutt, A.S., Perkins, G.C., Bateman, B.L., Perry, J.J., and Reside, A.E. 2013. Focus on poleward shifts in species' distribution underestimates the fingerprint of climate change. *Nature Climate Change* 3 (3): 239–43.

Van Dyck, H., Bonte, D., Puls, R., Gotthard, K., and Maes, D. 2015. The lost generation hypothesis: Could climate change drive ectotherms into a developmental trap? *Oikos* 124 (1): 54–61.

van Heerwaarden, B. and Kellermann, V. 2020. Does plasticity trade off with basal heat tolerance? *Trends in Ecology and Evolution* 35 (10): 874–85.

van Heerwaarden, B. and Sgrò, C.M. 2021. Male fertility thermal limits predict vulnerability to climate warming. *Nature Communications* 12 (1): 2214.

Van Kleunen, M., Weber, E., and Fischer, M. 2010. A meta-analysis of trait differences between invasive and non-invasive plant species. *Ecology Letters* 13 (2): 235–45.

Vessby, K. and Wiktelius, S. 2003. The influence of slope aspect and soil type on immigration and emergence of some northern temperate dung beetles. *Pedobiologia* 47 (1): 39–51.

Vidal, M.C., Anneberg, T.J., Curé, A.E., Althoff, D.M., and Segraves, K.A. 2021. The variable effects of global change on insect mutualisms. *Current Opinion in Insect Science* 47: 46–52.

Vilcinskas, A. 2019. Pathogens associated with invasive or introduced insects threaten the health and diversity of native species. *Current Opinion in Insect Science* 33: 43–48.

Violle, C., Reich, P.B., Pacala, S.W., Enquist, B.J., and Kattge, J. 2014. The emergence and promise of functional biogeography. *Proceedings of the National Academy of Sciences* 111 (38): 13690–13696.

Vitasse, Y., Ursenbacher, S., Klein, G., Bohnenstengel, T., Chittaro, Y., Delestrade, A., Monnerat, C. et al. J. 2021. Phenological and elevational shifts of plants, animals and fungi under climate change in the European Alps. *Biological Reviews* 96 (5): 1816–1835.

Vogel G. 2017. Where have all the insects gone? *Science* 356 (6338): 576–79.

Wagner, D.L., Fox, R., Salcido, D.M., and Dyer, L.A. 2021. A window to the world of global insect declines: Moth biodiversity trends are complex and heterogeneous. *Proceedings of the National Academy of Sciences* 118 (2): e2002549117.

Wagner, D.L. and Liebherr, J.K. 1992. Flightlessness in insects. *Trends in Ecology and Evolution* 7 (7): 216–20.

Walters, R.J., Hassall, M., Telfer, M.G., Hewitt, G.M., and Palutikof, J.P. 2006. Modelling dispersal of a temperate insect in a changing climate. *Proceedings of the Royal Society B: Biological Sciences* 273 (1597): 2017–2023.

Warren, M.S., Hill, J.K., Thomas, J.A., Asher, J., Fox, R., Huntley, B., Roy, D.B. et al. 2001. Rapid responses of British butterflies to opposing forces of climate and habitat change. *Nature* 414 (6859): 65–69.

Watts, M.J., Fordham, D.A., Akçakaya, H.R., Aiello-Lammens, M.E., and Brook, B.W. 2013. Tracking shifting range margins using geographical centroids of metapopulations weighted by population density. *Ecological Modelling* 269: 61–69.

Weis, J.S. 2010. The role of behavior in the success of invasive crustaceans. *Marine and Freshwater Behaviour and Physiology* 43 (2): 83–98.

Willig, M.R., Kaufman, D.M., and Stevens, R.D. 2003. Latitudinal gradients of biodiversity: Pattern, process, scale, and synthesis. *Annual Review of Ecology, Evolution, and Systematics* 34 (1): 273–309.

Wisz, M. S., Pottier, J., Kissling, W. D., Pellissier, L., Lenoir, J., Damgaard, C. F., Dormann, C.F. et al. 2013. The role of biotic interactions in shaping distributions and realised assemblages of species: Implications for species distribution modelling. *Biological Reviews* 88 (1): 15–30.

Wittman, S.E., Sanders, N.J., Ellison, A.M., Jules, E.S., Ratchford, J.S., and Gotelli, N.J. 2010. Species interactions and thermal constraints on ant community structure. *Oikos* 119 (3): 551–59.

Yang, L.H., Postema, E.G., Hayes, T.E., Lippey, M.K., and MacArthur-Waltz, D.J. 2021. The complexity of global change and its effects on insects. *Current Opinion in Insect Science* 47: 90–102.

Zeuss, D., Brunzel, S., and Brandl, R. 2017. Environmental drivers of voltinism and body size in insect assemblages across Europe. *Global Ecology and Biogeography* 26 (2): 154–65.

Zhang, S.N. and Kubota, K. 2021. Dispersal constraints on the potential distribution of cold-adapted stag beetles (genus *Platycerus*) in Japan and the implications of climate change. *Insect Conservation and Diversity* 14: 356–66.

Zhou, Y., Ge, X., Zou, Y., Guo, S., Wang, T., and Zong, S. 2021. Prediction of the potential global distribution of the Asian longhorned beetle Anoplophora glabripennis (Coleoptera: Cerambycidae) under climate change. *Agricultural and Forest Entomology* 23 (4): 557–68.

Ziter, C., Robinson, E.A., and Newman, J.A. 2012. Climate change and voltinism in Californian insect pest species: Sensitivity to location, scenario and climate model choice. *Global Change Biology* 18 (9): 2771–2780.

Zuckerberg, B., Woods, A.M., and Porter, W.F. 2009. Poleward shifts in breeding bird distributions in New York State. *Global Change Biology* 15 (8): 1866–1883.

Zurell, D., Thuiller, W., Pagel, J., Cabral, J.S., Münkemüller, T., Gravel, D., Dullinger, S. et al. 2016. Benchmarking novel approaches for modelling species range dynamics. *Global Change Biology* 22 (8): 2651–2664.

CHAPTER 12

Insect communities adapting to climate change

Using species' trajectories along elevation gradients in tropical and temperate zones

Genoveva Rodríguez-Castañeda and Anouschka R. Hof

12.1 Introduction: Threats to insect biodiversity

The biodiversity of insects is threatened worldwide (Sánchez-Bayo and Wyckhuys 2019; Wagner et al. 2021). Human activities and wealth generation are the main culprits of changes in land use, nitrification of the soils and climate change which in turn are linked to sharp declines in insect biomass (Wagner et al. 2021; summarized in Figure 12.1). Insects are keystone species for biodiversity and the economy. A large percentage of amphibians, reptiles, birds and mammals rely on insects for food. Moreover, plants' survival also depends on insects; for example, insects pollinate 90 percent of the flowering plant species on Earth (Hoshiba and Sasaki 2008). Humanity and its economy rely heavily on insect pollinators, median valuations for short-term pollination services are at 1 trillion US dollars (Lippert and Narjes 2021).

A review of the documentation behind insect extinctions and climate change describes how little we know about how climate change causes extinction. Our knowledge of the main drivers of insect extinction will improve when we focus on studies of direct relationships between species' tolerance to high temperatures and local extinctions. The current evidence suggests that a diverse suite of factors implicated in climate change has not been studied as much, such as precipitation, food abundance, plant phenology changes and the consequences of community loss on species survival (Cahill et al. 2013).

12.2 Climate change past and present

Insects have dominated the Earth since 325 million years ago and are not immune to mass decreases in diversity. Past climate change events have caused some of the largest turnovers of insect species communities (Schachat and Labandeira 2021; Chapter 2 of

Genoveva Rodríguez-Castañeda and Anouschka R. Hof, *Insect communities adapting to climate change*. In: *Effects of Climate Change on Insects*. Edited by: Daniel González-Tokman and Wesley Dáttilo, Oxford University Press. © Oxford University Press (2024). DOI: 10.1093/oso/9780192864161.003.0012

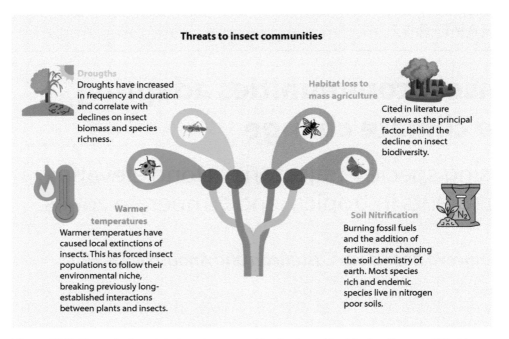

Figure 12.1 The main threats to insect communities (reviewed by Sánchez-Bayo and Wyckhuys 2019; Wagner et al. 2021).

this book). Climate fluctuates in cycles, a series of events that are repeated regularly in the same order. Life on Earth is governed by climatic cycles. Like all life on Earth, insects have adapted to short (e.g., circadian cycles, seasonal cycles) and midscale climatic cycles (e.g., El Nino-Southern oscillation). Longer climatic cycles, such as glaciations and post-glaciations, with changes in temperature and precipitation that span thousands to millions of years, are often correlated to extinctions (Bennett 1990). For example, during the Late Pennsylvanian (323–298 million years ago), the warmer temperatures and dryer conditions in coal swamps caused the demise of two-thirds of the coal swamp plant species, and all insects closely associated with them went extinct (Labandeira 2006), Yet this event is not considered a mass extinction of insects. Mass extinction events are defined as episodes in which 75 percent or more of the extant taxa go extinct (Barnosky et al. 2011). Recently, Schachat and Labandeira (2021) propose that these extreme insect extinction events are not considered mass extinctions but insect turnovers in which even though large numbers of genera and species go extinct, remaining clades evolve and diversify anew. Since the insect extinctions of the Late Pennsylvanian, insects evolved to an estimated 4.8 million insect species (Di Michelle and Phillips 1996; Stork 2018). The glacial episodes of the Quaternary (2.6 million years ago to the present) played a crucial role in shaping the distributions of extant flora and fauna. The expansions and contractions of the ice sheets that persisted during the summer, rendered large areas uninhabitable for most species (Provan and Bennett 2008). Climatic refugia have allowed taxa to rebound, recolonize and diversify to current biodiversity levels in temperate and tropical zones (Binney et al. 2009). Relative to temperate zones covered with ice sheets, tropical mountain climate has remained stable for more than 40,000 years. The large availability of refugial areas allowed species to resist past climate change. And the recolonization process following the climatic stress allowed for higher speciation rates in the tropics (Dynesius and Jansson 2000; Rodríguez-Castañeda et al. 2017).

When studying contemporary climate change, we need to determine whether this event will represent an episode where insects experience a significant turnover or if insects will face a mass extinction event for the first time. (Schachat and Labandeira 2021). While we have centered studies on how species will move in response to climate change, we need to spend more effort determining how and where insect communities will find refuge. Pleistocene refugia were pivotal in biodiversity maintenance and as sources for post-glacial recolonization (Keppel et al. 2012). Refugia are broadly defined as locations uniquely buffered from the intense cold and fluctuations of glacial climates, in which species could persist and had time to adapt to the new conditions. This concept has expanded to encompass past glaciations, high temperatures, droughts and fluctuating sea levels (Morelli et al. 2016).

Mountain ranges increase the likelihood that current climates will continue to exist within short migrating distances, enabling species to track their climatic niche. Moreover, mountains are ideal regions to look for refugia. The complex terrain creates a myriad of microclimates with the potential to buffer the effects of climate change. Deep snow drifts provide surface insulation and water later in the season, valleys promote cold air pools and inversions can decouple local climatic conditions from regional circulation patterns. Some mountains also provide shaded areas in the lowlands that buffer solar heating. Insect species have shown resilience. They manage to take advantage of habitats altered by humans (Dornelas et al. 2019). In the lowlands of northern California, habitats such as train tracks, rights of way and hedges provided refuge and contributed to the resilience of lowland butterflies to the megadrought in the southwest United States that started in the year 2000 (Forister et al. 2018; Williams et al. 2022).

The tropics, particularly tropical mountains, were crucial sources of climatic refugia during the Pleistocene (Flantua and Hooghiemstra 2018). The vast areas with climatic stability in the tropics contributed to the higher observed species richness in the tropics (Jansson et al. 2013). In fact, the tropics have acted as cradles and museums of tropical biodiversity (Stebbins 1974). However, changing climate in locations that have been climatically stable for long periods poses an unprecedented threat to tropical species. A recent literature review of twenty-two insect sampling publications with time series has shown losses of 30.0 percent of the unique interactions between plants, herbivores and parasitoids due to changes in precipitation and temperature regimes (Salcido et al. 2022). Multiyear studies of trophic interactions are scare with conclusions from them being dependent on the landscape matrix the studies are embedded in. This makes it challenging to generalize these conclusions to other tropical locations.

12.3 Explored and unexplored role of climate change on insect communities

Climate change is one of the main factors behind the loss of insect biodiversity in mountains and tropical areas (Sánchez-Bayo and Wyckhuys 2019). Alarmingly, the specific drivers of present and future extinctions are much less clear and difficult to quantify than those of habitat loss (Cahill et al. 2013). Documenting climate change-caused extinctions will require creative ways of disentangling mixed factors. Since climate change-driven extinctions have more to do with shifting communities and resulting net losses of interspecific interactions than with the changes in temperature and precipitation regimes (Cahil et al. 2013).

12.4 Less explored effects of climate change on insects

Temperature is the most studied climate change factor. Other factors, such as changes in precipitation and extreme weather events, are equally important climate change factors that are less studied and may be vital in determining the fate of insects and have yet to be explored.

Breakthrough studies have identified extreme weather events that caused declines in species richness and insect abundance. Extreme weather events, measured as the precipitation variability, have strong adverse effects on parasitism rates (Stireman et al. 2005). Extreme temperatures outside of historical ranges have been linked to reduced occupancy of sixty-six species of bumblebees in North America and Europe (Soroye et al. 2020). Finally, elevated carbon dioxide (CO_2) affects host plant quality that can cascade down to herbivores and predators (Cornelissen 2011). In fact, there is a strong correlation between insect biodiversity and host plant biodiversity (Lewinsohn et al. 2005); hence factors such as fossil fuel burning, the nitrification of soils and drought will affect plant species and the insect communities they host (Wagner et al. 2021). Climate change is pushing insect montane species between a rock and a hard place by forcing a shift to distributions upslope in areas that have fragmented forests and are heavily impacted by agriculture and the intense use of pesticides (Brühl and Zalle 2019; Lewis et al. 2015)

Needing to migrate and track environmental niches becomes impossible with fragmentation and the vast amount of pesticide-sprayed lands that are unsuitable habitats for insect species to move through. These conditions leave insect species to face either extinction or adaptation to the new conditions (Gonzalez-Tokman et al. 2020; Thomas et al. 2004). Dornelas et al. (2019) found that declining and increasing populations (winners and losers) are roughly equally balanced, but both groups are less common than populations showing little to no change. However, in all these calculations, we must not forget the role refugia may have on the fate of insect biodiversity. The resilience of insects is in the smaller areas of green they need to thrive. Halsch et al. (2021) found that insects in Northern California respond positively to hedgerows and backyard gardens.

12.5 Species inter-dependence along elevation gradients

When local populations go extinct there is a net reduction of relationships established in communities, which may cause populations of dependent species to go extinct. Losing interspecific interactions is particularly problematic in the tropics, where there is increased specialization of plant–insect interactions (Dyer et al. 2007). Tropical plant species form mesocosms, broadly defined as systems which contain two unique, interacting communities that coexist. It involves plant–plant, plant–insect and insect–insect interactions with a degree of specialization (Dyer and Letourneau 2003). No timeseries studies of how plant–insect interactions shift along elevation gradients exist, and few studies describe how these mesocosms change along elevation gradients. A few studies show how plant protection roles decrease with increasing elevation. Plowman et al. (2017) studied plant–ant interactions along an elevation gradient in Mount Wilhelm, Mandang province, Papua New Guinea. The team researched three dominant ant–plant species. They show how fewer partners at higher elevations (1600 metres (m)) decreased specialization between ant–plant interactions.

Furthermore, a decrease in herbivory protection was observed as elevation increased. Rodriguez-Castañeda et al. (2011) conducted manipulative studies of ant exclusion on three pepper plant species in the genus *Piper*. Two plant species were ant–plant mutualists with ants in the genus *Pheidole*, and one had occasional associations with various ant types. We found that ant protection against herbivory peaked at middle elevations (1200 m above sea level (m.a.s.l.)) and decreased at elevations where ants and plants do not occur naturally (1800 m.a.s.l.). Interestingly, when plants had active ant colonies inside, they could thrive outside their environmental niche (i.e., at 2200 m.a.s.l.). The plant–ant mutualists established at environmental conditions outside their elevational distribution only when ant colonies lived within the plant. In all instances where the plant was outside its elevational range and ants were excluded, the plants failed to survive. However, when planted within their elevation range, some plant–ant mutualists survived without ants. We found that ants create cemeteries within the plant stem. Add dead ants in a plant chamber, and the plant seals it and fungus grows on them. We hypothesize that the plant can absorb the nitrogen from the dead mutualist ants' bodies once broken down by the fungus. If true, these results suggest a double role of ants on plants. A nurturing role where the plant can draw nutrients from dead ants in cool temperatures and low light conditions. Moreover, a protective role in lowland locations where herbivory and competition are fierce (G. Rodríguez-Castañeda and M. Olson, unpublished data).

We need more manipulative studies to understand how the presence or absence of one species from a mesocosm affects the ability to adapt to the climatic changes of another species. We also need to establish how species interactions aid in colonizing new locations and adapting to climatic conditions.

12.6 Changes in precipitation cycles and extreme weather events

Effects of regional events, such as high precipitation and humidity on sites undergoing megadroughts, or the long-term effects of extreme weather events on insect communities can only be evaluated with multi-year sampling and time-series analysis. Moreover, evaluating whether mountains provide locations as a refuge for insect populations will require the inclusion of multiple locations and of elevations that include heterogeneous geological features (e.g., forests around stream beads, snowbanks, lagoons, valleys and alpine zones protected by rock crops).

Climate extremes can be placed into two broad groups: (i) those based on simple climate statistics, which include extremes such as extremely low or high daily temperatures or heavy daily or monthly rainfall amount, that occur every year; and (ii) more complex event-driven extremes, examples of which include drought, floods or hurricanes, which do not necessarily occur every year at a given location. Globally, both climatic extremes have increased with climate change (Easterling et al. 2014). Researchers have found that extreme weather events decrease insect abundance and diversity by analyzing time-series data sampled in different parts of the world. For example, annual variability in precipitation reduced herbivores' parasitism rates (Stireman et al. 2005). El Niño-Southern Oscillation (ENSO) was correlated with the altitudinal distributions of butterflies along the northern California mountains (Pardikes et al. 2015). For example, if one studies in mountainous areas subject to drought, it would be important to understand how species migrate and insect populations fare in patches with milder microclimates (i.e., higher humidity and closer access to water in aguadas and streambeds) along elevation gradients and contrast these results with locations that are more exposed to the climatic elements.

12.7 Effects of climate change on insect communities across elevation gradients

We categorize insects' response to climate change according to whether they remain in the location or disperse to find the climatic conditions that form their environmental niche. Within each of these responses, various researchers have recorded genetic, behavioral and ecological adaptations that project insect species resilience. When analyzing insect communities that remain on site, we must differentiate between species resistance and resilience. Species resistance is defined as the capacity of populations to survive the changes imposed by climate through persistence. Resilience, on the other hand, studies the populations' capacity to recover to pre-disturbance abundance levels instead of shifting their distribution.

A direct way to understand species' responses to global warming is to study changes over the past 100 to twenty-five years. During this period, mountains' mean temperatures have increased by 1–2 degrees Celsius. The expectation is that the species distribution range will move to track its environmental niche either poleward, along a latitudinal gradient or uphill, along an elevational gradient. Warmer conditions at higher latitudes and elevations will promote population growth and colonization (Thomas 2010), whereas rear edges of distribution will become increasingly unsuitable (Colwell et al. 2008). Chen et al. (2011) spearheaded a historical to current day contrast approach by resampling historic elevation records of insects in Mount Kinabalu, Malaysia. They sampled geometrid moths during 2007 and compared the rear and lead altitudinal distribution limits to what Jeremy D. Holloway had sampled in 1965 (i.e., forty-two years earlier). Their criteria included occurrences for a specific elevation and only species sampled three or more times in historical and current collections. They found that 80 percent of the species studied expanded their distribution and shifted on average 152 m uphill except for species at 2,700 m, that have suffered range contractions on both the leading and rear edge of their altitudinal distribution (Chen et al. 2011).

Differences in the velocity of the processes of colonizing at higher elevations and going extinct at lower elevations when these studies have been conducted determine whether total elevation ranges expand, contract, or remain stable. Thus, the pace of range shifting along with the direction of the shift determines species' extinction risk due to climate change (Lenoir and Svenning 2015). Rumpf et al. (2018) revised 3,446 studies contrasting the elevational distribution range of species. Of those, only twenty-two publications had primary data on historical and recent elevational limits. The studies included a total of 1,026 species including plants, vertebrates and invertebrates. They found that population extinctions at the rear edges of altitudinal distributions are at least as frequent as colonizations at the leading edges. Their results demonstrated that species of plants, vertebrates and invertebrates had had a true upslope range shift in response to the 1–2 degrees Celsius increases in the average temperature of the mountain ranges studied. Similarly, results from a meta-analysis that included eighteen taxonomic groups including plants, birds and insects among them geometrid moths, and that calculated the velocity at which species moved upslope revealed that, on average, species are moving upslope at a speed of 11 m per decade (Chen et al. 2011). This information has proven how most mountainous species respond to climate change. However, because these studies rely on primary data where species distribution was sampled along a single slope, they fail to represent how multiple populations of the species across its entire distribution range are responding to climate change. Studies solely looking for upward elevational range shifts limit what we can infer about how insect species respond to climate change. Getting partial information on an altitudinal distribution detracts from gaining essential information about how species that move downslope or remain onsite may be using refugia to resist the stressors posed by climatic change.

Ten possible states of a species' elevational range result from the contrast between historical and present distributions. A brief literature search by Lenoir et al. (2010) found that only 65 percent of the species had an upslope shift, expected with climate change. The remaining 45 percent was split between downslope range shifts (35 percent) and species that remained at their original location (i.e., stable range, 10 percent). We define and summarize these in Figure 12.2.

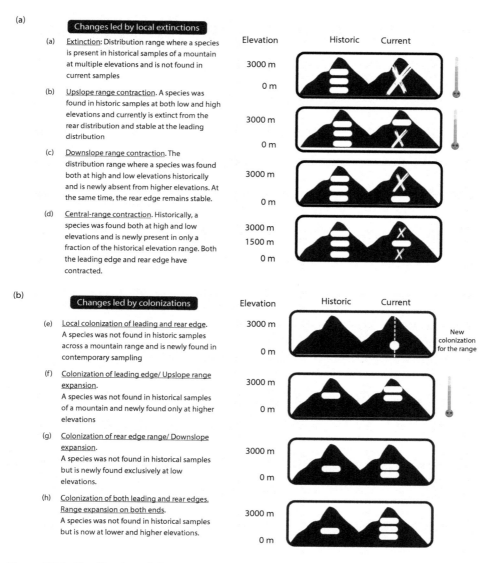

Figure 12.2 Classification of changes in species' distribution when historical and current samplings are contrasted. For the shifts that are expected with the recent increases in temperature, we depict them with a thermometer.

(c)

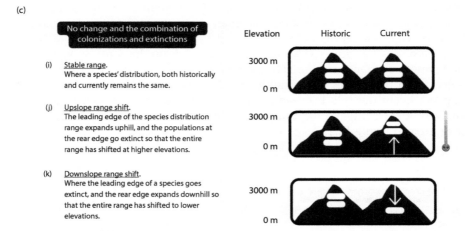

Figure 12.2 Continued

As powerful as this simple approach is, there are limitations to assuming insect species' changes in altitudinal distribution. When shifts in time and space are studied, a time-series approach in which multiple observations are included is much more informative than merely information on how species were distributed at the beginning and at the end.

Studies that contrast species distribution across elevations must make the following necessary assumptions:

1. Species were extensively collected in historical and current sampling so that an absence from a sampled elevation represents an actual distribution limit for the species at the time it was studied.

2. To guarantee that if a species is absent from a particular location, it no longer exists there, we assume that the current sampling is extensive and that the efforts and insect trap methods are comparable to how surveys were conducted during the historical sampling.

3. The sampling in both historical and current periods was conducted under comparable phenological seasons and in years with similar weather conditions as insects are irruptive in nature, with periods of low abundance followed by periods of high abundance (McCain and Garfinkel 2021).

4. That the sampling is such that catching a set number of individuals from a species in both historical/present collections represents that those species have viable populations at that elevation.

5. The species included in the study are not vagrants (i.e., a species outside its living/breeding area).

6. If the species is an altitudinal migrant, the leading and rear edges of their living and breeding distributions have been established.

7. Species known to have meta-population dynamics (i.e., a species that goes extinct upslope with multiple colonizations from downslope) are excluded. If a species

has multiple colonizations and extinctions along the gradient, it is impossible to determine its distribution limits with a historical and a current sampling event.

8. To establish that the two sampling time points represent the long-term distribution limits for the species, it assumes that there are no long-term cyclic patterns of colonization and extinctions along the elevation gradient (e.g., multivoltine species that reach higher elevations on ENSO years and are set back to the valleys in regular years).

9. Seeing the strong adverse effects that changes in land use have on insects and how much habitats have been degraded in the last century, studies must measure and note changes in land use along the elevation gradient studied.

10. Studies that contrast historical and current distributions only sometimes include complete distribution ranges. Being unaware of the entire altitudinal distribution of a species makes it difficult to discern whether a species is endangered. For example, a range expansion of the carabid beetle *Blennidus pichincha* must be evaluated not only based on the expansion of the leading edge but tracking whether there was an extinction of the populations at the rear edge. A good starting point is to discern between the responses of the leading and rear edges of a species' altitudinal distribution. Furthermore, it is crucial to know if species have had a complete upslope shift (i.e., species colonized uphill and their rear edge populations have gone locally extinct). Alternatively, when a species colonizes at the leading edge, but populations remain at the rear edge, it is considered an incomplete elevational range shift (McCain and Garfinkel 2021).

11. Considering the high interannual variability of insects is a challenge in studies contrasting two sampling events because determining occurrence depends on species abundance (McCain et al. 2016). The data collected in only two points in time are susceptible to be incorrect. Stuble et al. (2021) used simulations. They found that short-term snapshot resurveys based on modeled populations detected a 50 percent probability of erroneously detecting the opposite trend in population abundance change and nearly zero probability of detecting no change.

There are multiple literature reviews to understand how insects respond to climate change. However, we decided to include a literature review here to extend the search to publications available only in Spanish and Portuguese. We also wanted to present a global map with the studied locations of insects' elevational range shifts. In addition, we found the definition of insects' response to climate change too reductionist. Reviews on insect shifts in distribution mainly focus on local extinctions and shifts in distribution that goes upslope, neglecting to discuss the taxa that remain on-site or shift downslope. When studying insect responses to climate change, it is essential to include all responses to climate change. For example, Chen et al. (2011) found that downslope range shifts in high-altitude geometrid moths were related to tracking the changes in cloud permanence in a cloud forest. A refugium for a particular insect is a site where the effects of climate change are buffered to allow populations to persist. These conditions need not exclusively occur upslope. Finally, much has been published on differences between the dynamics in temperate mountains with seasonal changes in temperature and tropical mountains where temperatures remain constant throughout the years (Janzen 1967). In tropical Lepidoptera, there is higher specialization in host plant use (Dyer et al. 2007). In this review, we wanted to understand if tropical insects respond differently from temperate insect species. To evaluate if all studied species of insects are indeed shifting upward in elevation in response to climate change globally, we conducted an extensive literature review and classified insect species responses in the ten categories presented in Figure 12.2. Our hypotheses to test were:

1. Most of the studies have been conducted on European mountains, with a longer trajectory of studying insects' distribution.
2. Insect species are responding to climate warming at a global scale. Most insect species will respond to global warming by shifting to higher elevations through upslope range contractions, upslope range expansions, or complete upslope range shifts. As the thermal tolerance breadth of montane tropical species is smaller than that of temperate species (Ghalambor et al. 2006; Sheldon et al. 2011), we predict that temperate species will respond to global warming by expanding their niche, while tropical species will respond by either range contractions or upslope shifts in elevation.
3. Some species will remain within their altitudinal distribution. We expect higher proportions of tropical species to remain within the altitudinal distribution because there is higher diet breadth specialization (Dyer et al. 2007) and species mutualisms in the tropics that make it unlikely to colonize upslope independently.

12.8 Methods

To test these hypotheses, we extracted range shift data from the literature during years 2021–2022. Following a systematic review protocol (Page et al. 2021), we identified scientific papers from The Web of Science, Scopus and Google Scholar online databases. When we found previous meta-analyses or literature reviews, we screened their reference list to include data from those papers. We used the terms (ALL (communit* AND (mountain* OR alpine OR montane) AND (shift OR expan* OR move* or contrac*) AND (elevation OR altitude) AND (tropic* OR temperate)) AND TITLE- ABS- Key (Insect* Climate Change* Global Warming). When meta-analyses and literature reviews were found, we looked for references that included these criteria. Based on the inclusion criteria, we ended with a final set of nineteen publications containing 1,483 data points and a total of 1,287 species of arthropods (Insecta+Araneae).

We downloaded the primary data from supplementary materials or from the published graphs if the data were unavailable.

The following inclusion criteria were used in the screening process to assess the search results:

1. Included historical and current sampling along multiple locations of an elevational gradient.
2. Historical and current sampling was comparable in both sampling effort and methods.
3. The study subjects were arthropod species.
4. Only range shift studies that occurred more than twenty-five years ago and at present (2011–2022) were included. As other authors have published, our time window attempts to analyze range shifts over fifty years. However, in our effort to include as many tropical publications as possible, we lowered the period to twenty-five years.
5. The publication included their sampling methods and description of their sampling methods and decisions to include a species as present in a particular sampling location.

For each species included in the publications, we recorded the geographical coordinates of each elevation studied. We classified the distribution of each species as tropical if they were conducted in mountain ranges between 23 decimal degrees in the Southern Hemisphere and 23 decimal degrees in the Northern Hemisphere. Any study beyond 23 degrees (North or South) was considered a temperate study. We also recorded the trapping or collection method used in each study and the abundance criteria for considering a species present. For each study, we recorded the year(s) in which historical ($t1$) and current ($t2$) sampling occurred.

To evaluate the global representation of species' elevational range shift, we mapped each study by identifying the country in which each mountain range was studied and plotted the number of species per study within each country.

For each arthropod species, we also recorded the rear (i.e., the lowest elevation species occurred at) and leading edges (i.e., the highest elevation the species occurred at) during both time events (i.e., $t1$ and $t2$). How much of the elevation gradient a species occupies, or the species' altitudinal broadness, was calculated as leading (le) edge at $t1$ minus rear edge (re) at $t1$.

Equation 1

$$le.t1 - re.t1$$

Equation 2

$$le.t2 - re.t2$$

To understand the individual responses of insects' altitudinal distribution over the past twenty-five to sixty years, we used the 1,287 species we gathered leading and rear edge at $t1$ and $t2$ for and set the following rules for classification. We classified them as local extinctions, species that were recorded from anywhere in the mountain historically (i.e., at $t1$) and were not found at the latest sampling ($t2$). Similarly, we classified it as colonization when a species was not reported in the historical range ($t1$) and was newly found at any locations sampled during the resampling ($t2$).

The altitudinal distribution of species is not always a continuum. That is, some species are present or absent in the middle parts of the range at any point of time studied. We simplified the species distribution response by assuming a continuous altitudinal distribution from the rear edge to the leading edge of the elevation gradient. This simplification allowed us to make inferences about general patterns of the shifts in species altitudinal distribution over the past 25–100 years. We focused on the directionality (i.e., whether the species shifted in an upslope, neutral or downslope direction). We also focused on the broadness of the altitudinal distribution (i.e., whether the species' elevation range expanded or contracted). Rules to classify the altitudinal distribution when there are two points in time studied are summarized in Table 12.1.

Contingency tables were built based on the classification of the elevational range shift (Table 12.1) and whether the studies were conducted at tropical versus temperate zones. Including all responses and contrasting temperate versus tropical insect responses represent a novel framework as it expands our understanding of insects' non-conventional responses to climate change and may uncover characteristics of climate change refugia along the mountains.

Table 12.1 Set of rules programmed to classify the species' changes in altitudinal distribution by distinguishing their leading edge (*le*) from the rear edge (*re*) dynamics between historical (*t1*) and current sampling events (*t2*).

If condition	Classification
Species present at any surveyed elevation in the historical survey AND absent in the current survey	1. Locally extinct
re.t1 < *re.t2 AND le.t1* = le.t2	2. Upslope range contraction
re.t1 = *re.t2 AND le.t1* > le.t2	3. Downslope range contraction
re.t1 < *re.t2 AND le.t1* > le.t2	4. Central range contraction
Species absent from any surveyed elevation in the historical survey AND newly present in the current survey	5. Colonization
le.t1 < *le.t2 AND re.t1* = re.t2	6. Upslope range expansion
re.t1 > re.t2 AND *le.t1* = le.t2	7. Downslope range expansion
re.t1 > re.t2 AND *le.t1* < le.t2	8. Range expansion both ends
re.t1 = re.t2 AND *le.t1* = le.t2	9. Range stable
re.t1 < re.t2 AND *le.t1* < le.t2	10. Upslope range shift
re.t1 > re.t2 AND *le.t1* > le.t2	11. Downslope range shift

12.9 Most studies will be conducted on European mountains, with a longer trajectory of studying insect distribution

More than half our studies include species' range shifts in European mountains. Except for one study from New Zealand, all temperate studies of species' range shifts have been conducted in Northern temperate zones. Finally, the study includes only five publications in which tropical species' distribution range shifts were documented (Figure 12.3).

Completing the picture with the major mountain ranges of Earth is possible if there is a concerted effort to revisit and sample museum collections collected along elevational gradients. We propose making a database of globally studied mountain ranges. That includes the metadata explorers and museums holding those collections so that upcoming researchers can expand our knowledge of how insects with a mountainous distribution are coping with global warming. For example, the records of arachnids and insects along an elevational gradient after the 1980 Mount Saint Helens' eruption that have been curated by Rod Crawford and volunteers at the Burke Museum in Seattle, Washington would present an excellent opportunity to expand our knowledge to other insect orders. One could also resample seminal studies conducted along elevational gradients, such as the study of galling and free-living herbivores along the San Francisco peaks (3843 m) to the Sonoran Desert (305 m) in Arizona and Sierra do Cipó (350–1350 m), Minas Gerais, Brazil (Fernandes and Price 1988).

Arthropod orders have been sampled differently in tropical and temperate regions (Figure 12.4, number of species per order in temperate versus tropical zones: χ^2 = 313.94, $p < 0.0001$). Biases in taxa interests from 25–100 years ago are perpetuated in these studies since current studies rely on the baseline distributions set at that time. Not surprisingly, attractive groups such as butterflies and beetles dominate the literature. Other literature reviews that include analyses of multiple years have found similar biases (Halsch et al. 2021; McCain and Garfinkel 2021)

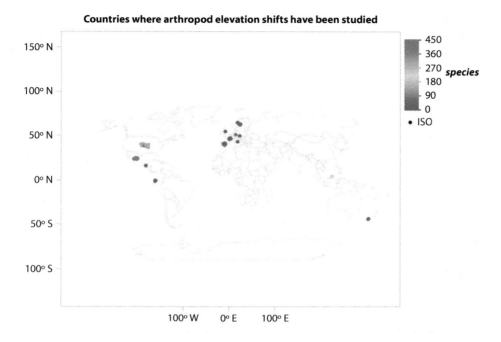

Figure 12.3 Countries for which published studies of elevational range shifts in distribution over the past 150 to 25 years exist. Red dots depict studies that include elevational information for more than 270 species, and dark-blue dots for studies that include less than 100 species.

Interestingly, while there is a wealth of research on how Hymenoptera pollinators respond to climate change along elevational gradients in temperate regions, we found no equivalent documentation on insect pollinators in tropical zones. From our perspective, finding shifts of distributions of pollinators in the tropics is of the highest priority because of the trends of specialization and narrow distributions of the species along the elevational gradient (Vizentin-Bugoni et al. 2018). Furthermore, assuming that we can draw similar conclusions from temperate zone studies could be seriously misleading.

12.10 Time interval studied

The average period covered by temperate studies (73.4 ± 0.99 years) was twofold that of tropical studies (31.8 ± 0.29 years) (Welch's T = 32.36, $P < 0.0001$).

Given the unbalance between the number of temperate ($n = 14$) versus tropical ($n = 5$) studies and the difference in time between historical and contemporary sampling studies in the temperate and tropical zones (Figure 12.5), we used a subset of the temperate studies that matched the timespan studied at tropical latitudes to evaluate insects' response to climate change along elevational gradients. In the end, we include 809 species and twelve studies.

There were significantly different shift responses in temperate and tropical regions. We excluded local extinctions and colonizations as those that did not represent shifts in elevation and did not have sufficient sample size in both temperate and tropical regions (Figure 12.6; $\chi^2 = 130.12$, df = 8, $p < 0.0001$, phi coefficient = 0.41)

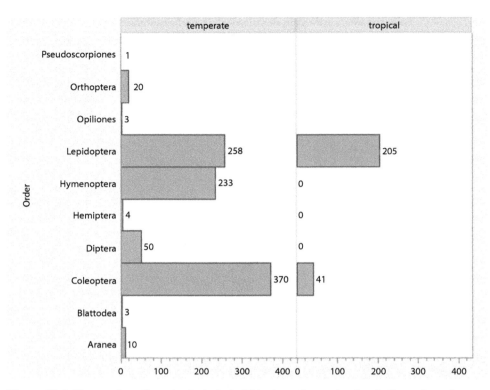

Figure 12.4 The number of species included within the arthropod orders in the reviewed studies of temperate versus tropical latitudes.

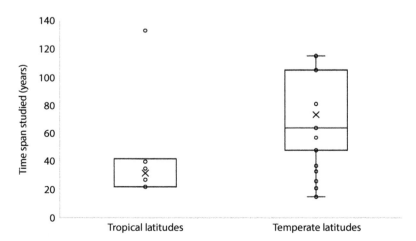

Figure 12.5 Boxplots with the time spanned between the historical and current sampling in tropical ($n = 5$) versus temperate ($n = 14$) publications.

12.11 Most insect species will respond to global warming by shifting to higher elevations

We did not find most species shifting to higher elevations (Figure 12.6). In the tropics, the largest proportion of species did move upslope, however it does not even represent half of the species studied (i.e., 43 percent). Remaining on site, or "stable distribution" was the dominant response in temperate insects (i.e., 41 percent). We found no evidence to support that tropical montane species are behind temperate species in responding to global warming (Figure 12.6). Differences between the proportion of tropical (33 percent) versus temperate species (41 percent) whose altitudinal distribution remained stable during the past forty-eight years were not significant (χ^2 = 1.321; df = 1; p = 0.25).

These results merit a more profound understanding of how range shifts are being studied. Chen et al. (2011) found that roughly 85 percent of the species shifted their distribution across elevation. In a similar literature search, Lenoir et al. (2010) found that 65 percent of the species studied had responded to global warming by shifting their altitudinal distribution upslope. In our study, only 43 percent of tropical and 37 percent of temperate species responded to climate change by shifting upslope. Nearly half of the species remained at the same altitudinal distribution as they were 48 years ago (Figure 12.6). This level of species permanence in their elevational niche is even more than the 20–40 percent of butterfly species in northern California with zero shift in elevation (Halsch et al. 2020).

In general, we can say that close to 50 percent of the species are shifting upslope and highlight the importance of studying species that either remain or move downslope.

One of the challenges with species that do not track their niche is to establish whether populations are thriving given the environmental stressors, if the species have not shifted upwards due to their lower dispersal ability or due to the interdependence with other species, or if their upslope shift is prevented by fragmentation. In other words, we need to determine if populations that fail to move upslope are doomed to extinction or if the microclimate is buffered where the species are, allowing species to thrive.

An upcoming stream of data analyses of insect abundance by species or genus link declines in insect biomass to climate change and habitat degradation (Hallman et al. 2017; Lister and García 2020). However, it is uncertain how these population declines payout along elevation gradients. It is important to find integrated ways of analyzing long-term data to correlate decreases in sampling frequency of caterpillars with net losses of intraspecific interactions (i.e., parasitoids). Salcido et al. (2020) studied in protected tropical forests, highlighting the extinction peril for higher trophic guilds. More complete studies that integrate complete elevation gradients to abundances are needed to understand what happens to the species that do not track their niche with climate change and identify potential areas within the mountain ranges that are acting as refugia for insect species.

12.12 Temperate species will respond to global warming by expanding their niche, while tropical species will respond by either range contractions or upslope shifts in elevation

We found statistical differences in how temperate versus tropical montane species respond to global warming (Figure 12.6). Temperate montane species tend to expand upslope

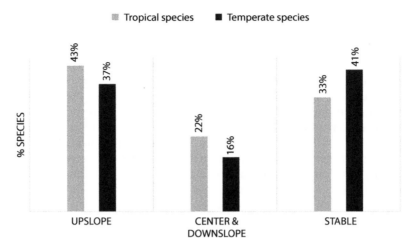

Figure 12.6 The proportion of temperate (*n* = 300) and tropical (*n* = 500) species that responded to global warming over the past 48 years as a net upslope response, a mixed central and downslope response, or remained at the historical elevational distribution.

through colonizations uphill by the populations at their leading edge that is not followed by the local extinction of the populations at the rear edge of their distribution, what McCain and Garfinkel (2021) call an incomplete response to climate change, but traditionally would be considered a range expansion. In contrast, tropical species responded to climate change by colonizing upslope on their leading edge and having local extinctions at the rear edge of their distribution. Thus, tropical species had a complete response to climate change. These responses support the idea that rear edges of distribution are becoming increasingly unsuitable for thriving populations, hence tropical species are colonizing upslope and going extinct at the rear edge of their distribution, making rainforest species attrition a real possibility (Colwell et al. 2008).

The higher proportion of upslope shifts in elevation for tropical species makes tropical lowland forests vulnerable to species attrition. In summarizing a plant and insect altitudinal distribution study, that included ants and geometrid moths in Costa Rica, the authors discussed the threats of extinction for keystone lowland species when it gets too hot to remain in the lowlands and either migrate upslope or go extinct (Colwell et al. 2008). No studies of tropical lowland attrition have been published in fourteen years, and the 1,330 publications cite this study in their references. Lister and García (2020) conclude that warmer temperatures and extreme weather over the last thirty years may be responsible for declines in 10–60 percent of the arthropod's biomass within two locations of intact rain forest (150–350 m) at Luquillo National Forest in Puerto Rico, getting close to a concept of lowland insect attrition. However, to answer the lowland biotic attrition question, one must study species' dispersal ability and colonizations upslope as responses to warmer temperatures and the regions and conditions that mitigate the effects of warmer temperatures. Rain forests' canopy may act as a buffer to the rise in temperatures. Montejo-Kovacevich et al. (2020) studied the buffering effect that forest canopy had on the temperature and humidity of tropical understory passion vine butterflies in the genus *Heliconius*. Just as forest canopy can mitigate the effects of climate warming, plant leaf modifications may be a refuge for multiple insect species. Romero et al. (2022) found a buffering effect of leaf rolls in preserving predator biomass. They concluded that as climate variability and aridity are likely to increase due to climate change, leaf rolls and leaf folds act as a biotic refugia in that they mitigate the dissection effects of climate. Most of these studies discuss mitigating factors in Lepidoptera herbivores. We need to know more about different taxa and guilds.

The differences in upslope response between temperate and tropical species to climate change could be explained by the narrower tolerance to higher temperatures tropical montane species have, compared to temperate taxa (Sheldon et al. 2011). As Janzen pointed out in the 1960s, temperatures in tropical mountains remain constant year-round in the tropics versus the typical seasonal fluctuations of temperatures at temperate latitudes. Constant temperatures select for narrower thermal tolerance and narrower thermal tolerances, in turn, exert stronger pressure on tropical species so that populations at rear edges of the distribution either migrate upslope or perish (Janzen 1967; Sheldon et al. 2011; Polato et al. 2015).

Unfortunately, because these studies do not typically include reliable measures of abundance and long-term data, we can't confirm if tropical insects at the rear edge of the mountain range are declining faster than those at the rear edges of temperate mountains. However, we do know that in general rainforest caterpillar genera and their parasitoids are declining. A time-series analysis of changes in frequency of capture of Lepidoptera in Costa Rica that spanned for twenty-two years found that the losses of diversity and density of tropical lowland Lepidoptera associated with a decline in parasitism (Salcido et al. 2020). For example, consider a mesocosm studied above ground that would include a plant species, one or two ant species, eighteen species of Lepidoptera herbivores that use this plant as a host and the twelve species of Tachinidae (Diptera), Bracconidae and Ichneumonide (Hymenoptera) parasitoids. When the local abundance of a rainforest caterpillar is lower or worse becomes locally extinct (whether or not populations upslope is thriving), there is more potential for local extinctions for the parasitoids that specialize on their hosts.

12.13 Some species will remain within their altitudinal distribution, with a higher proportion of tropical species remaining on site, compared to the shifts of temperate species

We found no evidence to support that tropical montane species are behind temperate species in responding to global warming (Figure 12.6). Differences between the proportion of tropical (33 percent) versus temperate species (41 percent) whose altitudinal distribution remained stable during the past 48 years were not significant ($\chi^2 = 1.321$; df = 1; $p = 0.25$). This means that globally, there is more than a third of species that are not shifting their distribution to track their environmental niche and that for this, there is no difference between temperate and tropical species. This warrants attention to studying the population dynamics of these taxa. One would need first to determine mobility of the species that shifted and remained to determine if that is a factor for their response to warmer temperatures

Because of the presence/absence nature of the data we collected and the interannual variability of insect abundances, we have no means of reviewing population trends along elevational gradients. Analyses of time-series samplings from different elevations of mountain ranges are needed to understand if the species that remain within their original range are thriving or declining. Understanding where and how populations that remain thrive under stressors of climate change is important to make accurate predictions of the future of these species that represent 33–41 percent of the insects on mountain ranges.

An upcoming stream of long-term data analysis linking decreases in local and landscape scale decreases of insect biomass (Hallman et al. 2017; Lister and García 2020). Integrated ways of analyzing long-term data have recently documented declines in diversity and density of Lepidoptera was associated with net losses of intraspecific interactions in protected tropical forests, highlighting the extinction peril for higher trophic guilds (i.e., parasitoids; Salcido et al. 2020).

12.14 Differences between the range dynamics along elevational gradients for tropical and temperate species

Except for Mount Kinabalu, Indonesia, we know how montane insects respond to global warming from temperate mountains. Here we show that a greater proportion of tropical montane insects respond to climate change through range shifts while in contrast, temperate montane insects respond by mainly extending their distribution upslope (Figures 12.7).

One would expect differences in the response to warmer temperatures in tropical versus temperate insects because tropical mountains lack temperature seasonality (Janzen 1967). Moreover, the climatic stability during the interglacial periods and gradual colonization after that has allowed specialized interactions to persist in tropical ecosystems. The smaller thermal tolerances of tropical species and complete upslope shifts in response

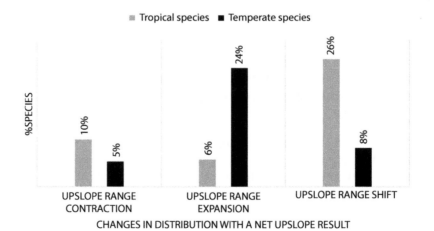

Figure 12.7 The proportion of temperate (*n* = 300) and tropical (*n* = 500) species that responded to global warming over the past 48 years as a net upslope response, a mixed central and downslope response, or remained at the historical distribution. No significant difference existed between the proportion of species that responded uphill, stable, or moved either toward the center of the mountain or downhill (Chi square = 5.79; df = 2, *p* = 0.06 phi coefficient = 0.09). * To contrast the contingency tables by type of shift in elevation, we excluded local extinctions and colonizations as those that did not represent shifts in elevation and did not have enough sample size in temperate and tropical regions.

to warmer temperatures lead to a crisis of disassembled communities that have been coevolving together for thousands of years (Sheldon et al. 2011). Consider disruptions to Species like the pepper plant *Piper kelleyi*, where even though the study area comprised 25 percent of its entire altitudinal distribution range, it hosted at least nineteen unique interactions (Tepe et al. 2014). Eleven *Piper* specialist geometrid moth species in the genus *Eois* are uniquely associated with it, and eight parasitoid species (Wilson et al. 2012). At least eleven specialized insect species were found to be uniquely associated with this plant species. Moreover, twenty to thirty species of Hymenoptera and Diptera parasitoids were found to be uniquely associated with the herbivores feeding from this plant (Tepe et al. 2014; Wilson et al. 2012). The high specialization is due to changes in plant chemistry along the elevational gradient (Glassmire et al. 2016)

12.15 Concluding remarks

Elevational gradients studied over time provide unique ways to understand species' responses to global warming and to identify potential regions that are mitigating the effects of global warming on communities. We show how poorly represented global mountains are in the studies of how insects have adapted their distribution over the past forty-eight years. Efforts must be made to resample expeditions conducted from the 1960s to the 1980s to improve our global coverage. It would be incorrect to assume that what we have learned from butterflies, bees and beetles can be extrapolated to all insect orders. The solution to this bias is to speak with global taxonomists, search digital data from entomological collections and compile a database of available information on elevational gradients. From there, one can assess whether enough data is systematically collected to recreate the sampling in current times.

As discussed earlier in this chapter, if it is a matter of priorities, understanding the distribution of Hymenopteran pollinators across elevation gradients in the tropics is vital as specialist interactions make it so that extinctions and elevational range shifts in pollinators may correlate with the demise of plant species. For example, orchid bees pollinate more than 700 neotropical orchids (Ramírez et al. 2010). We can start by working with entomologists in Latin America that have studied these groups systematically and support them and their collections so that their research groups conduct resamplings.

12.16 Perspectives

In the past two years, two research groups have warned us about the complexities of resampling studies to infer insect responses to climate change. These authors explain the need for data that (i) follow complete elevational gradients, (ii) are conducted at consistent phenological times year after year, and (iii) have sound inferences that have a good measure of abundance from the sampling method (McCain and Garfinkel 2021; Stuble et al. 2020). We argue that is not worth dwelling on what we don't have and instead we must focus on what can be done with the information that has been collected but due to lack of funding is sitting in collections without sorting or analyzing.

Literature reviews of elevation shift studies that include arthropods are becoming dominant in the literature (e.g., Chen et al. 2011; Halsch et al. 2021; Soroye et al. 2020). We need to develop a common framework where this information can reside, with descriptions of sampling methods, so that future captures and resampling efforts can build up on what we already know. Ideally, this database would include not only the shift in leading and rear edges of the species across a time span of thirty to fifty years but also information on the probability of capturing, a discussion on how seasons and weather compare to the original sampling and details of the sampling method used. Discussing all the necessary assumptions to study elevational range shifts provides new entomologists with a rich idea of the necessary data to publish. This way, the data obtained facilitates the modeling of how proportional abundances of species respond to climate change along elevational gradients (see Birkett et al. 2018; or methods used by McCain and Garfinkel 2021 to estimate shifts in abundance along elevation gradients).

We must also draw from historical information living at museum collections and herbariums and find genetic, modeling and time-series probabilistic solutions for infrequent sampling along elevation gradients. While wishing for better data and making calls for researchers to sample along complete elevation gradients is essential (e.g., McCain and Garfinkel 2021). The urgency of understanding the ecological and demographic processes of insects' response to climate change calls us to be creative and work with what we have.

There is no doubt that data availability and consensus on the framework under which to study elevational range shifts should spark diverse and essential work from regional to global scales. However, funding of the curation and publication of long-term data that include a complete elevational gradient with a good measure of abundance need to be made available to scientists and insect collections not only in Europe and North America but around the world. Moreover, information from tropical countries should be prioritized so that local scientists and parataxonomists are hired to curate and make this information available. We must also identify where historical sampling data are so that resampling trips can occur. This means talking to local entomologists to identify what happened to sampling efforts and work with the samples even when they are still sitting unsorted in jars of entomological cabinets. One of the great examples of the potential that lies in available data is the data compiled by the Global Biodiversity Information Facility (GBIF). In 1999, GBIF was founded to serve as a hub for curated data from museums and institutions worldwide. To date, more than 9,000 scientific articles used this curated information. Moreover, four scientific articles per day use the information from GBIF (<https://www.gbif.org/data-use>). However, having presence/absence data is the bare minimum for what we need to know, as data on abundances can provide much more about the population dynamics of the species that remain on site to understand if they are doomed and on their way to extinction or have found refugium where they are and represent stable populations insights in species' responses to changing climatic conditions.

It is frustrating to know that the information we need to answer important questions about the fate of biodiversity in the tropics confronting climate change sits unsorted in the jars of many museums across the globe. We could answer questions such as whether there are new refugia in tropical mountain ranges with such data. Alternatively, we could study how tropical highland species are adapting to the new conditions with information we collect now and contrast with complete collections that date more than fifty years that remained unsorted and stored in these institutions.

Once we explore these data sources we may also have to accept that we do not have all the needed data and find creative and probabilistic ways to move on. We can turn to paleontologists' methods to reconstruct entire ecosystems and events based on less data than we have (e.g., Schachat and Labandeira 2021).

The work ahead is an integrative approach. For example, we should conduct laboratory-based thermal tolerance studies on insects and use the results to model current thermal tolerance studies and distributions along elevation to predict range shifts. We can then verify the results from the model with current field collections. When modeled and actual distributions are not the same, there is usually an ecological reason such as interspecific competition, lack of preferred host plant or high predation. For example, in a study conducted along elevation gradients in Guatemala, we found that the altitudinal distribution of the leafminer fly *Liriomyza huidobrensis* (Diptera: Agromycidae) was limited by warm temperatures. However, the distribution of a close species, *L. sativae*, was limited by the presence of *L. huidobrensis* (Rodríguez-Castañeda et al. 2017). To truly understand how ecology shapes species' distribution and provides resilience to species, we need to conduct broad studies of plants, herbivores and parasitoids along elevation gradients. These studies would enable us to understand ecology's buffering effect on insects against climate change stressors. Moreover, to address community-wide effects, we need detailed trait studies of reciprocal transplants of microcosms along elevational gradients (Merill et al. 2008; Rodríguez-Castañeda et al. 2011), relationships between the upslope migration of host plants and pollinators (Inouye et al. 2020) and improve our understanding on the interactions between elevation, temperature plant defenses and insect communities (e.g., Fernández-Conradi et al. 2022; Pellisier et al. 2014) yet these studies are rare due to the logistical complexities to conduct manipulative experiments along elevation gradients. Further, we need to do more than just finding changes over the past thirty-plus years in the distribution of insects. The following steps should involve:

1. Determining population size from changes or capture frequencies at edge elevations.
2. Study the physiology of specimens collected over time and look for physiological modifications.
3. Identify conditions that buffer microclimates and allow species to resist changes imposed by global warming. This would allow us to invest conservation efforts in supporting insects' resilience to the inevitable changes imposed by global warming.

In conclusion, instead of looking at the errors in the published data we should focus on the potentially very valuable sources of data available that are currently not used or that need complementing in order to be useful for increasing our understanding of how climatic changes are and will affect insect communities worldwide. We call for funding and action to use these resources.

Key reflections

- Paleontologists argue that actual mass extinction has not occurred for insects. The unprecedented mass extinction of insects has considerable implications for the world's future. A large proportion of animals and plants and our economy depend on them.
- Even though climate change is not the most critical factor driving insect biomass losses is undoubtedly a significant stressor on insect biodiversity. The effects of climate change on insects are difficult to detect due to interactions with habitat degradation and the need for long-term monitoring of insect abundance across complete elevation gradients.
- Much has been reviewed and published on how insects respond to climate change along gradients. However, our review highlights understudied areas (e.g., African and Asian tropics). Furthermore, it discusses the undermined insect taxa (i.e., Most of what we know of shifts in elevation as a response to climate change comes from three out of twenty-six insect orders).
- We introduced a comprehensive framework analyzing the resampling of insects along altitudinal gradients. Added a temperate versus tropical comparison, finding that more tropical

species respond to climate change by shifting their altitudinal distribution entirely. In contrast, temperate species either remained on-site or expanded upslope.

- Unfortunately, we cannot conclude much about how insect communities will fare with climate change. Only a few studies address species interactions and how their dependencies change across elevation gradients. The conditions of these studies are unique. We, however, provide examples of studies that attempt to disentangle temperature changes along elevation gradients from ecological interactions.
- There is a need to join forces and compile a global public database with information about insects and elevation gradients.
- Finally, we stress the need to move forward by (i) determining population size from changes or capture frequencies at edge elevations, (ii) studying the physiological traits of speci-mens collected over time looking for adaptations, and (iii) identifying conditions that buffer microclimates and allow species to resist changes imposed by global warming.

Key further reading

- For the most up-to-date review of the threats to insect biodiversity, see the following:

  Sánchez-Bayo, F., and Wyckhuys, K.A. 2019. Worldwide decline of the entomofauna: A review of its drivers. *Biological Conservation 232* 8–27.

  Wagner, D.L., Grames, E.M., Forister, M.L., Berenbaum, M.R., and Stopak, D. 2021. Insects decline in the Anthropocene: Death by a thousand cuts. *Proceedings of the National Academy of Sciences 118* (2): p.e2023989118.

- For the available studies documenting insects' decline, see the following:

  Hallmann, C.A., Sorg, M., Jongejans, E., Siepel, H., Hofland, N., Schwan, H., Stenmans, W., Müller, A., Sumser, H., Hörren, T., and Goulson, D. 2017. More than 75 percent decline over 27 years in total flying insect biomass in protected areas. *PloS One 12* (10): e0185809.

  Lister, B.C., and Garcia, A. 2018. Climate-driven declines in arthropod abundance restruc-ture a rainforest food web. *Proceedings of the National Academy of Sciences 115* (44): E10397–E10406.

  Salcido, D.M., Forister, M.L., Garcia Lopez, H., and Dyer, L.A. 2020. Loss of dominant caterpillar genera in a protected tropical forest. *Scientific Reports 10* (1): 1–10.

- Classic studies contrasting historic and current records along altitudinal gradients include:

  Chen, I.C., Hill, J.K., Shiu, H.J., Holloway, J.D., Benedick, S., Chey, V.K., Barlow, H.S., and Thomas, C.D. 2011. Asymmetric boundary shifts of tropical montane Lepidoptera over four decades of climate warming. *Global Ecology and Biogeography 20* (1): 34–45.

  Merrill, R.M., Gutiérrez, D., Lewis, O.T., Gutiérrez, J., Díez, S.B., and Wilson, R.J. 2008. Combined effects of climate and biotic interactions on the elevational range of a phytophagous insect. *Journal of Animal Ecology 1*:145–55.

- For studies using and summarizing some of the best long-term data available across elevation gradients:

  Parkes, N.A., Shapiro, A.M., Dyer, L.A., and Forister, M.L. 2015 Global weather, and local butterflies: variable responses to a large-scale climate pattern along an elevational gradient. *Ecology 96* (11): 2891–2901.

  Forister, M.L., Fordyce, J.A., Nice, C.C., Thorne, J.H., Waetjen, D.P., and Shapiro, A.M. 2018. Impacts of millennium drought on butterfly faunal dynamics. *Climate Change Responses 5* (1): 1–9.

Halsch, C.A., Shapiro, A.M., Fordyce, J.A., Nice, C.C., Thorne, J.H., Waetjen, D.P., and Forister, M.L. 2021. Insects and recent climate change. *Proceedings of the National Academy of Sciences 118* (2): e2002543117.

- For studies of plant-ant interactions across elevation gradients, see:

Plowman, N.S., Hood, A.S., Moses, J., Redmond, C., Novotny, V., Klimes, P., and Fayle, T.M. 2017. Network reorganization and breakdown of an ant–plant protection mutualism with elevation. *Proceedings of the Royal Society B: Biological Sciences 284* (1850): 20162564.

Rodríguez-Castañeda, G., Forkner, R.E., Tepe, E.J., Gentry, G.L., and Dyer, L.A. 2011. Weighing defensive and nutritive roles of ant mutualists across a tropical altitudinal gradient. *Biotropica 43* (3): 343–50.

- For studies that explore the current role of spatial and ecological refugia:

Morelli, T.L., Daly, C., Dobrowski, S.Z., Dulen, D.M., Ebersole, J.L., Jackson, S.T., Lundquist, J.D., Millar, C.I., Maher, S.P., Monahan, W.B., and Nydick, K.R. 2016. Managing climate change refugia for climate adaptation. *PLoS One 11* (8): p.e0159909.

Montejo-Kovacevich, G., Martin, S.H., Meier, J.I., Bacquet, C.N., Monllor, M., Jiggins, C.D., and Nadeau, N.J. 2020. Microclimate buffering and thermal tolerance across elevations in a tropical butterfly. *Journal of Experimental Biology 223* (8): jeb220426.

References

Barnosky, A.D., Matzke, N., Tomiya, S., Wogan, G.O., Swartz, B., Quental, T.B., Marshall, C., McGuire, J.L., Lindsey, E.L., Maguire, K.C. and Mersey, B. 2011. Has the Earth's sixth mass extinction already arrived? Nature, 471 (7336): 51–7.

Bennett, K.D. 1990. Milankovitch cycles and their effects on species in ecological and evolutionary time. *Paleobiology 16* (1): 11–21.

Berger, A. 1988. Milankovitch theory and climate. *Reviews of Geophysics 26* (4): 624–57.

Binney, H.A., Willis, K.J., Edwards, M.E., Bhagwat, S.A., Anderson, P.M., Andreev, A.A., M. Blaauw et al. 2009. The distribution of late-Quaternary woody taxa in northern Eurasia: Evidence from a new macrofossil database. *Quaternary Science Reviews 28* (23–24): 2445–2464.

Brühl, C.A., and Zaller, J.G. 2019. Biodiversity decline as a consequence of an inappropriate environmental risk assessment of pesticides. *Frontiers in Environmental Science 7*: 177.

Busse, A., Bässler, C., Brandl, R., Friess, N., Hacker, H., Heidrich, L., . . . and Müller, J. 2022. Light and Malaise traps tell different stories about the spatial variations in arthropod biomass and method-specific insect abundance. *Insect Conservation and Diversity 15*(6): 655–65.

Cahill, A.E., Aiello-Lammens, M.E., Fisher-Reid, M.C., Hua, X., Karanewsky, C.J., Yeong Ryu, H., Sbeglia, G.C., et.al. 2013. How does climate change cause extinction? *Proceedings of the Royal Society B: Biological Sciences 280* (1750): 20121890.

Chen, I.C., Hill, J.K., Ohlemüller, R., Roy, D.B., and Thomas, C.D. 2011. Rapid range shifts of species associated with high levels of climate warming. *Science 333* (6045): 1024–1026.

Chen, I.C., Hill, J.K., Shiu, H.J., Holloway, J.D., Benedick, S., Chey, V.K., Barlow, H.S. et al. 2011. Asymmetric boundary shifts of tropical montane Lepidoptera over four decades of climate warming. *Global Ecology and Biogeography 20* (1): 34–45.

Colwell, R.K., Brehm, G., Cardelús, C.L., Gilman, A.C., and Longino, J.T. 2008. Global warming, elevational range shifts, and lowland biotic attrition in the wet tropics. *Science 322* (5899): 258–61.

Cornelissen, T. 2011. Climate change and its effects on terrestrial insects and herbivory patterns. *Neotropical Entomology 40*: 155–63.

Dahlhoff, E.P., Dahlhoff, V.C., Grainger, C.A., Zavala, N.A., Otepola-Bello, D., Sargent, B.A., Roberts, K.T. et al. 2019. Getting chased up the mountain: High elevation may limit performance and fitness characters in a montane insect. *Functional Ecology 33* (5): 809–18.

Diemer, M. 1996. Microclimatic convergence of high-elevation tropical páramo and temperate-zone alpine environments. *Journal of Vegetation Science* 7 (6): 821–30.

DiMichele, W.A. and Phillips, T.L. 1996. Climate change, plant extinctions and vegetational recovery during the Middle-Late Pennsylvanian transition: The case of tropical peat-forming environments in North America. *Geological Society, London, Special Publications* 102 (1): 201–21.

Dornelas, M., Gotelli, N.J., Shimadzu, H., Moyes, F., Magurran, A.E., and McGill, B.J. 2019. A balance of winners and losers in the Anthropocene. *Ecology Letters* 22 (5): 847–54.

Dyer, L.A. and Letourneau, D. 2003. Top-down and bottom-up diversity cascades in detrital vs. living food webs. *Ecology Letters* 6 (1): 60–68.

Dyer, L.A. and Letourneau, D.K. 2013. Can climate change trigger massive diversity cascades in terrestrial ecosystems?. *Diversity* 5 (3): 479–504.

Dyer, L.A., Singer, M. S., Lill, J.T., Stireman, J.O., Gentry, G.L., Marquis, R.J., Ricklefs, R.E. et al. 2007. Host specificity of Lepidoptera in tropical and temperate forests. *Nature* 448 (7154): 696–99.

Dynesius, M. and Jansson, R. 2000. Evolutionary consequences of changes in species' geographical distributions driven by Milankovitch climate oscillations. *Proceedings of the National Academy of Sciences* 97 (16): 9115–9120.

Easterling, D.R., Kunkel, K.E., Wehner, M.F. and Sun, L. 2016. Detection and attribution of climate extremes in the observed record. *Weather and Climate Extremes* 11: 17–27.

Fernandes, G.W. and Price, P.W. 1988. Biogeographical gradients in galling species richness: tests of hypotheses. *Oecologia* 76:161–7.

Fernandez-Conradi, P., Defossez, E., Delavallade, A., Descombes, P., Pitteloud, C., Glauser, G., Pellissier, et al. 2022. The effect of community-wide phytochemical diversity on herbivory reverses from low to high elevation. *Journal of Ecology* 110 (1): 46–56.

Flantua, S.G. and Hooghiemstra, H. 2018. Historical connectivity and mountain biodiversity. In *Mountains, Climate and Biodiversity*, edited by Hoorn, H., Perrigo, A., and Antonelli, A. New Jersey: Wiley-Blackwell, 595.

Forister, M.L., Fordyce, J.A., Nice, C.C., Thorne, J.H., Waetjen, D.P., and Shapiro, A.M. 2018. Impacts of a millennium drought on butterfly faunal dynamics. *Climate Change Responses* 5 (1): 1–9.

Franco, A.M., Hill, J.K., Kitschke, C., Collingham, Y.C., Roy, D.B., Fox, R., Huntley, B., et al. 2006. Impacts of climate warming and habitat loss on extinctions at species' low-latitude range boundaries. *Global Change Biology* 12 (8): 1545–1553.

Franzén, M. and Öckinger, E. 2012. Climate-driven changes in pollinator assemblages during the last 60 years in an Arctic mountain region in Northern Scandinavia. *Journal of Insect Conservation* 16: 227–38.

García-Robledo, C. and Baer, C.S. 2021. Demographic attritions, elevational refugia, and the resilience of insect populations to projected global warming. *The American Naturalist* 198 (1): 113–27.

Ghalambor, C.K., Huey, R.B., Martin, P.R., Tewksbury, J.J., and Wang, G. 2006. Are mountain passes higher in the tropics? Janzen's hypothesis revisited. *Integrative and Comparative Biology* 46 (1): 5–17.

Glassmire, A.E., Jeffrey, C.S., Forister, M.L., Parchman, T.L., Nice, C.C., Jahner, J.P., Wilson, J.F., et al. 2016. Intraspecific phytochemical variation shapes community and population structure for specialist caterpillars. *New Phytologist* 212 (1): 208–19.

González-Tokman, D., Córdoba-Aguilar, A., Dáttilo, W., Lira-Noriega, A.,Sánchez-Guillén, R.A., and Villalobos, F. 2020. Insect responses to heat: physiological mechanisms, evolution and ecological implications in a warming world. *Biological Reviews* 95 (3): 802–21.

Hallmann, C.A., Sorg, M., Jongejans, E., Siepel, H., Hofland, N., Schwan, H., Stenmans, W. et al. 2017. More than 75 percent decline over 27 years in total flying insect biomass in protected areas. *PloS One* 12 (10): e0185809.

Halsch, C.A., Shapiro, A.M., Fordyce, J.A., Nice, C.C., Thorne, J.H., Waetjen, D.P., and Forister, M.L. 2021. Insects and recent climate change. *Proceedings of the National Academy of Sciences* 118 (2): e2002543117.

Hoshiba, H., and Sasaki, M. 2008. Perspectives of multi-modal contribution of honeybee resources to our life. *Entomological Research* 38: S15–S21.

Inouye, D.W. 2020. Effects of climate change on alpine plants and their pollinators. *Annals of the New York Academy of Sciences* 1469 (1): 26–37.

Jansson, R., Rodríguez-Castañeda, G., and Harding, L.E. 2013. What can multiple phylogenies say about the latitudinal diversity gradient? A new look at the tropical conservatism, out of the tropics, and diversification rate hypotheses. *Evolution* 67 (6): 1741–1755.

Janzen, D.H. 1967. Why mountain passes are higher in the tropics. *The American Naturalist* 101 (919): 233–49.

Keppel, G., Van Niel, K.P., Wardell-Johnson, G.W., Yates, C.J., Byrne, M., Mucina, LA., and Schutt. 2012. Refugia: Identifying and understanding safe havens for biodiversity under climate change. *Global Ecology and Biogeography* 21 (4): 393–404.

Kerr, J.T., Pindar, A., Galpern, P., Packer, L., Potts, S.G., Roberts, S.M., and Rasmont, P. 2015. Climate change impacts on bumblebees converge across continents. *Science* 349 (6244): 177–80.

Konvicka, M., Maradova, M., Benes, J., Fric, Z., and Kepka, P. 2003. Uphill shifts in distribution of butterflies in the Czech Republic: Effects of changing climate detected on a regional scale. *Global Ecology and Biogeography* 12 (5): 403–10.

Labandeira, C.C. 2006. Silurian to Triassic plant and insect clades and their associations: New data, a review, and interpretations. *Arthropod Systematics and Phylogeny*.

Lenoir, J., Gégout, J.C., Guisan, A., Vittoz, P., Wohlgemuth, T., Zimmermann, N.E., and Svenning, J.C. 2010. Going against the flow: Potential mechanisms for unexpected downslope range shifts in a warming climate. *Ecography* 33 (2): 295–303.

Lenoir, J. and Svenning, J. C. 2015. Climate-related range shifts–a global multidimensional synthesis and new research directions. *Ecography* 38 (1): 15–28.

Lewinsohn, T.M., Novotny, V. and Basset, Y. 2005. Insects on plants: diversity of herbivore assemblages revisited. Annu. Rev. Ecol. Evol. Syst. 36: 597–620.

Lewis, S.L. and Maslin, M.A. 2015. A transparent framework for defining the Anthropocene Epoch. *The Anthropocene Review* 2 (2): 128–46.

Lippert, C., Feuerbacher, A., and Narjes, M. 2021. Revisiting the economic valuation of agricultural losses due to large-scale changes in pollinator populations. *Ecological Economics* 180: 106860.

Lister, B. C. and García, A. 2018. Climate-driven declines in arthropod abundance restructure a rainforest food web. *Proceedings of the National Academy of Sciences* 115 (44): E10397–E10406.

Lister, B.C. and García, A. 2018. Climate-driven declines in arthropod abundance restructure a rainforest food web. *Proceedings of the National Academy of Sciences* 115 (44): E10397–E10406.

McCain, C., Szewczyk, T., and Bracy Knight, K. 2016. Population variability complicates the accurate detection of climate change responses. *Global Change Biology* 22 (6): 2081–2093.

McCain, C.M. and Garfinkel, C.F. 2021. Climate change and elevational range shifts in insects. *Current Opinion in Insect Science* 47: 111–18.

Montejo-Kovacevich, G., Martin, S.H., Meier, J.I., Bacquet, C.N., Monllor, M., Jiggins, C. D., and Nadeau, N.J. 2020. Microclimate buffering and thermal tolerance across elevations in a tropical butterfly. *Journal of Experimental Biology* 223 (8): 1–12.

Morelli, T.L., Daly, C., Dobrowski, S.Z., Dulen, D.M., Ebersole, J.L., Jackson, S.T., Lundquist, J. D. 2016. Managing climate change refugia for climate adaptation. *PLoS One* 11 (8): 1–17.

Nice, C.C., Forister, M.L., Harrison, J.G., Gompert, Z., Fordyce, J.A., Thorne, J.H., Waetjen, D.P. et al. 2019. Extreme heterogeneity of population response to climatic variation and the limits of prediction. *Global Change Biology* 25 (6): 2127–2136.

Page, M.J., McKenzie, J.E., Bossuyt, P.M., Boutron, I., Hoffmann, T.C., Mulrow, C.D., Shamseer, L., et al. R. 2021. The PRISMA 2020 statement: An updated guideline for reporting systematic reviews. *Systematic Reviews* 10 (1): 1–11.

Pardikes, N.A., Shapiro, A.M., Dyer, L.A., and Forister, M.L. 2015. Global weather and local butterflies: Variable responses to a large-scale climate pattern along an elevational gradient. *Ecology* 96 (11): 2891–2901.

Parmesan, C. 2006. Ecological and evolutionary responses to recent climate change. Annual Review of Ecology, Evolution, and Systematics 37: 637–69.

Pellissier, L., Roger, A., Bilat, J., and Rasmann, S. 2014. High elevation Plantago lanceolata plants are less resistant to herbivory than their low elevation conspecifics: Is it just temperature? *Ecography* 37 (10): 950–59.

Plowman, N.S., Hood, A.S., Moses, J., Redmond, C., Novotny, V., Klimes, P., and Fayle, T. M. 2017. Network reorganization and breakdown of an ant–plant protection mutualism with elevation. *Proceedings of the Royal Society B: Biological Sciences* 284 (1850): 1–10.

Polato, N.R., Gill, B.A., Shah, A.A., Gray, M.M., Casner, K.L., Barthelet, A., Messer, P.W., Simmons, M.P., Guayasamin, J.M., Encalada, A.C. and Kondratieff, B.C. 2018. Narrow thermal tolerance and low dispersal drive higher speciation in tropical mountains. *Proceedings of the National Academy of Sciences* 115 (49): 12471–12476.

Ramírez, S., Dressle, R.L., and Ospina, M. 2002. Euglossine bees (Hymenoptera: Apidae) from the Neotropical Region: A species checklist with notes on their biology. *Biota Colombiana* 3: 7–118.

Ramirez, S.R., Roubik, D.W., Skov, C., and Pierce, N.E. (2010). Phylogeny, diversification patterns and historical biogeography of euglossine orchid bees (Hymenoptera: Apidae). *Biological Journal of the Linnean Society* 100 (3): 552–72.

Rodríguez-Castañeda, G., Forkner, R.E., Tepe, E.J., Gentry, G.L., and Dyer, L.A. 2011. Weighing defensive and nutritive roles of ant mutualists across a tropical altitudinal gradient. *Biotropica* 43 (3): 343–50.

Rodríguez-Castañeda, G., Hof, A.R., and Jansson, R. 2017. How bird clades diversify in response to climatic and geographic factors. *Ecology Letters* 20 (9): 1129–1139.

Rodríguez-Castañeda, G., MacVean, C., Cardona, C., and Hof, A.R. 2017. What limits the distribution of *Liriomyza huidobrensis* and its congener *Liriomyza sativae* in their native niche: When temperature and competition affect species' distribution range in Guatemala. *Journal of Insect Science* 17(4): 88; 1–13

Romero, G.Q., Gonçalves-Souza, T., Roslin, T., Marquis, R.J., Marino, N.A., Novotny, V., Cornelissen, T., et al. 2022. Climate variability and aridity modulate the role of leaf shelters for arthropods: A global experiment. *Global Change Biology* 28 (11), 3694–3710.

Rumpf, S.B., Hülber K., Klonner G., Moser D., Schütz M., Wessely J., Willner W., Zimmermann N.E., and Dullinger S. 2018. Range dynamics of mountain plants decrease with elevation. *Proceedings of the National Academy of Sciences.* 115 (8): 1848–1853.

Salcido, D.M., Forister, M.L., Garcia Lopez, H., and Dyer, L.A. 2020. Loss of dominant caterpillar genera in a protected tropical forest. *Scientific Reports* 10 (1): 1–10.

Salcido, D.M., Sudta, C., and Dyer, L.A. 2022. "Plant-caterpillar-parasitoid natural history studies over decades and across large geographic gradients provide insight into specialization, interaction diversity, and global change." In Marquis, R.J., and Koptur, S. *Caterpillars in the Middle: Tritrophic Interactions in a Changing World.* Springer International Publishing, 583–606.

Sánchez-Bayo, F. and Wyckhuys, K.A. 2019. Worldwide decline of the entomofauna: A review of its drivers. *Biological Conservation*, 232, 8–27.

Schachat, S.R. and Labandeira, C.C. 2021. Are insects heading toward their first mass extinction? Distinguishing turnover from crises in their fossil record. *Annals of the Entomological Society of America* 114 (2): 99–118.

Shapiro, A.M. 2022. Monitoring butterfly populations across Central California for 47 years. [Website] https://ucdavis.github.io/butterfly.ucdavis.edu/. (accessed September 2022).

Sheldon, K.S., Yang, S., and Tewksbury, J.J. 2011. Climate change and community disassembly: Impacts of warming on tropical and temperate montane community structure. *Ecology Letters* 14 (12): 1191–1200.

Soroye, P., Newbold, T., and Kerr, J. 2020. Climate change contributes to widespread declines among bumble bees across continents. *Science* 367 (6478): 685–88.

Stebbins, G.L. 1974. *Flowering Plants: Evolution above the Species Level.* Cambridge: Harvard University Press.

Stefanescu, C., Carnicer, J., and Penuelas, J. 2011. Determinants of species richness in generalist and specialist Mediterranean butterflies: The negative synergistic forces of climate and habitat change. *Ecography* 34 (3): 353–63.

Stireman III, J.O., Dyer, L.A., Janzen, D.H., Singer, M.S., Lill, J.T., Marquis, R.J., Ricklefs, R.E. et. al. 2005. Climatic unpredictability and parasitism of caterpillars: Implications of global warming. *Proceedings of the National Academy of Sciences* 102 (48): 17384–17387.

Stuble, K.L., Bewick, S., Fisher, M., Forister, M.L., Harrison, S.P., Shapiro, A.M., Latimer, A. M. and Fox, L.R., 2021. The promise and the perils of resurveying to understand global change impacts. *Ecological Monographs* 91 (2): e01435.

Stork, N.E. 2018. How many species of insects and other terrestrial arthropods are there on Earth? *Annual Review of Entomology* 63: 31–45.

Tepe, E.J., Rodríguez-Castañeda, G., Glassmire, A.E., and Dyer, L.A. 2014. Piper kelleyi, a hotspot of ecological interactions and a new species from Ecuador and Peru. *PhytoKeys* (34): 19–32.

Thomas, C.D., Cameron, A., Green, R.E., Bakkenes, M., Beaumont, L.J., Collingham, Y.C., Erasmus, B.F., et al. 2004. Extinction risk from climate change. *Nature* 427 (6970): 145–48.

Thomas, C.D. 2010. Climate, climate change and range boundaries. *Diversity and Distributions* 16 (3): 488–95.

Vizentin-Bugoni, J., Maruyama, P.K., de Souza, C.S., Ollerton, J., Rech, A.R., and Sazima, M. 2018. Plant-pollinator networks in the tropics: A review. In. *Ecological Networks in the Tropics: An Integrative Overview of Species Interactions from Some of the Most Species-Rich Habitats on Earth*, edited by Dattilo, W. and Rico-Gray, V. Springer International Publishing, 73–91.

Wagner, D.L., Grames, E.M., Forister, M.L., Berenbaum, M.R., and Stopak, D. 2021. Insect decline in the Anthropocene: Death by a thousand cuts. *Proceedings of the National Academy of Sciences* 118 (2), article e2023989118: 1–10.

Williams, A.P., Cook, B.I., and Smerdon, J.E. 2022. Rapid intensification of the emerging southwestern North American megadrought in 2020–2021. *Nature Climate Change* 12 (3): 232–34.

Wilson, J.S., Forister, M.L., Dyer, L.A., O'Connor, J.M., Burls, K., Feldman, C.R., Jaramillo, M.A., et al. 2012. Host conservatism, host shifts and diversification across three trophic levels in two Neotropical forests. *Journal of Evolutionary Biology* 25 (3): 532–46

Data for the literature review taken from the following.

Bässler, C., Hothorn, T., Brandl, R., and Müller, J. 2013. Insects overshoot the expected upslope shift caused by climate warming. *PLoS One* 8 (6):1–6 article e65842.

Beza-Beza, C., Schuster, J., and Cano, E. 2024. Submitted. Replicate studies separated by 40 years reveal changes in the altitudinal stratification of montane passalid beetle species (Passalidae) in Mesoamerica. Submitted for publication to Frontiers of Biogeography, March 2023.

Biella, P., Bogliani, G., Cornalba, M., Manino, A., Neumayer, J., Porporato, M., Rasmont, P. et al. 2017. Distribution patterns of the cold adapted bumblebee Bombus alpinus in the Alps and hints of an uphill shift (Insecta: Hymenoptera: Apidae). *Journal of Insect Conservation* 21: 357–66.

Birkett, A.J., Blackburn, G.A., and Menéndez, R. 2018. Linking species thermal tolerance to elevational range shifts in upland dung beetles. *Ecography* 41 (9): 1510–1519.

Chen, I.C., Hill, J.K., Shiu, H.J., Holloway, J.D., Benedick, S., Chey, V.K., Barlow, H.S. et al. 2011. Asymmetric boundary shifts of tropical montane Lepidoptera over four decades of climate warming. *Global Ecology and Biogeography* 20 (1): 34–45.

Chinn, W.G.H. and Chinn, T.J.H. 2020. Tracking the snow line: Responses to climate change by New Zealand alpine invertebrates. *Arctic, Antarctic, and Alpine Research* 52 (1): 361–89.

Fourcade, Y., Åström, S., and Öckinger, E. 2019. Climate and land-cover change alter bumblebee species richness and community composition in subalpine areas. *Biodiversity and Conservation* 28: 639–53.

Franzén, M., and Öckinger, E. 2012. Climate-driven changes in pollinator assemblages during the last 60 years in an Arctic mountain region in Northern Scandinavia. *Journal of Insect Conservation* 16: 227–38.

Konvicka, M., Maradova, M., Benes, J., Fric, Z., and Kepka, P. 2003. Uphill shifts in distribution of butterflies in the Czech Republic: Effects of changing climate detected on a regional scale. *Global Ecology and Biogeography* 12 (5): 403–10.

Marshall, L., Perdijk, F., Dendoncker, N., Kunin, W., Roberts, S., and Biesmeijer, J.C. 2020. Bumblebees moving up: Shifts in elevation ranges in the Pyrenees over 115 years. *Proceedings of the Royal Society B: Biological Sciences* 287 (1938): 1-10. Article 20202201.

Menéndez, R., González-Megías, A., Jay-Robert, P., and Marquéz-Ferrando, R. 2014. Climate change and elevational range shifts: Evidence from dung beetles in two European mountain ranges. *Global Ecology and Biogeography* 23 (6): 646–57.

Merrill, R.M., Gutiérrez, D., Lewis, O.T., Gutiérrez, J., Díez, S.B., and Wilson, R.J. 2008. Combined effects of climate and biotic interactions on the elevational range of a phytophagous insect. *Journal of Animal Ecology* 77: 145–55.

Molina-Martínez, A., León-Cortés, J.L., Regan, H.M., Lewis, O. T., Navarrete, D., Caballero, U., and Luis-Martínez, A. 2016. Changes in butterfly distributions and species assemblages on a Neotropical mountain range in response to global warming and anthropogenic land use. *Diversity and Distributions* 22 (11): 1085–1098.

Moret, P., Aráuz, M.D.L.Á., Gobbi, M., and Barragán, Á. 2016. Climate warming effects in the tropical Andes: First evidence for upslope shifts of Carabidae (Coleoptera) in Ecuador. *Insect Conservation and Diversity* 9 (4): 342–50.

Nufio, C.R., McGuire, C.R., Bowers, M.D., and Guralnick, R.P. 2010. Grasshopper community response to climatic change: Variation along an elevational gradient. *PLoS One* 5 (9): e12977.

Ploquin, E.F., Herrera, J.M., and Obeso, J.R. (2013). Bumblebee community homogenization after uphill shifts in montane areas of northern Spain. *Oecologia* 173: 1649–1660.

Provan, J. and Bennett, K.D. 2008. Phylogeographic insights into cryptic glacial refugia. *Trends in Ecology and Evolution* 23 (10): 564–71.

Shapiro, A. accessed 2022. <https://sites.google.com/view/westernbutterflies/data>.

CHAPTER 13

Impacts of climate change on insect pollinators and consequences for their ecological function

Laura A. Burkle and Shalene Jha

13.1 Introduction

One of the most urgent concerns in global biodiversity conservation is the potential impact of changing climate conditions on insect pollinator biology. Pollinators are one of the most ecologically and economically valuable ecosystem service providers, facilitating the reproduction of more than 80 percent of all flowering plant species (Ollerton et al. 2011) and contributing to more than $300 billion annually in global crop production (IPBES 2016). However, recent long-term studies have indicated that these insects may be particularly sensitive to altered climate conditions, especially with respect to direct impacts on their species distributions, abundance, phenology, morphology, physiology and behavior (Figure 13.1). Altered climate conditions can include warmer temperatures, increased drought, increased extreme weather events, including fire (IPCC 2007), and these conditions can also have indirect effects on pollinators via their interactions with plants, which provide pollinators with critical food resources (reviewed in Gérard et al. 2020; Hegland et al. 2009). These indirect effects include interaction mismatches, of which there are three main types: trait mismatches, which primarily result from physiological and behavioral changes; temporal mismatches, which primarily result from phenological changes; and spatial mismatches, which result from distribution and abundance changes (Figure 13.1; Glossary). While we acknowledge the important role of disturbance regimes and additional species interactions that are also likely impacted by climate change (e.g., fire, pathogen dynamics), in this chapter we focus on the effects of warmer temperatures, drought, enriched carbon dioxide and extreme weather events on insect pollinators, their plant food resources and potential shifts in their species interactions (gray sections of Figure 13.1). When possible, we provide details about the

Laura A. Burkle and Shalene Jha, *Impacts of climate change on insect pollinators and consequences for their ecological function*. In: *Effects of Climate Change on Insects*. Edited by: Daniel González-Tokman and Wesley Dáttilo, Oxford University Press. © Oxford University Press (2024). DOI: 10.1093/oso/9780192864161.003.0013

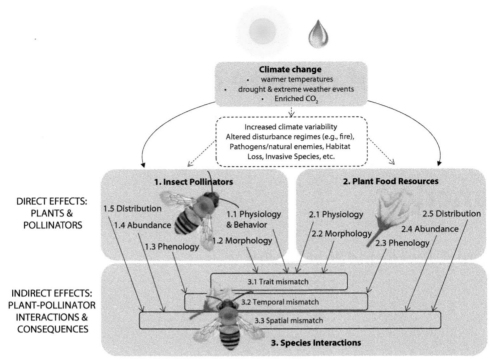

Figure 13.1 Visual representation of the impacts of altered climate conditions directly on species distributions, abundance, phenology, morphology, physiology and behavior of insect pollinators and their plant food resources (1 and 2), as well as indirectly on plant–pollinator interactions, potentially leading to trait mismatches, temporal mismatches and spatial mismatches (3). In this chapter, we focus on the effects of warmer temperature, drought, enriched carbon dioxide and extreme weather events on insect pollinators, their plant food resources, as well as potential shifts in their species interactions (gray sections).

Drawings by Ann Sanderson.

specific aspects of climate change being studied (e.g., increasing temperatures in winter); however, many past studies do not isolate the specific aspects of climate change underlying observed responses and thus we highlight this as an important future research direction.

13.2 Direct effects of climate change on insect pollinators

13.2.1 *Insect pollinator physiology and behavior*

The direct effects of climate change on insect pollinators occur via shifts in abiotic conditions including warmer temperatures, drought, extreme weather events and increased variability in components of climate, among others. In particular, insect pollinator physiology has been well studied under warmer temperatures, while the influence of other aspects of climate change on their physiology is poorly understood. When studied experimentally, the effects of increased temperatures on insect pollinator physiology generally show increases (up to a point, often followed by damage or mortality) in developmental rates, metabolism and activity levels, though populations and species follow temperature response curves depending on life histories and adaptations to local conditions (Higgins et al. 2014; Oyen et al. 2016). For instance, feeding and foraging rates of insect pollinators increase with temperature (Willmer and Stone 2004), but large-bodied pollinators (e.g., bumblebees) may overheat during flight at high temperatures (Dudley 2000). Nevertheless, experimentally manipulated air temperature up to 36 °C did not influence intraspecific patterns of short flights among workers of the bumblebee *Bombus impatiens* with different body sizes (Couvillon et al. 2010).

Presumably due to increased metabolic rates under warmer over-wintering or rearing temperatures, insect pollinators often exhibit increased developmental rates (Davidowitz et al. 2004; O'Neill et al. 2011; Radmacher and Strohm 2010; Stephen 1965), increased weight loss (Fründ et al. 2013; Radmacher and Strohm 2010; Slominski and Burkle 2019; Stuhldreher et al. 2014) and depletion of energetic reserves (e.g., fat content) (CaraDonna et al. 2018; Williams et al. 2012), which can be associated with increased over-wintering mortality (Bosch and Kemp 2003, 2004; CaraDonna et al. 2018; Sgolastra et al. 2010; Slominski and Burkle 2019; Stuhldreher et al. 2014) and reduced larval (Davidowitz et al. 2004) or adult body size (Radmacher and Strohm 2010). Adult pollinator mortality may increase in warmer temperatures, either as knock-on effects of warmer overwintering temperatures (solitary bee, *Osmia bicornis*; Radmacher and Strohm 2010) or as a result of exposure to warmer temperatures as adults (fig wasps; Jevanandam et al. 2013). In contrast, however, other species of insect pollinators, including solitary bees (*Megachile* spp) and butterflies (*Papilio* spp) have shown the opposite pattern, including reduced loss of energetic reserves (O'Neill et al. 2011), no effect on energetic reserves (Williams et al. 2012), or reduced weight loss (Slominski and Burkle 2019) in response to warmer rearing or over-wintering temperatures. Overall, there is variation in physiological and behavioral responses depending on species, with pollinators exhibiting elevated or reduced physiological rates under warmer temperatures.

13.2.2 *Insect morphology and other traits*

Multiple aspects of climate change, including warmer temperatures, can further influence morphology and other traits of insect pollinators. As seen in an observational study of the small carpenter bee, *Ceratina calcarata*, adult body size may decline with increasing temperatures over time (118 years; Kelemen and Rehan 2021). In addition to reductions in body size at warmer temperatures across taxa, experimentally high rearing temperatures reduced bumblebee wing size and altered wing shape (Gerard et al. 2018). Besides temperature, shifts in body color and body size have been observed in a tropical bee species in response to precipitation gradients, with larger body sizes at drier sites (Suni and Dela Cruz 2021). When scaling up the impacts of climate change on pollinator traits, there is limited but intriguing evidence that altered climate conditions may favor species with morphological traits related to resource acquisition. Specifically, in parts of the Rocky Mountains, short-tongued bumblebee species (generalists) appear to be replacing long-tongued species (specialists), possibly related to warming-induced reductions in flower availability (Miller-Struttmann et al. 2015). In general, shifts in or adaptations of insect morphology and other traits in response to climate change are poorly explored pathways through which climate impacts may be occurring.

13.2.3 *Insect pollinator phenology*

Related to effects of warmer temperatures on pollinator physiology are changes in pollinator phenology, the seasonal timing of life-cycle events. Many observational studies have documented earlier emergence of insect pollinators in the spring associated with warmer temperatures. This pattern has been found for butterflies (Forister and Shapiro 2003; Gordo and Sanz 2005, 2006; reviewed in Parmesan 2007), honey bees (Gordo and Sanz 2006; Sparks et al. 2010) and native wild bees (Bartomeus et al. 2011; Ovaskainen et al. 2013), though it is not known whether shifts in emergence phenology also hold for mating phenology or other life-cycle events. Some butterflies exhibit extended flight periods with warmer temperatures (Roy and Sparks 2000), which is associated with multi-voltinism (Altermatt 2010; Roy and Sparks 2000). Other observational studies and analyses based on biodiversity databases and museum collections have documented reductions in the flight periods of insect pollinators over many decades (Burkle et al. 2013; Duchenne et al. 2020), leading to interaction mismatches with their host plants and decreases in pollinator function (Burkle et al. 2013), though the degree to which these observational patterns are due specifically to warmer temperatures is often difficult to disentangle. Experimentally warmed winters or shortened winter duration mainly corroborate observational studies, resulting in earlier solitary bee emergence (Bosch and Kemp 2003, 2004; Slominski and Burkle 2019). However, solitary bees in systems with less strongly pronounced seasons (e.g., subtropical, tropical) may exhibit delayed spring emergence in response to warming (CaraDonna et al. 2018), possibly due to inadequate winter chilling (Sgolastra et al. 2010). Overall, phenological responses are driven by temperature in temperate areas, while inadequate winter chilling or precipitation may be a more important factor at lower latitudes, likely associated with regional drivers of seasonality (meta-analysis of Cohen et al. 2018).

13.2.4 *Insect pollinator abundance*

While declines in insect abundance have made headlines over the last several years, the effects of climate change on pollinator abundance are poorly understood relative to other factors implicated in pollinator declines (e.g., land-use change, pathogens). This knowledge gap is partially due to the challenge of quantifying changes in abundance over long time scales. However, recent analysis of long-term data for sixty-six bumblebee species has shown that the frequency of unusually hot days increases local extinction rates, reduces colonization and site occupancy and reduces species richness within a region, independent of land-use change (Soroye et al. 2020). Further, climate is one of the strongest forces driving wild bee abundance and richness, even when compared with land-use factors, and climate change conditions are associated with reduced abundances of wild bees (Kammerer et al. 2021). Such changes in abundance can also be reflected in species losses and gains. For example, Hofmann et al. (2018) documented an increase in warm-loving and a decrease in cool-adapted bee species in the Munich Botanical Garden (Germany), a protected area without pesticides, between 1997 and 2017. For some bee species that experience increased abundances under climate change, we may expect associated changes in life-history traits, such as the development of sociality (e.g., *Halictus rubicundus*) (Schürch et al. 2016).

13.2.5 *Insect pollinator distributions*

Ultimately, changes in insect pollinator abundance due to climate warming are reflected in species' distributions. There is ample evidence of species shifts to higher elevations and latitudes, particularly for butterflies (Forister et al. 2010; Hill et al. 2002; Konvicka et al. 2003; Mair et al. 2012; Parmesan 1996; Parmesan et al. 1999; Pöyry et al. 2009; White and Kerr 2006) and bumblebees (Pyke et al. 2016). However, there are also examples of insect pollinators failing to track climate warming. Northern range shifts of eighty-one butterfly species in Canada did not keep pace with climate change (Bedford et al. 2012), and North American and European bumblebees failed to shift distributions northward, while also exhibiting range contraction through retreating southern distributions (Kerr et al. 2015). These restrictions to insect pollinator species ranges are consistent with insights gained through predictive modeling in which half of the twelve studied South African bee species are predicted to experience range contractions in response to climate (Kuhlmann et al. 2012). In a few cases, however, range expansions are expected under climate change, most often for already widespread bee species (Dew et al. 2019).

13.3 Direct effects of climate change on plant food resources of insect pollinators

One critical way by which climate change can influence insect pollinators is via altered interactions with their food resources. Many insect pollinators, including bees, obtain the majority of their adult and brood food provision from pollen and nectar resources (Michener 2000). Some insect pollinators, like butterflies, consume leaves and other plant biomass during their larval stage. While other review papers have documented the broader impacts of climate change on plant species distributions, abundance, and phenology (e.g., Parmesan and Yohe 2003), in this section we focus on potential climate impacts on the aspects of plant biology—specifically physiology, morphology, phenology, abundance and distributions—that directly connect to pollinator food resources (Figure 13.1). This focus is critical, given that indirect effects of climate change on food resources may be even stronger than the direct effects of climate change on pollinators, as seen in some bumblebees (Ogilvie et al. 2017). In this section, we briefly summarize the impacts of climate change on floral traits related to pollen, nectar and pollinator attraction, as well as host-plant biomass for butterflies. Because some of these topics have been thoroughly reviewed elsewhere, we direct you to citations and Key Further Reading for more detailed information and helpful meta-analyses of this extensive body of knowledge.

13.3.1 *Plant physiology and traits related to plant development*

There is substantial evidence that climate change can alter plant physiology and development in ways that have major implications for pollinators. In a literature review of heat-stress impacts on pollen–pistil interactions, Snider and Ooserhuis (2011) describe numerous studies that indicate heat stress can inhibit male and female gametophyte development, effectively reducing pollen availability for a broad suite of insect pollinators. The study also found that heat stress can reduce pollen tube germination, limit pollen tube growth and generally alter anther and pistil tissue, potentially disrupting reproductive success after insect pollinator visitation. Butterflies in particular may be influenced by aspects of climate change through physiological effects on the quality of their larval host plants. Generally, increased temperatures and enriched carbon dioxide (CO_2) reduce host plant quality, while drought appears to increase it, with commensurate effects on larval butterfly performance. The mechanisms underlying these shifts in host plant quality are varied. Increased temperatures and CO_2 may reduce leaf nitrogen content for some species (Jeong et al. 2018) and drought stress can reduce plant investment in defenses (e.g., Gutbrodt et al. 2011). For instance, increased temperatures reduced host plant quality for a tropical butterfly species, especially for males, resulting in decreased body mass and increased development time (Kuczyk et al. 2021).

13.3.2 *Plant morphology and traits related to forage attractiveness and quality*

There is also substantial literature documenting climate impacts on floral traits specifically related to floral resource attractiveness and quality. For example, in an experimental study, Hoover et al. (2012) found that climate warming, elevated nitrogen and elevated CO_2 interacted to impact nectar quality, leading to changes in nectar concentration. In another experimental study, Wilson Rankin et al. (2020) found that reductions in water availability led to decreased nectar quality and pollen protein quality, which in turn led to decreased survival and productivity in bumblebees. Indeed, past reviews have indicated that nectar concentration and composition is mediated by temperature and can exhibit variation across a species' blooming period (reviewed in Pacini et al. 2003). This is particularly important to consider given that pollinator taxa differ in their nectar preferences and often forage in response to nectar concentrations and composition (reviewed in Vandelook et al. 2019). In addition to nectar, in a recent meta-analysis of water-deficit impacts on floral traits, Kuppler and Kotowska (2021) found that reduced water availability led to consistent decreases in floral size, among other traits (described below). Further, global change factors such as increases in CO_2, ozone and temperature have also been shown to impact the volatile organic compounds emitted by plants (reviewed in Yuan et al. 2009). For instance, by examining the floral emissions of various Mediterranean plants in response to temperature gradients, Farré-Armengol et al. (2014) found that floral emissions of volatile organic compounds (VOCs) generally increased with temperature to a point and then decreased, but that VOC compositional responses varied by species and by compound within each species, indicating a wide diversity of trait responses to climate, even for species from the same ecoregion. Drought and CO_2 have also been shown to influence floral VOC emissions and composition, with effects on pollinator attraction (Burkle and Runyon 2016; Campbell et al. 2019; Glenny et al. 2018).

13.3.3 *Plant phenology*

In their classic synthesis of hundreds of taxa across the globe, including plants, and spanning periods of 16 to 132 years, Parmesan and Yohe (2003) documented a strong shift to earlier dates for both first flowering and tree budburst. In a similar synthesis across 385 plant species in Britain, Fitter and Fitter (2002) found that the average first day of flowering in the last decade was advanced by 4.5 days, relative to the four previous decades. They also found that insect-pollinated species and annual species were more likely to flower earlier than wind-pollinated species and congeneric perennial species. In another long-term study, spanning fifty-eight years for eleven plant species and twenty years for an additional thirteen species within the British Isles, Sparks and colleagues (2000) found that the timing of all flowering events was significantly related to temperature, and suggested that future increases in temperature across the islands will continue to drive substantial change in flowering events. Using observations initiated by Henry David Thoreau in 1852 and spanning seasons until 2006, Miller-Rushing and Primack (2008) found not only that plants in the northeastern United States were flowering seven days earlier but also that spring-flowering species were more responsive to temperatures in the preceding months than summer-blooming species. Compared to studies focused on first flowering events, Menzel et al. (2006) focused on multiple temporal phases within a year, across a fifty-year span in Germany and found greater variation specifically for spring events for plants, with greater variability in warmer years, confirming the existing emphasis and continued need to focus on climate change risks posed to early phenological phases.

13.3.4 *Plant food resource abundance/quantity*

Though we reviewed climate impacts on pollinator forage-related plant traits above, in this section, we show that beyond qualitative trait changes, climate change can alter the abundance of key floral features. For example, long-term observational studies indicate that increasing spring temperatures may result in reduced flower abundance (Inouye 2008) though responses to temperature and rainfall can be species-specific and variable (Tyler 2001). Further, species that are not responsive to temperature are more likely to have declined in abundance over 150 years than species with flowering phenologies that correlate with temperature, and there is a phylogenetic signal in these patterns (i.e., closely related species have more similar patterns in temperature-induced abundance changes than distantly related species) (Willis et al. 2008). In the experiment conducted by Hoover and colleagues (2012) described earlier, climate warming, elevated nitrogen and elevated CO_2 not only impacted nectar concentration but also altered overall nectar volume. In another experimental study, Descamps et al. (2021) found that borage plants grown at warmer temperatures (26 degrees C) had fewer flowers, smaller flowers and lower nectar volume than those grown at cooler temperatures (21 degrees C), reducing bumblebee visitation and affecting foraging behavior. Similarly, in the meta-analysis conducted by Kuppler and Kotowska (2021) also described earlier, the authors found that reduced water availability not only decreased floral size but also reduced the number of flowers and nectar volume. Both these studies indicate that absolute differences in food resource quantity often exist alongside changes in aspects of floral quality. In addition to warming effects, CO_2 appears to have species-specific positive, neutral, or negative effects on the amount of nectar produced and the number of flowers produced, as evidenced by distinct responses for a suite of butterfly visited plants commonly found in calcareous grasslands (e.g., Rusterholz and Erhardt 1998). While the impacts of CO_2 on floral resources appear more mixed than temperature impacts, it represents an important pathway for further study, and one by which insect pollinators may be increasingly affected.

13.3.5 *Plant food resource distributions*

There are fewer studies that specifically examine floral resource response to climate change across spatial scales, and these tend to focus on elevational gradients. In one study comparing high elevation bumblebee and plant communities across a thirty-three-year period, Pyke et al. (2016) found that with increasing air temperature, community flowering shifted to earlier time periods across the elevation gradient. However, most plant species did not exhibit the expected altitude shift upward (ca. 317 meters) required to match average temperature, suggesting a possible time lag in the spatial response of plants to climate changes. In addition to this long-term study, some studies have also used a combination of transplant experiments across a latitudinal gradient and open-top chambers (e.g., De Frenne et al. 2011) to show that forest understory plant species have earlier emergence times and onset of flowering with increasing temperatures, irrespective of latitude origin. In another experimental transplant study, Wang et al. (2014) moved soil blocks containing the same plant communities to four different altitudes, where higher altitudes had lower temperatures than lower elevations. The authors found that transplanting the blocks to lower elevations advanced the first flowering date, while transplanting to higher elevation had the opposite effect (Wang et al. 2014), indicating the potential for rapid acclimation after spatial shifts or dispersal events. While a number of studies have documented elevational shifts for multiple plant species, including those known to depend on native pollinators (e.g., *Ceanothus*; Kelly and Goulden 2008), additional investigation of how climate change may dually impact spatial distribution as well as flowering dynamics is warranted.

13.4 Indirect effects of climate change: trait, temporal, and spatial mismatches

Given that climate conditions can influence the physiology, morphology, phenology, abundance and distribution of both insect pollinators and their food plants, there is increasing concern that changing climate may disrupt ecological interactions (reviewed in Hegland et al. 2009; Gérard et al. 2020), resulting in potential trait, temporal and spatial mismatches in the face of climate change (Figure 13.1). Mismatches take place when the original partners experience a reduction in shared habitat, usually through a reduction in shared time or space (*sensu* Hegland et al. 2009; Stenseth and Mysterud 2002). However, more recent syntheses have suggested additional trait-based axes where partners may also exhibit mismatches (Gérard et al. 2020). For plants and pollinators, climate-induced shifts in these mutualistic interactions may be driven by the fact that pollinators may respond more to temperature in their development and activity (e.g., Hegland et al. 2009) while forbs may use a broader suite of cues, such as photoperiod, temperature and snowmelt, among others (Inouye et al. 2003; Menzel and Fabian 1999). The examples in the sections that follow are from temperate and arctic systems in the Northern Hemisphere. Climate-induced interaction mismatches in tropical systems continue to be understudied (as noted first in Hegland et al. 2009), despite the fact that trait matching is expected to be more important (Vizentin-Bugoni et al. 2018) and that species may exhibit lower dispersal ability, higher specialization, and lower tolerance to environmental change within tropical systems (reviewed in Sheldon 2019), and thus represent an urgent focal area for future investigation.

13.4.1 *Trait mismatches*

One example of a trait mismatch is that documented by Miller-Struttmann et al. (2015) in a paper described above, which leveraged a combination of insect collections and herbarium specimens spanning forty years, revealing that two alpine bumblebee species have evolved shorter tongues indicative of more generalist foraging, even though their associated specialized flowers did not become shallower over the same time period. The authors posit that warmer summers have led to declining floral resources, which favors generalist foraging and drives a mismatch between the long-tubed plants and the now shorter-tongued species (Miller-Struttmann et al. 2015). Given the substantial morphological data, as well as long time-series data sets, required to document trait mismatches over time, the Miller-Struttmann study remains one of the few studies to empirically document trait mismatches.

In contrast, a number of studies have leveraged experimental setups to investigate trait *alterations* (trait changes that may or may not indicate mismatches) in response to climate conditions. In the study evaluating borage response to climate change described earlier (Descamps et al. 2021), not only did plants grown at higher temps have fewer flowers, smaller flowers and lower nectar volume, they also experienced one-quarter the bumblebee visitation as those plants grown at a cooler temperature. In a study of the temperate-zone butterfly *Pieris napi*, the authors document a direct impact of higher temperature on larvae, seen in accelerated development and lower adult body mass; they also document a weaker but significant indirect impact of temperature mediated through poor host-plant quality, in which larvae feeding on host plants (*Sinapis alba*, Brassicaceae) raised at higher temperatures exhibited longer larval development time and reduced adult body mass (Bauerfeind and Fischer 2013). In another experiment in the same study system, a host plant water-reduction experiment did not affect *P. napi* larval performance, while increased host plant temperatures decreased pupal mass and fat content, among other traits (Kuczyk et al. 2021). As described earlier in the chapter, in an experimental study where Hoover et al. (2012) increased temperature, nitrogen and CO_2, they found changes in nectar quality, and they additionally showed that bumblebees exhibit preference for nectar with higher sugar concentration, indicating that plant traits modified by climate conditions could indeed impact bee visitation, nectar consumption and even survival.

13.4.2 *Temporal mismatches*

A series of theoretical papers and examinations of historic data have suggested that changes in the interactions between insect pollinators and their food plants may lead to temporal mismatches between species in altered climate conditions. In a paper that simulated global warming impacts on plant-pollinator networks, Memmott and colleagues (2007) suggested that 17–80 percent of pollinators would experience reduced food supply, especially those that were highly specialized (i.e., used a small suite of plants). In a study leveraging historic data along with contemporary sampling, ca. 50 percent of bee species were lost from the system over 120 years, where diet specialists were lost more than generalists (Burkle et al. 2013). Further, the remaining bee species participated in fewer interactions with plant species, with over 20 percent of these unique interactions lost due to temporal mismatches from shifts in phenology of both bee and plant species, 17 percent lost due to spatial mismatches (see next section for climate-induced spatial mismatches) or a combination of both (Burkle et al. 2013). Similarly, taking advantage of historic data, a review of northeastern US bee museum collections over the past 140 years indicated that 56 percent of species exhibited a change in relative abundance over time, where declining species (species in the genus *Bombus*) were often those with small phenological breadth (Bartomeus et al. 2013). However, in a complementary study focused on ten bee species and their potential associated plants, the authors found that both bees and bee-pollinated plants had similarly advanced by ca. ten days in the last 130 years, suggesting that some species may be able to keep pace with shifts in their forage plants (Bartomeus et al. 2011). Likewise, using forty years of data from a Russian boreal forest, Ovaskainen et al. (2013) demonstrated substantial synchrony in the first appearance of bumblebees and early flowering plants. In contrast, another long-term study mentioned earlier by Pyke et al. (2016), conducted over thirty-three years and across high elevations, the authors not only found earlier flowering phenology, but they also documented unchanged bumblebee phenology, resulting in decreased synchrony over time. Differences across studies suggest that biogeographic regions may differ dramatically in temporal synchrony patterns and that elevation may play a key role in driving differences in local plant and pollinator phenology.

In addition to leveraging long-term (> ten year) historic data to understand temporal mismatch, experimental studies and those investigating a subset of focal plant and pollinator species have revealed greater variation in response to climate changes. In an experimental study delaying snowmelt date in an Arctic community, plant flowering was also delayed, reducing average overlap between floral availability and insect activity; however, despite this reduction in overlap, the composition of flowers and insect visitors was not significantly altered (Gillespie and Cooper 2021). Research on focal plant species from the field has also found that warmer temperatures have resulted in advanced flowering for *Pulsatilla* plants, earlier than their solitary bee (*Osmia* spp.) pollinators (Kehrberger and Holzschuh 2019). Similarly, *Corydalis ambigua* plants often bloom before their bumblebee pollinators, and repeated surveying suggests that this mismatch has been increasing over time (Kudo and Ida 2013). Likewise, cabbage white butterflies have been found foraging before their focal nectar plants bloom, largely in response to increasing temperatures (Gordo and Sanz 2005). In contrast, experimental studies manipulating the onset of flowering for a number of plant species, including some that have historically exhibited earlier shifts, revealed few mismatches between those plants and wild pollinators at the onset of flowering across groups (Rafferty and Ives 2011); these studies and others indicate that plasticity in wild pollinator response may reduce the potential for climate-induced temporal mismatch (Burkle et al. 2013).

13.4.3 *Spatial mismatches*

In contrast to the substantial observational and experimental literature describing temporal mismatches between plants and pollinators, there is limited but growing concern that spatial mismatch may also increase with altered climate conditions. For example, using surveys of bumblebees across an elevational gradient, conducted from 1974 and 2007, Pyke et al. (2016) found that most bumblebee species moved to higher altitudes while only a few plant species moved to higher latitudes, indicating reduced contemporary spatial synchrony. A study modeling multiple global climate change scenarios for a monophagous butterfly and its host plant indicates that all three distinct future scenarios, varying from the most modest to the maximum climate change estimated by the IPCC, result in spatial mismatch, where the extent of mismatch depends on the host plant's ability to fill its projected niche space (Schweiger et al. 2008). In another modeling paper focused on 150 high-mountain plant species, Dullinger and colleagues (2012) predict an average range reduction of 44–50 percent by the end of the twenty-first century, with species endemic to the Alps experiencing the highest range losses. While these modeling efforts are compelling, there is less field-based study and evidence of climate-induced spatial mismatches between plants and pollinators relative to temporal mismatches, suggesting greater need for research in this area (reviewed in Gérard et al. 2020; Hegland et al. 2009).

13.5 Functional consequences of climate change for pollination and next steps for research

While trait, temporal and spatial mismatches between insect pollinators and non-cultivated flowering plants are occurring, the ecological consequences of these mismatches are poorly understood. For plant–pollinator interactions that typically result in reciprocal benefits to the partners, we expect trait mismatches to disrupt the quality of ecosystem service delivery, temporal mismatches to alter the timing of delivery, and spatial mismatches to influence where ecosystem services are delivered. Such alterations to offspring fitness could generate co-evolutionary feedbacks between plants and pollinators, which is another important concept to consider in the light of climate change, but which is outside the scope of this review (see e.g., Bronstein et al. 2004). Although many studies have shown that climate change can result in trait *alterations*, there is a distinct need for researchers to investigate links between potential trait changes for individual species with actual trait and interaction *mismatches*. Further, in almost all cases, studies of "ecosystem services" are restricted to the pollination services that insect pollinators provide to plants. While equally important, almost nothing is known about how trait, temporal, and spatial mismatches affect the provisioning services that plants provide to insect pollinators. We hope that this review will serve as a call to action to investigate the ecological consequences of the effects of climate change on plants, pollinators and their functional, temporal and spatial synchrony.

13.5.1 *Functional consequences for pollination*

Even if many plants and pollinators exhibit synchronous response to climate change across both temporal and spatial scales, strong mismatch in a small subset of these species could still critically impact population vital rates (reviewed in Forrest 2015). Despite this fact, it has been challenging to link observed or predicted mismatches to demographic outcomes (Miller-Rushing et al. 2010). While there are few experimental examples of long-term fitness effects of temporal asynchrony, there is increasing field evidence of reduced plant reproductive success. For example, pollen limitation may be more likely for high elevation plant species, like glacier lily, due to temporal mismatch with bumblebee queens (Thomson 2010). Another bumblebee-dependent plant species, *Corydalis ambigua*, exhibited lowered seed set when there was phenological mismatch between flowering and bumblebee queens (Kudo and Ida 2013; Kudo et al. 2004). Cascading impacts of phenological mismatch on plant fitness also depend on other consumer-resource interactions, where experimentally advanced flowering may reduce pollinator visitation but also reduce herbivorous pollen beetles, in some cases resulting in increased seed set (Parsche et al. 2011). Despite increasing evidence that temporal mismatch may have negative fitness effects on plants and pollinators, more empirical and experimental study is required.

Investigating functional consequences of climate change for pollination involves defining "function" and making decisions about plant responses to measure. While fine-scale demographic approaches to documenting vital rates are time-consuming and labor intensive, they are critical to capturing fitness consequences. From the plant perspective, investigation of fitness consequences often invokes studies of pollen limitation of plant reproduction (Ashman et al. 2004; Knight et al. 2005). Because quantifying pollen limitation (or other demographic parameters) is often prohibitively labor intensive at the community level, tracking heterospecific and conspecific stigma pollen receipt (Arceo-Gómez and Ashman 2011; Ashman and Arceo-Gómez 2013; Ashman et al. 2020) for numerous plant species in a community may provide early insights to the functional patterns of fitness consequences.

13.5.2 *Functional consequences for insect pollinators and next steps for research*

Similar procedures for quantifying the functional consequences of climate change are rarely, if ever, applied to insect pollinators. In one experimental manipulation to understand the implications of temporal mismatches, temporal asynchrony between solitary bee emergence and flower availability resulted in reduced bee visitation rates to flowers and reduced offspring weight (Slominski and Burkle 2021). Traditionally, when population-level studies have been conducted on native species, they have primarily focused on bumblebees and threatened butterflies (e.g., monarchs), in part because some species in these groups are known targets of conservation concern. However, this singular focus on bumblebees perpetuates in-depth understanding of only a handful of species, while the status of thousands of other species of native insect pollinators is virtually unknown because we lack basic information on their life history traits and understand almost nothing about their population trends at one or more locations across their range. Investing the time to do so could uncover other taxa especially vulnerable to the effects of climate change.

An additional avenue for future research is to explore the potential effects of climate change on pollinators that may occur through changes in nesting habitat and resources. This chapter included a section on the impacts of climate change on plant food resources for pollinators. Plant responses to climate change represent a well-developed field, with several syntheses capturing the main responses of floral traits and other food resources of pollinators. What is missing beyond this work on plant food resources are studies investigating the impacts of climate change on the other key resource that pollinators require: nesting habitat. It seems likely that the quality and quantity of nesting habitat and resources (e.g., availability of mud, bare ground and stems and other plant materials) could also be influenced by climate change and related environmental shifts like wildfire and invasive species. Further, life-history traits are likely to help predict the direct effects of climate change on nesting success for insect pollinators. For example, solitary bee species that overwinter as adults (i.e., *Osmia* spp.) may be more vulnerable to negative effects of climate change compared to those that overwinter as prepupae (i.e., *Megachile* spp.) (Slominski and Burkle 2019). In temperate areas with distinct seasons and winter snowpack, ground-nesting bees might be more protected from over-wintering temperature changes than cavity or stem nesting bees, which may be less insulated from temperature extremes. Lastly, one aspect of climate change that has received little attention with respect to its effects on insect pollinators is that of climate variability (reviewed for terrestrial plants in Reyer et al. 2013), which addresses longer-term climate fluctuation (beyond averages) and could trigger changes in insect pollinator emergence and life-cycle events. Overall, while great gains have been made in understanding the effects of climate change on pollinators, their food resources and plant–pollinator interactions, there is much still unknown about the functional consequences for pollination and insect pollinators.

Key reflections

- Various aspects of climate change are directly influencing insect pollinator physiology, morphology and traits, phenology, abundance and distributions.
- Climate change is influencing insect pollinators through their food resources, including the quality, quantity and phenology of floral nectar, pollen and other plant parts.
- Impacts of climate change on insect pollinators and their food resources can further indirectly influence their species interactions, creating trait, spatial and temporal mismatches.
- The functional consequences of these interaction mismatches for pollination are poorly understood, and even less is known about population-level consequences for insect pollinators.

Key further reading

- Borghi, M., de Souza, L.P., Yoshida, T., and Fernie, A.R. 2019. Flowers and climate change: A metabolic perspective. *New Phytologist 224* (4): 1425–1441.
- Descamps, C., Quinet, M., and Jacquemart, A-L. 2021. The effects of drought on plant-pollinator interactions. *Environmental and Experimental Botany 182*: 104297.
- Hegland, S.J., Nielsen, A., Lazaro, A., Bjerknes, A-L., and Totland, O. 2009. How does climate warming affect plant-pollinator interactions?. *Ecology Letters 12*: 184–95.
- Gérard, M., Vanderplanck, M., Wood, T., and Michez, D. 2020. Global warming and plant–pollinator mismatches. *Emerging Topics in Life Sciences 4* (1): 77–86.
- Abarca, M. and Spahn, R. 2021. Direct and indirect effects of altered temperature regimes and phenological mismatches on insect populations. *Current Opinion in Insect Science 47*: 67–74.
- Forrest, J.R.K. 2017. Insect pollinators and climate change. In *Global Climate Change and Terrestrial Invertebrates*, edited by S.N. Johnson and T.H. Jones, John Wiley and Sons, Ltd., 69–91.
- Rafferty, N.E. 2017. Effects of global change on insect pollinators: Multiple drivers lead to novel communities. *Current Opinion in Insect Science 23*: 22–27

References

Altermatt, F. 2010. Climatic warming increases voltinism in European butterflies and moths. *Proceedings of the Royal Society B: Biological Sciences* 277 (1685): 1281–1287.

Arceo-Gómez, G. and Ashman, T.L. 2011. Heterospecific pollen deposition: Does diversity alter the consequences? *New Phytologist* 192 (3): 738–46.

Ashman, T.L. and Arceo-Gómez, G. 2013. Toward a predictive understanding of the fitness costs of heterospecific pollen receipt and its importance in co-flowering communities. *American Journal of Botany* 100 (6): 1061–1070.

Ashman, T.L., Arceo-Gómez, G., Bennett, J.M., and Knight, T.M. 2020. Is heterospecific pollen receipt the missing link in understanding pollen limitation of plant reproduction? *American Journal of Botany* 107 (6): 845–47.

Ashman, T.L., Knight, T.M., Steets, J.A., Amarasekare, P., Burd, M., Campbell, D.R., Dudash, M.R. et al. 2004. Pollen limitation of plant reproduction: Ecological and evolutionary causes and consequences. *Ecology* 85 (9): 2408–2421.

Bartomeus, I., Ascher, J.S., Gibbs, J., Danforth, B.N., Wagner, D.L., Hedtke, S.M., and Winfree, R. 2013. Historical changes in northeastern US bee pollinators related to shared ecological traits. *Proceedings of the National Academy of Sciences* 110 (12): 4656–4660.

Bartomeus, I., Ascher, J.S., Wagner, D., Danforth, B.N., Colla, S., Kornbluth, S., and Winfree, R. 2011. Climate-associated phenological advances in bee pollinators and bee-pollinated plants. *Proceedings of the National Academy of Sciences* 108 (51): 20645–20649.

Bauerfeind, S.S. and Fischer, K. 2013. Increased temperature reduces herbivore host-plant quality. *Global Change Biology* 19 (11): 3272–3282.

Bedford, F.E., Whittaker, R.J., and Kerr, J.T. 2012. Systemic range shift lags among a pollinator species assemblage following rapid climate change. *Botany* 90 (7): 587–97.

Bosch, J. and Kemp, W.P. 2003. Effect of wintering duration and temperature on survival and emergence time in males of the orchard pollinator Osmia lignaria (Hymenoptera: Megachilidae). *Environmental Entomology* 32 (4): 711–16.

Bosch, J. and Kemp, W.P. 2004. Effect of pre-wintering and wintering temperature regimes on weight loss, survival, and emergence time in the mason bee Osmia cornuta (Hymenoptera: Megachilidae). *Apidologie* 35 (5): 469–79. doi:10.1051/apido:2004035.

Bronstein, J.L., Dieckmann, U. and Ferriere, R. 2004. "Coevolutionary dynamics and the conservation of mutualisms." In *Evolutionary Conservation Biology*, edited by R. Ferriere, U. Dieckmann, and D. Couvet. Cambridge: Cambridge University Press, 305–26.

Burkle, L.A., Marlin, J.C., and Knight, T.M. 2013. Plant–pollinator interactions over 120 years: Loss of species, co-occurrence, and function. *Science* 339 (6127): 1611–1615. doi:10.1126/science.1232728.

Burkle, L.A. and Runyon, J.B. 2016. Drought and leaf herbivory influence floral volatiles and pollinator attraction. *Global Change Biology* 22 (4): 1644–1654.

Campbell, D.R., Sosenski, P., and Raguso, R.A. 2019. Phenotypic plasticity of floral volatiles in response to increasing drought stress. *Annals of Botany* 123 (4): 601–10.

CaraDonna, P.J., Cunningham, J.L., and Iler, A.M. 2018. Experimental warming in the field delays phenology and reduces body mass, fat content and survival: Implications for the persistence of a pollinator under climate change. *Functional Ecology* 32 (10): 2345–2356.

Cohen, J.M., Lajeunesse, M.J., and Rohr, J.R. 2018. A global synthesis of animal phenological responses to climate change. *Nature Climate Change* 8 (3): 224–28. doi:10.1038/s41558-018-0067-3.

Couvillon, M.J., Fitzpatrick, G., and Dornhaus, A. 2010. Ambient air temperature does not predict whether small or large workers forage in bumble bees (Bombus impatiens). *Psyche* Article ID 536430.

Davidowitz, G., D'Amico, L.J., and Nijhout, H.F. 2004. The effects of environmental variation on a mechanism that controls insect body size. *Evolutionary Ecology Research* 6 (1): 49–62.

De Frenne, P., Brunet, J., Shevtsova, A., Kolb, A., Graae, B.J., Chabrerie, O., Cousins, C. et al. 2011. Temperature effects on forest herbs assessed by warming and transplant experiments along a latitudinal gradient. *Global Change Biology* 17 (10): 3240–3253.

Descamps, C., Jambrek, A., Quinet, M., and Jacquemart, A.L. 2021. Warm temperatures reduce flower attractiveness and bumblebee foraging. *Insects* 12 (6): 493.

Dew, R.M., Silva, D.P., and Rehan, S.M. 2019. Range expansion of an already widespread bee under climate change. *Global Ecology and Conservation* 17: e00584.

Duchenne, F., Thébault, E., Michez, D., Elias, M., Drake, M., Persson, M., Rousseau-Piot, J.S. et al. 2020. Phenological shifts alter the seasonal structure of pollinator assemblages in Europe. *Nature Ecology and Evolution* 4 (1): 115–21.

Dudley, R. 2000. *The Biomechanics of Insect Flight: Form, Function, Evolution*. Princeton: Princeton University Press.

Dullinger, S., Gattringer, A., Thuiller, W., Moser, D., Zimmermann, N.E., Guisan, A., Willner, W., et al. 2012. Extinction debt of high-mountain plants under twenty-first-century climate change. *Nature Climate Change* 2 (8): 619–22. doi:10.1038/nclimate1514.

Farré-Armengol, G., Filella, I., Llusià, J., Niinemets, Ü., and Peñuelas, J. 2014. Changes in floral bouquets from compound-specific responses to increasing temperatures. *Global Change Biology* 20 (12): 3660–3669.

Fitter, A.H. and Fitter, R.S.R. 2002. Rapid changes in flowering time in British plants. *Science* 296 (5573): 1689–1691.

Forister, M.L., McCall, A.C., Sanders, N.J., Fordyce, J.A., Thorne, J.H., O'Brien, J., Waetjen, D.P., et al. 2010. Compounded effects of climate change and habitat alteration shift patterns of butterfly diversity. *Proceedings of the National Academy of Sciences* 107 (5): 2088–2092.

Forister, M.L. and Shapiro, A.M. 2003. Climatic trends and advancing spring flight of butterflies in lowland California. *Global Change Biology* 9 (7): 1130–1135.

Forrest, J.R. 2015. Plant–pollinator interactions and phenological change: What can we learn about climate impacts from experiments and observations? *Oikos* 124 (1): 4–13.

Fründ, J., Zieger, S.L., and Tscharntke, T. 2013. Response diversity of wild bees to overwintering temperatures. *Oecologia* 173, 1639–1648.

Gerard, M., Michez, D., Debat, V., Fullgrabe, L., Meeus, I., Piot, N., Sculfort, O., et al. 2018. Stressful conditions reveal decrease in size, modification of shape but relatively stable asymmetry in bumblebee wings. *Scientific Reports* 8 (1): 15169.

Gérard, M., Vanderplanck, M., Wood, T., and Michez, D. 2020. Global warming and plant–pollinator mismatches. *Emerging Topics in Life Sciences* 4 (1): 77–86.

Gillespie, M.A. and Cooper, E.J. 2021. The seasonal dynamics of a High Arctic plant–visitor network: Temporal observations and responses to delayed snow melt. *Arctic Science* 8 (3): 786–803.

Glenny, W.R., Runyon, J.B., and Burkle, L.A. 2018. Drought and increased CO 2 alter floral visual and olfactory traits with context-dependent effects on pollinator visitation. *New Phytologist* 220 (3): 785–98.

Gordo, O. and Sanz, J.J. 2005. Phenology and climate change: A long-term study in a Mediterranean locality. *Oecologia* 146: 484–95.

Gordo, O. and Sanz, J.J. 2006. Temporal trends in phenology of the honey bee Apis mellifera (L.) and the small white Pieris rapae (L.) in the Iberian Peninsula (1952–2004). *Ecological Entomology* 31 (3): 261–68.

Gutbrodt, B., Mody, K., and Dorn, S. 2011. Drought changes plant chemistry and causes contrasting responses in lepidopteran herbivores. *Oikos* 120 (11): 1732–1740.

Hegland, S.J., Nielsen, A., Lázaro, A., Bjerknes, A.L., and Totland, Ø. 2009. How does climate warming affect plant-pollinator interactions? *Ecology Letters* 12 (2): 184–95.

Higgins, J.K., MacLean, H.J., Buckley, L.B., and Kingsolver, J.G. 2014. Geographic differences and microevolutionary changes in thermal sensitivity of butterfly larvae in response to climate. *Functional Ecology* 28 (4): 982–89.

Hill, J.K., Thomas, C.D., Fox, R., Telfer, M.G., Willis, S.G., Asher, J., and Huntley, B. 2002. Responses of butterflies to twentieth century climate warming: Implications for future ranges. *Proceedings of the Royal Society of London. Series B: Biological Sciences* 269 (1505): 2163–2171.

Hofmann, M.M., Fleischmann, A., and Renner, S.S. 2018. Changes in the bee fauna of a German botanical garden between 1997 and 2017, attributable to climate warming, not other parameters. *Oecologia* 187 (3): 701–06.

Hoover, S.E., Ladley, J.J., Shchepetkina, A.A., Tisch, M., Gieseg, S.P., and Tylianakis, J.M. 2012. Warming, CO2, and nitrogen deposition interactively affect a plant-pollinator mutualism. *Ecology Letters* 15 (3): 227–34.

Inouye, D.W. 2008. Effects of climate change on phenology, frost damage, and floral abundance of montane wildflowers. *Ecology* 89 (2): 353–62.

Inouye, D.W., Saavedra, F., and Lee-Yang, W. 2003. Environmental influences on the phenology and abundance of flowering by Androsace septentrionalis (Primulaceae). *American Journal of Botany* 90 (6): 905–10.

IPBES 2016. *The Assessment Report on Pollinators, Pollination and Food Production: Summary for Policymakers.* Potts, S.G., Imperatriz-Fonseca, V., Ngo, H.T., Biesmeijer, J.C., Breeze, T.D., Dicks, L.V., Garibaldi, et al. Secretariat of the Intergovernmental Science-Policy Platform on Biodiversity and Ecosystem Services.

IPCC 2007. *Climate Change 2007—The Physical Science Basis: Working Group I Contribution to the Fourth Assessment Report of the IPCC* (Vol. 4). Solomon, S., Qin, D., Manning, M., Averyt, K., and Marquis, M. (eds). Cambridge: Cambridge University Press.

Jeong, H.M., Kim, H.R., Hong, S., and You, Y.H. 2018. Effects of elevated CO2 concentration and increased temperature on leaf quality responses of rare and endangered plants. *Journal of Ecology and Environment* 42: 1–11.

Jevanandam, N., Goh, A.G., and Corlett, R.T. 2013. Climate warming and the potential extinction of fig wasps, the obligate pollinators of figs. *Biology Letters* 9 (3): 20130041.

Kammerer, M., Goslee, S.C., Douglas, M.R., Tooker, J.F., and Grozinger, C.M. 2021. Wild bees as winners and losers: Relative impacts of landscape composition, quality, and climate. *Global Change Biology* 27 (6): 1250–1265.

Kehrberger, S. and Holzschuh, A. 2019. Warmer temperatures advance flowering in a spring plant more strongly than emergence of two solitary spring bee species. *PLoS One* 14 (6): e0218824.

Kelemen, E.P. and Rehan, S.M. 2021. Opposing pressures of climate and land-use change on a native bee. *Global Change Biology* 27 (5): 1017–1026.

Kelly, A.E. and Goulden, M.L. 2008. Rapid shifts in plant distribution with recent climate change. *Proceedings of the National Academy of Sciences* 105 (33): 11823–11826. Doi:10.1073/pnas.0802891105.

Kerr, J.T., Pindar, A., Galpern, P., Packer, L., Potts, S.G., Roberts, S.M., Rasmont, P., et al. 2015. Climate change impacts on bumblebees converge across continents. *Science* 349 (6244): 177–80. doi:10.1126/science.aaa7031.

Knight, T.M., Steets, J.A., Vamosi, J.C., Mazer, S.J., Burd, M., Campbell, D.R., Dudash, M.R., et al. 2005. Pollen limitation of plant reproduction: Pattern and process. *Annual Review of Ecology, Evolution, and Systematics* 36: 467–97.

Konvicka, M., Maradova, M., Benes, J., Fric, Z., and Kepka, P. 2003. Uphill shifts in distribution of butterflies in the Czech Republic: Effects of changing climate detected on a regional scale. *Global Ecology and Biogeography* 12 (5): 403–10.

Kuczyk, J., Müller, C., and Fischer, K. 2021. Plant-mediated indirect effects of climate change on an insect herbivore. *Basic and Applied Ecology* 53: 100–13.

Kuczyk, J., Raharivololoniaina, A., and Fischer, K. 2021. High temperature and soil moisture reduce host-plant quality for an insect herbivore. *Ecological Entomology* 46 (4): 889–97.

Kudo, G. and Ida, T.Y. 2013. Early onset of spring increases the phenological mismatch between plants and pollinators. *Ecology* 94 (10): 2311–2320.

Kudo, G., Nishikawa, Y., Kasagi, T., and Kosuge, S. 2004. Does seed production of spring ephemerals decrease when spring comes early? *Ecological Research* 19: 255–59.

Kuhlmann, M., Guo, D., Veldtman, R., and Donaldson, J. 2012. Consequences of warming up a hotspot: Species range shifts within a centre of bee diversity. *Diversity and Distributions* 18 (9): 885–97.

Kuppler, J. and Kotowska, M.M. 2021. A meta-analysis of responses in floral traits and flower–visitor interactions to water deficit. *Global Change Biology* 27 (13): 3095–3108.

Mair, L., Thomas, C.D., Anderson, B.J., Fox, R., Botham, M., and Hill, J.K. 2012. Temporal variation in responses of species to four decades of climate warming. *Global Change Biology* 18 (8): 2439–2447.

Memmott, J., Craze, P.G., Waser, N.M., and Price, M.V. 2007. Global warming and the disruption of plant–pollinator interactions. *Ecology Letters* 10 (8): 710–17.

Menzel, A., and Fabian, P. 1999. Growing season extended in Europe. *Nature* 397 (6721): 659–59.

Menzel, A., Sparks, T.H., Estrella, N., and Roy, D.B. 2006. Altered geographic and temporal variability in phenology in response to climate change. *Global Ecology and Biogeography* 15 (5): 498–504.

Michener, C.D. 2000. *The Bees of the World*. Baltimore: The Johns Hopkins University Press.

Miller-Rushing, A.J., Høye, T.T., Inouye, D.W., and Post, E. 2010. The effects of phenological mismatches on demography. *Philosophical Transactions of the Royal Society B: Biological Sciences* 365 (1555): 3177–3186.

Miller-Rushing, A.J. and Primack, R.B. 2008. Global warming and flowering times in Thoreau's Concord: A community perspective. *Ecology* 89 (2): 332–41.

Miller-Struttmann, N.E., Geib, J.C., Franklin, J.D., Kevan, P.G., Holdo, R.M., Ebert-May, D., Lynn, A.M., et al. 2015. Functional mismatch in a bumble bee pollination mutualism under climate change. *Science* 349 (6255): 1541–1544.

O'Neill, K.M., O'Neill, R.P., Kemp, W.P., and Delphia, C.M. 2011. Effect of temperature on post-wintering development and total lipid content of alfalfa leafcutting bees. *Environmental Entomology* 40 (4): 917–30.

Ogilvie, J.E., Griffin, S.R., Gezon, Z.J., Inouye, B.D., Underwood, N., Inouye, D.W., and Irwin, R.E. 2017. Interannual bumble bee abundance is driven by indirect climate effects on floral resource phenology. *Ecology Letters* 20 (12): 1507–1515.

Ollerton, J., Winfree, R. and Tarrant, S. 2011. How many flowering plants are pollinated by animals? *Oikos* 120 (3): 321–26. Doi:10.1111/j.1600-0706.2010.18644.x.

Ovaskainen, O., Skorokhodova, S., Yakovleva, M., Sukhov, A., Kutenkov, A., Kutenkova, N., Shcherbakov, A., et al. 2013. Community-level phenological response to climate change. *Proceedings of the National Academy of Sciences* 110 (33): 13434–13439.

Oyen, K.J., Giri, S., and Dillon, M.E. 2016. Altitudinal variation in bumble bee (Bombus) critical thermal limits. *Journal of Thermal Biology* 59: 52–57.

Pacini, E.N.M.V.J., Nepi, M., and Vesprini, J.L. 2003. Nectar biodiversity: A short review. *Plant Systematics and Evolution* 238 (1-4): 7–21.

Parmesan, C. 1996. Climate and species' range. *Nature* 382 (6594): 765–66. Doi:10.1038/382765a0.

Parmesan, C. 2007. Influences of species, latitudes and methodologies on estimates of phenological response to global warming. *Global Change Biology* 13 (9): 1860–1872.

Parmesan, C., Ryrholm, N., Stefanescu, C., Hill, J.K., Thomas, C.D., Descimon, H., Huntley, B., et al. 1999. Poleward shift of butterfly species' ranges associated with regional warming. *Nature* 399: 579–83.

Parmesan, C. and Yohe, G. 2003. A globally coherent fingerprint of climate change impacts across natural systems. *Nature* 421 (6918): 37–42.

Parsche, S., Fründ, J., and Tscharntke, T. 2011. Experimental environmental change and mutualistic vs. antagonistic plant flower–visitor interactions. *Perspectives in Plant Ecology, Evolution and Systematics* 13 (1): 27–35.

Pöyry, J., Luoto, M., Heikkinen, R.K., Kuussaari, M., and Saarinen, K. 2009. Species traits explain recent range shifts of Finnish butterflies. *Global Change Biology* 15 (3): 732–43.

Pyke, G.H., Thomson, J.D., Inouye, D.W., and Miller, T.J. 2016. Effects of climate change on phenologies and distributions of bumble bees and the plants they visit. *Ecosphere* 7 (3): e01267.

Radmacher, S. and Strohm, E. 2010. Factors affecting offspring body size in the solitary bee Osmia bicornis (Hymenoptera, Megachilidae). *Apidologie* 41 (2): 169–77. Doi:10.1051/apido/2009064.

Rafferty, N.E. and Ives, A.R. 2011. Effects of experimental shifts in flowering phenology on plant–pollinator interactions. *Ecology Letters* 14 (1): 69–74.

Reyer, C.P.O., Leuzinger, S., Rammig, A., Wolf, A., Bartholomeus, R.P., Bonfante, A., de Lorenzi, F., et al. 2013. A plant's perspective of extremes: Terrestrial plant responses to changing climatic variability. *Global Change Biology* 19 (1): 75–89. doi:10.1111/gcb.12023.

Roy, D.B. and Sparks, T.H. 2000. Phenology of British butterflies and climate change. *Global Change Biology* 6 (4): 407–16.

Rusterholz, H.P. and Erhardt, A. 1998. Effects of elevated CO_2 on flowering phenology and nectar production of nectar plants important for butterflies of calcareous grasslands. *Oecologia* 113: 341–49.

Schürch, R., Accleton, C., and Field, J. 2016. Consequences of a warming climate for social organisation in sweat bees. *Behavioral Ecology and Sociobiology* 70: 1131–1139.

Schweiger, O., Settele, J., Kudrna, O., Klotz, S., and Kühn, I. 2008. Climate change can cause spatial mismatch of trophically interacting species. *Ecology* 89 (12): 3472–3479.

Sgolastra, F., Bosch, J., Molowny-Horas, R., Maini, S., and Kemp, W.P. 2010. Effect of temperature regime on diapause intensity in an adult-wintering Hymenopteran with obligate diapause. *Journal of Insect Physiology* 56 (2): 185–94.

Sheldon, K.S. 2019. Climate change in the tropics: Ecological and evolutionary responses at low latitudes. *Annual Review of Ecology, Evolution, and Systematics* 50: 303–33.

Slominski, A.H. and Burkle, L.A. 2019. Solitary bee life history traits and sex mediate responses to manipulated seasonal temperatures and season length. *Frontiers in Ecology and Evolution* 7: 314.

Slominski, A.H. and Burkle, L.A. 2021. Asynchrony between solitary bee emergence and flower availability reduces flower visitation rate and may affect offspring size. *Basic and Applied Ecology* 56: 345–57.

Snider, J.L. and Oosterhuis, D.M. 2011. How does timing, duration, and severity of heat stress influence pollen-pistil interactions in angiosperms? *Plant Signaling and Behavior* 6 (7): 930–33.

Soroye, P., Newbold, T., and Kerr, J. 2020. Climate change contributes to widespread declines among bumble bees across continents. *Science* 367 (6478): 685–88.

Sparks, T.H., Jeffree, E.P., and Jeffree, C.E. 2000. An examination of the relationship between flowering times and temperature at the national scale using long-term phenological records from the UK. *International Journal of Biometeorology* 44: 82–87.

Sparks, T.H., Langowska, A., Głazaczow, A., Wilkaniec, Z., Bieńkowska, M., and Tryjanowski, P. 2010. Advances in the timing of spring cleaning by the honeybee Apis mellifera in Poland. *Ecological Entomology* 35 (6): 788–91.

Stenseth, N.C. and Mysterud, A. 2002. Climate, changing phenology, and other life history traits: Nonlinearity and match–mismatch to the environment. *Proceedings of the National Academy of Sciences* 99 (21): 13379–13381.

Stephen, W.P. 1965. Temperature effects on the development and multiple generations in the alkali bee, nomia melanderi cockerel 1. *Entomologia Experimentalis et Applicata* 8 (3): 228–40.

Stuhldreher, G., Hermann, G., and Fartmann, T. 2014. Cold-adapted species in a warming world–an explorative study on the impact of high winter temperatures on a continental butterfly. *Entomologia Experimentalis et Applicata* 151 (3): 270–79.

Suni, S.S. and Dela Cruz, K. 2021. Climate-associated shifts in color and body size for a tropical bee pollinator. *Apidologie* 52 (5): 933–45.

Thomson, J.D. 2010. Flowering phenology, fruiting success and progressive deterioration of pollination in an early-flowering geophyte. *Philosophical Transactions of the Royal Society B: Biological Sciences* 365 (1555): 3187–3199. Doi:10.1098/rstb.2010.0115.

Tyler, G. 2001. Relationships between climate and flowering of eight herbs in a Swedish deciduous forest. *Annals of Botany* 87 (5): 623–30.

Vandelook, F., Janssens, S.B., Gijbels, P., Fischer, E., Van den Ende, W., Honnay, O., and Abrahamczyk, S. 2019. Nectar traits differ between pollination syndromes in Balsaminaceae. *Annals of Botany* 124 (2): 269–79.

Vizentin-Bugoni, J., Maruyama, P.K., de Souza, C.S., Ollerton, J., Rech, A.R., and Sazima, M. 2018. "Plant-pollinator networks in the tropics: A review." In *Ecological Networks in the Tropics: An Integrative Overview of Species Interactions from Some of the Most Species-Rich Habitats on Earth*, edited by W. Dáttilo and V. Rico-Gray, Springer Publishing, 73–91.

Wang, S.P., Meng, F.D., Duan, J.C., Wang, Y.F., Cui, X.Y., Piao, S.L., Niu, H.S., et al. 2014. Asymmetric sensitivity of first flowering date to warming and cooling in alpine plants. *Ecology* 95 (12): 3387–3398.

White, P. and Kerr, J.T. 2006. Contrasting spatial and temporal global change impacts on butterfly species richness during the 20th century. *Ecography* 29 (6): 908–18.

Williams, C.M., Hellmann, J., and Sinclair, B.J. 2012. Lepidopteran species differ in susceptibility to winter warming. *Climate Research* 53 (2): 119–30.

Willis, C.G., Ruhfel, B., Primack, R.B., Miller-Rushing, A.J., and Davis, C.C. 2008. Phylogenetic patterns of species loss in Thoreau's woods are driven by climate change. *Proceedings of the National Academy of Sciences* 105 (44): 17029–17033.

Willmer, P.G. and Stone, G.N. 2004. Behavioral, ecological, and physiological determinants of the activity patterns of bees. *Advances in the Study of Behavior* 34 (34): 347–466.

Wilson Rankin, E.E., Barney, S.K., and Lozano, G.E. 2020. Reduced water negatively impacts social bee survival and productivity via shifts in floral nutrition. *Journal of Insect Science* 20 (5): 15.

Yuan, J.S., Himanen, S.J., Holopainen, J.K., Chen, F., and Stewart, C.N. 2009. Smelling global climate change: Mitigation of function for plant volatile organic compounds. *Trends in Ecology and Evolution* 24 (6): 323–31.

CHAPTER 14

Insect vectors of human pathogens in a warming world

Summarizing responses and consequences

Berenice González-Rete, Jesús Guillermo Jiménez-Cortés, Margarita Cabrera-Bravo, Paz María Salazar-Schettino, Any Laura Flores-Villegas, José Antonio de Fuentes-Vicente and Alex Córdoba-Aguilar

> *"Climate change is sometimes misunderstood as being about changes in the weather. In reality it is about changes in our very way of life."*
>
> *Paul Polman*

14.1 Introduction

Global warming is a major threat currently faced by humanity. The consensus is that the worldwide average temperature will increase by 1.1–5.8 °C in the next century if we continue human activities at prepandemic levels (IPCC 2007, 2014). Some of the main consequences of global climate change for insects include alterations of their life cycles, phenology, survival, immune response and modifications in their natural geographic distribution (Gage et al. 2008). From this perspective, we consider a "vector" as any host or object that transmits parasites, including intermediate, definitive and paratenic hosts, as well as water or wind. Specifically, and more commonly applied, this term refers to arthropods that transmit parasites among vertebrate hosts. In this case, the vector may be mechanical—a means of physical movement or biological—in which pathogens multiply, propagate or develop, or a vector in which maturation and increase in numbers occur (i.e., cyclopropagative transmission) (Botzler and Brown 2014).

In this chapter, we consider insect vectors of pathogens as a key group within the panorama of climate change, given their effects on human health. We consider a vector as any hematophagous arthropod that transports and transmits a parasite or other organism that constitutes a public health risk (Wilson et al. 2017). This includes holometabolous insects such as mosquitoes (Diptera: Culicidae), phlebotomines (Diptera: Psychodidae), the tsetse fly (Diptera: Glossinidae), blackflies (Diptera: Simuliidae), and hemimetabolous insects such as kissing bugs (Hemiptera: Triatominae), along with many species of lice (Phthiraptera) and fleas (Siphonaptera) (WHO 2016). In the widest sense of the definition,

Berenice González-Rete et al., *Insect vectors of human pathogens in a warming world*. In: *Effects of Climate Change on Insects*. Edited by: Daniel González-Tokman and Wesley Dáttilo, Oxford University Press. © Oxford University Press (2024). DOI: 10.1093/oso/9780192864161.003.0014

we considered a "parasite" any species of virus, bacteria, protozoa (microparasites) or nematode (macroparasite) that requires an insect host during its life cycle.

The objective of this chapter is to summarize existing information on how environmental temperature influences the physiology, ecology and evolution of insect vectors and their parasites. We have limited our review to temperature as the main factor in global warming for the following reasons: first, temperature is perhaps the most important ecological factor in ectotherms, including insects, as it determines their distribution and abundance (Harrison et al. 2012). Second, temperature determines much of the interaction between a host, in this case an insect vectors, and the species of parasite it harbors (Elliot et al. 2015; Thomas and Blanford, 2003). Finally, temperature affects some elements of the insect immune response, such as phenoloxidase (PO), which generally increases its activity at high temperatures (Adamo and Lovett 2011).

14.2 Climate effects on physiology

14.2.1 *Heat shock proteins*

The widest known physiological aspect of organisms that is affected by increased temperature are heat shock proteins (HSPs) (reviewed by González-Tokman et al. 2020; see also Chapter 3). HSPs are associated with biological processes of organisms including development, survival, aging and death. At a cellular level, HSPs participate in protein homeostasis, preventing denaturation under both normal and stressful conditions (González-Tokman et al. 2020). An increase in the gene expression of HSPs in response to increased temperature usually—but not always—maintains the general integrity and stability of the organism (e.g., Hunter et al. 1984).

In microorganisms such as the promastigotes of *Leishmania amazonensis* and *Leishmania braziliensis*, the protozoa that cause leishmaniasis, there is an immediate increase in the translation and synthesis of HSP70 and HSP83. In addition, an increased expression of polycistronic genes at temperatures of 26–37 °C, in temperature-generalist species (Shapira et al. 2001), which leads to increased parasitic genetic variability compared to parasites that are specialized to certain temperatures (Saxena et al. 2007). In this case, differential gene expression is beneficial to the parasite, conferring improved infective capacity, motility and cellular maintenance (Saxena et al. 2007), and improving its ability to evade the vertebrate immune response and successfully establish an infection (Papageorgiou and Soteriadou 2002). In the vector, *Lutzomyia longipalpis*, an increase in temperature from 37 °C to 40 °C causes stress and up to a fourfold increase in HSP90 expression, compared to a onefold increase at 37 °C, suggesting that expression levels may be temperature dependent (Martins et al. 2021).

14.2.2 *Infectivity, replication time and susceptibility*

As depicted in Figure 14.1, the infectivity and replication time of parasites transmitted by vectors are affected by temperature (Samuel et al. 2016). The infectivity of the parasite is related to the susceptibility of the vector, while the replication time corresponds to the time interval at which several parasites reproduce within the vector (Bonsall 2010). Moreover, vector susceptibility to infection depends on the stage in life cycle, nutritional state, genotype or even the time of year (Ewing et al. 2021).

While there are few studies that have focused on the impact of temperature on the susceptibility of vectors to infection by different parasites, vector populations present phenotypic and genotypic differences due to their adaptation to different environmental and parasitic conditions (Figure 14.1). In other words, in a certain vector species, susceptibility is determined by the genotype of the parasite, the genotype of the vector and the influence of the environment in which the interaction takes place (reviewed by Samuel et al. 2016). Some examples illustrate this point. In mosquitoes of the family Culicidae, such as *Anopheles* and *Aedes*, there is an increasing rate of infection by dengue virus, chikungunya virus and malaria parasites as temperature increases (Figure 14.1). This is due to the higher rate of replication of the parasites over a shorter time period than at lower temperatures; along the same line, increases in temperature leads to a decrease in the life expectancy of the mosquitoes (mainly explained for the reverse relationship between temperature and longevity in insects) (Harrison et al. 2012) and may drastically alter risk of transmission (Paaijmans et al. 2009).

In triatomines, the insect vectors of Chagas disease, the susceptibility to infection with *Trypanosoma cruzi* is closely related to genetic variability and the geographic distribution of the parasite. In fact, when triatomines were experimentally infected with strains of *T. cruzi* from outside their geographic ranges, their survival depended on the ambient temperature, so that an increase in temperature due to climate change and the different virulence of the strains circulating in the environment, would reflect drastic events in the replication of the parasite, as well as in the survival of the vector and its geographical distribution (Dworak et al. 2017). Finally, it is also known that the increase in temperature, expected due to climate change, could positively affect the release of metacyclic trypomastigotes (the infective stage in humans), as well as accelerate its appearance in triatomine feces (reviewed by Melo et al. 2020).

14.2.3 *Replication*

Temperature is important in the replication of parasites within their insect hosts. Several examples have shown that at elevations above 3000 meters (m), where temperatures are below 15 °C, the reproduction of different parasites in a vector stops (Beck-Johnson et al. 2013). A representative case is that of viruses transmitted by insects which are only found at latitudes below 50°N and 20°S, and their primary limitation for dispersal is temperature, because it has been shown that in these regions the cold temperature limits the vector development, the pathogen replication and its subsequent transmission (Kraemer et al. 2015). Still, not only low but any extreme temperature may affect parasitic replication.

In *Plasmodium vivax* and *Plasmodium falciparum*, predominantly found in the Americas and Africa, respectively, replication increases at temperatures above 27 °C but ceases when temperatures exceed 33 °C (Paaijmans et al. 2012). In the tsetse fly, *Glossina morsitans*, exposing adults to temperatures between 26 °C and 31 °C increases the replication rate of the trypanosomes. However, when the temperature exceeds 31 °C, the replication of *T. brucei* decreases, with an increase in mortality in *G. morsitans* associated with nutritional stress caused by high temperatures that induce rapid digestion of bloody meals, and consequently, the habitat is not suitable for the establishment of the parasite (Nnko et al. 2017). *Leishmania infantum* and *Leishmania braziliensis* develop slowly in the intestine of *Phlebotomus perniciosus* and *Lutzomyia longipalpis*, respectively at low temperatures (20 °C), but at 26 °C their replication rate increases (Hlavacova et al. 2013).

In addition to thermal tolerances, optimal temperatures for development of the parasite and vector have been studied. These are key variables for understanding success in vectorial capacity. One case is that of the optimal temperatures during the life cycle of different species of *Anopheles* mosquito, vectors of *P. falciparum* and *P. vivax*, which fall between 30 °C and 31 °C (Paaijmans et al. 2009), however *Anopheles spp.* mosquito populations decrease at 40 °C (Ngarakana-Gwasira et al. 2014). In *Simulium woodi*, vector of *Onchocerca volvulus* parasitic nematodes, the worms take sixteen days to reach the infective stage at 18 °C, but at 28 °C, the highest temperature at which the flies survived, they take four days (Wegesa 1966). In both cases, although parasites develop at an optimal temperature within their vectors, increases in temperature alter this interaction in such a way that the development and transmission success of the pathogen is modified.

14.2.4 *Extrinsic incubation period (EIP)*

The extrinsic incubation period (EIP)—that is, the interval between the entrance of the infectious agent into the vector and the moment that the vector becomes infectious—is one of the factors affected by increased temperature (e.g., Reisen et al. 2006). The EIP is highly variable and generally decreases with increasing temperatures. One of the main ways that temperature increase affects EIP is the rapid digestion of blood and the rapid defecation (Lafferty 2009), leading to a shorter period of maturation and oviposition, as well as increased frequency of feeding, increasing the probability of transmission of infection (e.g., Lafferty 2009). In *Culex* spp. mosquitos, after feeding on an infected human or other vertebrate with West Nile Virus, the EIP is approximately sixteen to twenty-five16 to 25 days at 20 °C, but this decreases to thirteen13 days when temperature is 25 °C (Anderson et al. 2008). In *Anopheles stephensi* infected by *Plasmodium yoelii*, the same pattern of decreased EIP with increasing temperature occurs, with a decrease from fifteen15 days at 20.8 °C to eight8 days at 26.8 °C (Paaijmans et al. 2012). However, one limiting factor is that transmission is not possible at temperatures below 20 °C, restricting its dispersal (Watts et al. 2021).

In fleas infected with *Yersinia pestis* bacteria, the EIP ranges from seven to thirty-one days and decreases with increasing temperature (Eisen et al. 2015). For triatomines and phlebotomines, the exact EIP is unknown. However, the development of parasites as *T. cruzi* within triatomines, as well as metacyclogenesis, are delayed from ten to fourteen days in insects at 26 °C compared to those maintained at 28 °C (Tamayo et al. 2018). In phlebotomines, the replication rate is higher over a shorter time interval at higher temperatures (Hlavacova et al. 2013).

14.2.5 *The gonotrophic cycle and the basic reproductive rate*

Another biological characteristic that is often altered in insect vectors by temperature increase is the duration of the gonotrophic cycle, defined as the interval between two consecutive blood meals or two consecutive ovipositions. For example, the duration of the gonotrophic cycle of *Aedes albopictus* decreases with increasing temperature, from 8.1 days at 20 °C to 3.5 days at 30 °C (Delatte et al. 2009). In *Simulium ochraceum*, the duration of the gonotrophic cycle is also modified by temperature: at 10 °C the cycle lasts nine days, and the time is reduced to two days as the temperature increases to 28 °C (Takaoka et al. 1982).

Another characteristic that is modified by temperature is the basic reproductive rate (R_0), defined as an epidemiological parameter that helps us understand how parasite populations, in this case vectors, are maintained and dispersed, considering aspects such as their fertility or mortality (Anderson and May 1978; Poulin 2007). In short, when

$R_0 > 1$, the insect vector population would be growing, while values of $R_0 < 1$ would mean that the population decreases and could even become extinct (Poulin 2007). Identifying this parameter will help us know if vector species that expand their distribution ranges in a warming world could be successful in these new environments. R_0 can vary depending on entomological parameters, such as the number of bites, EIP and insect population density (Couper et al. 2021). In a study using fixed parameters, R_0 for dengue virus was determined to be 12.02, thus the infection can expand and persist in the population. When the values of some of the parameters were modified to demonstrate the change in R_0, for example, $\beta m = 3.0 \times 10^{-1}$ (rate of transmission of the parasite from human to female mosquito) and $\beta h = 6.0 \times 10^{-2}$ (rate of transmission of the parasite from female mosquito to human) to 2.0×10^{-1} and 5.4×10^{-2}, respectively, R_0 was reduced to 0.721, indicating the infection would be low in the population and may be reduced (Yang et al. 2014).

14.2.6 *Immune response*

Immune system components (Figure 14.1) can be sensitive to increases in temperature (Ferguson and Sinclair 2017). In *An. stephensi* mosquitoes, temperature significantly affects the expression of antimicrobial peptides such as *defensin 1* (DEF1) and *cecropin 1* (CEC1), as well as the nitric oxide synthase (NOS) molecule when confronting bacteria and parasites (Murdock et al. 2012). Mosquitos maintained at 26 °C display increased expression of *DEF1* from six to twelve hours and this expression remains elevated at higher temperatures (30–34 °C), while *DEF1* expression decreases when mosquitos are incubated at 18 °C. In the case of *CEC1*, expression increases when mosquitos are maintained at higher temperatures (26 °C and 34 °C) than at lower temperatures (18 °C). NOS also presents similar values to *DEF1* when the temperature is increased (Murdock et al. 2012).

In triatomines, vector insects of Chagas disease, it has been demonstrated that the immune response is expressed differently depending on the strain or stage of the parasite, the infectious agent (viruses, bacteria, fungi or protist), the nutritional state of the vector and temperature (Castro et al. 2012). For example, in *Meccus pallidipennis*, a triatomine species, the immune enzyme phenoloxidase, used against infection by *T. cruzi*, was enhanced when the temperature increased from 20 °C to 30 °C, but this pattern changed dramatically at 34 °C, indicating a more efficient prophenoloxidase activity before a subtle increase in temperature (González-Rete et al. 2019).

In tsetse flies (*Glossina* spp.) infected with *T. brucei*, immune molecules are expressed, including antimicrobial peptides (attacins, cecropins and defensins) (Harrington 2011), glutamine/proline-rich proteins (EP), and reactive oxygen species, and the Imd pathway is also activated (Harrington 2011). Increased temperature leads to nutritional stress in the fly *G. morsitans*, mediated by a negative effect on the immune response of the vector when infected with *T. brucei rhodesiense* (Akoda et al. 2009).

The presented evidence suggests that under warming conditions, parasites will face a highly reactive immune system, potentially compromising parasite survival. A second possibility is that vectors generate higher tolerance, defined as the capacity to limit the negative effects of a given parasite load without fully controlling the infection (Graham et al. 2011). Either of these possibilities would lead to physiological trade-offs with other vital functions, such as reproduction, given the cost of the immune response (Kraaijeveld et al. 2002).

14.3 Ecological consequences

14.3.1 *The hard data*

Some effects of extreme temperature increases can lead to a reduction or even local extinction of insect vectors. For example, using simulations and weather records, the tropical zone of Darwin (Australia), local extinction of *Ae. aegypti* was predicted and associated with the increase in temperature in synergistic combination with the specialization for a reduced habitat type and the physiological vulnerability to dry conditions (Williams et al. 2010).

A fundamental step for understanding the viability of vectors under increasing temperatures is the investigation of thermal thresholds. For example, *Ae. albopictus* has higher survival rates under conditions of diapause than *Ae. aegypti*, even though *Ae. aegypti* can survive over a larger temperature range (Mordecai et al. 2019). The development of the immature stages (eggs, larvae and pupae) is also altered by the increase in temperature. Also, *Aedes aegypti* eggs can resist desiccation for up to a year at high temperatures, but when the eggs hatch the larvae are motionless and die within a couple of days; in other cases, the larvae do not develop well and this is reflected in the adult stage (Reinhold et al. 2018; Simoy et al. 2015).

Laboratory studies in triatomines have shown that increased temperature leads to an increase in the development and reproductive rates of the vector (Martínez-Ibarra et al. 2008). The populations of some species of triatomines (*Triatoma infestans*, *T. gerstaeckeri* and *T. sanguisuga*) are therefore expected to increase in the next years (Garza et al. 2014). However, for some species (e.g., *Meccus pallidipennis*) increased temperature had a negative effect on survival (González-Rete et al. 2019), indicating a cost in development and fitness of the vector.

It is also estimated that increasing temperatures in temperate climates could lead to an increase in insect populations, in some cases accompanied by geographic expansion of their territories (Chen et al. 2011). For example, *Plasmodium* spp. and its vector insects are shifting towards higher altitudes, including the African highlands where climate suitability for transmission increased by about 30 percent between 2012 and 2017 compared with 1950 (Giesen et al. 2020; Watts et al. 2019). Also, in another geographical region such as Nepal, an increase of 1 °C in the minimum temperature has resulted in a 27 percent increase in the incidence of malaria and transmission has expanded from thirty-eight to sixty-five districts throughout the country (including areas > 2000 m above sea level) (Dhimal et al. 2014, 2015). Another study in southeastern Brazil documented the presence of *Ae. aegypti* and *Ae. albopictus* at mid and high altitudes of tropical montane cities Mariana and Ouro Preto. From 1961 to 2014, the annual temperature increased due to the increases in winter temperatures, where the first autochthonous dengue cases were observed in 2007, which increased threefold and reached their maximum in 2013 (Pedrosa et al. 2021). In laboratory studies, *Culicoides* mosquitoes have greater sensitivity to Oropouche virus, as a result of increasing temperature, which explains the joint geographic expansion of *Culicoides* and Oropouche virus in recent years (Wittmann and Baylis 2000).

14.3.2 *From hard data to predictive models*

Predictive models are a tool that can be used together with laboratory studies. Currently, epidemiologic modeling incorporates different environmental variables and has focused almost completely on the impact of temperature changes. However, it is important to recognize the use of other environmental factors in modeling the transmission of parasites by vectors (Zhou et al. 2004). Regarding the intrinsic factors of parasites transmitted by vectors, predictive models have considered the strain of the parasite, the number of parasites and the stages of the parasite (Churcher et al. 2017). Based on these models, it has been observed that, in *An. stephensi*, there is a direct relationship between the number of parasites acquired when feeding on infected blood and the number of parasites inoculated into the host when they bite. Churcher et al. (2017) estimated that heavily infected mosquitoes will present higher rates of infection in less time than less infected mosquitoes, which could probably be exacerbated by increasing temperatures, according to results in other vector insects. For example, in blackflies, models scenarios estimate that indicators such as the annual transmission potential (number of infective larvae potentially received annually per person), the number of female nematodes per human host and the number of microfilariae per milligram of human skin will increase gradually with increasing of temperature from 19 °C, but a decrease in these indicators will be observed at temperatures above 30 °C (Cheke et al. 2015; Walsh et al. 1978). In *R. prolixus*, metacyclic trypomastigotes of *T. cruzi* were observed at six days post-infection at 30 °C, while at 28 °C infective forms were observed at fourteen days. The number of parasites increased over time (six to eight weeks) in infected triatomines at 30 °C, while at lower temperatures, the number decreased (Tamayo et al. 2018).

Models primarily consider the bite frequency, fecundity, vector density, and thermal preference regarding the vector (Beck-Johnson et al. 2013). For example, in *An. gambiae* and *An. funestus*, the bite frequency per gonotrophic cycle increased with increasing temperature from 21 °C to 34 °C (Shapiro et al. 2017).

Another tool for estimating the possible short- and medium-term panoramas of the distribution of vector species in predictive models is based on spatial estimations (Murdock et al. 2012). These models predict species distribution on a regional or global level using climate as a predictive variable to identify areas of risk (Kraemer et al. 2015). If any general conclusion can be drawn, it is that insect vector species have notably expanded their geographic distributions. For example, models of mosquito population of *An. gambiae* in Africa indicate that a 0.5 °C increase in mean temperature of that region could lead to a 30–100 percent increase in mosquito abundance (Pascual et al. 2006). However, if the temperature increases beyond the thermal tolerance threshold of the mosquito, there would be a reduction in survival (Thomas and Blanford, 2003).

Aedes aegypti and *Ae. albopictus* are widely distributed on all continents except Antarctica, and their distribution ranges are predicted to further expand in the coming years (Kraemer et al. 2015). These predictive data need to be reinforced by field observations in order to make the models more robust and realistic. For example, the culicid *Aedes japonicus* has been reported in recent years in regions with low temperatures (below 20 °C) at certain times of the year, such as Japan and Korea, and even in regions like Hawaii at higher temperatures, which exceeds the predictions based on its place of origin (Egizi et al. 2015). In *Simulium* spp., predictive models suggest that Liberia and Ghana will experience an increase in the number of onchocerciasis vectors in the next decades, due to the increasing acceleration of blackfly development rates and the accelerated development of larvae within them with the increase in temperature of up to 1 °C (Cheke et al. 2015). In triatomines, it is estimated that the geographic distributions of different species will be affected with range increases, so strong repercussions to increasing temperatures are expected in terms of physiological and behavioral processes, especially when infected by *T. cruzi* (Garza et al. 2014).

14.4 Evolutionary consequences

Some of the most fascinating aspects of the analysis of the interaction between vectors and the parasites they transmit are evolutionary. For example, we have not been able to completely uncover the degree of parasitism-commensalism, or even mutualism, that may be present in the vector–parasite relationship. While in some cases it is thought that the vector is simply a carrier of the parasite, today we know that this is not always the case (Wilson et al. 2017). In most cases it has been shown that parasites trigger an immune response, and therefore energetic costs to the vector, which suggests parasitism, not commensalism (Wilson et al. 2017). In the following paragraphs some important evolutionary trends are outlined.

Focusing first on the effects of the immune response in the context of climate change, some aspects of the immune response are expected to increase vector activity when faced with different pathogens in the environment. However, this does not mean that the vectors will be able to eliminate all the parasites they harbor in the coming years. First, the immune response is costly to activate and maintain (Schmid-Hempel 2011). Secondly, parasites are expected to respond to counteract the increased immune response using mechanisms that range from molecular processes to alterations of the entire organism such as the synthesis of many molecules with signaling and antimicrobial functions and the proliferation of immune cells (Dolezal et al. 2019; Loker and Hofkin 2015).

One aspect that should be monitored, given the widening distributions of several species of insect vectors, is the acquisition of new species or strains of parasites in the vectors' new environments, and the potential transmission of these new parasites to vertebrates including humans. There have recently been records of new pathogens in several species of insect vectors. These include *Rickettsia felis* in *An. gambiae* in Africa and different strains of *Rickettsia* in *Culex pipiens* in China; these pathogens are the etiologic agents of spotted fever in humans (Dieme et al. 2015). While transmission to mammals in the laboratory has been documented, transmission to humans has yet to be confirmed.

Some vector–parasite interactions result in changes in host behavior and some parasite-induced changes of vectors can increase the probability of infecting the vertebrate host (Lehane 2005). Some of the main alterations include physical blockages of the digestive tract, pathologies of the salivary glands and an increase in the number of vertebrate hosts bitten (Lehane 2005). Physical blockages have been observed in the flea *Xenopsylla cheopis*, vector of *Y. pestis*, and the protozoan *Leishmania mexicana*, which interferes with receptors in the phlebotomine *Lutzomyia longipalpis*, blocking blood flow. This interference affects the feeding and limits the volume of blood that a sandfly can obtain, and consequently, infected sandflies spend more time feeding, increasing parasite transmission (Rogers and Bates 2007). Interestingly, in the case of the flea vector of *Y. pestis*, this blockage decreases as temperature increases (Lehane 2005).

Several studies have shown how genetic variation increases both defense and virulence in vector–parasite interactions as temperature increases (Iriso et al. 2017). If we consider the increases in temperature, it is possible that this genetic variation will become subject to selection (Schade et al. 2014). Dengue virus has a substitution rate of approximately

7.6×10^{-4} substitutions per site per year (Mukhopadhyay et al. 2005). This mutation rate in genotypes and serotypes of the dengue virus is slow compared to non-vector-borne viruses. This is likely because Dengue virus alternates between two types of hosts: mosquitoes and vertebrates (Vasilakis et al. 2009). However, under the influence of new climatic conditions, the vectorial capacity of the mosquitoes is predicted to change, most likely increasing, and it is also likely that there will be increased propagation of the virus in new hosts, both vertebrates and mosquitoes. An additional risk is the ability of mosquitoes to transovarianly transmit dengue virus to their progeny, which would gradually increase the risk of transmission (Günther et al. 2007; Iriso et al. 2017).

Finally, climate change could affect not only the spatial distribution of insect vector species but also the biological interactions among species. An example of these interactions is the association of insects with their intestinal microbiota or symbionts, which in many cases have been present for hundreds of years. These symbionts are mainly bacteria, as far as has been studied, though fungi could also be important (Dobson 2009). Symbiotic bacteria are essential for vectors in that they provide B-complex vitamins and participate in immune pathways or in the establishment of some pathogens or parasites in their hosts (Jiménez-Cortés et al. 2018). These microorganisms are acquired by insects in their natural habitats, either by vertical or horizontal transmission and their composition can vary depending on factors including vector species, sex, life-cycle stage, geographic origin and feeding behavior (e.g., Valiente-Moro et al. 2013).

Temperature impacts the microbiota of vectors in several ways, mainly in their nutrition, digestion, metabolism, development and immunity (Kikuchi et al. 2016). One example is the effect of temperature on the intestinal microbiota of mosquitoes (Culicidae). For example, Jupatanakul et al. (2014) found that the *Wolbachia* strain *wAlbB* reduced infection by *Plasmodium yoelii* at 28 °C in *An. gambiae*, but at 20 °C had the opposite effect. Finally, in the bug *R. prolixus* kept at 37 °C and infected with *T. cruzi*, there was an increase in the diversity and population of the microbiota compared to non-infected bugs (Castro et al. 2012). There are few studies on the topic but it is nevertheless important to consider studying this effect in other vector species, such as phlebotomines, and even repeat studies on vectors that have already been studied, since microbiota communities are dynamic in space and time.

14.5 Conclusions

Perhaps the best general conclusion for this chapter is that temperature is the key to relationships between vectors and their parasites, although generalizations cannot be made yet to describe the complete picture. For example, while several species of insect vectors will expand their geographic ranges with current and predicted climate change, it is unknown if this will be accompanied by increased propagation of their parasites. While increasing temperature could negatively or positively affect parasite life cycles within their vectors, the diversity of abiotic (e.g., precipitation) and biotic (e.g., microbiota) factors reduce the predictive capacity of the available tools. The best source of information to date is mathematical models, which need to consider intrinsic factors of the vectors and their parasites. Several environmental parameters, such as relative humidity, will also vary with temperature, and these factors will need to be accounted for when using improved models and laboratory experiments. Further research should include manipulation of CO_2 levels along with temperature to simulate more realistic conditions.

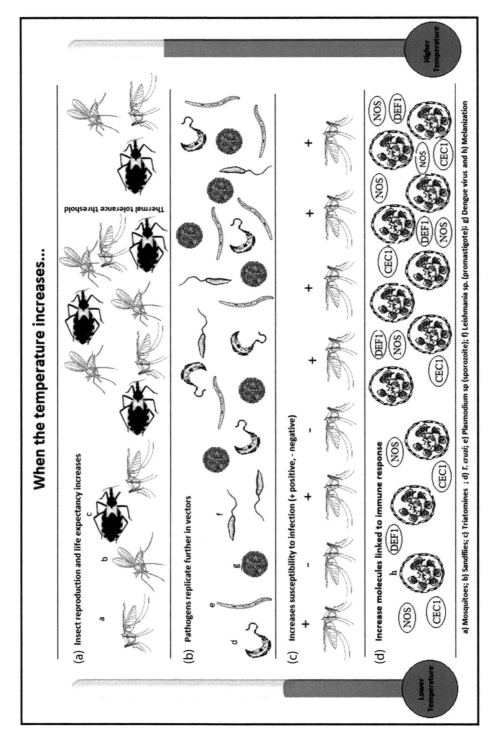

Figure 14.1 Temperature effects on: A) insect reproduction and life expectancy; B) parasite replication within insect vectors; C) susceptibility to parasite infection (mosquito cases) (See section 14.2.2 for further explanation); and D) molecules linked to insect immune responses (NOS = nitric acid synthase molecule; CEC1 and DEF1 = antimicrobial peptides *cecropin 1* and *defensin 1*, respectively; see section 14.2.6 for further explanation). a = mosquitoes; b = sandflies; c = kissing bugs; d = *Trypanosoma cruzi*; e = *Plasmodium sp.* (sporozoite); f = *Leishmania* sp. (promastigote); g = dengue virus (virion); h = melanization. Note that there are more infections and heavier infection as temperature increases from left to right.

Key reflections

- An increase in temperature via global warming has promoted adaptive changes in insect vectors at varying physiological and ecological levels.
- Insect vectors are susceptible to be infected by a plethora of pathogens. Since this process is temperature dependent, global warming will change current susceptibility dynamics.
- Some models have suggested non-adaptive responses by insect vectors which can render them to go extinct.
- We need new predictive models that include humidity and CO_2 variables to predict insect vectors' ecology and behavior.

Key further reading

- Cahill, A.E., Aiello-Lammens, M.E., Fisher-Reid, M.C., Hua, X., Karanewsky, C.J., Ryu, H.Y. et al. 2013. How does climate change cause extinction? *Proceedings of the Royal Society Series B: Biological Sciences 280*: 20121890.
- Leger, R.J.S. 2021. Insects and their pathogens in a changing climate. *Journal of Invertebrate Pathology 184* 1–11.
- Koltz, A.M., and Culler, L.E. 2021. Biting insects in a rapidly changing Arctic. *Current Opinion in Insect Science* 47: 75–81.

References

Adamo, S.A., and Lovett, M.M. 2011. Some like it hot: The effects of climate change on reproduction, immune function and disease resistance in the cricket *Gryllus texensis*. *Journal of Experimental Biology*, 214 (12): 1997–2004.

Akoda, K., Van den Bossche, P., Marcotty, T., Kubi, C., Coosemans, M., De Deken, R., and Van den Abbeele, J. 2009. Nutritional stress affects the tsetse fly's immune gene expression. *Medical and Veterinary Entomology* 23: 195–201.

Anderson, J.F., Main, A.J., Delroux, K., and Fikrig, E. 2008. Extrinsic incubation periods for horizontal and vertical transmission of West Nile virus by *Culex pipiens pipiens* (Diptera: Culicidae). *Journal of Medical Entomology* 45: 445–51.

Anderson, R.M. and May, R.M. 1978. Regulation and stability of host-parasite population interactions. *Journal of Animal Ecology* 47 (1): 219–47.

Beck-Johnson, L.M., Nelson, W.A., Paaijmans, K.P., Read, A.F., Thomas, M.B., and Bjørnstad, O.N. 2013. The effect of temperature on Anopheles mosquito population dynamics and the potential for malaria transmission. *PLoS One* 8 (11): e79276.

Bonsall, M.B. 2010. Parasite replication and the evolutionary epidemiology of parasite virulence. *PloS One* 5 (8): e12440.

Botzler, R.G. and Brown, R.N. 2014. *Foundations of Wildlife Diseases*. California: University of California Press.

Castro, D.P., Moraes, C.S., Gonzalez, M.S., Ratcliffe, N.A., Azambuja, P., and Garcia, E.S. 2012. *Trypanosoma cruzi* immune response modulation decreases microbiota in *Rhodnius prolixus* gut and is crucial for parasite survival and development. *PLoS One* 7 (5): e36591.

Cheke, R.A., Basánez, M.G., Perry, M., White, M.T., Garms, R., Obuobie, E., Lamberton, P.H.L. et al. 2015. Potential effects of warmer worms and vectors on onchocerciasis transmission in West Africa. *Philosophical Transactions of the Royal Society B: Biological Sciences* 370 (1665): 20130559.

Chen, I.C., Hill, J.K., Ohlemüller, R., Roy, D.B., and Thomas, C.D. 2011. Rapid range shifts of species associated with high levels of climate warming. *Science* 333 (6045): 1024–1026.

Churcher, T.S., Sinden, R.E., Edwards, N.J., Poulton, I.D., Rampling, T.W., Brock, P.M., Griffin, J.T., et al. 2017. Probability of transmission of malaria from mosquito to human is regulated by mosquito parasite density in naïve and vaccinated hosts. *PLoS Pathology* 13: e1006108.

Couper, L.I., Farner, J.E., Caldwell, J.M., Childs, M.L., Harris, M.J., Kirk, D.G., Nova, N., et al. 2021. How will mosquitoes adapt to climate warming? *Elife* 10: e69630.

Delatte, H., Gimonneau, G., Triboire, A., and Fontenille, D. 2009. Influence of temperature on immature development, survival, longevity, fecundity, and gonotrophic cycles of *Aedes albopictus*, vector of chikungunya and dengue in the Indian Ocean. *Journal of Medical Entomology* 46 (1): 33–41.

Dhimal, M., Ahrens, B., and Kuch, U. 2015. Climate change and spatiotemporal distributions of vector-borne diseases in Nepal–a systematic synthesis of literature. *PloS One* 10 (6): e0129869.

Dhimal, M., O'Hara, R.B., Karki, R., Thakur, G.D., Kuch, U., and Ahrens, B. 2014. Spatio-temporal distribution of malaria and its association with climatic factors and vector-control interventions in two high-risk districts of Nepal. *Malaria Journal* 13: 1–14.

Dieme, C., Bechah, Y., Socolovschi, C., Audoly, G., Berenger, J.M., Faye, O., Raoult, D., et al. 2015. Transmission potential of *Rickettsia felis* infection by *Anopheles gambiae* mosquitoes. *Proceedings of the National Academy of Sciences* 112 (26): 8088–8093.

Dobson, A. 2009. Climate variability, global change, immunity, and the dynamics of infectious diseases. *Ecology* 90 (4): 920–27.

Dolezal, T., Krejcova, G., Bajgar, A., Nedbalova, P., and Strasser, P. 2019. Molecular regulations of metabolism during immune response in insects. *Insect Biochemistry and Molecular Biology* 109: 31–42.

Dworak, E.S., Araújo, S.M.D., Gomes, M.L., Massago, M., Ferreira, É.C., and Toledo, M.J.D.O. 2017. Sympatry influence in the interaction of *Trypanosoma cruzi* with triatomine. *Revista da Sociedade Brasileira de Medicina Tropical* 50: 629–37.

Egizi, A., Fefferman, N.H., and Fonseca, D.M. 2015. Evidence that implicit assumptions of "no evolution" of disease vectors in changing environments can be violated on a rapid timescale. *Philosophical Transactions of the Royal Society B: Biological Sciences* 370 (1665): 20140136.

Eisen, R.J., Dennis, D.T., and Gage, K.L. 2015. The role of early-phase transmission in the spread of Yersinia pestis. *Journal of Medical Entomology* 52 (6): 1183–1192.

Elliot, S.L., Rodrigues, J.D.O., Lorenzo, M.G., Martins-Filho, O.A., and Guarneri, A.A. 2015. *Trypanosoma cruzi*, etiological agent of Chagas disease, is virulent to its triatomine vector *Rhodnius prolixus* in a temperature-dependent manner. *PLoS Neglected Tropical Diseases* 9 (3): e0003646.

Ewing, D.A., Purse, B.V., Cobbold, C.A., and White, S.M. 2021. A novel approach for predicting risk of vector-borne disease establishment in marginal temperate environments under climate change: West Nile virus in the UK. *Journal of the Royal Society Interface* 18 (178): 20210049.

Ferguson, L.V. and Sinclair, B.J. 2017. Insect immunity varies idiosyncratically during overwintering. *Journal of Experimental Zoology Part A: Ecological and Integrative Physiology* 327 (5): 222–34.

Gage, K.L., Burkot, T.R., Eisen, R.J., and Hayes, E.B. 2008. Climate and vectorborne diseases. *American Journal of Preventive Medicine* 35 (5): 436–50.

Garza, M., Feria Arroyo, T.P., Casillas, E.A., Sanchez-Cordero, V., Rivaldi, C.L., and Sarkar, S. 2014. Projected future distributions of vectors of *Trypanosoma cruzi* in North America under climate change scenarios. *PLoS Neglected Tropical Diseases* 8 (5): e2818.

Giesen, C., Roche, J., Redondo-Bravo, L., Ruiz-Huerta, C., Gomez-Barroso, D., Benito, A., and Herrador, Z. 2020. The impact of climate change on mosquito-borne diseases in Africa. *Pathogens and Global Health* 114 (6): 287–301.

González-Rete, B., Salazar-Schettino, P.M., Bucio-Torres, M.I., Córdoba-Aguilar, A., and Cabrera-Bravo, M. 2019. Activity of the prophenoloxidase system and survival of triatomines infected with different *Trypanosoma cruzi* strains under different temperatures: Understanding Chagas disease in the face of climate change. *Parasites and Vectors* 12 (1): 1–12.

González-Tokman, D., Córdoba-Aguilar, A., Dáttilo, W., Lira-Noriega, A., Sánchez-Guillén, R.A., and Villalobos, F. 2020. Insect responses to heat: Physiological mechanisms, evolution and ecological implications in a warming world. *Biological Reviews* 95 (3): 802–21.

Graham, A.L., Shuker, D.M., Pollitt, L.C., Auld, S.K., Wilson, A.J., and Little, T.J. 2011. Fitness consequences of immune responses: Strengthening the empirical framework for ecoimmunology. *Functional Ecology* 25 (1): 5–17.

Günther, J., Martínez-Muñoz, J.P., Pérez-Ishiwara, D.G., and Salas-Benito, J. 2007. Evidence of vertical transmission of dengue virus in two endemic localities in the state of Oaxaca, Mexico. *Intervirology* 50 (5): 347–52.

Harrington, J.M. 2011. Antimicrobial peptide killing of African trypanosomes. *Parasite Immunology* 33 (8): 461–69.

Harrison, J.F., Woods, H.A., and Roberts, S.P. 2012. *Ecological and Environmental Physiology of Insects*. Oxford: Oxford University Press.

Hlavacova, J., Votypka, J., and Volf, P. 2013. The effect of temperature on Leishmania (Kinetoplastida: Trypanosomatidae) development in sand flies. *Journal of Medical Entomology* 50 (5): 955–58.

Hunter, K., Cook, C., and Hayunga, E.G. 1984. Leishmanial differentiation in vitro: Induction of heat shock proteins. *Biochemical and biophysical research communications*, 125(2): 755–60.

IPCC. 2007. 4th Assessment Report "Climate Change 2007: Synthesis Report." <http://www.ipcc.ch/ipccreports/ar4-syr.htm>.

IPCC. 2014. *Climate Change 2014: Synthesis Report Contributions of Working Groups I, II and III to the Fifth Assessment Report of the IPCC*, edited by Core Writing Team, R.K. Pachauri, and L.A. Meyer. Geneva: Intergovernmental Panel on Climate Change.

Iriso, C.A., Bueno, M.R., De las Heras, E., Lucientes, J., and Molina, R. 2017. Climate change in Spain and its influence on vector-transmitted diseases. *Revista de Salud Ambiental* 17: 70–86.

Jiménez-Cortés, J.G., García-Contreras, R., Bucio-Torres, M.I., Cabrera-Bravo, M., Córdoba-Aguilar, A., Benelli, G., and Salazar-Schettino, P.M. 2018. Bacterial symbionts in human blood-feeding arthropods: Patterns, general mechanisms and effects of global ecological changes. *Acta Tropica* 186: 69–101.

Jupatanakul, N., Sim, S., and Dimopoulos, G. 2014. The insect microbiome modulates vector competence for arboviruses. *Viruses* 6 (11): 4294–4313.

Kikuchi, Y., Tada, A., Musolin, D.L., Hari, N., Hosokawa, T., Fujisaki, K., and Fukatsu, T. 2016. Collapse of insect gut symbiosis under simulated climate change. *MBio* 7 (5): e01578–16.

Kraaijeveld, A.R., Ferrari, J., and Godfray, H.C.J. 2002. Costs of resistance in insect-parasite and insect-parasitoid interactions. *Parasitology* 125 (7): S71–S82.

Kraemer, M.U., Sinka, M.E., Duda, K.A., Mylne, A.Q., Shearer, F.M., Barker, C.M., Moore, C.G., et al. 2015. The global distribution of the arbovirus vectors *Aedes aegypti* and *Ae. albopictus*. *Elife* 30 (4): e08347.

Lafferty, K.D. 2009. The ecology of climate change and infectious diseases. *Ecology* 90 (4): 888–900.

Lehane, M.J. 2005. *The Biology of Blood-Sucking Insects*. 2nd edn, Cambridge: Cambridge University Press.

Loker, E.S. and Hofkin, B.V. 2015. *Parasitology. A Conceptual Approach*. New York: Garland Science.

Martínez-Ibarra, J.A., Grant-Guillén, Y., Morales-Corona, Z.Y., Haro-Rodríguez, S., Ventura-Rodríguez, L.V., Nogueda-Torres, B., and Bustos-Saldaña, R. 2008. Importance of species of Triatominae (Heteroptera: Reduviidae) in risk of transmission of *Trypanosoma cruzi* in western Mexico. *Journal of Medical Entomology* 45: 476–82.

Martins, K.A., Morais, C.S., Broughton, S.J., Lazzari, C.R., Bates, P.A., Pereira, M.H., and Dillon, R.J. 2021. Response to thermal and infection stresses in an American vector of visceral leishmaniasis. *Medical and Veterinary Entomology* 37(2): 238–51.

Melo, R.D.F.P., Guarneri, A.A., and Silber, A.M. 2020. The influence of environmental cues on the development of *Trypanosoma cruzi* in triatominae vector. *Frontiers in Cellular and Infection Microbiology* 10: 27.

Miller, E. and Huppert, A. 2013. The effects of host diversity on vector-borne disease: The conditions under which diversity will amplify or dilute the disease risk. *PLoS One* 8 (11): e80279.

Mordecai, E.A., Caldwell, J.M., Grossman, M.K., Lippi, C.A., Johnson, L.R., Neira, M., Rohr, J.R., et al. 2019. Thermal biology of mosquito-borne disease. *Ecology Letters* 22 (10): 1690–1708.

Mukhopadhyay, S., Kuhn, R.J., and Rossmann, M.G. 2005. A structural perspective of the flavivirus life cycle. *Nature Reviews Microbiology* 3 (1): 13–22.

Murdock, C.C., Paaijmans, K.P., Bell, A.S., King, J.G., Hillyer, J.F., Read, A.F., and Thomas, M.B. 2012. Complex effects of temperature on mosquito immune function. *Proceedings of the Royal Society B: Biological Sciences* 279 (1741): 3357–3366.

Ngarakana-Gwasira, E.T., Bhunu, C.P., and Mashonjowa, E. 2014. Assessing the impact of temperature on malaria transmission dynamics. *Afrika Matematika* 4 (25):1095–1112.

Nnko, H.J., Ngonyoka, A., Salekwa, L., Estes, A.B., Hudson, P.J., Gwakisa, P.S., and Cattadori, I.M. 2017. Seasonal variation of tsetse fly species abundance and prevalence of trypanosomes in the Maasai Steppe, Tanzania. *Journal of Vector Ecology* 42 (1): 24–33.

Paaijmans, K.P., Blanford, S., Chan, B.H., and Thomas, M.B. 2012. Warmer temperatures reduce the vectorial capacity of malaria mosquitoes. *Biology Letters* 8 (3): 465–68.

Paaijmans, K.P., Read, A.F., and Thomas, M.B. 2009. Understanding the link between malaria risk and climate. *Proceedings of the National Academy of Sciences* 106 (33): 13844–13849.

Papageorgiou, F.T. and Soteriadou, K.P. 2002. Expression of a novel Leishmania gene encoding a histone H1-like protein in Leishmania major modulates parasite infectivity in vitro. *Infection and Immunity* 70 (12): 6976–6986.

Pascual, M., Ahumada, J.A., Chaves, L.F., Rodo, X., and Bouma, M. 2006. Malaria resurgence in the East African highlands: Temperature trends revisited. *Proceedings of the National Academy of Sciences* 103 (15): 5829–5834.

Pedrosa, M.C., Borges, M.A.Z., Eiras, Á.E., Caldas, S., Cecílio, A.B., Brito, M.F., and Ribeiro, S.P. 2021. Invasion of tropical montane cities by *Aedes aegypti* and *Aedes albopictus* (Diptera: Culicidae) depends on continuous warm winters and suitable urban biotopes. *Journal of Medical Entomology* 58 (1): 333–42.

Poulin, R. 2007. *Evolutionary Ecology of Parasites*. Princeton: Princeton University Press.

Reinhold, J.M., Lazzari, C.R., and Lahondère, C. 2018. Effects of the environmental temperature on Aedes aegypti and Aedes albopictus mosquitoes: A review. *Insects* 9 (4): 158.

Reisen, W.K., Fang, Y., and Martinez, V.M. 2006. Effects of temperature on the transmission of West Nile virus by *Culex tarsalis* (Diptera: Culicidae). *Journal of Medical Entomology* 43 (2): 309–17.

Rogers, M.E. and Bates, P.A. 2007. Leishmania manipulation of sand fly feeding behavior results in enhanced transmission. *PLoS Pathogens* 3 (6): e91.

Samuel, G.H., Adelman, Z.N., and Myles, K.M. 2016. Temperature-dependent effects on the replication and transmission of arthropod-borne viruses in their insect hosts. *Current Opinion in Insect Science* 16: 108–13.

Saxena, A., Lahav, T., Holland, N., Aggarwal, G., Anupama, A., Huang, Y., Volpin, H., et al. 2007. Analysis of the *Leishmania donovani* transcriptome reveals an ordered progression of transient and permanent changes in gene expression during differentiation. *Molecular and Biochemical Parasitology* 152 (1): 53–65.

Schade, F.M., Shama, L.N., and Wegner, K.M. 2014. Impact of thermal stress on evolutionary trajectories of pathogen resistance in three-spined stickleback (*Gasterosteus aculeatus*). *BMC Evolutionary Biology* 14 (1): 1–12.

Schmid-Hempel, P. 2011. *Evolutionary Parasitology: The Integrated Study of Infections, Immunology, Ecology, and Genetics*, 2nd edn, Oxford: Oxford University Press.

Shapira, M., Zilka, A., Garlapati, S., Dahan, E., Dahan, I., and Yavesky, V. 2001. Post transcriptional control of gene expression in *Leishmania*. *Medical Microbiology and Immunology* 190: 23–26.

Shapiro, L.L., Whitehead, S.A., and Thomas, M.B. 2017. Quantifying the effects of temperature on mosquito and parasite traits that determine the transmission potential of human malaria. *PLoS Biology* 15 (10): e2003489.

Simoy, M.I., Simoy, M.V., and Canziani, G.A. 2015. The effect of temperature on the population dynamics of Aedes aegypti. *Ecological Modelling* 314: 100–110.

Takaoka, H., Ochoa, J.O., Juarez, E.L., and Hansen, K.M. 1982. Effects of temperature on development of *Onchocerca volvulus* in *Simulium ochraceum*, and longevity of the simuliid vector. *The Journal of Parasitology* 68: 478–83.

Tamayo, L.D., Guhl, F., Vallejo, G.A., and Ramírez, J.D. 2018. The effect of temperature increase on the development of *Rhodnius prolixus* and the course of *Trypanosoma cruzi* metacyclogenesis. *PLoS Neglected Tropical Diseases* 12 (8): e0006735.

Thomas, M.B. and Blanford, S. 2003. Thermal biology in insect-parasite interactions. *Trends in Ecology and Evolution* 18 (7): 344–50.

Valiente Moro, C., Tran, F.H., Nantenaina Raharimalala, F., Ravelonandro, P., and Mavingui, P. 2013. Diversity of culturable bacteria including Pantoea in wild mosquito *Aedes albopictus*. *BMC Microbiology* 13: 1–11.

Vasilakis, N., Deardorff, E.R., Kenney, J.L., Rossi, S.L., Hanley, K.A., and Weaver, S.C. 2009. Mosquitoes put the brake on arbovirus evolution: Experimental evolution reveals slower mutation accumulation in mosquito than vertebrate cells. *PLoS Pathogens* 5 (6): e1000467.

Walsh, J.F., Davies, J.B., Le Berre, R., and Garms, R. 1978. Standardization of criteria for assessing the effect of Simulium control in onchocerciasis control programmes. *Transactions of the Royal Society of Tropical Medicine and Hygiene* 72 (6): 675–76.

Watts, M.J., Monteys, V.S., Mortyn, P.G., and Kotsila, P. 2021. The rise of West Nile Virus in Southern and Southeastern Europe: A spatial–temporal analysis investigating the combined effects of climate, land use and economic changes. *One Health* 13: 100315.

Watts, N., Amann, M., Arnell, N., Ayeb-Karlsson, S., Belesova, K., Boykoff, M., Byass, P., et al. 2019. The 2019 report of The Lancet Countdown on health and climate change: Ensuring that the health of a child born today is not defined by a changing climate. *The Lancet* 394 (10211): 1836–1878.

Wegesa, P. 1966. Variation of microfilarial densities of Onchocerca volvulus in the skin with the time of day. *Annual Report of East African Institute of Malaria and Vector-Borne Diseases* 31–32.

WHO. 2016. *Reglamento Sanitario Internacional 2005*. 3rd edn. Ginebra: OMS.

Williams, C.R., Bader, C.A., Kearney, M.R., Ritchie, S.A., and Russell, R.C. 2010. The extinction of dengue through natural vulnerability of its vectors. *PLoS Neglected Tropical Diseases* 4 (12): e922.

Wilson, A.J., Morgan, E.R., Booth, M., Norman, R., Perkins, S.E., Hauffe, H.C., Mideo, N., et al. 2017. What is a vector? *Philosophical Transactions of the Royal Society B: Biological Sciences* 372 (1719): 20160085.

Wittmann, E.J. and Baylis, M. 2000. Climate change: Effects on Culicoides-transmitted viruses and implications for the UK. *The Veterinary Journal* 160 (2): 107–17.

Yang, H.M., Boldrini, J.L., Fassoni, A.C., de Lima, K.K.B., Freitas, L.F.S., Gomez, M.C., Andrade, V.R., et al. 2014. Abiotic effects on population dynamics of mosquitoes and their influence on dengue transmission. In *Ecological Modelling Applied to Entomology*, edited by Ferreira, C.P., Godoy, W.A.C. Berlin: Springer.

Zhou, G., Minakawa, N., Githeko, A.K., and Yan, G. 2004. Association between climate variability and malaria epidemics in the East African highlands. *Proceedings of the National Academy of Sciences* 101 (8): 2375–2380.

CHAPTER 15

Climate change disrupts insect biotic interactions

Cascading effects through the web of life

Pedro Luna and Wesley Dáttilo

15.1 Introduction

In the last decade, ecological research has made major advances in understanding the dynamics and properties of species interactions, particularly of those comprised of insects and their biotic interactions with other organisms (Dáttilo and Rico-Gray 2018; Del-Claro and Torezan-Silingardi 2021). We know that the ecological and evolutionary dynamics of insect biotic interactions can be explained by the spatial and temporal variation of current and historical climate in the same way that climate can explain the diversity of insects and how they are distributed in space (Classen et al. 2020; Dáttilo and Vasconcelos 2019; Hargreaves et al. 2019; Luna et al. 2021). The fact that climate can influence species and their biotic interactions is relevant when considering insects because climate change has been identified as one of the main drivers of insect declines and consequential cascading extinctions worldwide (Kehoe et al. 2021).

Basic research on the effects of climate change on insect biodiversity has mainly focused on how future climate change will disrupt biological systems scaling from species to populations and communities; for example, how warmer temperatures may affect insect physiology, demography or distribution patterns (González-Tokman et al. 2020; Halsch et al. 2021; Renner and Zohner 2018; Wilson and Fox 2021). However, we must consider that biodiversity is more than species richness and that there are other components of biodiversity such as those involving biotic interactions that are fundamental for the main-tenance and functioning of Earth's ecosystems (Andresen et al. 2018). Considering that insects are the most diverse group of animals and are capable of linking multiple trophic levels (Dáttilo and Rico-Gray 2018; Del-Claro and Torezan-Silingardi 2021; Herrera and Pellmyr 2002), understanding the effects of climate change on insect trophic interactions is crucial for identifying sensitive ecological interactions and forecasting the local per-sistence of species (Kehoe et al. 2021). This is relevant because whenever a species starts declining toward extinction its closest interaction partners may follow, which triggers further extinctions that propagate through the web of life through direct and indirect

Pedro Luna and Wesley Dáttilo, *Climate change disrupts insect biotic interactions*. In: *Effects of Climate Change on Insects*. Edited by: Daniel González-Tokman and Wesley Dáttilo, Oxford University Press. © Oxford University Press (2024). DOI: 10.1093/oso/9780192864161.003.0015

interactions jeopardizing whole populations, communities and ecosystems (Bascompte and Jordano 2007; Montoya et al. 2006; Rezende et al. 2007; Tylianakis et al. 2010). In light of current and global insect declines (Wagner 2020) it is imperative for ecologists to develop conceptual frameworks to understand how insect declines may affect the ecological and evolutionary dynamics of species interactions.

In this chapter, we focus on biotic interactions of insects with other trophic levels such as plants or other types of partners (e.g., invertebrates or vertebrates) to summarize how climate change can drive cascading effects through their interactions. Under an ecological approach, a cascading effect is triggered by the decline, gain or extinction of a species that generates an effect on other species, either positive or negative. We will focus on the direct effects that climate change may impose on antagonisms (i.e., herbivory, granivory and host-parasite systems) and mutualisms (i.e., pollination and seed dispersal). Finally, we will identify knowledge gaps and discuss future directions in the study of insects and their biotic interactions under a climate change approach.

15.2 Direct effects of climate change on insect biotic interactions

Insects comprise one of the most diverse groups of organisms that inhabit Earth and they establish many kinds of interactions with other organisms ranging from antagonisms to mutualisms (Herrera and Pellmyr 2002). These biotic interactions involving insects are historically thought as isolated and paired systems (e.g., plant-pollinator systems); however, in most cases insects are embedded in complex multi-trophic and non-trophic systems, scaling from microorganisms and plants to vertebrates (e.g., humans) (Kehoe et al. 2021; Medone et al. 2015; Reverchon and Méndez-Bravo 2021). In this sense, interactions between insects and plants are among the most diverse on Earth and under a climate change perspective it is worth noting that almost all insect species depend on plants throughout their life cycles (e.g., butterflies depend on plants in their larval and adult stages) (Del-Claro and Torezan-Silingardi 2021; Herrera and Pellmyr 2002). Moreover, insects provide many services that are fundamental for plant life cycles, such as pollination, seed dispersal or protection against herbivores (Del-Claro and Torezan-Silingardi 2021; Herrera and Pellmyr 2002). The interdependence between plants and animals indicates that any negative or positive effect on either trophic level can affect both groups through their direct or indirect interactions. Thus, cascading effects affecting insect populations can be triggered by other insects or by any other trophic level with which they interact (Kehoe et al. 2021; Petsopoulos et al. 2021). Because of the long coevolutionary history of animal and plant interactions, we should consider that cascading effects due to climate change can also trigger evolutionary responses in insect populations in the short and long term (e.g., Miller-Struttmann et al. 2015). In the face of climate change accelerating local extinctions of insects and their interactions, here we review how cascading effects are triggered and buffered by several mechanisms. We also discuss how cascading effects are affecting the ecological and evolutionary dynamics of insects and the species with which they interact.

15.3 Cascading effects through antagonistic insect interactions

15.3.1 *Herbivory and granivory*

When we talk about antagonisms, we refer to an association between organisms in which one benefits at the expense of the other, in this case insects benefiting at the expense of plants and animals. One type of antagonism is herbivory, which is the consumption of plants by animals and encompasses many types of interactions that range from eating specific parts of plants (e.g., leaf chewers), to eating complete plants (e.g., granivores), to facultative herbivores (i.e., floral and extrafloral nectar consumers; Herrera and Pelmyr 2002). Insects that feed on plants are highly diverse and mainly include the orders Lepidoptera (~160,000 species), Orthoptera (~20,000 species), Coleoptera (~350,000 species), Hymenoptera (~150,000 species) and Hemiptera (~112,000 species) (Herrera and Pellmyr 2002). In this sense, any population decline, extinction or distribution shift of plant species directly threatens a great number of insect populations.

The current temperature increase due to climate change is leading to insect and plant population declines and shifts in their spatial and temporal distributions (Thuiller et al. 2005; Wilson and Fox 2021). This is because native habitats no longer present ideal conditions for living and therefore the co-occurrence of herbivorous insects and their host plants is changing, which thus affects their relationships (González-Tokman et al. 2020; Schweiger et al. 2012; Wilson and Fox 2021). This is important because herbivorous insects often have specialized evolutionary relationships with their host plants (Strauss and Zangerl 2002), therefore, the local extinction of host plants would directly affect their associated insect herbivores as they would not have enough time to develop new mechanisms to consume alternative plants that may have different defensive systems (e.g., trichomes and chemical defenses) compared to those of their actual host plants (Pelini et al. 2010). In fact, expected plant extinctions due to warmer climates are likely to drive more animal coextinctions than the reverse scenario (Figure 15.1a and b) mainly because plants are the primary resource in almost all types of trophic chains. Therefore, the loss of animals is less detrimental to plants as plants have alternative mechanisms to complete their life cycles that can replace animals (e.g., mechanical chemical defenses, wind pollination, and self-cloning) (Schleuning et al. 2016). One way to assess how the extinction of plants causes secondary animal extinctions (or vice-versa) is by estimating and simulating the consequences of removing a species from the species interaction network; this approach is known as coextinction modeling. Evidence obtained by using such coextinction models has shown that plant extinctions can predict, for example, lepidopteran coextinctions (Pearse and Altermatt 2013). In particular, the loss of larval host plants has been identified as one of the factors explaining lepidopteran losses (i.e., when the loss of a primary species leads to the loss of its consumer, often a specialist partner) (Pearse and Altermatt 2013). The vulnerability of insects to plant extinctions is not merely hypothetical because empirical data have shown similar results. For instance, the extinction of butterflies and wasps on tropical islands can be the result of the loss of their host plants (Koh et al. 2004). In other words, the loss of host plants on which herbivores feed (particularly specialized ones) is one path to cascading effects in insect herbivore communities. Generalist insect herbivores are in less danger when host plants are lost, mainly due to their wider diet breadth (Figure 15.1c). Unlike specialist herbivores, generalist herbivores may have increased tolerance to plant extinction as they can persist by interacting with alternative host plants; however, they are less abundant compared with the vulnerable and diverse specialized insect herbivores (Bridle et al. 2013; Janzen 1988). It is worth mentioning that most herbivorous insects and biodiversity itself is made up of many rare and specialized species that depend on generalized species to persist. Thus, the loss of single generalist species (animal or plant) could lead to the loss of many rare and specialized species (Bascompte and Jordano 2007; Dáttilo et al. 2016; Montoya et al. 2006).

The loss of host plants must be evaluated carefully since herbivorous insects also serve as a primary resource for predators. In this sense, herbivorous insects can also trigger cascading effects over third trophic level species such as insect predators. An effect on the third trophic level is expected since herbivore abundance can be regulated by predators and as a consequence, changes in herbivore abundances can be linked to increases or decreases in plant species richness (Marquis 2010). In fact, there is evidence showing that when insect predators are excluded, herbivore abundance can increase, leading to higher plant consumption (Marquis 2010; Rico-Gray and Oliveira 2007). In another example, Terborgh et al. (2006) showed that the absence of predators of leaf-cutting ants located on small islands in Venezuela led to reduced plant diversity as the diverse closed-canopy tropical forests were transformed into species-poor habitats due to the absence of ant predators. We must also consider that the loss of species from a single trophic level, such as predators, has the potential to generate several cascading effects. For example, as previously mentioned, the loss of a predator can lead to an increase in herbivore populations, which in turn can affect plant populations through increased herbivory, and if the reproductive organs of plants are compromised by the increased herbivory we can also expect negative effects in plant mutualistic interactions like pollination and seed dispersal. Although the previous types of cascading effects can be expected, there is little evidence of how cascading effects are propagated through multiple interaction types. Thus, mainly because herbivorous insects are highly diverse, we should increase our efforts of understanding how climate will continue to disrupt herbivore interactions to propose strategies to understand the effects of climate change over a larger scale.

Although ecologists have made advances in understanding the effects of climate change on insect herbivory, the effects of climate change on granivory have received less attention. Granivory is a type of herbivory in which animals eliminate potential plant individuals by eating their seeds. Granivorous insects can have negative and positive impacts on plants communities, the former by decreasing plant populations and the latter by decreasing competition between young plants (Hulme and Benkman 2002). One of the major threats that climate change imposes on granivores is in phenological events, mainly because fruiting and flowering are highly synchronized to climatic conditions (Lewis and Gripenberg 2008). Moreover, invasions and extinctions caused by shifts in phenology and geographical distributions should also be considered to understand potential cascading effects that climate change may impose to insect granivory. In this case, we must consider that granivorous insects can range from highly generalized (e.g., harvester ants) to highly specialized (e.g., parasitic wasps) seed predators and therefore, cascading effects may depend on which type of insect is involved (Hulme and Benkman 2002; Rico-Gray and Oliveira 2007).

Granivorous insects such as harvester ants are dominant in some ecosystems and can influence plant communities and populations by reducing seed banks and by allowing rare species to survive (Anjos et al. 2019; Luna et al. 2018; Pirk et al. 2009). Due to the ecological role of harvester ants, one might expect that the loss of a dominant insect granivore may have cascading impacts on plant communities as its ecosystem service could be lost (Lewis and Gripenberg 2008). However, experimental evidence based on species interaction networks have shown that the removal of dominant seed predators does not drive drastic changes to network organization. Instead, the removal of a dominant species showed that networks have structural plasticity because the species that remained undisturbed rearranged themselves and established new interactions replacing the lost species (a phenomenon known as network rewiring) and expanding their diets (Timóteo et al.

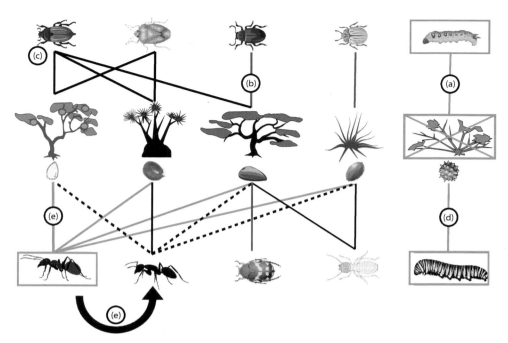

Figure 15.1 A hypothetical species interaction network showing the potential cascading effects on insect herbivore interactions. Species interactions in the top section represent those of insect chewers, leaf miners, or suckers. The bottom section represents insect seed predators. As plants can have both leaves and seeds we show how they can interact with herbivores and seed predators. Lines represent interactions between plants and insects, red lines represent interactions lost due to the extinction of either a host plant or an insect (a and d). Blue lines denote specialized interactions that are at risk if the host plants go extinct or suffer spatial shifts imposed by climate change (b). The loss of a generalist seed predator can be buffered by submissive insect species; dashed lines represent potential new interactions that could be established if generalist insects are excluded (e).

2016). By structural plasticity, we are referring to the ability of species interaction networks to withstand changes in their organization (Dehling 2018). The loss of a generalist species can imply community collapse because generalist species allow the persistence of rare species by interacting with them (Bascompte and Jordano 2007; Saavedra et al. 2011). Moreover, the loss of a generalist species can undermine other processes such as seed removal, displacement, and manipulation and may lead to the loss of other kinds of services provided to plants by seed predators (Lewis and Gripenberg 2008). Although the loss of generalists is worrying, there is growing evidence showing that increased functional redundancy (i.e., the ability of two or more species to perform similar functional roles within a network) can buffer against the loss of other species (Sanders et al. 2018). However, given the alarming extinction of insects that has occurred in recent decades, the vulnerability of ecosystems has also been increasing as biodiversity loss leads to a reduction in redundant interactions and consequently less robustness in coextinction cascades (Sanders et al. 2018). If we focus on specialized seed predator interactions, we could expect that the responses of granivorous insects to climate change might be similar to those of other specialist herbivores (Figure 15.1d). Accordingly, the loss of the plants (due to extinction or spatial-temporal shift) on which a specialist granivore feeds could exert negative effects on granivore populations as they would not have adaptations to consume alternative resources (Pelini et al. 2010). In fact, evidence shows that species with narrower diets also have narrow environmental niches (Slatyer et al. 2013), making them more sensitive to climate change because any change in climate may affect the distribution of their food sources and, ultimately, their survival (Schweiger et al. 2012).

In the case of phenological mismatches among granivores and plants, climate change has the potential to alter the synchronization of masting species as the main clue for masting is climate (Pearse et al. 2016). Mast seeding is the event in which plants of the same species and population synchronize to produce flowers and consequently seeds (Silvertown 1980). It is known that mast seeding is a strategy that benefits plants by increasing pollination efficiency and satiating predators due to increased floral and seed abundance (Pearse et al. 2016). Masting is triggered by weather cues (e.g., temperature and precipitation) and therefore it is sensitive to increasing temperatures or delayed rainy seasons driven by climate change (McKone et al. 1998). As insect granivores are highly diverse and depend on seed production, changes in masting could drive cascading effects in their populations. Accordingly, current evidence already indicates that mast frequency has increased during the last decades due to Earth's warming (Hacket-Pain and Bogdziewicz 2021), affecting insect granivores and their interactions with plants. Moreover, changes in mast frequency can lead to rises in resource availability for insect granivores (e.g., more frequent interannual flowering) allowing insect populations to grow and devastate annual seed crops (McKone et al. 1998). In another case, it was documented that increased fluctuations in seed production due to climate changes has decreased the impact of seed predators (Diptera and Hemiptera) as the resource has become less predictable, and therefore insect populations fluctuate more, increasing seed survival (Solbreck and Knape 2017). Although it is clear that seed production can be disrupted over time by climate change, there are only a few studies that directly evaluated the effects of climate change in the interactions between insect predators and mast seeding plants, and this topic is still limited and fragmented in the literature (Hacket-Pain and Bogdziewicz 2021; McKone et al. 1998; Pearse et al. 2016; Solbreck and Knape 2017).

15.3.2 *Parasitoids*

Besides insect–plant interactions, we must consider that there are other types of antagonistic interactions in which insects are embedded. One of these cases is animal–animal interactions, particularly those comprising antagonistic parasitoid interactions. Parasitoids are insects whose larvae feed and develop within or on the bodies of other arthropods, ultimately killing their hosts. Parasitoids depend on their hosts to complete their life cycles, so their seasonal activities are synchronized (i.e., phenology) and depend on environmental conditions; thus, any change in climate may desynchronize their activity and biotic interactions (Hance et al. 2007). For example, insect parasitoids that are distributed in polar and temperate regions have mechanisms like diapause to cope with unfavorable environmental conditions (e.g., cold climate) (Tougeron et al. 2020). Early or late entry into diapause may affect parasitoids because their populations may not co-occur in space or time with food (host) or sexual partners or simply die due to bad environmental conditions (Tougeron et al. 2020). The previous could lead to potential cascading effects by allowing the increase in host populations and parasitoid population declines because many intraspecific interactions (e.g., sexual interactions) will be lost. Global warming is already changing the spatial and temporal distribution of parasitoids since many habitats in temperate regions are becoming warmer and are affecting insect diapause dynamics and the co-occurrence of species (Damien and Tougeron 2019). For example, Furlong and Zalucki (2017) showed that future temperature increases will have a greater impact on the distribution of the parasitoid wasp *Diadegma semiclausum* rather than on its host and the exclusion of *D. semiclausum* from its native distribution will leave many regions without its pest control services. For generalist insect parasitoids, phenological mismatches or future changes in spatial distributions should not represent the same hazard as for specialists because generalists can switch to alternative hosts if their phenology or distribution is modified (Furlong and Zalucki 2017). In contrast, specialist parasitoids have constrained diets and narrower environmental tolerances, thus any temporal or spatial shift can directly affect their survival since switching hosts is not possible due to their lack of adaptations (Tougeron et al. 2020). In the previous paragraphs we have made clear that there are different responses to climate change among generalist and specialist species, suggesting that generalist species should be more resistant to climate change. However, generalists are not always the winners in climate change mainly because the roles of generalists and specialists depend on the number of species in the community (i.e., network size) (Gilman et al. 2010). For example, it has been shown that generalist species that contribute the most to network stability are also those that are the most vulnerable to extinction (Saavedra et al. 2011). This is possibly because species that play structural roles in ecological networks (i.e., highly interactive species) are subject to greater constraints than species that have intimate specialized interactions (e.g., generalist species interacting with a greater number of less rewarding hosts) (Saavedra et al. 2011).

15.3.3 *Disease vectors*

Another type of insect antagonistic interaction involves disease vectors such as mosquitoes or Chagas bugs. In this type of interaction insects act as vectors to transmit diseases (caused by viruses, bacteria, protozoa, etc.) to different kinds of hosts (particularly vertebrates). Insect-borne diseases are of particular concern as they include diseases that infect millions of humans and other vertebrates every year (e.g., dengue, malaria, chikunguya, Chagas disease, Zika, leishmaniasis, among others). Because of their medical importance insect disease vectors are a group of concern and their responses to climate change should be studied in detail. In the case of insect-borne diseases the prevalence of warmer temperatures due to climate change is worrying, particularly for mosquitoes since their populations proliferate faster and they bite more as temperatures become warmer (Epstein 2000). Accordingly, studies have shown that climate change has the potential to alter the distribution of insect disease vectors since many of them are originally from the tropics and warmer temperatures are expanding their optimal environmental conditions toward highlands and temperate regions (Epstein 2000). Indeed, a recent analytical review showed that 54 percent of studies agree that climate change will increase mosquito-borne diseases due to an increased vector abundance driven by variations in climate (Franklinos et al. 2019). In another case it was shown that increasing temperatures (projected until the year 2100) will increase the suitability of highlands for malaria (Ebi et al. 2005). In the case of insect vectors, the simple arrival of a vector to a new area already implies a cascading effect since the disease they bring with them meets new hosts, producing diseases where they did not exist earlier (Franklinos et al. 2019). Such is the case in northern Europe since expected increases in humidity and temperature due to climate change are likely to increase the environmental suitability for mosquitoes (*Aedes albopictus* and *Aedes aegypti*) which can transmit dengue and chikungunya (Caminade et al. 2012; Fischer et al. 2011; Liu-Helmersson et al. 2016). However, drier and warmer conditions will limit mosquito populations in southern Europe (Caminade et al. 2012). Beyond mosquitoes, a study with Chagas bugs (Hemiptera—Triatomine) found that increasing temperatures is likely to decrease the geographical distribution of two vector species in South America (Medone et al. 2015), which suggests that not all insect vectors will increase their distributional ranges due to current climate changes. A reduction in insect vectors is positive for human populations but not for the organisms that depend on those vectors to complete their life cycles (e.g., bacteria and protozoans). Under this view the loss of vectors will drive a cascading effect by affecting, for example, bacterial diversity. Other vector-borne diseases like leishmaniasis, which is vectored by sandfly species (Phlebotominae), are of interest since climate change will likely affect their distributions and transmission (Ready 2008). Leishmaniasis requires a reservoir host which adds a third trophic level to its interaction dynamics, and therefore, shifts in the distribution of its vectors due to climate change will likely produce cascading effects on hosts and reservoirs (Ready 2008). For example, climate will increase the risk of human exposure to leishmaniasis in northern America, bringing into contact populations of vectors and hosts producing disease where it historically did not exist (González et al. 2010). In another study, the authors showed that leishmaniasis incidence is associated with El Niño cycles (Franke et al. 2002), because to the increase in El Niño events (Trenberth and Hoar 1997), it is expected that leishmaniasis cases can increase. In summary, climate change is driving shifts in the distributions of many insect vectors as the potential for cold habitats to become warmer is increasing, because of this many insects are taking advantage of this environmental change by migrating to new areas, bringing with them new diseases, and creating biotic interactions with new partners.

Overall, insect antagonistic interactions are highly diverse and include interactions with many different taxa from different kingdoms (e.g., animals, plants, bacteria and protozoans). Due to this diversity and the sensitivity of insects to environmental changes, climate change is triggering many cascading effects and many antagonistic interactions that must be studied in detail. This is because empirical evidence is poor and most of what we know comes from simulations and predictions into the future using current data. Despite these limitations in available information, we know that cascading effects are propagating throughout the web of life by insect antagonistic interactions, particularly changing interaction dynamics by geographical shifts and the extinction of species.

15.4 Mutualisms: Seed dispersal, pollination and ant–plant protective systems

15.4.1 *Seed dispersal by insects*

Current estimates show that 57 percent of Earth's plants will go extinct due to climate change if they do not have the capacity or opportunity to migrate (Warren et al. 2013). This implies that animal seed dispersal is a fundamental ecosystem service to preserve plant populations under climate change. In fact, the transportation of seeds by animals is key for the persistence of plant populations as many plant species rely on animals to effectively disperse their seeds (Jordano 2000; Rico-Gray and Oliveira 2007). Accordingly, the great majority of seed dispersal studies have focused on studying vertebrates while giving little or no attention to seed dispersal by invertebrates (Corlett 2021; Rogers et al. 2021). Insects are remarkable seed dispersers and can regulate this fundamental ecosystem service in many environments (Corlett 2021; Herrera and Pellmyr 2002; Rico-Gray and

Oliveira 2007; Rogers et al. 2021). For example, more than 3,000 plant species belonging to more than sixty families can be dispersed by ants (Handel and Beattie 1990). Ants can be important seed dispersers across entire biomes, such as in the Brazilian Caatinga where ants can disperse seeds from more than 100 plant species (Crespo-Pérez et al. 2020; Leal et al. 2007, 2014). In addition, the widely distributed group of harvester ants (*Pogonomyrmex* spp.; commonly perceived as seed predators) can also disperse seeds from many plant species from different families (Anjos et al. 2020; Luna et al. 2018). Dispersal of seeds by ants is particular as their coevolutionary history with plants has led to the development of adaptations called elaisomes (i.e., fleshy structures rich in lipids and proteins that are attached to seeds) that are attractive food sources to many ant species (Beattie and Hughes 2002; Rico-Gray and Oliveira 2007). Ants carry seeds and eat the elaisome, then discard the seeds in their nest dumps which allows seeds to be transported some meters away from the mother (1 m to ~25 m, Anjos, Leal, et al. 2020). Evidence has shown that ants are effective dispersers of small seeds (< 5 millimeters (mm)), and although ants are able to move large seeds (> 5 mm), their seed dispersal capacity cannot compensate the loss of vertebrates, which usually move large seeds (Rogers et al. 2021). Under a climate change scenario, studies on defaunation have shown that the dynamics of seed dispersal can be jeopardized by the loss of large-bodied insects. For instance, in undisturbed remnants of Atlantic forests where large (~1.5 centimeters (cm)) and small ants (< 1 cm) coexist, seed removal and dispersal distances are greater when compared to highly fragmented and disturbed forests. This occurs because large ants are not present in disturbed forests, resulting in low seed removal and short dispersal distances (Bieber et al. 2014). According to current declines in vertebrate and insect diversity, previous studies suggest that plants with larger seeds are more threatened by the loss of seed dispersers including ants (Rogers et al. 2021). Moreover, ants are among of the most dominant terrestrial organisms linked to many organisms by diverse biotic interactions and are highly sensitive to temperature since they are ectotherms (Parr and Bishop 2022). For example, it is expected that climate may affect nocturnal more than diurnal granivore ants (i.e., lower seed removal at nights), which is because nocturnal ants seem to be more sensitive to warming temperatures than diurnal ants (Garcia-Robledo et al. 2018). Additionally, aridity as a consequence of climate change can impact ant seed dispersal by reducing rates of seed removal and distance of removal in Brazilian Caatinga (Oliveira et al. 2019). In other words, with increasing aridity imposed by climate change many myrmecochorous species are at risk of losing effective seed dispersal.

Beyond ants, dung beetles can also move and promote seed establishment, particularly in tropical environments (Andresen 2002; Andresen and Levey 2004). Dung beetles are secondary seed dispersers of many plant species by removing and transporting dung of frugivorous vertebrates (i.e., they disperse seeds that were primarily dispersed by vertebrates) (Andresen 2002). The loss of dung producers (i.e., vertebrates) could generate cascading effects in the multi-trophic interaction among vertebrates (e.g., howler monkeys), dung beetles and seeds (Portela Salomão et al. 2018). Although dung beetles are indicator species that provide many ecosystem services, it is not clear how the future climate will alter their distribution and their biotic interactions, particularly in the tropics. For temperate regions, we know that the warming will increase its suitability for dung beetles in Europe, while southern regions may experience species impoverishment (Dortel et al. 2013). It has also been observed that the warming of lowlands is causing lowland dung beetles to expand their distribution upwards. This migration may possibly leave many plant species without secondary seed dispersers (Menéndez et al. 2014). The potential upward migration of dung beetles seems to depend on the location of the mountain and its pool of species, affecting each montane community differently. In another example in South America, it was shown that the geographic distributions of many dung beetle species will decrease due to warmer temperatures, highlighting that species exposed to recent deforestation are the most sensitive to changes in temperature. This means that regions with a lot of deforestation (e.g., the Amazon region) are at higher risk of losing the seed dispersal services provided by dung beetles (Maldaner et al. 2021), possibly triggering secondary plant extinctions. In general, the loss of insect seed dispersers can drive cascading effects in plant populations in several ways, starting with reducing the number of plants dispersed since dispersal frequency can depend on seed disperser richness (Rumeu et al. 2017). Another aspect to consider is that seed disperser loss not only affects the reproduction of adult plants but also affects the diversity of seed banks (i.e., seed richness and abundance; Donoso et al. 2017). For example, reduced seed removal can increase seed mortality due to increased seed density under parent plants (Caughlin et al. 2014). In the case of the extinction of seed dispersers, plants seem to be the most affected group because even the loss of a few species can lead to a significant loss of this service, greatly reducing gene flow between plant populations (Donoso et al. 2020).

From what we know about herbivory, we could expect that for seed dispersal, plant species that establish a small number of interactions (i.e., specialists) are the ones at higher risk due to climate change as they have narrower climate niches and rely on few animal species to disperse their seeds (Pelini et al. 2010; Slatyer et al. 2013). However, evidence has shown that plants and animals that establish a great number of interactions (i.e., generalists) rely more on mutualistic interactions such as seed dispersal. Therefore, if one generalist species becomes extinct the risk of triggering cascading effects is low as there are other species that can reduce the impact of its loss (Clemente et al. 2013; Fricke et al. 2017). In contrast, plant species with a small number of interaction partners tend to show less dependence to their mutualistic partners as they can have alternative mechanisms to cope with the lack of mutualistic partners (e.g., vegetative reproduction, anemochory) (Dáttilo. 2012; Fricke et al. 2017). In another case, researchers showed that the loss of generalist mutualists on islands (seed dispersers and pollinators) cannot be compensated by other species since the species pool is limited, ultimately affecting the species with few interactions (Fricke et al. 2018). In summary, the effect of the degree of specialization in seed dispersal interactions seems be more variable when compared to antagonistic herbivore interactions where specialists are the species at higher risk of extinction or coextinction.

We must keep in mind that plants are embedded in many interactions at the same time and if plants lose the ability to move their seeds through interactions with animals it could initiate cascading effects in other types of interactions. For instance, mutualistic interactions such as seed dispersal and pollination depend on each other. In this case, the plants that produce seeds must first flower and be pollinated (Figure 15.2). Despite this, we often ignore the fact that there are interactions that precede others and we lack the theoretical and empirical knowledge to understand the cascading effects across mutualisms and between mutualism and antagonisms. Pioneer studies that investigated species interaction networks involving several types of mutualisms have shown that insects play fundamental roles in maintaining the robustness (i.e., the capacity to maintain network structure due to the loss of species) of networks that include multiple types of interactions. For example, Dáttilo et al. (2016) found that ant species that establish a large number of interactions can link different interaction types (e.g., seed dispersal, pollination and ant-defensive systems) and therefore their extinction could lead to cascading coextinctions that propagate over the different trophic levels. In the same study, when simulating coextinctions, the authors observed that the overlap of generalist species over

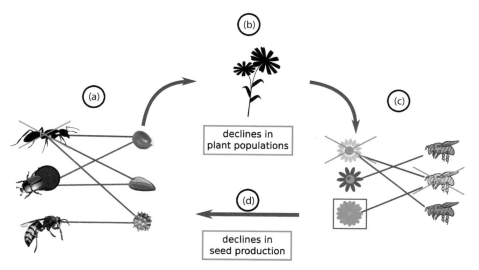

Figure 15.2 Feedback among pollination and seed dispersal. a) The loss of a generalist seed disperser affects the plant populations of the species that depended on the extinct animal. b) Reductions in seed dispersal reduce the abundances of plant populations. c) Reductions in plant abundances lead to reductions in floral production that can leave animals without food resources. d) Fewer flowers to pollinate produce fewer fruits/seeds for dispersers, threatening the populations of seed dispersers.

multiple interaction types increases network robustness (Dáttilo et al. 2016). In other words, the loss of generalist species always implies the extinction of their interactions, leaving many species without mutualistic partners. Under a climate change scenario, if the goal is to conserve species and their interactions, efforts should be focused on identifying the species that establish more interactions and link multiple interaction types (i.e., keystone mutualists) (Figure 15.3).

In the previous sections, we highlighted that the loss of generalist species and their interactions can be mitigated by other submissive species that rearrange and establish new interactions (i.e. rewiring) when keystone species are lost. In conventional models where species coextinctions are simulated, the rewiring of species interactions is neglected which has led to a possible overestimation of the effect of an extinction on other species (Vizentin-Bugoni et al. 2020). Although interaction rewiring has been identified for some time, it was not until recent years that it was considered an important driver of the dynamism and variation of biotic interactions (Luna and Dáttilo 2021; Poisot et al. 2012; Vizentin-Bugoni et al. 2020). Rewiring is a measure of variation of species interactions and indicates how dynamic a species can be in different scenarios; for example, between day–night periods, over weeks or years and even across spatial gradients (CaraDonna et al. 2017; Carstensen et al. 2014; Luna et al. 2018). This means that a single species or a group of species can establish many different pairwise interactions depending on the context, which may have consequences for community and population dynamics. Moreover, despite the high dynamism of biotic interactions we do not understand how the rapid turnover of interaction partners can affect evolutionary or coevolutionary patterns. For example, if the majority of the insects that visit a flower change every day it could be relevant to know which insect is the actual pollinator of the visited plant, since much of the variation in floral visitors could be generated only by species that do not

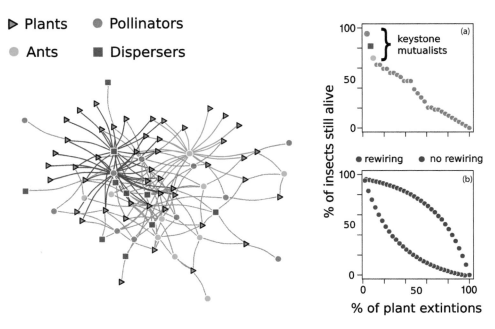

Figure 15.3 Hypothetical network showing different types of mutualistic interactions (pollination, seed dispersal and ant–plant defensive systems). Each node (triangle, circle, or square) represents one species of different trophic levels (plants, ants, pollinators or seed dispersers), and the lines represent the interactions between them. Colored interactions (blue, purple and orange) denote the interactions of keystone mutualists (species with many interactions that link different types of interactions). a) When the extinction of a keystone mutualist is simulated, we can observe that just by losing three species the plants that depend on them go extinct. b) When interaction rewiring is implemented (blue points) in coextinction simulation models, we can observe that networks are more robust to the loss of species (generalist or specialist). However, when rewiring (i.e., interaction rearrangement) is not considered, networks tend to be less robust to extinctions and consequent coextinctions (red points) occur rapidly.

pollinate the plant (reviewed by Burkle and Alarcon 2011). Conversely, if the majority of insects that visit the plant contribute to its reproduction, then its high dynamism may also have consequences for trait matching, for example. The capacity of a species to rewire depends on many factors (e.g., climate and species traits), but until now this has remained a topic yet to be elucidated in detail. Because interaction rewiring allows us to measure how species interact in different contexts, its measurement could be useful to understand further how species interactions are being affected by climate change. One development that has fostered our understanding of coextinction cascades is the inclusion of rewiring in coextinction simulation models. This is worth mentioning because more dynamic models (i.e., those including rewiring as a property of species interactions) show less catastrophic scenarios compared to the coextinction models that do not include it. Interaction rewiring as an ecological process therefore can increase the robustness of mutualistic networks (seed dispersal and pollination) since the more dynamic species are, the more robust the networks are against species extinctions (Schleuning et al. 2016; Figure 15.3). Models that include rewiring to simulate species losses are promising, however the data to develop such models are scarce (e.g., species abundances, morphological traits, phenological data);

by implementing these kinds of models we can build more accurate predictions of how species coextinctions can propagate cascading effects through the web of life.

As previously highlighted, plants face obstacles to dispersing their genes since they are sessile organisms. However, plants have developed adaptations to attract animals and ameliorate their lack of movement over the evolutionary time. Such adaptations involve the development of mutualistic interactions with animals, by either consuming and transporting their seeds or in the case of pollination by transporting pollen grains between flowers when looking for food rewards (e.g., fruits, nectar and pollen). Estimations show that insects are the main group of animals that visit flowers in search of food, mates, or to acquire other resources, and Lepidoptera (~141,000 species), Coleoptera (~77,000 species), Hymenoptera (~70,000 species) and Diptera (~54,000) are the most dominant flower-visiting species and potential pollinators (Wardhaugh 2015). Although we are aware of the high diversity of insect pollinators, most studies dealing with the effects of climate change have focused on current bee declines and on phenological mismatches; however, there are other effects that climate change can exert on insect biotic interactions (Wilson and Fox 2021). In this sense, increases in temperature and precipitation are contributing to declines in bumblebee diversity worldwide (affecting approximately sixty-six species), which is accompanied by the loss of their pollination services in many environments, particularly in the Northern Hemisphere (Soroye et al. 2020). Climate change can also affect the invasive honeybee (*Apis mellifera*) by influencing its behavior and physiology and by driving shifts in its geographical distribution, giving rise to competitive relationships with native floral visitors in different regions of the globe (Cruz et al. 2022). In the case of honeybees reaching new habitats, we are referring to the introduction of a highly generalist species into a native community, which can drive cascading effects in both plants and floral visitors. For instance, the honeybee can change native communities by monopolizing many interactions, leaving native species without food resources, decreasing overall community pollination effectiveness, and native bee abundances (Santos et al. 2012). The approach used by Santos et al. (2012) should be interpreted with caution since in their study they remove *A. mellifera* from the networks without having any control treatment (i.e., separate the effect of removing a species from the network regardless of its identity) (but see Dáttilo et al. 2022 where the authors used a standardized framework based on species removal). In another study, by controlling the removal of species from a network it was found that alien and native species have similar roles maintaining the structure and robustness of plant-pollinator networks (Parra-Tabla et al. 2019). It has also been shown that honeybees are not always detrimental to the communities they invade, as some communities have their own highly competitive bees that buffer the gain of an invasive species, which can even promote pollination in pollinator-poor habitats (Giannini et al. 2015) (Figure 15.4).

One of the main effects that climate change exerts on insect pollinators and floral visitors are shifts in their seasonal activities (i.e., phenology). Shifts in insect phenology are related to rising temperatures since temperature directly influences insect and plant development (Settele et al. 2016). Notably, rising temperatures are already influencing the first flowering and the emergence date of insects, leading to an increasing desynchronization between insect and plant life cycles with consequences for both trophic levels. Studies are already showing that several butterfly species are appearing earlier in the year (~ sixteen days earlier per decade) in response to shorter winters caused by warmer weather (Parmesan 2007). The desynchronization of insect and plant life cycles implies that many insect pollinators are no longer co-occurring with their food resources in time and space while many plants will lose potential pollinators. For example, in mountain

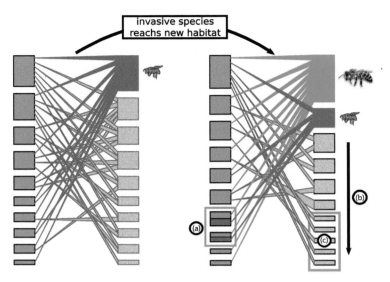

Figure 15.4 Diagram showing the gain of an invasive species in a native community. The left panel denotes a native community with its own dominant species (red bees and blue interactions). In the right panel we show how an invasive species reaches a native community (red interactions). There are several cascading effects: a) invasive species compete with a dominant native species for food resources; b) submissive species interact less due to the higher competition imposed by the invasive species; and c) submissive species are in danger of losing their food sources as they compete with an invasive and dominant species.

systems where we find habitats located in cold climates and high elevations, warming is advancing the snowmelt date, which allows early flowering and leaves many bumblebee species without food as insect and plant life cycles are desynchronized (Miller-Struttmann et al. 2015). Similar trends have been reported for flower-visiting syrphids (Diptera) since early snowmelt is threatening many fly populations because it is related to less temporal overlap between syrphids and flowering plants (Iler et al. 2013). In another case, it was documented that shifts in floral resources (data spanning 130 years) may not threaten generalist bee species since they can keep up with changes in plant flowering (Bartomeus et al. 2011). On the other hand, butterfly declines have been linked to changes in environmental conditions as butterfly populations can decrease either by exposure to extremely warm or extremely cold temperatures (Roland and Matter 2013). Moreover, due to increasing temperatures in the Northern Hemisphere, the distribution of butterflies is shifting poleward and to higher elevations, such changes in the distribution of butterflies would leave many plants without pollinators. However, evidence has shown that if butterflies can expand their diet by interacting with species phylogenetically close to those with which they are adapted to interact, the risk of butterfly cascading extinctions due to the lack of host plants will be reduced (Descombes et al. 2016). Although it is clear that changing climate is threatening the phenology of insects and their host plants, most data come from temperate mountain systems. Therefore, we lack empirical evidence of how climate change affects insect pollinators in tropical and dry environments. Note that the loss of synchrony between insect pollinators and plant life cycles can lead to cascading effects in other trophic systems like seed dispersal; as we already mentioned pollination precedes seed production. It is worth mentioning that the loss of synchrony between insects and plants can also have evolutionary consequences for insects. For instance, long-tongued bumblebees have experienced a reduction in tongue length due to a reduction in the abundance of the floral resources in which they specialize (Miller-Struttmann et al. 2015). The reduction in tongue length have led bumblebees to develop a more generalist diet, which suggests that some species can undergo rapid evolution to cope with current climate change.

As mentioned for other types of interactions (herbivory, granivory and seed dispersal), one way to estimate and measure possible coextinctions due to the loss of a species and its interactions is using coextinction models. In the case of insect pollinators, coextinction models provide novel information since researchers have done some empirical experiments. In these studies, the extirpation of species has been done in the field, providing opportunities to compare results of simulations with actual empirical evidence. Following this, in one experiment, researchers tested how the removal of the most visited plants from experimental plots affected a plant-pollinator system (Biella et al. 2019). They found that by losing highly visited plants other plants lost flower visitors, suggesting that highly visited plants benefit the ones that are not frequently visited. Additionally, the authors found that flower traits limit pollinator foraging as floral visitor does not switch between different floral shapes. This means that the loss of a generalist plant species can decrease pollinator abundance and affect its foraging strategies. By applying a network approach, the same research group found that the accumulation of local extinctions (of both species and interactions) increases faster when plant generalist species are lost, which is also faster than predicted with coextinction models. Rewiring seemed to be a component that was able to maintain network structure to some extent, as the network became less nested (i.e., less stable) and more modular (i.e., more isolated sub-networks within the main network) (Biella et al. 2020). Experiments seem to be congruent to simulated coextinction models, however they provide more detailed information of the processes triggered by the loss of interacting species.

Estimations show that around 90 percent of flowering plants depend on animals for pollination, of which most are insects (Ollerton 2017). Worryingly, current knowledge shows that many insect floral visitors are at risk, compromising the pollination of many plant species, affecting human crops in particular. In this sense, we should focus our conservation efforts on the preservation of generalist species as it has been determined that these species can maintain the interactions of less abundant species and buffer the loss of another generalist. The ideal scenario would be to stop all extinctions to avoid extinction cascades, but conservation efforts are limited by budgets, and thus we need to be realistic and try preserve the largest number of generalist species that we can.

15.5 Scientometric overview of insect biotic interactions and climate change

In the previous sections we reviewed current literature by synthesizing what we know about insects and their interactions. In this final section, we performed a scientometric review to identify different knowledge gaps. Initially, we used the following keywords: "insects," "climate," "change," "biotic" and "interactions" in the Scopus database to assess the current state of knowledge of insect biotic interactions in the era of climate change. In this review, we did not include the term "cascading effects" since it reduced our search

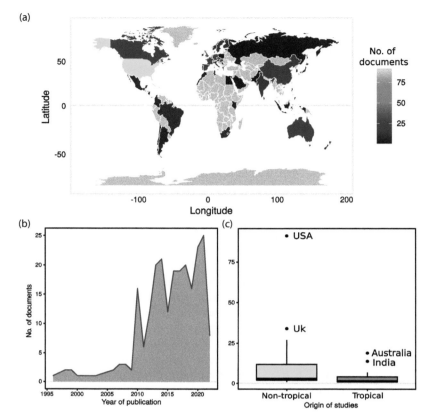

Figure 15.5 a) Map showing the number of published documents by country. b) Frequency of published documents by year, starting with the first traced article (1996) to the most recent (2022). c) Boxplot showing the number of published documents according to the latitudinal location of the country that produced it (non-tropical or tropical region).

to only twenty-one published documents. Our search included the use of the selected keywords in article titles, abstracts and keywords.

We found that there are 235 published documents (69% journal articles, 22% reviews, and 6% book chapters, books and short surveys) published from all around the world (53 countries) that have our selected keywords (Figure 15.5a). To have a point of comparison of how much we know of insect biotic interactions under a climate change approach, we compared our findings with another search focusing only on the keywords "insects" and "climate change" where we found 6,443 documents. When comparing the number of documents focusing on biotic interactions with the overall knowledge regarding insects and climate change, we observed that biotic interaction research represents only 3 percent of the research effort. Moreover, in this case, the first article assessing biotic interactions and climate change involving insects can be traced back to 1996 and was published 63 years after the first article assessing insects and climate change (year 1933). From 1996 to 2009 there was a very low number of published documents (mean ± SD = 1.8 ± 0.78), which started to increase from 2010 to 2022 (16.69 ± 5.73) (t = 9.2, df = 12, P = 0.0001). The highest number of published documents was in 2021 (n = 25 documents) and the lowest in 1996, 2000 and 2002 (one document in each year). Most published documents were produced in non-tropical countries (10.13 ± 16.09) compared with tropical countries (4 ± 5.2) (t = 2.07, df = 49, P = 0.04).

Our scientometrics analyses indicate that our understanding of how insect biotic interactions are influenced by climate change is relatively poor and biased towards non-tropical regions. Although in recent years there has been growing interest in understanding species biotic interactions, the interest in insect research is recent, particularly when assessing its biotic interactions. It is important to note that climate change is a topic of current interest, but the lack of interest in insect research might be because this group is not as charismatic as other taxonomic groups (e.g., vertebrates). However, the recent growing interest in insect biotic interactions could also be related to the development and implementation of new tools to understand species interactions, for example genetic and network analyses. Bias towards non-tropical regions is expected since there is a higher number of developed countries out of the tropics, which in turn invest more economic resources in science. The lack of studies in tropical regions could be due the lack of investment, but also indicates that we do not know how biotic interactions will react to climate change in the regions where diversity peaks! Thus, almost all our knowledge is probably from regions where diversity is low and therefore climate change effects on insect biotic interactions (i.e., cascading effects) could differ between tropical and non-tropical regions. Thus, some cascading effects mentioned in our different sections could only represent the dynamics of non-tropical regions.

15.6 Future perspectives and concluding remarks

One aspect that has been almost ignored in this chapter is indirect interactions and how such interactions may affect whole ecological networks by climate-induced effects. Indirect effects refer to those effects that are not triggered by direct interactions but rather by alternative paths that link two species within a network (Figure 15.6a). Indirect effects driven by indirect interactions are often overlooked since we do not have enough tools and theoretical frameworks to measure and explain their effects through

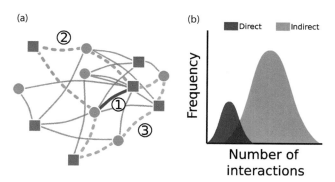

Figure 15.6 a) Hypothetical network showing interactions between insects (blue squares) and plants (green circles); the links between the nodes represent the interactions between insects and plants. (1) denotes a direct interaction between two species, and (2) and (3) denote alternative pathways through which the same focal species may interact indirectly. b) Hypothetical expected frequency of indirect and direct interactions that can be observed with an ecological network according to Pires et al. (2017).

ecological networks in an intuitive way and with biological sense. In fact, most measures to characterize ecological networks focus on direct pairwise interactions (e.g., coextinction models), limiting our ability to predict how climate change can affect insects and their indirect interactions with other trophic levels. This is relevant as after perturbations, indirect effects can be responsible for changes in population and community composition (Novak et al. 2016). The propagation of effects by indirect interactions also has coevolutionary consequences as the convergence of traits in ecological networks can be driven by such diffuse interactions (Guimarães et al. 2011, 2017). To understand indirect interactions and their role in network stability, Pires et al. (2020) proposed a framework to study indirect interactions by including them in coextinction models. They showed that species have more impact on others through indirect interactions rather than by direct pairwise interactions, probably because focal species can be linked to virtually whole network trough indirect interactions (Figure 15.6b). Moreover, they also show that the importance of indirect effects grows as the dependence between species increases because indirect paths provide alternative routes for perturbations to travel through a network (Pires et al. 2020). Because of the ecological and evolutionary relevance of indirect effects to network dynamics we postulate that future directions in the study of cascading effects should focus on elucidating the role of indirect effects in all type of interactions (from antagonisms to mutualisms).

In summary, we can highlight that insects are threatened by climate change through cascading effects mainly driven by shifts in species temporal (i.e., phenological shifts) and spatial (i.e., geographical shifts) distributions, and by the extinction of trophic levels with which they interact (e.g., plants, predators and hosts). Finally, our chapter represents an effort to join evidence involving the most diverse interactions in which insects are entangled and provides a base for future studies.

Acknowledgments

This research was supported by the Consejo Nacional de Ciencia y Tecnología (CONACyT, Mexico) in 2022 under Grant No. FOP16-2021-01-319,227. P. Luna has a scholarship provided by CONACyT (CVU 771,366). P. Luna thanks Claudia Suarez for her support during the writing of this chapter.

Key reflections

- Insects are highly threatened by cascading effects triggered by the loss of interacting partners, particularly plants.
- Different types of interactions are linked through direct and indirect interactions, thus cascading effect can be propagated from one system to another (e.g., from pollination to seed dispersal and granivory).
- Predictions provided by coextintion models seem to agree with empirical experiments of species removal from ecological networks.
- Including interaction rewiring and indirect interactions into coextinction models could provide new and more realistic predictions of how species coextinctions can be propagated in ecological networks.

Key further reading

- Herrera, C.M., and Pellmyr, O. (eds). 2009. *Plant–Animal Interactions: An Evolutionary Approach*. John Wiley and Sons, Malde, USA.
- Dáttilo, W., and Rico-Gray, V. (eds). 2018. *Ecological Networks in the Tropics: An Integrative Overview of Species Interactions from Some of the Most Species-Rich Habitats on Earth*. Cham: Springer.
- Del-Claro, K., and Torezan-Silingardi, H.M. (eds). 2021. *Plant-Animal Interactions: Source of Biodiversity*. Springer Nature, Cham, Switzerland.

References

Andresen, E. 2002. Dung beetles in a Central Amazonian rainforest and their ecological role as secondary seed dispersers. *Ecological Entomology* 27 (3): 257–70.

Andresen, E., Arroyo-Rodríguez, V., and Escobar, F. 2018. "Tropical biodiversity: The importance of biotic interactions for its origin, maintenance, function, and conservation." In *Ecological Networks in the Tropics: An Integrative Overview of Species Interactions from Some of the Most Species-Rich Habitats on Earth*, edited by W. Dáttilo and V Rico-Gray. Cham, Switzerland: Springer Nature, pp. 1–13.

Andresen, E. and Levey, D.J. 2004. Effects of dung and seed size on secondary dispersal, seed predation, and seedling establishment of rain forest trees. *Oecologia* 139: 45–54.

Anjos, D.V., Andersen, A.N., Carvalho, R.L., Sousa, R.M., and Del-Claro, K. 2020. Switching roles from antagonist to mutualist: A harvester ant as a key seed disperser of a myrmecochorous plant. *Ecological Entomology* 45 (5): 1063–1070.

Anjos, D.V., Leal, L.C., Jordano, P., and Del-Claro, K. 2020. Ants as diaspore removers of non-myrmecochorous plants: A meta-analysis. *Oikos* 129 (6): 775–86.

Anjos, D.V., Luna, P., Borges, C.C., Dáttilo, W., and Del-Claro, K. 2019. Structural changes over time in individual-based networks involving a harvester ant, seeds, and invertebrates. *Ecological Entomology* 44 (6): 753–61.

Bartomeus, I., Ascher, J.S., Wagner, D., Danforth, B.N., Colla, S., Kornbluth, S., and Winfree, R. 2011. Climate-associated phenological advances in bee pollinators and bee-pollinated plants. *Proceedings of the National Academy of Sciences* 108 (51): 20645–20649.

Bascompte, J. and Jordano, P. 2007. Plant-animal mutualistic networks: The architecture of biodiversity. *Annual Review of Ecology, Evolution, and Systematics* 38 (1): 567–93. doi:10.1146/annurev.ecolsys.38.091206.095818.

Beattie, A. and Hughes, L. 2002. "Ant-plant interactions." In *Animal-plant Interaction: An Evolutionary Approach* edited by C. M. Herrera and O. Pellmyr. Oxford: Blackwell Publishing, pp. 211–53.

Bieber, A.G.D., Silva, P.S., Sendoya, S.F., and Oliveira, P.S. 2014. Assessing the impact of deforestation of the Atlantic Rainforest on ant-fruit interactions: A field experiment using synthetic fruits. *PLoS One* 9 (2): id90369.

Biella, P., Akter, A., Ollerton, J., Nielsen, A., and Klecka, J. 2020. An empirical attack tolerance test alters the structure and species richness of plant–pollinator networks. *Functional Ecology* 34 (11): 2246–2258.

Biella, P., Akter, A., Ollerton, J., Tarrant, S., Janeček, Š., Jersáková, J., and Klecka, J. 2019. Experimental loss of generalist plants reveals alterations in plant-pollinator interactions and a constrained flexibility of foraging. *Scientific Reports* 9 (1): 7376.

Bridle, J.R., Buckley J., and Bodsworth E. 2013. Evolution on the move: Specialization on widespread resources associated with rapid range expansion in response to climate change, *Proceedings of the Royal Society B: Biological Sciences* 281 (1776). doi:10.1098/rspb.2013.1800.

Burkle, L.A. and Alarcón, R. 2011. The future of plant–pollinator diversity: Understanding interaction networks across time, space, and global change. *American Journal of Botany* 98 (3): 528–38.

Caminade, C., Medlock, J.M., Ducheyne, E., McIntyre, K.M., Leach, S., Baylis, M., and Morse, A.P. 2012. Suitability of European climate for the Asian tiger mosquito Aedes albopictus: Recent trends and future scenarios. *Journal of the Royal Society Interface* 9 (75): 2708–2717.

CaraDonna, P.J., Petry, W.K., Brennan, R.M., Cunningham, J.L., Bronstein, J.L., Waser, N.M., and Sanders, N.J. 2017. Interaction rewiring and the rapid turnover of plant–pollinator networks. *Ecology Letters* 20 (3): 385–94.

Carstensen, D.W., Sabatino, M., Trøjelsgaard, K., Morellato, L.P.C. 2014. Beta diversity of plant-pollinator networks and the spatial turnover of pairwise interactions. *PloS One* 9 (11). doi:10.1371/journal.pone.0112903.

Caughlin, T.T., Ferguson, J.M., Lichstein, J.W., Zuidema, P.A., Bunyavejchewin, S., and Levey, D.J. 2015. Loss of animal seed dispersal increases extinction risk in a tropical tree species due to pervasive negative density dependence across life stages. *Proceedings of the Royal Society B: Biological Sciences*, 282(1798): id20142095.

Classen, A., Eardley, C.D., Hemp, A., Peters, M.K., Peters, R.S., Ssymank, A., and Steffan-Dewenter, I. 2020. Specialization of plant–pollinator interactions increases with temperature at Mt. Kilimanjaro. *Ecology and Evolution* 10 (4): 2182–2195.

Clemente, M.A., Lange, D., Dáttilo, W., Del-Claro, K., and Prezoto, F. 2013. Social wasp-flower visiting guild interactions in less structurally complex habitats are more susceptible to local extinction. *Sociobiology* 60, 337–44.

Corlett, R.T. 2021. "Frugivory and seed dispersal." In *Plant-Animal Interactions: Source of Biodiversity*, edited by K. Del-Claro, H.M. Torezan-Silingardi, Cham: Springer International Publishing, pp. 175–204.

Crespo-Pérez, V. et al. 2020. The importance of insects on land and in water: A tropical view. *Current Opinion in Insect Science* 40: 31–38. doi:10.1016/j.cois.2020.05.016.

Cruz, C.P., Luna, P., Guevara, R., Hinojosa-Díaz, I.A., Villalobos, F., and Dáttilo, W. 2022. Climate and human influence shape the interactive role of the honeybee in pollination networks beyond its native distributional range. *Basic and Applied Ecology* 63: 186–95.

Damien, M. and Tougeron, K. 2019. Prey–predator phenological mismatch under climate change. *Current Opinion in Insect Science* 35: 60–68.

Dáttilo, W. 2012. Different tolerances of symbiotic and nonsymbiotic ant-plant networks to species extinctions. *Network Biology* 2 (4): 127–38.

Dáttilo, W., Lara-Rodríguez, N., Jordano, P., Guimarães Jr, P.R., Thompson, J.N., Marquis, R.J., . . . and Rico-Gray, V. 2016. Unravelling Darwin's entangled bank: Architecture and robustness of mutualistic networks with multiple interaction types. *Proceedings of the Royal Society B: Biological Sciences* 283 (1843): 20161564.

Dáttilo, W. and Rico-Gray, V. 2018. *Ecological Networks in the Tropics: An Integrative Overview of Species Interactions from Some of the Most Species-Rich Habitats on Earth*. Cham: Springer. doi:10.1007/978-3-319-68228-0.

Dáttilo, W., and Vasconcelos, H.L. 2019. Macroecological patterns and correlates of ant–tree interaction networks in Neotropical savannas. *Global Ecology and Biogeography* 28 (9): 1283–1294.

Dehling, D.M. 2018. The structure of ecological networks. *Ecological Networks in the Tropics: An Integrative Overview of Species Interactions from Some of the Most Species-Rich Habitats on Earth.* 29–42.

Del-Claro, K. and Torezan-Silingardi, H.M. (eds.) 2021. *Plant-Animal Interactions*. Cham: Springer International Publishing.

Descombes, P., Pradervand, J. N., Golay, J., Guisan, A., and Pellissier, L. 2016. Simulated shifts in trophic niche breadth modulate range loss of alpine butterflies under climate change. *Ecography* 39 (8): 796–804. doi:10.1111/ecog.01557.

Descombes, P., Pradervand, J.N., Golay, J., Guisan, A., and Pellissier, L. 2016. Simulated diet expansion in alpine butterflies reduces range loss under climate change. *Ecography* 39 (8): 796–804.

Donoso, I., Schleuning, M., García, D., and Fründ, J. 2017. Defaunation effects on plant recruitment depend on size matching and size trade-offs in seed-dispersal networks. *Proceedings of the Royal Society B: Biological Sciences* 284 (1855): id20162664.

Donoso, I., Sorensen, M.C., Blendinger, P.G., Kissling, W.D., Neuschulz, E.L., Mueller, T., and Schleuning, M. 2020. Downsizing of animal communities triggers stronger functional than structural decay in seed-dispersal networks. *Nature Communications* 11 (1): 1582.

Dortel, E., Thuiller, W., Lobo, J.M., Bohbot, H., Lumaret, J.P., and Jay-Robert, P. 2013. Potential effects of climate change on the distribution of Scarabaeidae dung beetles in Western Europe. *Journal of Insect Conservation* 17: 1059–1070.

Ebi, K. L., Hartman, J., Chan, N., Mcconnell, J., Schlesinger, M., Weyant, J., et al. 2005. Climate suitability for stable malaria transmission in Zimbabwe under different climate change scenarios', *Climatic Change* 73 (3): 375–93. doi:10.1007/s10584-005-6875-2.

Epstein, P.R. 2000. Is global warming harmful to health? *Scientific American* 283 (2): 50–57.

Fischer, D., Thomas, S.M., Niemitz, F., Reineking, B., and Beierkuhnlein, C. 2011. Projection of climatic suitability for Aedes albopictus Skuse (Culicidae) in Europe under climate change conditions. *Global and Planetary Change* 78 (1–2): 54–64.

Forrest, J.R.K. 2017. "Insect pollinators and climate change." In *Global Climate Change and Terrestrial Invertebrates*, edited by S. N. Johnson and T. H. Jones, Hoboken, NJ: Wiley-Blackwell, 69–91.

Franke, C.R., Ziller, M., Staubach, C., and Latif, M. 2002. Impact of El Niño/southern oscillation on visceral Leishmaniasis, Brazil. *Emerging Infectious Diseases* 8 (9): 914.

Franklinos, L. H., Jones, K. E., Redding, D. W., Abubakar, I., et al. 2019. The effect of global change on mosquito-borne disease. *The Lancet Infectious Diseases* 19 (9): e302–e312. doi: 10.1016/S1473-3099(19)30161-6.

Fricke, E.C., Tewksbury, J.J., and Rogers, H.S. 2018. Defaunation leads to interaction deficits, not interaction compensation, in an island seed dispersal network. *Global Change Biology* 24 (1): e190–e200.

Fricke, E.C., Tewksbury, J.J., Wandrag, E.M., and Rogers, H.S. 2017. Mutualistic strategies minimize coextinction in plant–disperser networks. *Proceedings of the Royal Society B: Biological Sciences* 284 (1854): 20162302.

Furlong, M.J. and Zalucki, M.P. 2017. Climate change and biological control: The consequences of increasing temperatures on host–parasitoid interactions. *Current Opinion in Insect Science* 20: 39–44.

Garcia-robledo, C., Chuquillanqui, H., Kuprewicz, E.K., and Escobar-sarria, F. 2018. Lower thermal tolerance in nocturnal than in diurnal ants: A challenge for nocturnal ectotherms facing global warming. *Ecological Entomology* 43 (2): 162–67.

Gilman, S.E., Urban, M.C., Tewksbury, J., Gilchrist, G.W., and Holt, R.D. 2010. A framework for community interactions under climate change. *Trends in Ecology and Evolution* 25 (6): 325–31.

Giannini, T.C., Garibaldi, L.A., Acosta, A.L., Silva, J.S., Maia, K.P., Saraiva, A.M., Guimarães-Jr, P., et al. 2015. Native and non-native supergeneralist bee species have different effects on plant–bee networks. *PloS One* 10 (9): e0137198.

Gilman, S.E., Urban, M.C., Tewksbury, J., Gilchrist, G.W., and Holt, R.D. 2010. A framework for community interactions under climate change. *Trends in Ecology and Evolution* 25 (6): 325–31.

González, C., Wang, O., Strutz, S.E., and González-Salazar, C. Sánchez-Cordero, V., Sarkar, S. 2010. Climate change and risk of leishmaniasis in North America: Predictions from ecological niche models of vector and reservoir species. *PLoS Neglected Tropical Diseases* 4 (1): e585.

González-Tokman, D., Córdoba-Aguilar, A., Dáttilo, W., Lira-Noriega, A., Sánchez-Guillén, R.A., and Villalobos, F. 2020. Insect responses to heat: Physiological mechanisms, evolution and ecological implications in a warming world. *Biological Reviews* 95 (3): 802–21.

Guimaraes Jr, P.R., Pires, M.M., Jordano, P., Bascompte, J., and Thompson, J.N. 2017. Indirect effects drive coevolution in mutualistic networks. *Nature* 550 (7677): 511–14.

Guimarães Jr, P.R., Jordano, P., and Thompson, J.N. 2011. Evolution and coevolution in mutualistic networks. *Ecology Letters* 14 (9): 877–85.

Hacket-Pain, A. and Bogdziewicz, M. 2021. Climate change and plant reproduction: Trends and drivers of mast seeding change. *Philosophical Transactions of the Royal Society B: Biological Sciences* 376 (1839): id20200379.

Halsch, C.A., Shapiro, A.M., Fordyce, J.A., Nice, C.C., Thorne, J.H., Waetjen, D.P., and Forister, M.L. 2021. Insects and recent climate change. *Proceedings of the National Academy of Sciences* 118 (2): id2002543117.

Hance, T., van Baaren, J., Vernon, P., Boivin, G. 2007. Impact of extreme temperatures on parasitoids in a climate change perspective. *Annual Review of Entomology* 52: 107–26. doi:10.1146/annurev.ento.52.110405.091333.

Handel, S.N. and Beattie, A.J. 1990. Seed dispersal by ants. *Scientific American* 263 (2): 76–83B.

Hargreaves, A.L., Suárez, E., Mehltreter, K., Myers-Smith, I., Vanderplank, S.E., Slinn, H.L., . . . and Thomas, H.J.D.M, PAM. 2019. Seed predation increases from the Arctic to the Equator and from high to low elevations. *Science Advances* 5 (2). eaau4403.

Herrera, C.M. 2002. "Seed dispersal by vertebrates." In *Plant–Animal iInteractions: An Evolutionary Approach*, edited by C. M. Herrera and O. Pellmyr. Oxford: Blackwell Publishing, pp. 185–208.

Herrera, C.M. and Pellmyr, O. 2002. *Plant-Animal Interactions: An Evolutionary Approach*. Oxford: Blackwell Publishing.

Hulme, P.E. and Benkman, C.W. 2002. "Granivory." In *Animal-Plant Interaction: An Evolutionary Approach*, edited by C. M. Herrera and O. Pellmyr. Oxford: Blackwell Publishing, pp. 132–54.

Iler, A. M., Inouye, D. W., Høye, T. T., Miller-Rushing, A. J., Burkle, L. A., Johnston, E. B. 2013. Maintenance of temporal synchrony between syrphid flies and floral resources despite differential phenological responses to climate. *Global Change Biology* 19 (8): 2348–2359. doi:10.1111/gcb.12246.

Janzen, D.H. 1988. Ecological characterization of a Costa Rican dry forest caterpillar fauna. *Biotropica*, 120–35.

Jordano, P. 2000. Fruits and frugivory. *Seeds: The Ecology of Regeneration in Plant Communities* 2: 125–66.

Kehoe, R., Frago, E., and Sanders, D. 2021. Cascading extinctions as a hidden driver of insect decline. *Ecological Entomology* 46 (4): 743–56.

Koh, L.P., Dunn, R.R., Sodhi, N.S., Colwell, R.K., Proctor, H.C., and Smith, V.S. 2004. Species coextinctions and the biodiversity crisis. *Science* 305 (5690): 1632–1634.

Leal, I.R., Wirth, R., and Tabarelli, M. 2007. Seed dispersal by ants in the semi-arid Caatinga of north-east Brazil. *Annals of Botany* 99 (5): 885–94.

Leal, L.C., Andersen, A.N., and Leal, I.R. 2014. Anthropogenic disturbance reduces seed-dispersal services for myrmecochorous plants in the Brazilian Caatinga. *Oecologia*, 174, 173–81.

Lewis, O.T. and Gripenberg, S. 2008. Insect seed predators and environmental change. *Journal of Applied Ecology* 45 (6): 1593–1599.

Liu-Helmersson, J., Quam, M., Wilder-Smith, A., Stenlund, H., Ebi, K., Massad, E., Rocklöv, J. 2016. Climate Change and Aedes Vectors: 21st Century Projections for Dengue Transmission in Europe. *EBioMedicine* 7: 267–77. doi:10.1016/j.ebiom.2016.03.046.

Luna, P., Villalobos, F., Escobar, F., Neves, F. S., Dáttilo, W. 2021. Global trends in the trophic specialisation of flower-visitor networks are explained by current and historical climate. *Ecology Letters* (October): 113–24. doi:10.1111/ele.13910.

Luna, P. and Dáttilo, W. 2021. "Disentangling plant-animal interactions into complex networks: A multi-view approach and perspectives." In *Plant-Animal Interactions: Source of Biodiversity*, edited by W. Dáttilo and V. Rico-Gray, Cham: Springer International Publishing, pp. 261–81.

Luna, P., García-Chávez, J.H., and Dáttilo, W. 2018. Complex foraging ecology of the red harvester ant and its effect on the soil seed bank. *Acta Oecologica* 86: 57–65.

Luna, P., Peñaloza-Arellanes, Y., Castillo-Meza, A.L., García-Chávez, J.H., and Dáttilo, W. 2018. Beta diversity of ant–plant interactions over day-night periods and plant physiognomies in a semiarid environment. *Journal of Arid Environments* 156: 69–76.

Maldaner, M.E., Sobral-Souza, T., Prasniewski, V.M., and Vaz-de-mello, F.Z. 2021. Effects of climate change on the distribution of key native dung beetles in South American grasslands. *Agronomy* 11 (10): 2033.

Marquis, R.J. 2010. "The role of herbivores in terrestrial trophic cascades." In *Trophic Cascades: Predators, Prey and the Changing Dynamics of Nature*, edited by J. Terborgh & J.A. Estes. Washington, DC: Island Press, pp. 109–123.

McKone, M.J., Kelly, D., and Lee, W.G. 1998. Effect of climate change on mast-seeding species: Frequency of mass flowering and escape from specialist insect seed predators. *Global Change Biology* 4 (6): 591–96.

Medone, P., Ceccarelli, S., Parham, P.E., Figuera, A., and Rabinovich, J.E. 2015. The impact of climate change on the geographical distribution of two vectors of Chagas disease: Implications for the force of infection. *Philosophical Transactions of the Royal Society B: Biological Sciences* 370 (1665): id20130560.

Menéndez, R. González–Megías, A., Jay–Robert, P., and Marquéz–Ferrando, R. 2014. Climate change and elevational range shifts of dung beetles. *Global Ecology and Biogeography* 23: 646–57.

Miller-Struttmann, N.E., Geib, J.C., Franklin, J.D., Kevan, P.G., Holdo, R.M., Ebert-May, D., Lynn, A.M. 2015. Functional mismatch in a bumble bee pollination mutualism under climate change. *Science* 349 (6255): 1541–1544.

Montoya, J.M., Pimm, S.L., and Solé, R.V. 2006. Ecological networks and their fragility. *Nature* 442 (7100): 259–64.

Novak, M., Yeakel, J.D., Noble, A.E., Doak, D.F., Emmerson, M., Estes, J.A., Jacob, U. 2016. Characterizing species interactions to understand press perturbations: What is the community matrix? *Annual Review of Ecology, Evolution, and Systematics* 47: 409–32.

Oliveira, F.M., Andersen, A.N., Arnan, X., Ribeiro-Neto, J.D., Arcoverde, G.B., and Leal, I.R. 2019. Effects of increasing aridity and chronic anthropogenic disturbance on seed dispersal by ants in Brazilian Caatinga. *Journal of Animal Ecology* 88 (6): 870–80.

Ollerton, J. 2017. Pollinator diversity: Distribution, ecological function, and conservation. *Annual Review of Ecology, Evolution, and Systematics* 48 (1): 353–76.

Parmesan, C. 2007. Influences of species, latitudes and methodologies on estimates of phenological response to global warming. *Global Change Biology* 13 (9): 1860–1872.

Parr, C.L. and Bishop, T.R. 2022. The response of ants to climate change. *Global Change Biology* 28 (10): 3188–3205.

Parra-Tabla, V., Angulo-Pérez, D., Albor, C., Campos-Navarrete, M.J., Tun-Garrido, J., Sosenski, P., Alonso, C. et al. 2019. The role of alien species on plant-floral visitor network structure in invaded communities. *PLoS One* 14 (11): id0218227.

Pearse, I.S. and Altermatt, F. 2013. Extinction cascades partially estimate herbivore losses in a complete Lepidoptera–plant food web. *Ecology* 94 (8): 1785–1794.

Pearse, I.S., Koenig, W.D., and Kelly, D. 2016. Mechanisms of mast seeding: Resources, weather, cues, and selection. *New Phytologist* 212 (3): 546–62.

Pelini, S.L., Keppel, J.A., Kelley, A.E., and Hellmann, J.J. 2010. Adaptation to host plants may prevent rapid insect responses to climate change. *Global Change Biology* 16 (11): 2923–2929.

Petsopoulos, D., Lunt, D.H., Bell, J.R., Kitson, J.J., Collins, L., Boonham, N., Morales-Rojas, R., et al. 2021 Using network ecology to understand and mitigate long-term insect declines. *Ecological Entomology* 46 (4): 693–8. doi:10.1111/een.13035.

Pires, M. M., O'Donnell, J. L., Burkle, L. A., Díaz-Castelazo, C., Hembry, D. H., Yeakel, J. D., Newman, E.A. et al. 2020. The indirect paths to cascading effects of extinctions in mutualistic networks. *Ecology* 101 (7): 1–8. doi:10.1002/ecy.3080.

Pirk, G.I., De Casenave, J.L., Pol, R.G., Marone, L., and Milesi, F.A. 2009. Influence of temporal fluctuations in seed abundance on the diet of harvester ants (Pogonomyrmex spp.) in the central Monte desert, Argentina. *Austral Ecology* 34 (8): 908–19.

Poisot, T., Canard, E., Mouillot, D., Mouquet, N., and Gravel, D. 2012. The dissimilarity of species interaction networks. *Ecology Letters* 15 (12): 1353–1361.

Salomão, R.P., González-Tokman, D., Dáttilo, W., López-Acosta, J.C., and Favila, M.E. 2018. Landscape structure and composition define the body condition of dung beetles (Coleoptera: Scarabaeinae) in a fragmented tropical rainforest. *Ecological Indicators* 88 (January): 144–51.

Ready, P.D. 2008. Leishmaniasis emergence and climate change. *Revue scientifique et Technique* 27 (2): 399–412.

Renner, S.S. and Zohner, C.M. 2018. Climate change and phenological mismatch in trophic interactions among plants, insects, and vertebrates. *Annual Review of Ecology, Evolution, and Systematics* 49: 165–82.

Reverchon, F. and Méndez-Bravo, A. 2021. "Plant-mediated above-belowground interactions: A phytobiome story." In *Plant-Animal Interactions: Source of Biodiversity*, edited by K. Del-Claro, H.M. Torezan-Silingardi. Cham: Springer International Publishing, pp. 205–31.

Rezende, E.L., Lavabre, J.E., Guimarães, P.R., Jordano, P., Bascompte, J. 2007 Non-random coextinctions in phylogenetically structured mutualistic networks. *Nature*, 448 (7156): 925–8. doi:10.1038/nature05956.

Rico-Gray, V. and Oliveira, P.S. 2007. *The Ecology and Evolution of Ant-Plant Interactions*. Chicago: University of Chicago Press. doi:10.1017/CBO9781107415324.004.

Rogers, H.S., Donoso, I., Traveset, A., and Fricke, E.C. 2021. Cascading impacts of seed disperser loss on plant communities and ecosystems. *Annual Review of Ecology, Evolution, and Systematics* 52: 641–66.

Roland, J. and Matter, S.F. 2013. Variability in winter climate and winter extremes reduces population growth of an alpine butterfly. *Ecology* 94 (1): 190–99.

Rumeu, B., Devoto, M., Traveset, A., Olesen, J.M., Vargas, P., Nogales, M., and Heleno, R. 2017. Predicting the consequences of disperser extinction: Richness matters the most when abundance is low. *Functional Ecology* 31 (10): 1910–1920.

Saavedra, S., Stouffer, D.B., Uzzi, B., and Bascompte, J. 2011. Strong contributors to network persistence are the most vulnerable to extinction. *Nature* 478 (7368): 233–35.

Sanders, D., Thébault, E., Kehoe, R., and Frank van Veen, F.J. 2018. Trophic redundancy reduces vulnerability to extinction cascades. *Proceedings of the National Academy of Sciences* 115 (10): 2419–2424.

Santos, G. M. M., Aguiar, C. M., Genini, J., Martins, C. F., Zanella, F. C., and Mello, M. A. 2012. Invasive Africanized honeybees change the structure of native pollination networks in Brazil. *Biological Invasions* 14 (11): 2369–2378. doi:10.1007/s10530-012-0235-8.

Schleuning, M., Fründ, J., Schweiger, O., Welk, E., Albrecht, J., Albrecht, M., Beil, M., et al. 2016. Ecological networks are more sensitive to plant than to animal extinction under climate change. *Nature Communications* 7 (1): 13965.

Schweiger, O., Heikkinen, R. K., Harpke, A., Hickler, T., Klotz, S., Kudrna, O., Kühn, I., et al. 2012. Increasing range mismatching of interacting species under global change is related to their ecological characteristics. *Global Ecology and Biogeography* 88–99. doi:10.1111/j.1466-8238.2010.00607.x.

Settele, J., Bishop, J., and Potts, S.G. 2016. Climate change impacts on pollination. *Nature Plants* 2 (7). id16092. doi:10.1038/NPLANTS.2016.92.

Silvertown, J.W. 1980. The evolutionary ecology of mast seeding in trees. *Biological Journal of the Linnean Society* 14 (2): 235–50.

Slatyer, R.A., Hirst, M., and Sexton, J.P. 2013. Niche breadth predicts geographical range size: A general ecological pattern. *Ecology Letters* 16 (8): 1104–1114. doi: 10.1111/ele.12140.

Solbreck, C. and Knape, J. 2017. Seed production and predation in a changing climate: New roles for resource and seed predator feedback? *Ecology* 98 (9): 2301–2311.

Soroye, P., Newbold, T., and Kerr, J. 2020. Climate change contributes to widespread declines among bumble bees across continents. *Science* 367 (6478): 685–88.

Strauss, S.Y. and Zangerl, A.R. 2002. "Plant–insect interactions in terrestrial ecosystems." In *Plant-Animal Interactions: An Evolutionary Approach*, edited by C. M. Herrera and O. Pellmyr. Oxford: Blackwell Publishing, pp.77–106.

Terborgh, J., Feeley, K., Silman, M., Nuñez, P., and Balukjian, B. 2006. Vegetation dynamics of predator-free land-bridge islands. *Journal of Ecology* 94, 253–63.

Thuiller, W., Lavorel, S., Araújo, M. B., Sykes, M. T., and Prentice, I. C. 2005. Climate change threats to plant diversity in Europe. *Proceedings of the National Academy of Sciences* 102 (23): 8245–8250. doi:10.1073/pnas.0409902102.

Timóteo, S., Ramos, J.A., Vaughan, I.P., and Memmott, J. 2016. High resilience of seed dispersal webs highlighted by the experimental removal of the dominant disperser. *Current Biology* 26 (7): 910–15.

Tougeron, K., Brodeur, J., Le Lann, C., and van Baaren, J. 2020. How climate change affects the seasonal ecology of insect parasitoids. *Ecological Entomology* 45 (2): 167–81.

Trenberth, K.E. and Hoar, T.J. 1997. El Niño and climate change. *Geophysical Research Letters* 24 (23): 3057–3060.

Tylianakis, J.M., Laliberté, E., Nielsen, A., and Bascompte, J. 2010. Conservation of species interaction networks. *Biological Conservation* 143 (10): 2270–2279.

Vizentin-Bugoni, J., Debastiani, V.J., Bastazini, V.A., Maruyama, P.K., and Sperry, J.H. 2020. Including rewiring in the estimation of the robustness of mutualistic networks. *Methods in Ecology and Evolution* 11 (1): 106–16.

Wagner, D.L. 2020. Insect declines in the Anthropocene. *Annual Review of Entomology* 65: 457–80.

Wardhaugh, C.W. 2015. How many species of arthropods visit flowers? *Arthropod– Plant Interactions* 9 (6): 547–65.

Warren, R., VanDerWal, J., Price, J., Welbergen, J.A., Atkinson, I., Ramirez-Villegas, J., . . . and Lowe, J. 2013. Quantifying the benefit of early climate change mitigation in avoiding biodiversity loss. *Nature Climate Change* 3 (7): 678–82.

Wilson, R.J. and Fox, R. 2021. Insect responses to global change offer signposts for biodiversity and conservation. *Ecological Entomology* 46 (4): 699–717.

CHAPTER 16

Refugia from climate change, and their influence on the diversity and conservation of insects

Guim Ursul, Mario Mingarro, Juan Pablo Cancela, Helena Romo and Robert J. Wilson

16.1 Introduction

Climate change is one of the main threats driving recent changes to global insect abundance, distributions and diversity (Goulson 2019; Halsch et al. 2021; Outhwaite et al. 2022; Wilson and Fox 2021). There are widespread examples of shifts in insect latitudinal (Hickling et al. 2006; Lenoir et al. 2020; Parmesan et al. 1999) and altitudinal (Marshall et al. 2020; Rödder et al. 2021; Scalercio et al. 2014; Wilson et al. 2005) distributions that are consistent with climate change, and these individual species range shifts in turn have impacts on the composition of insect communities (Nieto-Sánchez et al. 2015; Santorufo et al. 2021; Wilson et al. 2007). Increases in drought, or in temperature beyond the upper limits of thermal tolerance, have been implicated in declines in insect abundance and diversity especially at lower latitudes (Halsch et al. 2021; Janzen and Hallwachs 2021; Outhwaite et al. 2022; Soroye et al. 2020), while the ability of species to track newly suitable climatic conditions beyond existing poleward or high-elevation range margins is constrained by land-use change and a lack of suitable habitat (Lenoir et al. 2020; Platts et al. 2019). One approach that has been advocated for conservation to adapt to the rapid and increasing threat of climate change to biodiversity, is to prioritize particular geographic areas to function as refugia for species against climate change. In this chapter, we explore the theory behind refugia, and the evidence for whether they might play an important role in adapting the conservation of insects to ongoing climate change.

The frequent observation that changes to the composition of ecological communities are lagging behind those expected from observed rates of warming, creating a "climatic debt" (Devictor et al. 2012; Fourcade et al. 2021; Menéndez et al. 2006; Mingarro et al. 2021), implies an effect of landscape and habitat structure on ecological responses to climate change. Delays in species' local extinctions or colonizations as ambient conditions cross apparent climatic tolerance thresholds (Wilson 2022) could result from the

Guim Ursul et al., *Refugia from climate change, and their influence on the diversity and conservation of insects*. In: *Effects of Climate Change on Insects*. Edited by: Daniel González-Tokman and Wesley Dáttilo, Oxford University Press. © Oxford University Press (2024). DOI: 10.1093/oso/9780192864161.003.0016

effects of local adaptation, biotic interactions or habitat condition on physiological and demographic responses to climate change (Alexander et al. 2018; Brambilla and Gobbi 2014). But certain features of the landscape might also reduce rates of local extinction by creating fine-resolution variation in local climatic conditions and rates of climate change, potentially helping populations withstand broader changes to the climate. For example, landscapes with heterogeneous topography create mosaics of climatic conditions and rates of climatic change, promoting the appearance of refugia for taxa threatened by changes to prevailing conditions (Dobrowski 2011). Structural heterogeneity in habitats like old-growth forests, can also buffer temperature fluctuations, reducing the exposure of organisms to extremes (Wolf et al. 2021). As small, ectothermic organisms often with specialist habitat requirements, the performance of insects is tied to very fine-scale variation in climatic conditions (Pincebourde and Woods 2012, 2020; Potter et al. 2013). Exploring how landscape and habitat structure could create refugia from climate change is therefore potentially an area of vast importance for conserving insect diversity.

In this review, we summarize the physical and biotic characteristics of proposed climate change refugia and explore evidence for the effects of refugial landscapes on historical and contemporary responses of insects to climate change. We then focus on how this understanding can be applied in conservation to buffer insect diversity against climate change. An increasing literature focused on the identification of climate change refugia has identified several main physical and biotic characteristics (Figure 16.1): the stability and spatial heterogeneity of local climatic conditions, and the climatic and geographic isolation of refugial areas from their surroundings (Abellán and Svenning 2014; Ashcroft et al. 2012; Harrison and Noss 2017). The size, quality and connectivity of habitats within refugia also influence the persistence or potential for fine-scale redistribution in response to climate change. Larger and more heterogeneous refugia are likely to host a greater number of

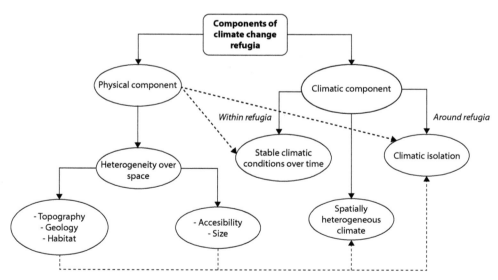

Figure 16.1 Schematic representation of the main physical and climatic components of climate change refugia. Macrorefugia are characterized by regions or landscapes with temporally stable climates, or spatially heterogeneous climatic conditions created by variation in topography, geology or habitat. Microrefugia are environments in which topography, geology, or habitat lead to rare or distinct climates from their surroundings.

species differing in their ecological requirements (Keppel et al. 2015). Refugia have been defined in many ways, but broadly can be separated into macrorefugia and microrefugia (Ashcroft 2010). Macrorefugia refer to regions or landscapes with temporally stable climates, or spatially heterogeneous climatic conditions created by variation in topography, geology or habitat. These landscapes often represent centers of endemism because over long time periods they have allowed populations to withstand or adapt to changing climatic conditions (Harrison and Noss 2017; Hewitt 2000). Microrefugia represent more localized environments in which topography, geology or habitat lead to rare or distinct climates from the surrounding areas (Dobrowski 2011), and may also facilitate long-term persistence in otherwise climatically hostile environments (Mee and Moore 2014; Patsiou et al. 2014; Rull 2009; Stewart et al. 2010).

There is strong biogeographic, molecular and fossil evidence for the role of stable climate refugia for insect species persistence during the Last Glacial Maximum (LGM), with subsequent range expansions shaping current patterns of intra- and inter-specific diversity (Dincă et al. 2021; Hernández-Roldán et al. 2011; Maresova et al. 2021). Several studies demonstrate how climatic stability, heterogeneity or isolation in refugia have influenced insect biodiversity patterns (Table 16.1). However, the potential of refugial landscapes to maintain insect diversity in the face of ongoing climate change is a matter for debate. The ability of climatic stability or heterogeneity in refugia to provide safe havens for biodiversity (Keppel et al. 2012) could be threatened not only by rapid rates of anthropogenic warming, but by the ecological characteristics (e.g., local adaptation to hitherto stable conditions, or proximity to low-latitude range margins) that make species in proposed refugia vulnerable to climate change, and which risk refugia becoming highly threatened "hospices for biodiversity" (Brown et al. 2020; Kocsis et al. 2021).

16.2 Physical characteristics of climate change refugia

Variation across landscapes both in physical factors and human land use cause fine-scale variation in the climatic conditions to which organisms are exposed (Maclean et al. 2017; Oldfather et al. 2020). Common climatic gradients such as from low to high latitudes or elevations lead to regional-scale gradients in diversity (Laiolo et al. 2018), but insect diversity can vary at much finer resolutions related to local topographic variation and associated variation in climate and microhabitat, for example between valley bottoms, slopes and ridges (Rabl et al. 2020). Heterogeneity in topography, geology and land cover (Doxa et al. 2022; Graae et al. 2018) all create fine-resolution variation in the microclimates experienced by organisms, with a considerable effect on small ectotherms like insects (Pincebourde and Woods 2020; Pincebourde et al. 2016; Potter et al. 2013) because of their often lower dispersal ability, narrower thermal niche and more complex life cycles than larger and endothermic species (Calosi et al. 2010; Koot et al. 2022; Ma et al. 2021).

In a changing climate, microclimatic variation influences a range of factors important for population dynamics and persistence: for example, where, when and for how long individual organisms are exposed to conditions beyond their tolerance; the magnitude of climatic change and the scope for behavioral thermoregulation to cope with it; and the distances that need to be traveled during dispersal or foraging to maintain climatically favorable surroundings. For example, Lawson et al. (2014) found that climatic variation generated by topography influences fine-scale egg-site selection by a temperate butterfly (*Hesperia comma* Linnaeus, 1758), with greater selection of hot microhabitats (increased bare ground cover) on poleward-facing than equatorward-facing habitats. In this case, the

Table 16.1 Evidence of how physical components of climate change refugia have influenced insect diversity and biogeography.

Taxon/Region	Component	Main result	Reference
Insect responses to Quaternary ice ages	Stability	Stable low latitude refugia allowed species to persist through glacial maxima	Coope 2004
Endemics including butterflies and water beetles, Iberian Peninsula	Stability and heterogeneity	Long-term climatic velocity explains Iberian endemic richness patterns	Abellán and Svenning 2014
Endemic insects, South America	Stability	Long-term stability determines endemic persistence	Montemayor et al. 2017
Ice crawlers (*Grylloblatta*), North America	Stability, isolation and, heterogeneity	Pleistocene elevation shifts explain patterns of diversification	Schoville et al. 2019
Hemiptera, China	Heterogeneity	Mountains support high diversity through persistent lineages and new divergence	Li et al. 2021
Short-horned scorpionfly (*Cerapanorpa brevicornis*)	Isolation	Interglacial mountain refugia facilitated speciation	Gao et al. 2022
Patagonian bee (*Centris cineraria*), South America	Heterogeneity, isolation	Uphill postglacial expansion from glacial refugia beside the Andes	Sosa-Pivatto et al. 2020
Nebria germari (Coleoptera), Eastern Alps	Isolation	Persistence in isolated climates on supraglacial debris	Valle et al. 2020
Xylotrupes siamensis Minck, 1920 (Coleoptera), South East Asia	Isolation	Isolation by distance of historical refugia explains phenotypic variation	Morgan and Huang 2021
Pearly heath butterfly (*Coenonympha arcania* Linnaeus, 1761), Europe	Isolation	Wing morphology reveals historical isolation in refugia	Cassel-Lundhagen et al. 2020

local effects of aspect greatly outweighed the effects of approximately 200 kilometers (km) of latitude and longitude on the climates experienced by individual egg-laying butterflies. Climate-driven changes in butterfly microhabitat selection also occur over elevation gradients, with cooler, more shaded microhabitats selected at hotter, low elevations both by egg-laying *Aporia crataegi* (Linnaeus, 1758) (Merrill et al. 2008) and foraging larvae of *Parnassius apollo* (Linnaeus, 1758) (Ashton et al. 2009). Topography thus drives changes in local and microclimatic conditions that can influence the behavior, population dynamics, distributions and diversity of insects at a fine spatial resolution.

In this context, local variation in both climate and rates of climate change can lead to the appearance of microrefugia (Dobrowski 2011): habitats or landscapes whose localized climatic conditions buffer species against broader-scale changes to prevailing climates (Lenoir et al. 2017). Flat landscapes such as plains and plateaux have less spatial variation in climatic conditions and rates of climate change compared to mountainous areas (e.g., see Figure 16.2 for the Iberian Peninsula). Species that are tracking favorable climatic conditions by range-shifting (Lenoir and Svenning 2015) therefore have less far to move in more heterogeneous landscapes, by moving uphill (e.g., at temperate latitudes, 1 °C temperature change occurs roughly over 167 meters (m) of elevation, in contrast with 145 km of latitude; Jump et al. 2009) or from equatorward to poleward facing sides of hills. This concept of *climate change velocity* refers to the speed required for a species to keep pace with climate change by shifting its distribution, and can be used to quantify regional exposure to climate change (Dobrowski and Parks 2016; Loarie et al. 2009) being higher in flat landscapes and lower in mountain landscapes (Ackerly et al. 2010; Brito-Morales et al. 2018). Mountainous landscapes with low climate change velocities since the LGM (21,000 years before present (BP)) have high rates of endemism among weakly dispersing and small-range vertebrates, suggesting that topographic variation has allowed such species to withstand climatic change by undertaking range shifts over short distances (Sandel et al. 2011), and the same factors are likely to explain patterns of endemism in sedentary and habitat specialist insects (Kenyeres et al. 2009; Massa and Fontana 2020; Schoville et al. 2019). Topographic variation slows down the velocity of climate change to which insects are exposed (Loarie et al. 2009). As a consequence, in 10 × 10 km grid cells in England with greater topographic heterogeneity, extirpation rates since 1990 have been reduced for climate-change sensitive species of Lepidoptera and Coleoptera compared with flatter, more homogeneous landscapes (Suggitt et al. 2018).

The responses of individual species to topographic variation in climate and climate change have consequences for insect communities. In the case of the Iberian Peninsula, changes before and after 1980 to the climatic associations of butterfly communities (measured by the community temperature index (CTI) and community precipitation index (CPI)) were reduced in 10 × 10 km grid cells with greater local spatial variation in climatic conditions (Mingarro et al. 2021). Although mean annual temperature had increased on average by 1.14 °C between 1900–1979 and 1980–2016, this rate of warming is exceeded by fine spatial-scale variation in the temperatures to which species are exposed (e.g., Figure 16.2a for maximum monthly temperatures at 1 km resolution in 1979–2019). The steeper spatial temperature gradients in more topographically variable regions increase the opportunities for species to cope with climate change by local redistribution (Figure 16.2b). Temporal rates of climatic change are also spatially variable, and are reduced by elevational heterogeneity (Mingarro et al. 2021). Thus, whilst the western part of the Iberian Peninsula has warmed more slowly since 1979 (related to effects of the Atlantic Ocean in contrast with more rapid warming in the Mediterranean basin), warming has been reduced on north-facing compared with south-facing slopes (Figure 16.2c).

Microclimates are influenced not only by topography but also by vegetation. For example, the effects of climate change could be buffered under forest canopies, where climatic conditions are cooler, more spatially variable at a fine resolution and have warmed less compared with open habitats (De Frenne et al. 2021; De Lombaerde et al. 2022; Kašpar et al. 2021; Wolf et al. 2021). Considering these effects of topography or habitat on climatic variation and rates of climate change at an appropriately fine scale may be needed to properly understand climate change impacts on insects (Bütikofer et al. 2020; Pincebourde and Woods 2020) because, as small-bodied ectotherms, microclimatic differences at the scale of mm to cm (e.g., above and below a single leaf) can affect insect performance (Pincebourde and Woods 2012). In central Spain, increases in the butterfly CTI between 1967–1973 and 2006–2012 were reduced in sample sites with greater cover of forest and meadow habitats, suggesting a possible role for natural vegetation cover to buffer the

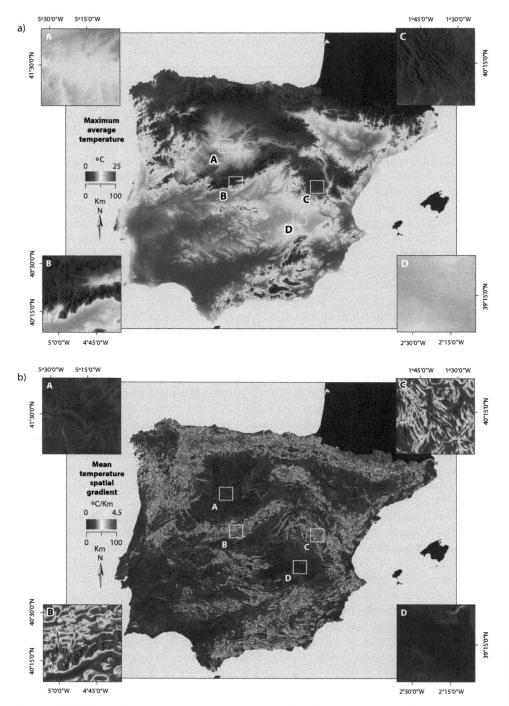

Figure 16.2 Effects of topography on temperature and spatiotemporal rates of temperature change in the Iberian Peninsula. a) Maximum monthly average temperature (°C), b) mean temperature spatial gradient (°C/km) and c) mean temperature temporal trend (°C/year) between 1979 and 2019 using CHELSA climate data (Karger et al. 2017) with a resolution of 1 km². Inset maps (50 × 50 km, locations indicated by white squares) show two flat landscapes in Castilla y León (A) and Castilla-La Mancha (D) and two mountain landscapes in Gredos (B) and Albarracín (C).

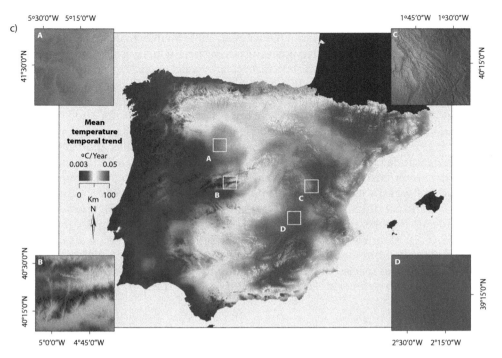

Figure 16.2 Continued

effects of climate change on insect communities (Nieto-Sánchez et al. 2015). In northeast Spain, declines since 1994 have been greatest among butterflies of dry lowland habitats in contrast with those of forests or mountains (Herrando et al. 2019). For one exemplar species in this system (*Pieris napi* (Linnaeus, 1758), Lepidoptera: Pieridae) open habitats exacerbated the negative effects of high temperatures and drought on food plant quality and larval survival (Carnicer et al. 2019).

Geology can also influence climatic heterogeneity across a landscape through its effects on fine-scale variation in topography and vegetation (Opedal et al. 2015), or via direct effects on drainage, humidity and temperatures near the ground (e.g., because of variation in albedo). Bátori et al. (2020) found that hollows in karst landscapes provide warm/dry microclimates on equatorward-facing slopes, and cool/moist microclimates on poleward-facing slopes and depression bottoms, that each provided localized climatically suitable habitats for specialized ants and plants. Mountain environments are exposed to extreme temperatures but have many components that can buffer these extremes and provide localized refugial habitats, such as rocky cliffs (García et al. 2020) and cold rocky landforms like moraines (Brighenti et al. 2021). Valle et al. (2020) found that a cold-adapted Coleoptera species (*Nebria germari* Heer, 1837) which might otherwise be in danger in the eastern European Alps because of climate warming, has persistent populations in supraglacial debris that provides more stable, colder and wetter conditions than the surroundings.

Topographic heterogeneity has had an important role in shaping today's insect diversity because it has provided safe havens to withstand unfavorable climatic conditions in the past. In Europe, species distributions contracted southwards during the last ice age, with populations surviving and sometimes speciating in isolated climatically favorable mountain ranges in the Mediterranean basin (Dincă et al. 2021; Menchetti et al. 2021). Since the LGM, species ranges have expanded polewards from these refugia, leading to current patterns of insect species richness and endemism (Maresova et al. 2021). Lepidoptera species richness and endemism across the Iberian Peninsula demonstrate how topography may influence where and how insect diversity is exposed to ongoing effects of climate change: mountain areas host the highest species richness both overall and for endemic butterflies (Figure 16.3 and Table 16.2). The notable decline in species richness from north to south reflects an underlying pattern of western Palearctic butterflies approaching their current low-latitude range margins in the Mediterranean basin, related to an increasingly hot and dry water/energy balance (Hawkins and Porter 2003). Mountain ranges in the center and south of the Iberian Peninsula provide localized enclaves of cool and spatially variable climatic conditions (Figure 16.2), supporting higher species richness and (especially) endemic richness than more lowland regions at equivalent latitudes (Figure 16.3 and Table 16.2). Areas of high and variable altitude in southern Europe also host very high levels of endemism for other insects such as the Orthoptera, including many highly sedentary or apterous species with restricted distributions that have evolved and persisted in these regions (Kenyeres et al. 2009; Massa and Fontana 2020).

16.3 Biotic components of refugia

Historical refugia have been studied across many taxa (Gavin et al. 2014; Selwood and Zimmer 2020) and can be identified using a combination of fossil records (Sommer and Nadachowski 2006), species distribution models (Montemayor et al. 2017) and phylogeographic surveys (Provan and Bennett 2008; Tang et al. 2022). A common method to identify past climate refugia is by combining genetic analysis with species distribution models based on current and past climates to corroborate phylogeographic results. Such approaches reveal evidence about the nature and distribution of historical climate change refugia, and how they relate to current patterns of insect diversity and endemism.

The current geographic ranges and genetic structure of species have been greatly influenced by the Quaternary ice ages of the past 2.4 million years (Hewitt 2000; Ohlemüller et al. 2012). As ice sheets advanced and retreated, species became isolated in places where climatic conditions were still favorable. The equatorward contraction of warm-adapted species to refugia in the Mediterranean basin during glacial periods, followed by interglacial recolonization of higher latitudes, has been demonstrated in a wide range of taxa including vertebrates, invertebrates and plants (e.g., Hewitt 1999, 2000; Harrison and Noss 2017; Schmitt 2007). Isolation in geographically separate refugia (e.g., in the Iberian Peninsula, Italy and the Balkans) caused genetic differentiation among lineages in these taxa, with possible eventual speciation and the appearance of endemics (Harrison and Noss 2017; Keppel et al. 2018; Morales-Barbero et al. 2021; Schmitt 2007). Small sedentary and ectothermic animals like insects can be especially prone to speciation following population isolation in climatically favorable areas during glacial-interglacial cycles, as Ribera and Vogler (2004) found for endemic Iberian diving beetles (Coleoptera: Dytiscidae) (see also Calosi et al. 2010). This history of contraction to lower-latitude refugia and subsequent postglacial range expansion is closely related to current patterns of inter- and intra-specific European insect diversity (e.g., Dincă et al. 2021; Hernández-Roldán et al. 2011; Pinkert et al. 2018).

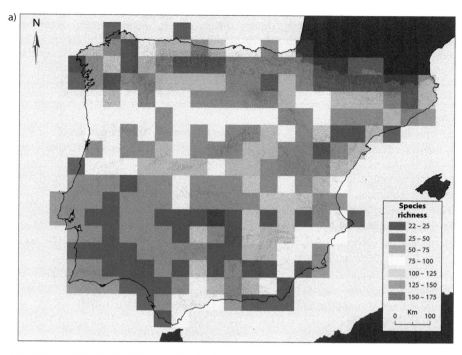

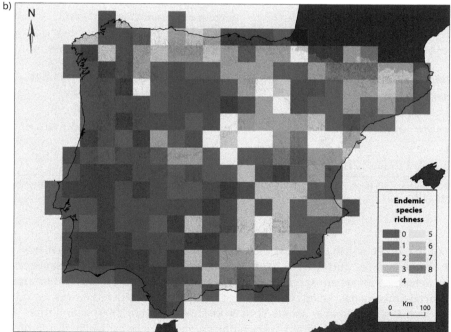

Figure 16.3 Butterfly species richness (a) and endemic species richness (b) in the Iberian Peninsula, based on distribution records since 1900 (García-Barros et al. 2004, 2013) at a resolution of 50 × 50 km. Squares with fewer than 50 overall butterfly occurrence records are shown in gray (see Table 16.2 for the endemic species included).

Table 16.2 Most endemic butterfly species in the Iberian Peninsula are restricted to mountains. The species listed were used to plot endemic richness for Figure 16.3b. Geographic range is summarized from García-Barros et al. (2004) and Vila et al. (2018), and IUCN category from The European Red List of Butterflies (van Swaay et al. 2010). Species names are listed as in Wiemers et al. (2018) followed by the naming authority.

Species	Geographic range and IUCN category
Lycaenidae	
Aricia morronensis (Ribbe, 1910)	Mountain ranges > 1000 m elevation/LC
Agriades pyrenaicus (Boisduval, 1840)	Pyrenees, Cantabrian mountains (N Spain) > 1500 m/LC
Agriades zullichi Hemming, 1933	Sierra Nevada mountains (S Spain) > 2400 m/EN
Kretania hesperica (Rambur, 1839)	C, E and S Spain 500–2000 m/LC
Lycaena bleusei (Oberthür, 1884)	C and W Iberian Peninsula 500–2000 m/LC
Lysandra caelestissima (Verity, 1921)	Sistema Ibérico mountains (E Spain) 1000–1800 m/LC
Polyommatus fabressei (Oberthür, 1910)	Sistema Ibérico mountains (E Spain) 1000–2000 m/LC
Polyommatus fulgens (Sagarra, 1925)	NE Spain 400–1400 m/LC
Polyommatus golgus (Hübner, [1813])	Sierra Nevada mountains (S Spain) > 2500 m/VU
Polyommatus nivescens (Keferstein, 1851)	E Spain mountains 500–2000 m/NT
Polyommatus violetae (Gómez-Bustillo, Expósito and Martínez, 1979)	S Spain mountains 700–2100 m/VU
Pseudophiloptes panoptes (Hübner, [1813])	Widespread < 1500 m/NT
Nymphalidae	
Erebia gorgone Boisduval, 1833	Pyrenees mountains 1100–2800 m/LC
Erebia hispania Butler, 1868	Sierra Nevada mountains (S Spain) 1800–3000 m/LC
Erebia lefebvrei (Boisduval, 1828)	Pyrenees, Cantabrian mountains (N Spain) 900–2800 m/LC
Erebia palarica (Chapman, 1905)	Cantabrian mountains (N Spain) > 1000 m/LC
Erebia rondoui (Oberthür, 1908)	Pyrenees mountains 1600–2500 m/LC
Erebia sthennyo (Graslin, 1850)	Pyrenees mountains 1200–2800 m/LC
Erebia zapateri (Oberthür, 1875)	Sistema Ibérico mountains (E Spain) > 1200 m/LC
Pieridae	
Euchloe bazae (Fabiano, 1993)	Isolated arid sites NE and SE Spain/VU

Abbreviations: Central (C), North (N), South (S), East (E), West (W); Least Concern (LC), Near Threatened (NT), Vulnerable (VU), Endangered (EN).

Recent research has also examined the capacity of species to persist in smaller or cryptic microrefugia (Mee and Moore 2014; Rull 2009; Stewart et al. 2010). For example, there is debate about where species currently occurring in and around mountains survived glacial–interglacial cycles, when parts of their current restricted ranges would have been climatically unfavorable. Holderegger and Thiel-Egenter (2009) define three different types of refugia for mountain species during glacial periods: isolated microrefugia on mountain peaks above the glaciers in the center of mountain systems (*nunataks*); peripheral refugia that maintained some snow-free periods towards the borders of mountain systems; and lowland refugia outside mountain systems and beyond the limits of ice shields. In Europe, Alpine Lepidoptera undertook elevation range shifts to persist during glacial periods in lowland refugia (Haubrich and Schmitt 2007; Schmitt and Hewitt 2004), but some less dispersive alpine plant species (potentially with associated insects) are inferred to have survived in high-altitude nunataks (Holderegger and Thiel-Egenter 2009).

Insect species now restricted to mountains can show evidence of persistence during glacial–interglacial cycles in multiple refugia. For example, distinct genetic lineages in different parts of the European distributions of the flightless ground beetle *Carabus sylvestris* (Panzer, 1793) (Drees et al. 2016) and the butterfly *Erebia epiphron* (Knoch, 1783) (Minter et al. 2020) suggest both species survived in several topographically diverse regions that allowed persistence in low-elevation foothills during glacial periods. In South America, the Patagonian bee (*Centris cineraria* Smith, 1854) shows evidence of several low-elevation glacial refugia beside the Andes (Sosa-Pivatto et al. 2020). In western North America, narrow-range endemic species of flightless, cold-specialized ice crawlers (*Grylloblata* Walker, 1914) tracked expanding and retreating glaciers up and down multiple mountain slopes over the past approximately 2 million years (Schoville et al. 2019).

The locations where cold-adapted species survived warm interglacial periods are particularly pertinent in the context of ongoing warming. Gao et al. (2022) found two distinct lineages in the Chinese endemic scorpionfly *Cerapanorpa brevicornis* (Hua and Li, 2007), corresponding to a large interglacial macrorefugium in the Qinling mountains, as well as a more isolated "sky-island archipelago" providing multiple microrefugia in the eastern Bashan Mountains. In contrast, a study of the mountain butterfly *Parnassius apollo* in France suggests a single high-mountain refugium in the Auvergne region during the Last Interglacial (120,000 years ago), when temperatures were 2 °C warmer than nowadays, with subsequent colonization of the western Alps when conditions cooled (Kebaïli et al. 2022). Since the LGM, *P. apollo* has been contracting its range towards the peaks of both the Auvergne and western Alps (Kebaïli et al. 2022).

Long-term climatic stability, or at least the maintenance of favorable climatic conditions when prevailing climates are most restrictive, is likely to leave important legacies in insect diversity, especially for dispersal-limited organisms (Table 16.1). In the Australian Wet Tropics, rainforest refugia persisted in high elevation areas throughout periods of low temperature and rainfall around the LGM (18 ka), and later warmer and wetter periods (Boyer et al. 2016). In this region, current species richness and phylogenetic diversity in the mite harvestmen *Austropurcellia* (Juberthie, 1988) are most closely related to climatic suitability during the LGM, when the distribution of rainforest was most restricted by climatic conditions, illustrating how bottlenecks in the spatial distribution of suitable climate has long-term effects on patterns of diversity (Boyer et al. 2016).

However, apparently similar current patterns of diversity and endemism can reflect contrasting evolutionary histories. In Italy, butterfly endemicity is highest in the mountain ranges of the Alps (in the north) and Apennines (running down the peninsula). But Menchetti et al. (2021) found that the origins of the mountain endemics in the two regions are distinct. Endemics in the Apennines have evolved *in situ* in isolated high-altitude refugia as conditions have warmed since the Pleistocene. In contrast, endemics in the Alps are occupying an ecological center of endemicity, which currently provides a large area of favorable cool conditions, but these species evolved *ex situ* (e.g., during glacial periods) and later colonized the Alps as their distributions moved uphill when conditions warmed.

As we have seen for the Iberian Peninsula, many European butterfly species either approach their current low-latitude range margins in Mediterranean mountains, or are geographically isolated endemics in the high elevation areas that provide the coolest conditions and greatest topographic variation in southern Europe (Figures 16.2 and 16.3) (see also Ribera and Vogler 2004 for diving beetles). These locations represent regions where populations have persisted throughout glacial–interglacial periods owing to stable climatic conditions (e.g., Kebaïli et al. 2022), either because climate change is slower or because a wide range of climatic conditions are represented in an area accessible to dispersing individuals. Persistence over a long time period leads to high genetic diversity (Dincă et al. 2021; Hewitt 2000), potentially with intra-specific adaptations both to cool and warm conditions. In this case, the genetic diversity in rear-edge populations, occurring in the locations of past glacial refugia yet also persisting through interglacials, may make such populations a priority for conservation, because the range of thermal adaptations could make them resilient to future climatic changes (Hampe and Petit 2005).

Perhaps paradoxically, the climatic stability of historical refugia and centers of endemism could make their biotas especially vulnerable to anthropogenic climate change, for several ecological and evolutionary reasons. Climatic stability promotes species richness and endemism not only because of reduced extinction rates but because opportunities for greater specialization increase speciation rates (Fine 2015). In the tropics, for example, climatic stability permits narrower physiological tolerance, which in turn can increase the cost of a species dispersing over a climatic gradient such as elevation, and allowing allopatric speciation of specialists over short geographic distances (Janzen 1967; Sheldon et al. 2018). Species able to persist in geographically restricted climatically stable regions may in turn evolve traits such as reduced vagility. Small geographic ranges, weak dispersal ability and high levels of specialization could in turn increase species sensitivity to a given magnitude of climatic change or could reduce their adaptive capacity (Table 16.3).

Globally, regions with the highest species richness of terrestrial vertebrates (amphibians, birds and mammals) correspond to locations where temperatures have been most stable since the LGM (Brown et al. 2020), with the proviso that the most climatically stable regions per century have rarely stayed in fixed geographic locations over this period of time (Fordham et al. 2019). Nevertheless, the increases in average temperatures projected for the twenty-first century are likely to exceed the acclimation capacities of many of the species occurring in these hitherto stable refugia (Brown et al. 2020): ectothermic insects with small geographic ranges and specialist habitat requirements also occurring in tropical centers of diversity and endemism may be similarly vulnerable. Nevertheless, Kocsis et al. (2021) found that global terrestrial and freshwater locations classified as rich in biodiversity have experienced somewhat reduced warming compared with global averages, emphasizing their potential importance for conservation both in terms of their biodiversity value and relative climatic stability. It is important to note that these two studies on the global vulnerability of taxa in historical climate change refugia focused on relatively coarse scale information on climate and species richness. Both studies acknowledge that fine-scale variation in topography, climate and species distributions are likely to influence ecological vulnerability to future climate change, such that understanding how fine-scale climatic variation could buffer restricted-range, specialist species against climate change in historical refugia is a priority for conservation.

Table 16.3 Life-history traits of insects in proposed refugia that could increase sensitivity to climate change, and the resulting constraints, opportunities and contingency plans for their conservation.

Trait	Features in refugia	Constraints for conservation	Opportunities and contingency plans	Key references
Geographic range size	Small, restricted range; many endemics	Restricted range leads to high extinction risk. Small range size limits heterogeneity and overlap between *in situ* and *ex situ* refugia.	Opportunity: many species can be conserved in a small area. Contingency: *ex situ* translocations for species lacking *in situ* refugia (e.g., mountaintop species).	Ohlemüller et al. 2008 Morelli et al. 2016 Beaumont et al. 2019 Graham et al. 2019
Geographic range position	Low-latitude, low-elevation "rear-edge" margins	Greater effects of competition or natural enemies at low latitude/elevation margins. Climate change compounded by greater direct impact of human land-use at lower elevations. Relative lack of research and conservation capacity at lower latitudes.	Opportunity: Restriction at low latitude margins to mountains provides high heterogeneity of habitat, topography and microclimate, allowing fine-resolution climatic tracking, species coexistence, and acting as a barrier to intensive land-use. Contingency: Monitor and conserve species elsewhere throughout geographic range.	Hampe and Petit 2005
Thermal niche	Endemics: narrow thermal niche adapted to stable conditions Rear edge: close to upper thermal limits	High likelihood of exceeding thermal limits. Risk that species exceed limits simultaneously, driving ecosystem collapse.	Adaptations in genetically diverse populations that persisted through glacial and interglacial periods may provide wider thermal niche. Contingency: Monitor population dynamic responses to microclimatic variation.	Calosi et al. 2010 Trisos et al. 2020 Dincă et al. 2021 Weaving et al. 2022
Habitat specialization	Stability promotes high specialization	Restricted habitat availability and heterogeneity in *in situ* and *ex situ* refugia; reduced connectivity to enable climatic tracking.	Narrow specialism increases opportunity to target habitat protection and management. Contingency: Habitat restoration or management in potential *ex situ* refugia.	Janzen 1967 Thomas et al. 2009 Sheldon et al. 2018
Dispersal ability	Stability and habitat specialization select for reduced vagility	Reduced ability to track changing climatic conditions.	Possible increased opportunity for very fine-resolution behavioral thermoregulation or microclimatic tracking. Contingency: Habitat restoration to increase connectivity to *ex situ* refugia; translocations.	Sandel et al. 2011 Sahlean et al. 2014 Koot et al. 2022

16.4 The conservation of insects in climate change refugia

The life history of insects makes them potentially responsive to the role of refugia in adapting conservation to climate change. Insects are small and respond to the fine-resolution microclimatic variation provided by variation in topography and vegetation (Pincebourde and Woods 2020). They are mobile to a greater or lesser degree and have short life cycles and fast population dynamics, allowing them to adjust their local distributions and abundance quickly in response to changing climate (Halsch et al. 2021). In combination, the small size, mobility and rapid population dynamics of insects allow their populations or metapopulations to persist in relatively small areas and then expand from these if conditions improve (Bennie et al. 2013; Karban and Huntzinger 2021). Evidence-based management of abiotic and biotic conditions can therefore allow conservation programs to arrest declines and implement recoveries of threatened insect species from highly geographically restricted populations (Thomas et al. 2009). To adapt the conservation of highly threatened insect species, or those that are important as flagships or umbrella species for conservation, climate change refugia could therefore be prioritized by incorporating understanding of their responses to climatic variation (such as based on eco-physiological research or species distribution modelling) to prioritize sites in existing conservation programs and protected area networks. In addition, species distribution modeling can be used to identify *in situ* refugia for focal taxa, that will retain suitability under projected climate change, and *ex situ* refugia, that currently lie outside the geographic range but have a high chance of providing persistent suitable climate and habitat in the future.

A more general strategy for the many insects lacking detailed information about climate change sensitivity is to focus on regions, landscapes or habitats that are more resistant to climate change (Michalak et al. 2018; Stralberg et al. 2020a), or that maintain climatic conditions that are increasingly rare (Ohlemüller et al. 2008). Likely terrain-mediated refugia can be identified based on palaeoecological evidence and climate modeling, and have been classified according to their spatial scale of climate buffering. Macrorefugia formed by coastal regions or topographically variable uplands reduce climate change velocity by providing slower rates of climate change or greater intra-regional climatic variation (Haight and Hammill 2020). Microrefugia formed by local features such as hollows, narrow valleys or poleward-facing slopes can maintain distinctly cooler temperatures than their surroundings (Ashcroft et al. 2012; Dobrowski 2011), while lakes, rivers and locally impeded drainage may reduce warming or drying in terrestrial habitats (Stralberg et al. 2020a). Ecological resistance to external climate change provided by factors such as shading in forests, persistent moisture in peat soils, or eco-hydrological feedbacks caused by ecosystem engineers such as beavers can also lead to "ecosystem-protected" refugia (Stralberg et al. 2020a).

Recent research on climate change refugia has built on the increasing evidence and guidance regarding their identification to propose frameworks and remaining challenges for their practical incorporation in conservation (Keppel et al. 2015; Morelli et al. 2016; Selwood and Zimmer 2020). Designation of climate change refugia as protected areas is important to minimize impacts of additional anthropogenic threats (Monsarrat et al. 2019). In practice, many potential refugia are already incorporated in protected area networks in topographically complex uplands (Haight and Hammill 2020; Stralberg et al. 2020b), and the protection of new sites could be prioritized in areas of high environmental diversity, rare climatic conditions or slow projected climate change. Under-representation of sites in existing protected area networks can be identified based on modeled *in situ* and *ex situ* refugia for individual taxa under a range of climate change scenarios (e.g., Beaumont et al. 2019; Graham et al. 2019). However, additional protection of sites based on individual taxa, as well as the necessary protection of sites to provide connectivity between these, can incur greater additional cost than prioritization based more generally on terrain-based refugial capacity (Lawler et al. 2020).

Evidence of physiological, genetic, demographic and diversity responses is needed to validate the role of proposed refugia in conservation (Barrows et al. 2020), and is a vital part of monitoring and managing biodiversity in refugia (Morelli et al. 2016). To date, there is only limited evidence of ecological responses in proposed refugia at the landscape scales over which conservation management operates (Selwood and Zimmer 2020), although several studies have demonstrated evidence for insect persistence over the course of years to decades in potential climatic refugia. In England, local topographic variability has reduced rates of extirpation for warming-sensitive Coleoptera and Lepidoptera (Suggitt et al. 2018). The regional distribution of meadow spittlebugs in California contracts towards local environments providing protection from desiccating winds during dry years, expanding again in wet years (Karban and Huntzinger 2021). Restricted human-modified local environments allow ground beetles (Volf et al. 2018) or odonates (Husband and McIntyre 2021) to persist in landscapes providing otherwise unfavorable hydrological conditions. However, longer-term validation of the ongoing role of refugia is rare, and long-term work in California shows that extreme drought can threaten mountain butterfly communities that had for several decades been much more stable than nearby lowland communities (Halsch et al. 2021). Because refugia may not be permanent, both ongoing monitoring and contingency planning are essential, and *ex situ* approaches such as translocations may need to be considered for highly isolated mountaintop populations for which *in situ* refugia are not available (Morelli et al. 2016) (see Table 16.3 for additional possible contingency plans).

16.5 Conclusion

The threat of climate change to biodiversity is global and its effects are inherently difficult to manage. In this context, the identification, protection and management of refugia could play an important role in adapting conservation, both to changes in prevailing climates and to the incidence and severity of extreme events such as droughts, fires and heat waves. Climate change refugia provide environments where species are able to withstand deteriorating or extreme conditions, and that can act as foci for recovery if conditions improve (Keppel et al. 2012). Explicit incorporation of climate change refugia in conservation programs therefore provides a potential means of unifying existing *in situ* measures such as protected area networks, with an increasing need to account for the dynamic changes to biodiversity that are taking place in response to global change.

For insects, recent climate change has already caused widespread changes in abundance, species distributions and diversity (Lenoir et al. 2020; Outhwaite et al. 2022). The effective identification and management of climate change refugia could help conservation adapt to these dynamic and accelerating changes. The paleoecological and evolutionary effects of historical climate change provide strong evidence for the long-term importance of localized climatic refugia for insect biodiversity. What remains is to apply our increasing understanding of the mechanisms influencing (micro-)climatic variation and its ecological effects on insects, into practical programs to identify safe havens from climate change and prioritize these for protection and conservation management.

Key reflections

- Climate change refugia are landscapes, topographic features or ecosystems that allow taxa to persist through unfavorable climatic conditions and act as foci for subsequent recovery or range expansion.
- Macrorefugia are landscapes, often in mountains, providing climatic conditions that are spatially heterogeneous or have been stable over time, permitting local-scale redistributions and the evolution of endemics.
- Microrefugia provide distinct climatic conditions or rates of climatic change from their surroundings and could therefore buffer species against prevailing changes in the climate.
- Because insects are small and show rapid dynamic responses to fine-scale variation in the climate, the protection, monitoring and management of insects in refugia represent priorities for adapting their conservation to climate change.

Key further reading

- Hampe, A., and Petit, R.J. 2005. Conserving biodiversity under climate change: The rear edge matters. *Ecology Letters 8* (5): 461–67.
- Kebaïli, C., Sherpa, S., Rioux, D., and Després, L. 2022. Demographic inferences and climatic niche modelling shed light on the evolutionary history of the emblematic cold-adapted Apollo butterfly at regional scale. *Molecular Ecology 31* (2): 448–66.
- Lenoir, J., Hattab, T., and Pierre, G. 2017. Climatic microrefugia under anthropogenic climate change: Implications for species redistribution. *Ecography 40* (2): 253–66.
- Menchetti, M., Talavera, G., Cini, A., Salvati, V., Dincă, V., Platania, L., Bonelli, S., *et al.* 2021. Two ways to be endemic. Alps and Apennines are different functional refugia during climatic cycles. *Molecular Ecology 30* (5): 1297–1310.
- Mingarro, M., Cancela, J.P., Burón-Ugarte, A., García-Barros, E., Munguira, M.L., Romo, H., and Wilson, R.J. 2021. Butterfly communities track climatic variation over space but not time in the Iberian Peninsula. *Insect Conservation and Diversity 14* (5): 647–60.
- Morelli, T.L., Daly, C., Dobrowski, S.Z., Dulen, D.M., Ebersole, J.L., Jackson, S.T., Lundquist, J.D., *et al.* 2016. Managing climate change refugia for climate adaptation. *PLoS One 11* (8): 1–17.
- Oldfather, M.F., Kling, M.M., Sheth, S.N., Emery, N.C., and Ackerly, D.D. 2020. Range edges in heterogeneous landscapes: Integrating geographic scale and climate complexity into range dynamics. *Global Change Biology 26* (3): 1055–1067.
- Pincebourde, S., and Woods, H.A. 2020. There is plenty of room at the bottom: microclimates drive insect vulnerability to climate change. *Current Opinion in Insect Science 41*: 63–70.
- Suggitt, A.J., Wilson, R.J., Isaac, N.J.B., Beale, C.M., Auffret, A.G., August, T., Bennie, J.J., *et al.* 2018. Extinction risk from climate change is reduced by microclimatic buffering. *Nature Climate Change 8* (8): 713–17.

References

Abellán, P. and Svenning, J.C. 2014. Refugia within refugia–patterns in endemism and genetic divergence are linked to Late Quaternary climate stability in the Iberian Peninsula. *Biological Journal of the Linnean Society* 113 (1): 13–28.

Ackerly, D.D., Loarie, S.R., Cornwell, W.K., Weiss, S.B., Hamilton, H., Branciforte, R., and Kraft, N.J.B. 2010. The geography of climate change: Implications for conservation biogeography. *Diversity and Distributions* 16 (3): 476–87.

Alexander, J.M., Chalmandrier, L., Lenoir, J., Burgess, T.I., Essl, F., Haider, S., Kueffer, C., et al. 2018. Lags in the response of mountain plant communities to climate change. *Global Change Biology* 24 (2): 563–79.

Ashcroft, M.B. 2010. Identifying refugia from climate change. *Journal of Biogeography* 37 (8): 1407–1413.

Ashcroft, M.B., Gollan, J.R., Warton, D.I., and Ramp, D. 2012. A novel approach to quantify and locate potential microrefugia using topoclimate, climate stability, and isolation from the matrix. *Global Change Biology* 18 (6): 1866–1879.

Ashton, S., Gutierrez, D., and Wilson, R.J. 2009. Effects of temperature and elevation on habitat use by a rare mountain butterfly: Implications for species responses to climate change. *Ecological Entomology* 34 (4): 437–46.

Barrows, C.W., Ramirez, A.R., Sweet, L.C., Morelli, T.L., Millar, C.I., Frakes, N., Rodgers, J., et al. 2020. Validating climate-change refugia: Empirical bottom-up approaches to support management actions. *Frontiers in Ecology and the Environment* 18 (5): 298–306.

Bátori, Z., Lőrinczi, G., Tölgyesi, C., Módra, G., Juhász, O., Aguilon, D.J., Vojtkó, A., et al. 2020. Karst dolines provide diverse microhabitats for different functional groups in multiple phyla. *Scientific Reports* 9 (1): 1–13.

Beaumont, L.J., Esperón-Rodríguez, M., Nipperess, D.A., Wauchope-Drumm, M., and Baumgart-ner, J.B. 2019. Incorporating future climate uncertainty into the identification of climate change refugia for threatened species. *Biological Conservation* 237: 230–37.

Bennie, J., Hodgson, J.A., Lawson, C.R., Holloway, C.T.R., Roy, D.B., Brereton, T., Thomas, C.D., et al. 2013. Range expansion through fragmented landscapes under a variable climate. *Ecology Letters* 16 (7): 921–29.

Boyer, S.L., Markle, T.M., Baker, C.M., Luxbacher, A.M., and Kozak, K.H. 2016. Historical refugia have shaped biogeographical patterns of species richness and phylogenetic diversity in mite harvestmen (Arachnida, Opiliones, Cyphophthalmi) endemic to the Australian Wet Tropics. *Journal of Biogeography* 43 (7): 1400–1411.

Brambilla, M. and Gobbi, M. 2014. A century of chasing the ice: Delayed colonisation of ice-free sites by ground beetles along glacier forelands in the Alps. *Ecography* 37 (1): 33–42.

Brighenti, S., Hotaling, S., Finn, D.S., Fountain, A.G., Hayashi, M., Herbst, D., Saros, J.E., et al. 2021. Rock glaciers and related cold rocky landforms: Overlooked climate refugia for mountain biodiversity. *Global Change Biology* 27 (8): 1504–1517.

Brito-Morales, I., García Molinos, J., Schoeman, D.S., Burrows, M.T., Poloczanska, E.S., Brown, C.J., Ferrier, S., et al. 2018. Climate velocity can inform conservation in a warming world. *Trends in Ecology & Evolution* 33 (6): 441–57.

Brown, S.C., Wigley, T.M., Otto-Bliesner, B.L., Rahbek, C., and Fordham, D.A. 2020. Persistent quaternary climate refugia are hospices for biodiversity in the Anthropocene. *Nature Climate Change* 10 (3): 244–48.

Bütikofer, L., Anderson, K., Bebber, D.P., Bennie, J.J., Early, R.I., and Maclean, I.M.D. 2020. The problem of scale in predicting biological responses to climate. *Global Change Biology* 26 (12): 6657–6666.

Calosi, P., Bilton, D.T., Spicer, J.I., Votier, S.C., and Atfield, A. 2010. What determines a species' geographical range? Thermal biology and latitudinal range size relationships in European diving beetles (Coleoptera: Dytiscidae). *Journal of Animal Ecology* 79 (1): 194–204.

Carnicer, J., Stefanescu, C., Vives-Ingla, M., López, C., Cortizas, S., Wheat, C., Vila, R., et al. 2019. Phenotypic biomarkers of climatic impacts on declining insect populations: A key role for decadal drought, thermal buffering and amplification effects and host plant dynamics. *Journal of Animal Ecology* 88 (3): 376–91.

Cassel-Lundhagen, A., Schmitt, T., Wahlberg, N., Sarvašová, L., Konvička, M., Ryrholm, N., and Kaňuch, P. 2020. Wing morphology of the butterfly Coenonympha arcania in Europe: Traces of both historical isolation in glacial refugia and current adaptation. *Journal of Zoological Systematics and Evolutionary Research* 58 (4): 929–43.

Coope, G.R. 2004. Several million years of stability among insect species because of, or in spite of, Ice Age climatic instability? *Philosophical Transactions of the Royal Society of London. Series B: Biological Sciences* 359 (1442): 209–14.

Devictor, V., Van Swaay, C., Brereton, T., Brotons, L., Chamberlain, D., Heliölö, J., Herrando, S., et al. 2012. Differences in the climatic debts of birds and butterflies at a continental scale. *Nature Climate Change* 2 (2): 121–24.

Dincă, V., Dapporto, L., Somervuo, P., Vodă, R., Cuvelier, S., Gascoigne-Pees, M., Huemer, P. et al., 2021. High resolution DNA barcode library for European butterflies reveals continental patterns of mitochondrial genetic diversity. *Communications Biology* 4 (1): 1–11.

Dobrowski, S.Z. 2011. A climatic basis for microrefugia: The influence of terrain on climate. *Global Change Biology* 17 (2): 1022–1035.

Dobrowski, S.Z. and Parks, S.A. 2016. Climate change velocity underestimates climate change exposure in mountainous regions. *Nature Communications* 7 (1): 1–8.

Doxa, A., Kamarianakis, Y., and Mazaris, A.D. 2022. Spatial heterogeneity and temporal stability characterize future climatic refugia in Mediterranean Europe. *Global Change Biology* 28 (7): 2413–2424.

Drees, C., Husemann, M., Homburg, K., Brandt, P., Dieker, P., Habel, J.C., von Wehrden, H., et al. 2016. Molecular analyses and species distribution models indicate cryptic northern mountain refugia for a forest-dwelling ground beetle. *Journal of Biogeography* 43 (11): 2223–2236.

Fine, P.V. 2015. Ecological and evolutionary drivers of geographic variation in species diversity. *Annual Review of Ecology, Evolution, and Systematics* 46: 369–92.

Fordham, D.A., Brown, S.C., Wigley, T.M., and Rahbek, C. 2019. Cradles of diversity are unlikely relics of regional climate stability. *Current Biology* 29 (10): R356–R357.

Fourcade, Y., WallisDeVries, M.F., Kuussaari, M., van Swaay, C.A., Heliölä, J., and Öckinger, E. 2021. Habitat amount and distribution modify community dynamics under climate change. *Ecology Letters* 24 (5): 950–57.

De Frenne, P., Lenoir, J., Luoto, M., Scheffers, B.R., Zellweger, F., Aalto, J., Ashcroft, M.B., et al. 2021. Forest microclimates and climate change: Importance, drivers and future research agenda. *Global Change Biology* 27 (11): 2279–2297.

Gao, K., Hua, Y., Xing, L.X., and Hua, B.Z. 2022. Speciation of the cold-adapted scorpionfly Cerapanorpa brevicornis (Mecoptera: Panorpidae) via interglacial refugia. *Insect Conservation and Diversity* 15 (1): 114–27.

García, M.B., Domingo, D., Pizarro, M., Font, X., Gómez, D., and Ehrlén, J. 2020. Rocky habitats as microclimatic refuges for biodiversity. A close-up thermal approach. *Environmental and Experimental Botany* 170 (103886): 1–10.

García-Barros, E., Munguira, M., Stefanescu, C., and Vives, A. 2013. "*Lepidoptera: Papilionoidea.*" In *Fauna Iberica*, 37, edited by M.A. Ramos, J. Alba, X. Bellés, J. Gosálbez, A. Guerra, E. Macpherson, J. Serrano, et al. Madrid: Museo Nacional de Ciencias Naturales, 1213.

García-Barros, E., Munguira, M.L., Martín, J., Romo, H., Garcia-Pereira, P., and Maravalhas, E.S. 2004. *Atlas de las mariposas diurnas de la Península Ibérica e islas Baleares (Lepidoptera: Papilionoidea and Hesperioidea)*. Monografias de la SEA, vol 11 edn. Zaragoza: Sociedad Entomológica Aragonesa.

Gavin, D.G., Fitzpatrick, M.C., Gugger, P.F., Heath, K.D., Rodríguez-Sánchez, F., Dobrowski, S.Z., Hampe, A. et al. 2014. Climate refugia: Joint inference from fossil records, species distribution models and phylogeography. *New Phytologist* 204 (1): 37–54.

Goulson, D. 2019. The insect apocalypse, and why it matters. *Current Biology* 29 (19): R967–R971.

Graae, B.J., Vandvik, V., Armbruster, W.S., Eiserhardt, W.L., Svenning, J.C., Hylander, K., Ehrlén, J., et al. 2018. Stay or go—how topographic complexity influences alpine plant population and community responses to climate change. *Perspectives in Plant Ecology, Evolution and Systematics* 30: 41–50.

Graham, V., Baumgartner, J.B., Beaumont, L.J., Esperón-Rodríguez, M., and Grech, A. 2019. Prioritizing the protection of climate refugia: Designing a climate-ready protected area network. *Journal of Environmental Planning and Management* 62 (14): 2588–2606.

Haight, J. and Hammill, E. 2020. Protected areas as potential refugia for biodiversity under climatic change. *Biological Conservation* 241: 1–11.

Halsch, C.A., Shapiro, A.M., Fordyce, J.A., Nice, C.C., Thorne, J.H., Waetjen, D.P., and Forister, M.L. 2021. Insects and recent climate change. *Proceedings of the National Academy of Sciences* 118 (2): 1–9.

Hampe, A. and Petit, R.J. 2005. Conserving biodiversity under climate change: The rear edge matters. *Ecology Letters* 8 (5): 461–67.

Harrison, S. and Noss, R. 2017. Endemism hotspots are linked to stable climatic refugia. *Annals of Botany* 119 (2): 207–14.

Haubrich, K. and Schmitt, T. 2007. Cryptic differentiation in alpine-endemic, high-altitude butterflies reveals down-slope glacial refugia. *Molecular Ecology* 16 (17): 3643–3658.

Hawkins, B.A. and Porter, E.E. 2003. Water–energy balance and the geographic pattern of species richness of western Palearctic butterflies. *Ecological Entomology* 28 (6): 678–86.

Hernández-Roldán, J.L., Murria, C., Romo, H., Talavera, G., Zakharov, E., Hebert, P.D., and Vila, R. 2011. Tracing the origin of disjunct distributions: A case of biogeographical convergence in Pyrgus butterflies. *Journal of Biogeography* 38 (10): 2006–2020.

Herrando, S., Titeux, N., Brotons, L., Anton, M., Ubach, A., Villero, D., García-Barros, E., et al. 2019. Contrasting impacts of precipitation on Mediterranean birds and butterflies. *Scientific Reports* 9 (1): 1–7.

Hewitt, G. 1999. Post-glacial re-colonization of European biota. *Biological Journal of the Linnean Society* 68 (1–2): 87–112.

Hewitt, G.M. 2000. The genetic legacy of the Pleistocene ice ages. *Nature* 405: 907–13.

Hickling, R., Roy, D.B., Hill, J.K., Fox, R., and Thomas, C.D. 2006. The distributions of a wide range of taxonomic groups are expanding polewards. *Global Change Biology* 12 (3): 450–55.

Holderegger, R. and Thiel-Egenter, C. 2009. A discussion of different types of glacial refugia used in mountain biogeography and phylogeography. *Journal of Biogeography* 36 (3): 476–80.

Husband, D.M. and McIntyre, N.E. 2021. Urban areas create refugia for odonates in a semi-arid region. *Insects* 12 (5): 1–15.

Janzen, D.H. 1967. Why mountain passes are higher in the tropics. *The American Naturalist* 101 (919): 233–49.

Janzen, D.H. and Hallwachs, W. 2021. To us insectometers, it is clear that insect decline in our Costa Rican tropics is real, so let's be kind to the survivors. *Proceedings of the National Academy of Sciences* 118 (2): 1–8.

Jump, A.S., Mátyás, C., and Peñuelas, J. 2009. The altitude-for-latitude disparity in the range retractions of woody species. *Trends in Ecology & Evolution* 24 (12): 694–701.

Karban, R. and Huntzinger, M. 2021. Spatial and temporal refugia for an insect population declining due to climate change. *Ecosphere* 12 (11): 1–9.

Karger, D.N., Conrad, O., Böhner, J., Kawohl, T., Kreft, H., Soria-Auza, R.W., Zimmermann, N.E., et al. 2017. Climatologies at high resolution for the earth's land surface areas. *Scientific Data* 4 (1): 1–20.

Kašpar, V., Hederová, L., Macek, M., Müllerová, J., Prošek, J., Surový, P., Wild, J., et al. 2021. Temperature buffering in temperate forests: Comparing microclimate models based on ground measurements with active and passive remote sensing. *Remote Sensing of Environment* 263: 1–10.

Kebaïli, C., Sherpa, S., Rioux, D., and Després, L. 2022. Demographic inferences and climatic niche modelling shed light on the evolutionary history of the emblematic cold-adapted Apollo butterfly at regional scale. *Molecular Ecology* 31 (2): 448–66.

Kenyeres, Z., Rácz, I.A., and Varga, Z. 2009. Endemism hot spots, core areas and disjunctions in European Orthoptera. *Acta Zoologica Cracoviensia—Series B: Invertebrata* 52 (1–2): 189–211.

Keppel, G., Mokany, K., Wardell-Johnson, G.W., Phillips, B.L., Welbergen, J.A., and Reside, A.E. 2015. The capacity of refugia for conservation planning under climate change. *Frontiers in Ecology and the Environment* 13 (2): 106–12.

Keppel, G., Ottaviani, G., Harrison, S., Wardell-Johnson, G.W., Marcantonio, M., and Mucina, L. 2018. Towards an eco-evolutionary understanding of endemism hotspots and refugia. *Annals of Botany* 122 (6): 927–34.

Keppel, G., Van Niel, K.P., Wardell-Johnson, G.W., Yates, C.J., Byrne, M., Mucina, L., Schut, A.G.T., et al. 2012. Refugia: Identifying and understanding safe havens for biodiversity under climate change. *Global Ecology and Biogeography* 21 (4): 393–404.

Kocsis, Á. T., Zhao, Q., Costello, M.J., and Kiessling, W. 2021. Not all biodiversity rich spots are climate refugia. *Biogeosciences* 18 (24): 6567–6578.

Koot, E.M., Morgan-Richards, M., and Trewick, S.A. 2022. Climate change and alpine-adapted insects: Modelling environmental envelopes of a grasshopper radiation. *Royal Society Open Science* 9 (3): 1–15.

Laiolo, P., Pato, J., and Obeso, J.R. 2018. Ecological and evolutionary drivers of the elevational gradient of diversity. *Ecology Letters* 21 (7): 1022–1032.

Lawler, J.J., Rinnan, D.S., Michalak, J.L., Withey, J.C., Randels, C.R., and Possingham, H.P. 2020. Planning for climate change through additions to a national protected area network: Implications for cost and configuration. *Philosophical Transactions of the Royal Society B: Biological Sciences* 375 (1794): 1–8.

Lawson, C.R., Bennie, J., Hodgson, J.A., Thomas, C.D., and Wilson, R.J. 2014. Topographic microclimates drive microhabitat associations at the range margin of a butterfly. *Ecography* 37 (8): 732–40.

Lenoir, J. and Svenning, J.C. 2015. Climate-related range shifts–a global multidimensional synthesis and new research directions. *Ecography* 38 (1): 15–28.

Lenoir, J., Hattab, T., and Pierre, G. 2017. Climatic microrefugia under anthropogenic climate change: Implications for species redistribution. *Ecography* 40 (2): 253–266.

Lenoir, J., Bertrand, R., Comte, L., Bourgeaud, L., Hattab, T., Murienne, J., and Grenouillet, G. 2020. Species better track climate warming in the oceans than on land. *Nature Ecology and Evolution* 4 (8): 1044–1059.

Li, J., Li, Q., Wu, Y., Ye, L., Liu, H., Wei, J., and Huang, X. 2021. Mountains act as museums and cradles for hemipteran insects in China: Evidence from patterns of richness and phylogenetic structure. *Global Ecology and Biogeography* 30 (5): 1070–1085.

Loarie, S.R., Duffy, P.B., Hamilton, H., Asner, G.P., Field, C.B., and Ackerly, D.D. 2009. The velocity of climate change. *Nature* 462 (7276): 1052–1055.

De Lombaerde, E., Vangansbeke, P., Lenoir, J., Van Meerbeek, K., Lembrechts, J., Rodríguez-Sánchez, F., Luoto, M., et al. 2022. Maintaining forest cover to enhance temperature buffering under future climate change. *Science of the Total Environment* 810: 1–9.

Ma, C.S., Ma, G., and Pincebourde, S. 2021. Survive a warming climate: Insect responses to extreme high temperatures. *Annual Review of Entomology* 66: 163–84.

Maclean, I.M., Suggitt, A.J., Wilson, R.J., Duffy, J.P., and Bennie, J.J. 2017. Fine-scale climate change: Modelling spatial variation in biologically meaningful rates of warming. *Global Change Biology* 23 (1): 256–68.

Maresova, J., Suchackova Bartonova, A., Konvicka, M., Høye, T.T., Gilg, O., Kresse, J.C., Shapoval, N.A., et al. 2021. The story of endurance: Biogeography and the evolutionary history of four Holarctic butterflies with different habitat requirements. *Journal of Biogeography* 48 (3): 590–602.

Marshall, L., Perdijk, F., Dendoncker, N., Kunin, W., Roberts, S., and Biesmeijer, J.C. 2020. Bumblebees moving up: Shifts in elevation ranges in the Pyrenees over 115 years. *Proceedings of the Royal Society B: Biological Sciences* 287 (1938): 1–10.

Massa, B. and Fontana, P. 2020. Endemism in Italian Orthoptera. *Biodiversity Journal* 11 (2): 405–34.

Mee, J.A. and Moore, J.S. 2014. The ecological and evolutionary implications of microrefugia. *Journal of Biogeography* 41 (5): 837–41.

Menchetti, M., Talavera, G., Cini, A., Salvati, V., Dincă, V., Platania, L., Bonelli, S. et al. 2021. Two ways to be endemic. Alps and Apennines are different functional refugia during climatic cycles. *Molecular Ecology* 30 (5): 1297–1310.

Menéndez, R., Megías, A.G., Hill, J.K., Braschler, B., Willis, S.G., Collingham, Y., Fox, R., et al. 2006. Species richness changes lag behind climate change. *Proceedings of the Royal Society B: Biological Sciences* 273 (1593): 1465–1470.

Merrill, R.M., Gutiérrez, D., Lewis, O.T., Gutiérrez, J., Díez, S.B., and Wilson, R.J. 2008. Combined effects of climate and biotic interactions on the elevational range of a phytophagous insect. *Journal of Animal Ecology* 77 (1): 145–55.

Michalak, J.L., Lawler, J.J., Roberts, D.R., and Carroll, C. 2018. Distribution and protection of climatic refugia in North America. *Conservation Biology* 32 (6): 1414–1425.

Mingarro, M., Cancela, J.P., Burón-Ugarte, A., García-Barros, E., Munguira, M.L., Romo, H., and Wilson, R.J. 2021. Butterfly communities track climatic variation over space but not time in the Iberian Peninsula. *Insect Conservation and Diversity* 14 (5): 647–60.

Minter, M., Dasmahapatra, K.K., Thomas, C.D., Morecroft, M.D., Tonhasca, A., Schmitt, T., Siozios, S., et al. 2020. Past, current, and potential future distributions of unique genetic diversity in a cold-adapted mountain butterfly. *Ecology and Evolution* 10 (20): 11155–11168.

Monsarrat, S., Jarvie, S., and Svenning, J.C. 2019. Anthropocene refugia: Integrating history and predictive modelling to assess the space available for biodiversity in a human-dominated world. *Philosophical Transactions of the Royal Society B: Biological Sciences* 374 (1788): 1–10.

Montemayor, S.I., Melo, M.C., Scattolini, M.C., Pocco, M.E., Río, M.G. del, Dellapé, G., Scheibler, E.E., et al. 2017. The fate of endemic insects of the Andean region under the effect of global warming. *PloS One* 12 (10): 1–17.

Morales-Barbero, J., Gouveia, S.F., and Martinez, P.A. 2021. Historical climatic instability predicts the inverse latitudinal pattern in speciation rate of modern mammalian biota. *Journal of Evolutionary Biology* 34 (2): 339–51.

Morelli, T.L., Daly, C., Dobrowski, S.Z., Dulen, D.M., Ebersole, J.L., Jackson, S.T., Lundquist, J.D., et al. 2016. Managing climate change refugia for climate adaptation. *PLoS One* 11 (8): 1–17.

Morgan, B. and Huang, J.P. 2021. Isolation by geographical distance after release from Pleistocene refugia explains genetic and phenotypic variation in Xylotrupes siamensis (Coleoptera: Scarabaeidae). *Zoological Journal of the Linnean Society* 192 (1): 117–29.

Nieto-Sánchez, S., Gutiérrez, D., and Wilson, R.J. 2015. Long-term change and spatial variation in butterfly communities over an elevational gradient: Driven by climate, buffered by habitat. *Diversity and Distributions* 21 (8): 950–61.

Ohlemüller, R., Anderson, B.J., Araujo, M.B., Butchart, S.H., Kudrna, O., Ridgely, R.S., and Thomas, C.D. 2008. The coincidence of climatic and species rarity: High risk to small-range species from climate change. *Biology Letters* 4 (5): 568–72.

Ohlemüller, R., Huntley, B., Normand, S., and Svenning, J.C. 2012. Potential source and sink locations for climate-driven species range shifts in Europe since the Last Glacial Maximum. *Global Ecology and Biogeography* 21 (2): 152–63.

Oldfather, M.F., Kling, M.M., Sheth, S.N., Emery, N.C., and Ackerly, D.D. 2020. Range edges in heterogeneous landscapes: Integrating geographic scale and climate complexity into range dynamics. *Global Change Biology* 26 (3): 1055–1067.

Opedal, Ø.H., Armbruster, W.S., and Graae, B.J. 2015. Linking small-scale topography with microclimate, plant species diversity and intra-specific trait variation in an alpine landscape. *Plant Ecology and Diversity* 8 (3): 305–15.

Outhwaite, C.L., McCann, P., and Newbold, T. 2022. Agriculture and climate change are reshaping insect biodiversity worldwide. *Nature* 605 (7908): 97–102.

Parmesan, C., Ryrholm, N., Stefanescu, C., Hill, J.K., Thomas, C.D., Descimon, H., Huntley, B., et al. 1999. Poleward shifts in geographical ranges of butterfly species associated with regional warming. *Nature* 399 (6736): 579–83.

Patsiou, T.S., Conti, E., Zimmermann, N.E., Theodoridis, S., and Randin, C.F. 2014. Topo-climatic microrefugia explain the persistence of a rare endemic plant in the Alps during the last 21 millennia. *Global Change Biology* 20 (7): 2286–2300.

Pincebourde, S., Murdock, C.C., Vickers, M., and Sears, M.W. 2016. Fine-scale microclimatic variation can shape the responses of organisms to global change in both natural and urban environments. *Integrative and Comparative Biology* 56 (1): 45–61.

Pincebourde, S. and Woods, H.A. 2012. Climate uncertainty on leaf surfaces: The biophysics of leaf microclimates and their consequences for leaf-dwelling organisms. *Functional Ecology* 26 (4): 844–53.

Pincebourde, S. and Woods, H.A. 2020. There is plenty of room at the bottom: Microclimates drive insect vulnerability to climate change. *Current Opinion in Insect Science* 41: 63–70.

Pinkert, S., Dijkstra, K.D.B., Zeuss, D., Reudenbach, C., Brandl, R., and Hof, C. 2018. Evolutionary processes, dispersal limitation and climatic history shape current diversity patterns of European dragonflies. *Ecography* 41 (5): 795–804.

Platts, P.J., Mason, S.C., Palmer, G., Hill, J.K., Oliver, T.H., Powney, G.D., Fox, R., et al. 2019. Habitat availability explains variation in climate-driven range shifts across multiple taxonomic groups. *Scientific Reports* 9 (1): 1–10.

Potter, K.A., Woods, H.A., and Pincebourde, S. 2013. Microclimatic challenges in global change biology. *Global Change Biology* 19 (10): 2932–2939.

Provan, J. and Bennett, K.D. 2008. Phylogeographic insights into cryptic glacial refugia. *Trends in Ecology and Evolution* 23 (10): 564–71.

Rabl, D., Gottsberger, B., Brehm, G., Hofhansl, F., and Fiedler, K. 2020. Moth assemblages in Costa Rica rain forest mirror small-scale topographic heterogeneity. *Biotropica* 52 (2): 288–301.

Ribera, I. and Vogler, A.P. 2004. Speciation of Iberian diving beetles in Pleistocene refugia (Coleoptera, Dytiscidae). *Molecular Ecology* 13 (1): 179–93.

Rödder, D., Schmitt, T., Gros, P., Ulrich, W., and Habel, J.C. 2021. Climate change drives mountain butterflies towards the summits. *Scientific Reports* 11 (1): 1–12.

Rull, V. 2009. Microrefugia. *Journal of Biogeography* 36 (3): 481–84.

Sahlean, T.C., Gherghel, I., Papeş, M., Strugariu, A., and Zamfirescu, S.R. 2014. Refining climate change projections for organisms with low dispersal abilities: A case study of the Caspian whip snake. *PLoS One* 9 (3): 1–12.

Sandel, B., Arge, L., Dalsgaard, B., Davies, R.G., Gaston, K.J., Sutherland, W.J., and Svenning, J.C. 2011. The influence of Late Quaternary climate-change velocity on species endemism. *Science* 334 (6056): 660–64.

Santorufo, L., Ienco, A., and Scalercio, S. 2021. Climate warming drives divergence of montane butterfly communities in Southern Italy. *Regional Environmental Change* 21 (56): 1–13.

Scalercio, S., Bonacci, T., Mazzei, A., Pizzolotto, R., and Brandmayr, P. 2014. Better up, worse down: Bidirectional consequences of three decades of climate change on a relict population of Erebia cassioides. *Journal of Insect Conservation* 18: 643–50.

Schmitt, T., 2007. Molecular biogeography of Europe: Pleistocene cycles and postglacial trends. *Frontiers in Zoology* 4 (11): 1–13.

Schmitt, T. and Hewitt, G.M. 2004. Molecular biogeography of the arctic-alpine disjunct burnet moth species *Zygaena exulans* (Zygaenidae, Lepidoptera) in the Pyrenees and Alps. *Journal of Biogeography* 31 (6): 885–93.

Schoville, S.D., Bougie, T.C., Dudko, R.Y., and Medeiros, M.J. 2019. Has past climate change affected cold-specialized species differentially through space and time? *Systematic Entomology* 44 (3): 571–87.

Selwood, K.E. and Zimmer, H.C. 2020. Refuges for biodiversity conservation: A review of the evidence. *Biological Conservation* 245: 1–9.

Sheldon, K.S., Huey, R.B., Kaspari, M., and Sanders, N.J. 2018. Fifty years of mountain passes: A perspective on Dan Janzen's classic article. *The American Naturalist* 19 1 (5): 553–65.

Sommer, R.S. and Nadachowski, A. 2006. Glacial refugia of mammals in Europe: Evidence from fossil records. *Mammal Review* 36 (4): 251–65.

Soroye, P., Newbold, T., and Kerr, J. 2020. Climate change contributes to widespread declines among bumble bees across continents. *Science* 367 (6478): 685–88.

Sosa-Pivatto, M., Camps, G.A., Baranzelli, M.C., Espíndola, A., Sérsic, A.N., and Cosacov, A. 2020. Connection, isolation and reconnection: Quaternary climatic oscillations and the Andes shaped the phylogeographical patterns of the Patagonian bee Centris cineraria (Apidae). *Biological Journal of the Linnean Society* 131 (2): 396–416.

Stewart, J.R., Lister, A.M., Barnes, I., and Dalén, L. 2010. Refugia revisited: Individualistic responses of species in space and time. *Proceedings of the Royal Society B: Biological Sciences* 277 (1682): 661–71.

Stralberg, D., Arseneault, D., Baltzer, J.L., Barber, Q.E., Bayne, E.M., Boulanger, Y., Brown, C.D., et al. 2020. Climate-change refugia in boreal North America: What, where, and for how long? *Frontiers in Ecology and the Environment* 18 (5): 261–70.

Stralberg, D., Carroll, C., and Nielsen, S.E., 2020b. Toward a climate-informed North American protected areas network: Incorporating climate-change refugia and corridors in conservation planning. *Conservation Letters* 13 (4): 1–10.

Suggitt, A.J., Wilson, R.J., Isaac, N.J.B., Beale, C.M., Auffret, A.G., August, T., Bennie, J.J., et al. 2018. Extinction risk from climate change is reduced by microclimatic buffering. *Nature Climate Change* 8 (8): 713–17.

Tang, X.T., Lu, M.X., and Du, Y.Z. 2022. Molecular phylogeography and evolutionary history of the pink rice borer (Lepidoptera: Noctuidae): Implications for refugia identification and pest management. *Systematic Entomology* 47 (2): 371–83.

Thomas, J.A., Simcox, D.J., and Clarke, R.T. 2009. Successful conservation of a threatened Maculinea butterfly. *Science* 325 (5936): 80–83.

Trisos, C.H., Merow, C., and Pigot, A.L. 2020. The projected timing of abrupt ecological disruption from climate change. *Nature* 580 (7804): 496–501.

Valle, B., Ambrosini, R., Caccianiga, M., and Gobbi, M. 2020. Ecology of the cold-adapted species *Nebria germari* (Coleoptera: Carabidae): The role of supraglacial stony debris as refugium during the current interglacial period. *Acta Zoologica Academiae Scientiarum Hungaricae* 60: 199–220.

van Swaay, C., Cuttelod, A., Collins, S., Maes, D., López Munguira, M., Šašic, M., Settele, J., et al. 2010. European Red List of Butterflies. https://policycommons.net/artifacts/1375275/european-red-list-of-butterflies/1989536/

Vila, R., Stefanescu, C., and Sesma, J.M. 2018. *Guia de les papallones diürnes de Catalunya*. Bellaterra: Lynx Edicions.

Volf, M., Holec, M., Holcová, D., Jaroš, P., Hejda, R., Drag, L., Blízek, J. et al. 2018. Microhabitat mosaics are key to the survival of an endangered ground beetle (Carabus nitens) in its post-industrial refugia. *Journal of Insect Conservation* 22 (2): 321–28.

Weaving, H., Terblanche, J.S., Pottier, P., and English, S. 2022. Meta-analysis reveals weak but pervasive plasticity in insect thermal limits. *Nature Communications* 13 (5292): 1–11.

Wiemers, M., Balletto, E., Dincă, V., Fric, Z.F., Lamas, G., Lukhtanov, V., Munguira, M.L., et al. 2018. An updated checklist of the European Butterflies (Lepidoptera, Papilionoidea). *ZooKeys* 811: 9–45.

Wilson, R.J. 2022. Northern wildlife feels the heat. *Nature Climate Change* 12 (6): 506–07.

Wilson, R.J. and Fox, R. 2021. Insect responses to global change offer signposts for biodiversity and conservation. *Ecological Entomology* 46 (4): 699–717.

Wilson, R.J., Gutiérrez, D., Gutiérrez, J., and Monserrat, V.J. 2007. An elevational shift in butterfly species richness and composition accompanying recent climate change. *Global Change Biology* 13 (9): 1873–1887.

Wilson, R.J., Gutiérrez, D., Gutiérrez, J., Martínez, D., Agudo, R., and Monserrat, V.J. 2005. Changes to the elevational limits and extent of species ranges associated with climate change. *Ecology Letters* 8 (11): 1138–1146.

Wolf, C., Bell, D.M., Kim, H., Nelson, M.P., Schulze, M., and Betts, M.G. 2021. Temporal consistency of undercanopy thermal refugia in old-growth forest. *Agricultural and Forest Meteorology* 307: 1–11.

CHAPTER 17

Improving our understanding of insect responses to climate change

Current knowledge and future perspectives

Daniel González-Tokman, Ornela De Gasperin and Wesley Dáttilo

17.1 Introduction

Several insect populations are declining or have gone extinct due to climate change caused by the increase of atmospheric CO_2 driven by human activity (summarized by Dáttilo and González-Tokman in Chapter 1). Global warming has been ten times faster in the last decades than during the last aberrantly rapid warming event on Earth, 56 million years ago, when atmospheric CO_2 increased two to five times and temperature raised up to 8 °C in 10,000 years, with extreme weather events also increasing (Currano in Chapter 2). As summarized in this book, insects vary greatly in how they respond to climate change, and the responses of insects at the individual level impact phenology, population sizes, community composition, biotic interactions, ecosystem functioning, and human wellbeing. However, despite multiple alarming reports, our knowledge of how insects respond to climate change, and how in turn these responses have altered population, community and ecosystem dynamics is very limited, and related to a few taxa, geographic areas and fundamental aspects (Figure 17.1).

The first main gap in our knowledge of how insects respond to climate change is the low quantity of available information emerging from long-term ecological data. Furthermore, very little long-term ecological data have been collected in the tropics, despite these being the most biological diverse regions in the world. Long-term field studies are the only source that can provide real evidence of the role of recent climate change (and of other human-induced processes, like chemical pollution and deforestation) on local and global insect extinctions (Boggs in Chapter 9; Del Claro et al. in Chapter 10). Moreover, the available long-term data are also scarce documenting changes in phenology (Ma et al. in Chapter 6), species distributions (Rodríguez-Castañeda and Hof in Chapter

Daniel González-Tokman, Ornela De Gasperin and Wesley Dáttilo, *Improving our understanding of insect responses to climate change*. In: *Effects of Climate Change on Insects*. Edited by: Daniel González-Tokman and Wesley Dáttilo, Oxford University Press. © Oxford University Press (2024). DOI: 10.1093/oso/9780192864161.003.0017

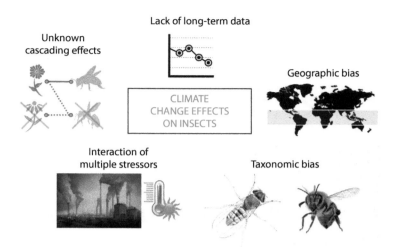

Figure 17.1 Our current knowledge about how insects respond to climate change is still limited by five key aspects. i) The lack of long-term data indicating population declines or extinctions. ii) The effects of climate change in some geographical regions, mainly the tropics, remain largely unknown despite their high diversities. iii) A few insect orders have been studied in detail whereas most remain unstudied. iv) The effect of climate change interacting with other stressors such as habitat loss, pollution and exotic species remains largely unexplored. v) The effects of species losses or gains on ecosystem functioning are constrained by cascading effects, which have been overlooked in insects (Figure from Wesley Dáttilo and freepik.com).

12; Aulus-Giacosa et al. in Chapter 11) population dynamics (Boggs in Chapter 9) and community compositions (Del Claro et al. in Chapter 10), all of which might impact the persistence of species, biotic interactions and overall ecosystem functioning. This information would contribute to the detection and validation of the role of specific sites (i.e., refugia) considered priority for conservation (Ursul et al. in Chapter 16). Moreover, long-term information would be valuable to identify changes in insect morphology with time, as climate affects body size and the expression of sexual traits through sexual selection, sexual conflict and other processes (Buzatto et al. in Chapter 7). Therefore, long-term studies re-sampling specific sites would be highly informative to understand changes in evolutionary and ecological processes driven by climate change, and to undertake conservation actions, particularly in the tropics. As carrying out long-term field monitoring studies is a difficult and time-intensive task, we encourage researchers to implement more Community Science (CS) projects to monitor insect abundance and diversity across space and time. CS is a research method that actively involves members of the public in the data collection process. This tool is very useful for documenting large-scale patterns, particularly when economic resources are limited. CS projects have been successful in monitoring long-term wildlife abundance and diversity in both industrialized and poor nations (e.g., Braschler 2009; Saunders et al. 2018), in evaluating ecosystem changes, such as water quality (e.g., Seibert and van Meerveld 2022) and changes in coral reef ecosystem structure (e.g., Gouraguine et al. 2019), and in detecting alien species (e.g., Johnson et al. 2020). We consider CS to be a promising and untapped research tool that can aid scientists in monitoring the role of climate change on wild insect populations.

A second gap in our knowledge of how insects respond to climate change is the geographic bias of where most studies are being carried out. Indeed, most field and laboratory evidence comes from insects that inhabit temperate latitudes, leaving tropical insects remarkably understudied (Rocha-Ortega et al. 2021). This is particularly worrying not just because the tropics are highly diverse but also because insects in the tropics are physiologically more sensitive to increases in temperature than temperate ones (González-Tokman and Villada-Bedoya in Chapter 4). This bias, mainly caused by economic and social factors limiting research in some regions (Juárez-Juárez et al. 2023; Khelifa and Mahdjoub 2022), starts with evidence from the fossil record, which is scarce per se, but particularly limited for tropical and polar regions, mainly in the southern hemisphere (Currano in Chapter 2). The tropics are also understudied regarding insect extinctions due to current climate change, which have been reported in longitudinal studies sampled across years, but less than 10 percent of such studies come from the tropics (Boggs in Chapter 9). Surprisingly, no studies have been carried out in the tropics evaluating changes in insect populations across elevation gradients through time (Rodríguez-Castañeda and Hof in Chapter 12), despite the relevance of tropical mountains as (i) experimental setups to monitor distribution changes of important pests and vectors (González-Rete et al. in Chapter 14) and (ii) as high diversity refugia against climate change (Ursul et al. in Chapter 16). Moreover, biotic interactions involving insects, such as those between plants and pollinators, are far less described in the tropics than in non-tropical latitudes (Luna and Dáttilo in Chapter 15; Burkle and Jha in Chapter 13). Therefore, evidence-based research in highly biodiverse regions in the tropics, particularly in mountainous ecosystems, are highly needed and desirable in the future.

Taxonomic bias in insect research is a third main gap in our knowledge of insect responses to climate change. This bias is also evident in the fossil record as only a few insect orders have been studied extensively, probably for being spectacular (such as giant but extremely rare Odonata), easy to preserve (like Coleoptera but not Lepidoptera) or simply preferred by researchers, independently of their relative abundance in the studied fauna (Currano in Chapter 2; Schachat and Labandeira 2021). A few taxa (notably *Drosophila*) have also dominated studies of physiological, genetic and plastic responses to climate change, leaving an important taxonomic gap in all these topics. Moreover, little attention has been paid to the effects that climate change has on systems with a wide range of variation in life-history traits and strategies, such as considering how the same stressors impact social, sub-social and solitary species, aquatic and terrestrial insects, migratory and non-migratory species, sexually dimorphic and monomorphic and haplodiploid and diplodiploid species. This is important, because even within a single population, insects sometimes display large variation in how well they cope with environmental stressors according to specific life-history traits, like colony size (Jones, and Oldroyd 2006) or colony genetic diversity (Jones et al. 2004). Therefore, climate change may have very different effects on species according to specific life-history traits, even within a given taxon. However, important evidence is emerging in non-model species, largely made available by genomic tools that allow us to uncover the mechanistic genes regulating heat tolerance and adaptation (Fuller and Wellenreuther in Chapter 3; González-Tokman and Villada-Bedoya in Chapter 4; Rohner in Chapter 5). A few insect orders, notably Diptera, Hymenoptera, Lepidoptera, Odonata and Orthoptera, also dominate the literature about hybridization processes resulting from changes in distributions, despite negative outcomes of hybridization being expected in endemic or threatened taxa (Sánchez-Guillén et al. in Chapter 8). Studying non-model insects is an important consideration for further research, that would largely benefit from genomic, transciptomic and epigenomic emerging evidence (Fuller and Wellenreuther in Chapter 3; Sánchez-Guillén et al. in Chapter 8).

A fourth gap in our knowledge relates to how insects respond to other environmental processes linked to climate change, and how the effects that these processes have on insects may interact with the effects of climate change. The Anthropocene (i.e., the currently new geological epoch shaped by human activities) is characterized by warming, changes in precipitation and extreme weather events, like heavy rainfalls and hurricanes, but also by other pressures such as habitat loss, increased ocean level and ocean acidification, pollution, urbanization, and introduction of exotic predators, competitors and parasites into ecosystems. All these factors may potentiate their negative effects with climate change and have profound consequences for insects (Del Claro et al. in Chapter 10). In terms of physiology and evolution, the mechanisms involved in response to combined stressors might differ from those involved in response to single stressors acting independently, with fitness consequences for stressed insects (González-Tokman and Villada-Bedoya in Chapter 4; Rohner in Chapter 5; Buzatto et al. in Chapter 7). At the population level, synergistic effects of increasing temperature with habitat loss might cause local extinction of insects (Boggs in Chapter 9; Del Claro et al. in Chapter 10; González-Rete et al. in Chapter 14). Colonization of new suitably climatic habitats, a common response to climate change, will also be limited by the presence of other stressors such as lack of habitat or pollution (Aulus-Giacosa et al. in Chapter 11). Therefore, evaluating responses to combined stressors in a wide range of taxa becomes urgent to understand how insects respond to dangerous combinations.

Finally, as individuals are entities interacting with conspecifics but also with competitors, predators, prey, symbionts and parasites, indirect cascading effects are of utmost importance to define community composition and maintain ecosystem functioning but are overlooked in response to past and current climate change (Currano in Chapter 2; Luna and Dáttilo in Chapter 15), partly due to insufficient analytical tools and theoretical frameworks (Luna and Dáttilo in Chapter 15). By affecting phenologies, abundances, diversities and distributions, cascading effects caused by climate change impact at the population and community levels (Ma et al. in Chapter 6; Boggs in Chapter 9; Aulus-Giacosa et al. in Chapter 11; Del Claro et al. in Chapter 10) and ultimately affect ecosystem functioning and human wellbeing (González-Rete et al. in Chapter 14; Burkle and Jha in Chapter 13). The importance of indirect effects at different organizational levels still needs to be considered in climate change studies with insects.

Insects have slowly come out ahead of several catastrophic events in the history of our planet, including extreme weather conditions, changes in the atmosphere composition and radiation, high vulcanism and meteorite impacts. For example, it took herbivorous insects 4–9 million years to recover the diversity of pre-extinction levels after the meteorite impacted the Earth 66 million years ago (Donovan et al. 2016). Even after these catastrophes of global magnitude, complete insect families have rarely become extinct, and new species have also originated, counteracting species losses, leaving that insect diversity has not necessarily declined worldwide due to low extinction rather than by high diversification rates (Briggs 2017). Even though insects have so far proven to be resilient to dramatic global changes, the main problem today is likely to be related to reductions in abundance and biomass of insect populations, which are the basis of ecosystem functioning and essential for human wellbeing (Cardoso et al. 2020; Sánchez-Bayo and Wyckhuys, 2021).

Key reflections

- Five key aspects still limit our knowledge of how insects respond to climate change: lack of long-term data monitoring populations, geographic and taxonomic bias in studies of climate change, limited studies evaluating combinations of climate and other stressors and few studies considering cascading effects.
- Conservation of insect abundance and biomass is needed to maintain ecosystem functioning and human wellbeing.

Key further reading

- Briggs, J.C. 2017. Emergence of a sixth mass extinction? *Biological Journal of the Linnean Society 122* (2): 243–48.
- Cardoso, P., Barton, P.S., Birkhofer, K., Chichorro, F., Deacon, C., Fartmann, T., Fukushima, C.S. et al. 2020. Scientists' warning to humanity on insect extinctions. Biological Conservation 242: 108426.
- Rocha-Ortega, M., Rodriguez, P., and Córdoba-Aguilar, A. 2021. Geographical, temporal and taxonomic biases in insect GBIF data on biodiversity and extinction. *Ecological Entomology 46* (4): 718–28.
- Schachat, S.R., and Labandeira, C.C. 2021. Are insects heading toward their first mass extinction? Distinguishing turnover from crises in their fossil record. *Annals of the Entomological Society of America 114* (2): 99–118.

References

Braschler, B. 2009. Successfully implementing a citizen-scientist approach to insect monitoring in a resource-poor country. *BioScience* 59: 103–04.

Briggs, J.C. 2017. Emergence of a sixth mass extinction? *Biological Journal of the Linnean Society* 122 (2): 243–48.

Cardoso, P., Barton, P.S., Birkhofer, K., Chichorro, F., Deacon, C., Fartmann, T., Fukushima, C.S. et al. 2020. Scientists' warning to humanity on insect extinctions. *Biological Conservation* 242: 108426.

Donovan, M.P., Iglesias, A., Wilf, P., Labandeira, C.C., and Cúneo, N.R. 2016. Rapid recovery of Patagonian plant–insect associations after the end-Cretaceous extinction. *Nature Ecology and Evolution* 1 (1): 1–5.

Gouraguine, A., Moranta, J., Ruiz-Frau, A., Hinz, H., Reñones, O., Ferse, S. C., Jompa, J. et al. 2019. Citizen science in data and resource-limited areas: A tool to detect long-term ecosystem changes. *PLoS One* 14 (1), e0210007.

Johnson, B.A., Mader, A.D., Dasgupta, R., and Kumar, P. 2020. Citizen science and invasive alien species: An analysis of citizen science initiatives using information and communications technology (ICT) to collect invasive alien species observations. *Global Ecology and Conservation* 21: e00812.

Jones, J.C., Myerscough, M.R., Graham, S., and Oldroyd, B. P. 2004. Honey bee nest thermoregulation: diversity promotes stability. *Science* 305: 402–04.

Jones, J.C. and Oldroyd, B.P. 2006. Nest thermoregulation in social insects. *Advances in Insect Physiology* 33: 153–91.

Juárez-Juárez, B., Dáttilo, W., and Moreno, C.E. 2023. Synthesis and perspectives on the study of ant-plant interaction networks: A global overview. *Ecological Entomology*. 48: 269–83.

Khelifa, R. and Mahdjoub, H. 2022. An intersectionality lens is needed to establish a global view of equity, diversity and inclusion. *Ecology Letters* 25 (5): 1049–1054.

Rocha-Ortega, M., Rodriguez, P., and Córdoba-Aguilar, A. 2021. Geographical, temporal and taxonomic biases in insect GBIF data on biodiversity and extinction. *Ecological Entomology* 46 (4): 718–28.

Sánchez-Bayo, F. and Wyckhuys, K.A. 2021. Further evidence for a global decline of the entomo-fauna. *Austral Entomology* 60 (1): 9–26.

Saunders, M.E., Roger, E., Geary, W. L., Meredith, F., Welbourne, D.J., Bako, A., Canavan, E., Herro, F., Herron, C., and Hung, O. 2018. Citizen science in schools: Engaging students in research on urban habitat for pollinators. *Austral Ecology* 43: 635–42.

Schachat, S.R. and Labandeira, C.C. 2021. Are insects heading toward their first mass extinction? Distinguishing turnover from crises in their fossil record. *Annals of the Entomological Society of America* 114 (2): 99–118.

Seibert, J. and van Meerveld. I. 2022. Bridge over changing waters–Citizen science for detecting the impacts of climate change on water. *PLoS Climate* 1: e0000088.

Glossary

Allopatry: when the distribution ranges of different species or populations are completely isolated.

Alternative reproductive tactics: discrete phenotypes (morphology or behavior) expressed by individuals within a sex to obtain reproductive success.

Altitudinal distribution: defined as the distribution of a species in mountain ranges that is determine through sampling along complete altitudinal gradients.

Anthropocene (our geological epoch): the last 200–250 years of profound change combining economic, social and political forces that have improved human welfare over-exploiting everything the Earth has to offer, like water, soils, plants, animals.

Assembly: DNA sequencing technology cannot read whole genomes in one go; instead it reads short pieces of bases from a genomic sequence. Sequence assembly is aligning and merging fragments from a longer DNA sequence to reconstruct the original sequence.

Biotic specialization: measure of the dependency between species by describing the range of resources used by a species, which is directly related to its survival and reproduction.

Biological invasions: the successful establishment and spread of non-native species outside their natural range.

Centers of endemism: geographic locations that support a high proportion of species found nowhere else in the world, often in mountain ranges, oceanic islands and the tropics.

Climate change velocity: the speed required for a species to keep pace with climate change by shifting its distribution, which is typically higher in flat landscapes because mountains provide more local-scale climatic gradients over elevation or topography.

Climate debt: changes to the geographic distributions of species or to community composition that do not appear to be keeping pace with rates of climate change.

Community Temperature Index (CTI): an index quantifying the composition of an ecological community based on the thermal associations of the constituent species across their geographic ranges.

Cryptic genetic variation: a form of genetic variation that is not expressed under normal conditions but is released in stressful environments.

Damage type (DT): discrete morphologies of damage on plant organs. DTs are defined based on size, shape and placement on the plant organ, as well as using other characters specific to the different functional feeding groups (hole feeding, margin feeding, skeletonization, surface feeding, mining, galling, and piercing and sucking).

Diapause: a dynamic process of arrested development/growth induced by a combination of several environmental cues where insects can adjust the timing of development and reproduction to adapt to seasonal changes. Insects progress through a series of physiological phases, including diapause induction, maintenance and termination, and post-diapause development during this process. Diapause may occur at any stage during insect life cycle, but many species diapause at a fixed developmental stage. Not all species/populations must undergo this process.

DNA barcoding: identification of an insect (in this case) using sequencing of a short sequence of mitochondrial DNA, usually the cytochrome c oxidase I gene (COI or COX1), followed by comparison to a sequence reference library. Multiple species can be tested at once, in which case this is termed meta-barcoding.

Downslope range shift: when the species altitudinal distribution range is found, on average, at lower elevations than it was found historically.

Ecological network: conceptual framework to study ecological interactions with the use of graph theory and matrices A, where A_{ij} = number of interactions between a species of a trophic level i (e.g., a plant) with a species of the other trophic level species j (e.g., an insect).

Ecosystem services: direct and indirect contributions of ecosystems to human wellbeing, which have an impact on human survival and quality of life.

Elevational range shift: occurs when the distribution of a species along an altitudinal gradient differs between historic and a current sampling events.

Epigenetics: the influence of non-genetic factors on the genome, which affects gene expression. Non-genetic factors include the environment like diet, gut microbiota, toxin and drug exposure, psychological and physical stressors, and levels of activity throughout life. Measuring the epigenetic changes that occur in diseases, including cancer and heart disease, can provide understanding of the underlying mechanisms.

Evolutionary rescue: The restoration of a declining population by adaptive evolutionary changes that restores the positive growth rate and prevents extinction.

Fossil: the preserved remains, impression or trace of ancient organisms.

Genetic accommodation: evolutionary modification of an ancestrally plastic response in a novel environment.

Genetic admixture: Process by which individuals from two or more populations (or species) interbreed, giving place to a new one. The previous populations are usually known as ancestral, while the new formed population is known as admixed.

Genetic assimilation: a form of genetic accommodation whereby phenotypes that were originally plastic become genetically fixed (i.e., environment-insensitive).

Genetic variance or variation: in quantitative genetics, the amount of *genetic variation* indicates the amount of *heritable* variation. It is, in a sense, a misnomer because genetic variation is my no means less shaped by genes and gene products than other variance components (e.g., plasticity, V_E).

Genome: the sum total of the genetic material of a cell or an organism.

Genomics: the study of genes and their function.

Genotype-by-environment interactions (G × E): G × E-interactions refer to differences in the way genotypes respond to environmental conditions. The term is considered problematic by some because it is the organism, not its genotype, that interacts with the environment.

Heat shock proteins: molecular chaperones that bind to other proteins and avoid their denaturation. They are named according to their molecular weight, in kDa (e.g., HSP70 and HSP90 have molecular weights of 70 and 90 kDa, respectively). They are the best studied mechanism of insect heat tolerance.

Heat tolerance: any measurement of individual performance at high temperature, the most common being upper lethal temperature, critical thermal maximum and heat knockdown time.

Heritability: it refers to the proportion of the total phenotypic variation that is due to genetic factors. It reflects the degree to which an offspring's phenotype can be predicted from its parents' phenotypes.

Hybrid lineage: an evolutionary lineage that originated from hybridization, which is at least partially reproductively isolated from both parental lineages and evolve regardless of parent species.

Hybrid zone: a region where two species or populations overlap their ranges and cross-fertilize to produce offspring of mixed ancestry.

Insect damage census: Method used to quantify herbivory in the fossil record in which every identifiable leaf from a fossil site that is > 50 percent complete is scored for the presence/absence of every damage type.

Interaction rewiring: property of species to rearrange their ecological interactions over time and space. It is used as a measure of dynamism and variation; for example, high rewiring between two ecological networks (i.e., a high proportion of rearrangement of interactions) means that the interactions between them vary greatly. It also means that the same species interact in different ways depending on the context where interactions are taking place.

Introgression: exchange of genes from one species or population into another by hybridization and backcrossing. Introgression can be unidirectional (in only one species), or bidirectional (in both species) depending on whether backcrossing occurs with one or both parental species.

Konservat-Lagerstätten: a fossil deposit with extraordinary preservation, including features like soft parts, cellular detail, color. Rapid burial, anoxial, microbial coverings are all thought to be important controls on formation. Insect Konservat-Lagerstätten include amber (e.g., Eocene Baltic Amber and Miocene Dominican Amber) and lacustrine (e.g., Eocene Messel and Florissant sites) deposits.

Landscape genomics: refers to analyses to identify relationships between environmental factors and the genetic adaptation of organisms in response to these factors.

Leading edge: refers to the highest elevation record for the species at a given sampling event.

Macroecology: the study of mechanisms generating ecological patterns across scales.

Macrorefugia: regions with temporally stable or spatially heterogeneous climatic conditions that often represent centers of endemism because over long time periods they have allowed populations to withstand or adapt to changing climatic conditions.

Mate choice: whenever the effects of traits expressed in one sex leads to nonrandom matings with members of the opposite sex (= intersexual selection).

Mating system: how males and females pair when choosing a mate, including how many mating partners each sex has.

Metapopulation: a set of demes (sub-populations) distributed across space, with migration among them but with independent population dynamics. Metapopulations are characterized by the rates of extinction and colonization of their demes.

Microclimate: fine-scale variation in the climate that deviates from prevailing climatic conditions, and which represents the conditions to which small organisms such as insects are exposed.

Microrefugia: localized environments where topography, geology or habitat create distinct climates that facilitate long-term persistence within otherwise climatically hostile surroundings.

Modularity: describes a non-random structure of ecological networks in which a subgroup of species in one trophic level interacts more frequently with a group of species from another trophic level, with few or no interactions with other groups of species in the network.

Nestedness: non-random structure of ecological networks, where species that engage in few interactions (i.e., potentially specialists) interact with a subset of highly interactive species, while relationships between species with few interactions rarely occur among them.

Neurohormone: chemical messenger that is released from a neuron (neurosecretory cell) to the hemolymph and functions as a hormone.

Neuromodulator: chemical messenger that is released in the vicinity of the synapse, modifying synaptic transmission.

Neurotransmitter: molecule used by the nervous system to transmit a signal from a pre-synaptic neuron to a post-synaptic neuron or a muscle cell. It can influence a cell in inhibitory, excitatory or modulatory way.

Non-anthropogenic (natural) hybridization: interbreeding that occurs in natural hybrid zones.

Omics: collective term for a range of new high-throughput biological research methods (e.g., transcriptomics, proteomics and metabolomics) that systematically investigate entire networks of genes, proteins and metabolites within cells.

Oxidative stress: damage caused to lipids, proteins and nucleic acids by reactive oxygen species that could not be counteracted by antioxidants. Causes lipid peroxidation and protein oxidation and leads to cell death.

Paleocene-Eocene Thermal Maximum (PETM):   the geologically rapid global warming event that occurred 56 million years ago and was caused by massive release of isotopically light carbon into the atmosphere and ocean. The PETM is considered to be the best ancient analog for modern-day climate change.

Parapatry: when the distribution ranges of two species or populations overlap (the overlapping range is less than either range), allowing the opportunity to interbreed.

Phenological shift: a change in the timing of life-history events.

Phenological mismatch: shifts in the timing of life events among two or more species, such that life stages that previously overlapped no longer do to the same extent.

Phenology: the seasonal timing of life events of a species or population.

Phenotypic plasticity: the ability of genotypes to produce different phenotypes when exposed to different environmental conditions.

Pollen limitation: occurs when insufficient conspecific pollen is deposited on a flower's stigma, resulting in sub-maximal plant reproductive success.

Range stable: refers to species that remain at the same altitudinal distribution they were in historical records.

Reaction norm: reaction norms (also referred to as *norms of reaction*) describe patterns of phenotypic changes of across several environments. Estimation usually requires experimental rearing of a genetic group (e.g., genotype, full sib family, population or species) under a wide range of values, ideally including extreme environments.

Rear edge: refers to the lowest elevation record for the species at a given sampling event.

Reduced Representation Approaches: sequencing approaches that use restriction enzymes to decrease genome complexity before sequencing (e.g., RADseq, GBS, ddRAD and MIG-seq).

RNA: Ribonucleic acid is the molecule that takes information from DNA to make protein and has many other activities. RNA is only made from genes in the DNA that are turned on.

Robustness: refers to the lack of phenotypic responses to environmental variation. It is often portrayed as the absence of plasticity, although this is not necessarily the case.

Sexual conflict: the evolutionary arms race between males and females when the interests of the different sexes are not aligned.

Sexual ornament: a morphological, acoustic, chemical or behavioral feature that is expressed by individuals of one sex and is the target of mate choice by the other sex.

Sexual selection: the result of differential reproduction of individuals of the same sex due to mating and fertilization success.

Single Nucleotide Polymorphism/SNP: Genetic variation that involves just one nucleotide, or base. About 90 percent of human genome variation can be accounted for by single nucleotide polymorphisms, or SNPs (pronounced "snips").

Spatial mismatch: reduced or eliminated opportunity for plant and pollinator species to interact due to changes in their spatial distributions.

Stable hybrid zone: region where hybridization between the involved lineages persist for numerous generations due to reduced hybrid fitness and it is maintained by immigrants from the parental species or populations.

Stress response: adaptive physiological response that allows living organisms to deal with adverse environmental conditions including extreme temperatures and oxidative stress. In insects, involves the synthesis and use of heat shock proteins, biogenic amines and neuroendocrine factors.

Sympatry: when the distribution range of two species or populations overlap partially or totally, allowing the opportunity to interbreed.

Temporal mismatch: reduced or eliminated opportunity for plant and pollinator species to interact due to changes in phenologies or reductions in their temporal overlap.

Thermal performance curve: describes the effect of temperature on vital rates (e.g., enzyme activity, metabolism, development).

Trait mismatch: reduced or eliminated opportunity for plant and pollinator species to interact due to changes in their morphological traits and/or pollinator foraging behavior (e.g., bee tongue length or flower corolla depth).

Transcription factor: protein needed to initiate or regulate transcription in eukaryotes.

Transcriptomics: sequencing RNA from a tissue or cell to measure the set of active genes.

Upslope range shift: when the species altitudinal distribution range is found, on average, at higher elevations than it was found historically.

Vector: an insect that transmits a disease to other organisms.

Vital rates: measures of age-specific survival and reproduction. These elements are used in matrix models of population dynamics.

Voltinism: the number of generations per year. Insect populations can be either multivoltine or univoltine. Multivoltine populations have more than one generation per year while univoltine populations have only one generation in a year. Some insect species exhibit only one type of voltinism (e.g., obligately univoltine), while others vary in their voltinism (functionally univoltine and multivoltine) according to local environmental conditions (e.g., latitude).

Weapon: a morphological trait that is directly used in male–male fights.

Whole genome sequencing (WGS): the process of determining the complete DNA sequence of an organism's genome.

Subject Index

Notes
Tables, figures and boxed material are indicated by an italic *t, f* or *b* after the page number.